MATHEMATICS RESEARCH DEVELOPMENTS

INTRODUCTION TO GEOMETRY AND RELATIVITY

MATHEMATICS RESEARCH DEVELOPMENTS

Additional books in this series can be found on Nova's website under the Series tab.

Additional e-books in this series can be found on Nova's website under the e-books tab.

PHYSICS RESEARCH AND TECHNOLOGY

Additional books in this series can be found on Nova's website under the Series tab.

Additional e-books in this series can be found on Nova's website under the e-books tab.

MATHEMATICS RESEARCH DEVELOPMENTS

INTRODUCTION TO GEOMETRY AND RELATIVITY

DAVID C. MELLO

nova publishers
New York

For permission to use material from this book please contact us:
Telephone 631-231-7269; Fax 631-231-8175
Web Site: http://www.novapublishers.com

Library of Congress Cataloging-in-Publication Data

Mello, David C. (David Cabral), 1950-
Introduction to geometry and relativity / David C. Mello.
pages cm
Includes bibliographical references and index.
ISBN: 978-1-62618-542-5 (hardcover)
1. Geometry, Differential. 2. Relativity (Physics) 3. Space and time. I. Title.
QA641.M455 2013
530.15'6--dc23

2013014127

Published by Nova Science Publishers, Inc. † New York

To my wife, Gracie

Contents

Preface

There can be no doubt that the Theory of Relativity is a great masterpiece of intellectual achievement. It provides, within the framework of differential geometry, a logical unification of the realms of electro-dynamics, mechanics, and gravitation.

While many excellent texts have been written on this important topic, this book is an effort to make relativity accessible to a broader audience of readers. This book is intended to provide an introduction to both modern geometry and relativity which is suitable not only for advanced undergraduate and first-year graduate students of physics, but also for students in related areas such as applied mathematics, astronomy, and astrophysics.

The author makes only modest assumptions on the academic background of the book's intended reader. Anyone who has a solid understanding of basic physics, and has completed the standard calculus sequence - including vector calculus - and has had some exposure to differential equations, should be able to profit from reading the book.

The book begins with a review of key concepts of Newtonian mechanics in Chapter 1. Here, the geometric viewpoint of the subject is introduced immediately, and the basic assumptions of Newtonian mechanics are clearly presented. Chapter 2 continues along the same lines, providing an examination of the structure of Newtonian spacetime. Particular emphasis is placed upon the covariance of Newtonian mechanics with respect to the group of general Galilean transformations.

Chapter 3 provides a solid introduction to Maxwell's equations. This chapter will be useful to those students who have not had a formal course in electrodynamics, or who may need a detailed review of this subject. The chapter concludes with a discussion of electromagnetic waves, and a demonstration of the fact that Maxwell's theory is not covariant with respect to the group of Galilean transformations.

A basic introduction to special relativity is provided in Chapter 4. At the outset, the electrodynamic origin of special relativity is stressed, a geometrically-inspired derivation of the Lorentz transformation is provided, and its standard consequences are discussed. In addition, a careful derivation of the transformation equations corresponding to a general Lorentz boost is provided.

Chapters 5 and 6 provide a detailed introduction to modern differential geometry. These chapters contain an introduction to the notions of topological spaces, differentiable manifolds, and tensor calculus. Particular care is taken to introduce new concepts using both modern coordinate-free notation, and the older component-based notation.

After having been equipped with the basic tools of differential geometry, the reader is invited to revisit special relativity in Chapter 7. Here, the idea of Minkowski space is introduced as a special type of semi-Riemannian space, and a rigorous discussion of the light-cone structure of this space is provided. Four-vectors are naturally presented as tangent vectors in Minkowski space, and their properties are closely examined. This chapter concludes with a treatment of relativistic collisions.

In Chapter 8, we examine Maxwell's equations in Minkowski space. This chapter begins with the potential formulation of Maxwell's equations and then proceeds with a rigorous derivation of the covariant form of these equations. The covariant form of the Lorentz force law is motivated by an appeal to the requirement of manifest covariance, and the energy-momentum tensor of the electromagnetic field is derived and studied. The chapter concludes with a discussion of the resulting transformation laws for electric and magnetic fields.

In preparation for the study of general relativity, a detailed discussion of relativistic hydrodynamics is provided in Chapter 9. The reader is first introduced to the standard mathematical treatment of Newtonian fluids, and is provided with a derivation of the energy-momentum tensors for relativistic dust, and the relativistic perfect fluid. The chapter concludes with a presentation of the relativistic continuity equation, and the relativistic version of Euler's equation.

Chapter 10 provides a discussion of the physical and philosophical motivations which guided Einstein in his formulation of general relativity. After a careful review of Newtonian gravitation, the chapter proceeds with a discussion of the equivalence principle, and some of its ramifications. Following some previous authors, including Einstein himself, the behavior of non-inertial frames in Minkowski space are used to motivate the non-Euclidean nature of spacetime in the vicinity of a gravitational field.

The role of the spacetime metric, and the phenomenon of geodesic deviation are also discussed in Chapter 10. The chapter continues with a discussion of the principles of general relativity, general covariance, minimal gravitational coupling, along with a heuristic derivation of Einstein's field equations. The chapter concludes with a discussion of the Newtonian limit of the full field equations, including a derivation of geodesic motion from the field equations themselves, and a careful statement of the formal postulates of general relativity.

Chapter 11 treats the major predictions of general relativity. A derivation of the spacetime metric for spherically symmetric fields is first presented, and is then followed by a derivation of Schwarz-child's exterior solution. The basic aspects of classical Keplerian motion are presented, along with the predicted precession of planetary orbits.

In addition, Chapter 11 proceeds with a discussion of the predicted deflection of light rays, and the gravitational red shift. The chapter concludes with a derivation of the linearized vacuum field equations and their resulting solutions leading to the prediction of gravitational waves.

Unfortunately, space limitations prohibited a discussion of relativistic cosmology and black holes; however, the author strongly feels that the detailed presentation of differential geometry will provide the reader with the necessary tools to take up the study of these interesting topics in any one of the many standard references.

The book is unique in that a great number of exercises, along with their solutions, are interpersed throughout the text. It is hoped that the detailed solutions will help to smooth the way for the less prepared readers, and provide a wealth of examples of the application of course material.

In short, the purpose of this book is to *explain* modern geometry and relativity; it has been intentionally designed to be a student-friendly introduction to these important fields. As such, the book contains "nothing new" in terms of subject matter, but it is hoped that the reader will find the presentation to be pedagogically satisfying, and the solutions to the exercises to be genuinely helpful.

In the preparation of the manuscript, the author has greatly relied upon the works of others in the field. A detailed list of the references used has been provided at the end of the text, and wherever appropriate, footnotes have been provided to specific references.

The author would like to acknowledge the valuable assistance of Professor David Capaldi, who reviewed the manuscript for errors, provided encouragement, and often made valuable suggestions to help improve the clarity of various discussions. Needless to say, any errors that remain are naturally the responsibility of the author.

David C. Mello
Providence, Rhode Island
March, 2013

Chapter 1

Introduction to Newtonian Spacetime

"Nature and Nature's laws lay hid in the night, God said Let Newton be! and all was light."

Alexander Pope

1.1. Introduction

Before we venture into the realm of relativity, let's review some of the more basic aspects of Newtonian mechanics. In this manner, we will be in a position to better appreciate the radical changes that Einstein found necessary to make when he first set forth the Special Theory of Relativity in 1905.

Newton's laws of motion, along with his law of universal gravitation, were set forth in his monumental work, *The Mathematical Principles of Natural Philosophy*, which was first published in 1687. This was truly a remarkable intellectual achievement in which Newton, almost single-handedly, created the mathematical science of mechanics, and brought a piercing bright light to bear upon the once hidden laws of nature.

Following in the foot steps of Galileo, Newton demonstrated that the universe has a well-defined structure, and what is equally important, that the laws of nature are best expressed in the language of mathematics.

Now, Newton's three laws of motion, formed the backbone of his mechanics. So, a close examination of these laws will provide us with a better understanding of the nature of the assumptions and the mode of thinking that Newton employed in his mechanics.

In deference to its original creator, Galileo, the first law of motion is sometimes called *Galileo's Law of Inertia*, and it first appeared as follows[1]:

"Everybody perseveres in its state of rest, or of uniform motion in a right [straight] line, unless it is compelled to change that state by forces impressed thereon."

Stated differently, this tells us that in the absence of external forces, a material particle must either remain in a state of rest, or move in a straight line at a constant speed.

Assuming the truth of the first law, it is apparent that in order to do physics, we must introduce a reference frame, along with appropriate measuring devices, such as accurate clocks and rulers, to serve as a backdrop which allows us to gauge the state of motion of a material particle.

Consequently, we are led to introduce the concept of a reference frame. A **reference frame** consists of three mutually perpendicular rigid rods so as to form an orthogonal Cartesian coordinate system. Such a reference frame is shown below, and it allows us to specify the position of a particle at any instant of time.

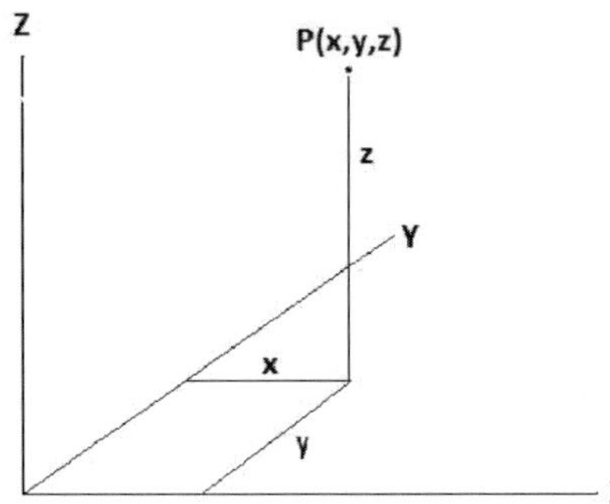

Figure 1.1. A Cartesian reference frame.

Although a time axis has been suppressed in Figure 1.1, it is apparent that **Newtonian spacetime** is inherently *four dimensional* in that we must use four coordinates - one time coordinate, and three space coordinates - to fully describe the motion of a particle. This may seem rather unusual to the reader who is used to doing physics in three dimensional space, but any driver who has narrowly escaped serious injury after experiencing a "close call" on a high-speed thoroughfare, will readily attest that not only was the location of his motor vehicle important, but also *when* it was at that location.

Each point in Newtonian spacetime is called an **event**, and each event is specified by a set of four coordinates: (t,x,y,z). A particle moving relative to our reference frame traces out a **path** or **trajectory** which may be described by the position vector

$$r = r(t) = x(t)i + y(t)j + z(t)k \tag{1.1}$$

where i, j,and k are unit vectors in the X,Y, and Z directions respectively. For each value of the time t, the tip of the vector $r(t)$ coincides with the instantaneous position of the particle.

It is apparent that Newton's first law also implies the existence of a family of reference frames called **inertial reference frames**. *Stated simply, an inertial reference frame, is one in which Newton's first law holds true.* If an observer sees an object at rest in an inertial reference frame, then in the absence of outside forces, that object simply remains at rest. Additionally, if the same observer sees any freely moving object, then that object will move in a straight line, and at a constant speed.

For Newton, the ideal inertial reference frame is one which is stationary with respect to the system of fixed stars. Experiment has shown that such a frame is inertial to a high degree.

On the other hand, a reference frame which is fixed to the surface of the earth is only approximately inertial, owing to the continual rotation of the earth.

For the same reason, a reference frame attached to the center of a rotating merry-go-round cannot be inertial either. If we imagine that the surface of the merry-go-round is perfectly smooth and frictionless, and if we place a small ball at any point on its surface, not coinciding with its center of rotation, then ball will appear to move in a circle (and not a straight line) relative to such an observer.

1.2. Newton's Laws

Newton's second law is that time rate of change of the momentum of a material particle is equal to the net force which acts upon it. In other words,

$$\mathbf{F} = \frac{d}{dt}\mathbf{p} = \dot{\mathbf{p}} \tag{1.2}$$

where the particle's **momentum p** is simply the product of its mass m, which is *assumed to be constant*, and its velocity $\mathbf{v}$. So, the last equation may be rewritten as

$$\mathbf{F} = m\frac{d\mathbf{v}}{dt} = m\frac{d^2\mathbf{r}}{dt^2} = m\ddot{\mathbf{r}}(t) \tag{1.3}$$

Newton's second law is logically compatible with his first law. To see this, note that in the absence of external forces, we have $\mathbf{F} = \mathbf{0}$, so that equation (1.3) may be readily integrated to give

$$\mathbf{r}(t) = \dot{\mathbf{r}}(0)t + \mathbf{r}(0) \tag{1.4}$$

where the constant vectors $\dot{\mathbf{r}}(0)$ and $\mathbf{r}(0)$ are simply the initial velocity and initial position of the particle. Equation (1.4) indicates that in the absence of external forces, a particle moves in a straight line. If it so happens that the initial velocity of the particle $\dot{\mathbf{r}}(0) = \mathbf{0}$, then (1.4) tells us that $\mathbf{r}(t) = \mathbf{r}(0)$ for all time, i.e., the particle remains at rest.

Newton's third law governs the forces of interaction between pairs of particles. Assume that we are given a pair of particles with masses m_1 and m_2. Let us suppose further, that as a result of this interaction (e.g., a collision), the first particle exerts a force $\mathbf{F}_{12}$ on the second particle. Then, according to Newton's third law, the second particle will exert a force $\mathbf{F}_{21}$ on the first particle which is given by

$$\mathbf{F}_{21} = -\mathbf{F}_{12} \tag{1.5}$$

In other words, "for every action there is an equal and opposite reaction."

1.3. Galilean Covariance

Now that are reacquainted with Newton's laws, and are equipped with the concept of an inertial reference frame, a question which naturally arises is how do we relate the space and time coordinates of two different inertial reference frames?

In order to answer this question, let's assume that there are two inertial reference frames, $S(t,x,y,z)$ and $\bar{S}(\bar{t},\bar{x},\bar{y},\bar{z})$, where the second frame $\bar{S}$ moves in the direction of the positive $x-$axis, at a *constant velocity* $\mathbf{v}$ relative to the first frame.

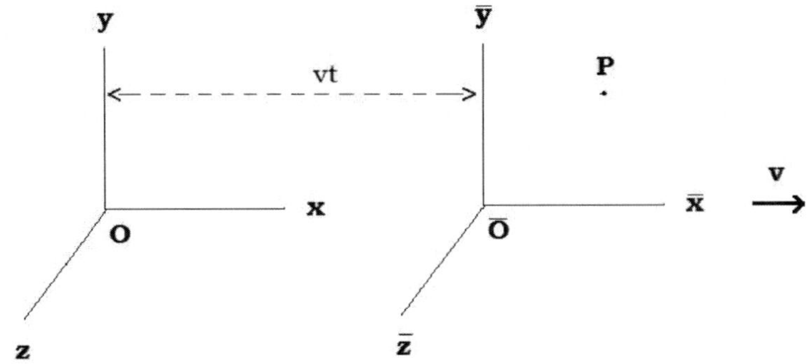

Figure 1.2. The Galilean transformation.

For the sake of simplicity, we also assume that the coordinate axes of $\bar{S}$ are parallel to the corresponding axes of S, and that the origins of each system coincide at $t = \bar{t} = 0$.

This general arrangement is shown in Figure 1.2.

Since the y and z axes of both reference frames are perpendicular to the direction of motion, then we must have $y = \bar{y}$ and $z = \bar{z}$. On the other hand, if $\bar{x}$ is first space coordinate of the point P which is fixed in the frame $\bar{S}$, and x is the x-coordinate of the *same point P* in the frame S, then it is clear that

$$\bar{x} = x - vt$$

so the space and time coordinates in both systems are related by the equations

$$\begin{aligned} \bar{t} &= t \\ \bar{x} &= x - vt \\ \bar{y} &= y \\ \bar{z} &= z \end{aligned} \tag{1.6}$$

We shall refer to equations (1.6) as the **Galilean transformation** equations, or alternatively, we shall say this set of equations represents a **Galilean boost** $\mathbf{v} = v\mathbf{i}$ in the positive x-direction of S. In order to obtain the inverse transformation, we simply solve for the unbarred coordinates in terms of the barred coordinates to obtain

$$\begin{aligned} t &= \bar{t} \\ x &= \bar{x} + vt \\ y &= \bar{y} \\ z &= \bar{z} \end{aligned} \tag{1.7}$$

Observe that equations (1.7) may be formally obtained from (1.6) simply by interchanging barred and unbarred coordinates in (1.6) and replacing v by $-v$; this is no accident, since if the barred frame moves at a speed v relative to the unbarred frame, then the unbarred frame must move at a speed $-v$ relative to the barred frame. This supports the idea that all inertial frames are equivalent, and no one frame can enjoy a privileged status.

In order to find out how velocities transform under a Galilean transformation, let's assume that a particle has velocity $\mathbf{u}$ relative to the unbarred frame; consequently,

$$\mathbf{u} = \frac{d}{dt}\mathbf{r}(t) = \frac{dx}{dt}\mathbf{i} + \frac{dy}{dt}\mathbf{j} + \frac{dz}{dt}\mathbf{k}$$

From equations (1.6), we find that the velocity of the same particle relative to the barred frame is given by

$$\begin{aligned}\bar{\mathbf{u}} &= \frac{d}{dt}\bar{\mathbf{r}}(t) = \frac{d\bar{x}}{dt}\mathbf{i} + \frac{d\bar{y}}{dt}\mathbf{j} + \frac{d\bar{z}}{dt}\mathbf{k} = \frac{d}{dt}(x - vt)\,\mathbf{i} + \frac{dy}{dt}\mathbf{j} + \frac{dz}{dt}\mathbf{k} \\ &= \mathbf{u} - \mathbf{v}\end{aligned}$$

where $\mathbf{v} = v\mathbf{i}$. So, the Galilean law for the **transformation of velocities** may be summarized as

$$\bar{\mathbf{u}} = \mathbf{u} - \mathbf{v} \tag{1.8}$$

Now, let's see what happens during a collision of two particles in the unbarred system. Suppose two particles, with masses m_1 and m_2 and with velocities $\mathbf{u}_1$ and $\mathbf{u}_2$ collide. By Newton's third law, we know that the first particle will exert a force $\mathbf{F}_{12}$ on the second particle, and in turn, the second particle will exert a force $\mathbf{F}_{21} = -\mathbf{F}_{12}$ on the first particle. Then, using Newton's second law, we have

$$\begin{aligned}\mathbf{F}_{21} &= \frac{d}{dt}(m_1\mathbf{u}_1) = -\mathbf{F}_{12} = -\frac{d}{dt}(m_2\mathbf{u}_2) \\ &\Longrightarrow \frac{d}{dt}(m_1\mathbf{u}_1 + m_2\mathbf{u}_2) = 0 \\ &\Longrightarrow \frac{d}{dt}(\mathbf{p}_1 + \mathbf{p}_2) = 0 \\ &\Longrightarrow \mathbf{p}_1 + \mathbf{p}_2 = \text{Constant}\end{aligned}$$

Thus, we see that, relative to the unbarred system, the total momentum $\mathbf{p}_1 + \mathbf{p}_2$ of the two particles is constant, or **conserved**. This, of course, is nothing more that a special case of the **law of the conservation of momentum.** If we now analyze the same state of affairs relative to the barred system, recalling equation (1.8), and remembering that $\mathbf{v}$ is constant, we obtain

$$\begin{aligned}&\Longrightarrow \frac{d}{dt}(m_1\bar{\mathbf{u}}_1 + m_2\bar{\mathbf{u}}_2) = \frac{d}{dt}\left[m_1(\mathbf{u}_1 - \mathbf{v}) + m_2(\mathbf{u}_2 - \mathbf{v})\right] \\ &= \frac{d}{dt}(m_1\mathbf{u}_1 + m_2\mathbf{u}_2) - (m_1 + m_2)\frac{d\mathbf{v}}{dt} \\ &\Longrightarrow \frac{d}{dt}(\mathbf{p}_1 + \mathbf{p}_2) = 0 \quad (\text{ since } \frac{d\mathbf{v}}{dt} = 0\,) \\ &\Longrightarrow \bar{\mathbf{p}}_1 + \bar{\mathbf{p}}_2 = \text{Constant}\end{aligned}$$

In other words, the law of the conservation of momentum also holds true in the barred reference system. This is an example of a general principle, often called the **postulate of Galilean Covariance**.

Postulate of Galilean Covariance

The laws of physics are **covariant** with respect to Galilean transformations.

That is, while observers attached to two different inertial reference frames may obtain *different* numerical values for various physical quantities, they will both arrive at the same laws of physics because the equations that describe those laws have the *same* mathematical form; this is the meaning of the term **covariance**. Thus, for example, although the total momentum of the two particles will usually be *different* in each inertial frame, we still can say that the total momentum is conserved in each frame.

We sometimes find confusion in the literature, where some authors use the terms *covariance* and *invariance* interchangeably, but one must be careful. The concept of invariance has a more restricted quality than that of covariance.

If we assert that the laws of mechanics are *invariant* under a Galilean tranformation, then we are not only asserting that all inertial observers will agree on the mathematical form of those laws, but they will also obtain the *same numerical values* for the relevant physical quantities that enter into those laws. Thus, although the "invariance" implies "covariance", it is easy to see that the converse is not true in general.

In addition, physical quantities that remain the same under a Galilean transformation are called **invariants**. For example, equations (1.6) indicate that the time t is an *invariant* quantity in Newton's mechanics. Newton supported the notion of **absolute time** in which he assumed that time flows at the same rate in all inertial systems, and it is therefore *independent* of the state of motion of an observer. Later on, we shall see that Einstein found it necessary to radically alter this idea in his special theory.

Another example of an invariant in Newton's theory is the mass, or more properly, the inertial mass of a body. The **inertial mass** of an object is considered to be a measure of its resistance to any change in its state of motion. Newton considered inertial mass to be an intrinsic property of an object, a property which remains independent of the motion of an inertial observer.

Now, it so happens that *Newton's laws are invariant under a Galilean transformation*. This is a direct consequence of the fact that under a Galilean transformation, the acceleration of a particle is invariant. To see this, note that we can take the time derivative of both sides of equation (1.8) to obtain

$$\begin{aligned} \bar{\mathbf{a}} &= \frac{d}{dt}\bar{\mathbf{u}} = \frac{d}{dt}(\mathbf{u}-\mathbf{v}) = \mathbf{a} - \frac{d\mathbf{v}}{dt} \\ &= \mathbf{a} \ \ (\text{since } \frac{d\mathbf{v}}{dt} = \mathbf{0}) \end{aligned} \tag{1.9}$$

So, $\bar{\mathbf{a}} = \mathbf{a}$, or in other words, under a Galilean tranformation, the acceleration of a particle is *invariant*. Consequently, each inertial observer will measure the same force on a moving particle

$$\bar{\mathbf{F}} = m\bar{\mathbf{a}} = m\mathbf{a} = \mathbf{F} \tag{1.10}$$

and thus, Newton's laws are invariant under a Galilean transformation.

In the examples that follow, the reader will see how these ideas can be generalized to systems of particles, and how the Galilean transformation can be generalized as well.

1.4. Exercises

1. Consider a system of particles with masses $m_1, m_2, ..., m_n$ and with corresponding position vectors $\mathbf{r}_1, \mathbf{r}_2, ..., \mathbf{r}_n$. Recall that the **center of mass** $\mathbf{R}_{CM}$ of the system is given by

$$\mathbf{R}_{CM} = \frac{1}{M}\sum_k m_k \mathbf{r}_k \tag{1.11}$$

where M is the total mass of the system. Show that the net external force acting on the system is equal to the total mass of the system times the acceleration of the center of mass:

$$\mathbf{F} = M\,\ddot{\mathbf{R}}_{CM} \tag{1.12}$$

(Solution)

Let $\mathbf{F}_k$ be the total force on the k-th particle, then

$$\mathbf{F}_k = \mathbf{F}_k^{INT} + \mathbf{F}_k^{EXT}$$

where $\mathbf{F}_k^{INT}$ and $\mathbf{F}_k^{EXT}$ are the internal and external forces acting on the k-th particle. If we let $\mathbf{F}_{ik}^{INT}$ represent the internal force exerted by the i-th particle on the k-th particle, then by Newton's third law we have

$$\mathbf{F}_{ik}^{INT} = -\mathbf{F}_{ki}^{INT} \text{ and } \mathbf{F}_{kk}^{INT} = \mathbf{0} \tag{1.13}$$

Consequently, the net force on the entire system of particles is given by

$$\mathbf{F} = \sum_k \mathbf{F}_k = \sum_k \left(\mathbf{F}_k^{INT} + \mathbf{F}_k^{EXT}\right) = \sum_k\sum_i \mathbf{F}_{ik}^{INT} + \sum_k \mathbf{F}_k^{EXT}$$

But, by virtue of equation (1.13), we must have

$$\sum_k\sum_i \mathbf{F}_{ik}^{INT} = \mathbf{0}$$

so the last equation simply becomes

$$\mathbf{F} = \sum_k \mathbf{F}_k^{EXT} \tag{1.14}$$

On the other hand, if we differentiate (1.11) twice, and use the previous result, we obtain

$$M\,\ddot{\mathbf{R}}_{CM} = \sum_k m_k \ddot{\mathbf{r}}_k = \sum_k \mathbf{F}_k^{EXT} = \mathbf{F}$$

as required.

2. Assume there are two inertial frames, $S(t,x,y,z)$ and $\bar{S}(\bar{t},\bar{x},\bar{y},\bar{z})$, where the second frame $\bar{S}$ moves at a *constant velocity*

$$\mathbf{v} = v_x\mathbf{i} + v_y\mathbf{j} + v_z\mathbf{k}$$

relative to the first frame. Once again, we also assume that the coordinate axes of $\bar{S}$ are parallel to the corresponding axes of S, and that the origins of each system coincide at $t = \bar{t} = 0$. This situation is called a **general Galilean boost v**, since we don't restrict ourselves to the special case where $\mathbf{v} = v\mathbf{i}$. Show that under a general Galilean boost v the transformation equations become

General Galilean Boost

$$\bar{t} = t, \ \ \bar{x} = x - v_x t, \ \ \bar{y} = y - v_y t, \ \ \bar{z} = z - v_z t \tag{1.15}$$

(Solution)

Let P be a fixed but arbitrary point whose position vectors are $\mathbf{r}$ and $\bar{\mathbf{r}}$ in the frames $S(t,x,y,z)$ and $\bar{S}(\bar{t},\bar{x},\bar{y},\bar{z})$, respectively.

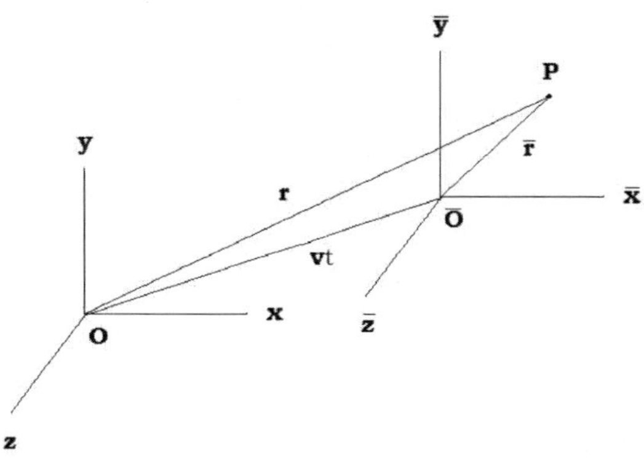

Figure 1.3. A general Galilean boost, **v**.

From Figure 1.3, it is apparent that $\mathbf{r} = \bar{\mathbf{r}} + vt$, so that

$$\begin{aligned} \bar{\mathbf{r}} &= \mathbf{r} - \mathbf{v}t \\ &\Longrightarrow \bar{x}\mathbf{i} + \bar{y}\mathbf{j} + \bar{z}\mathbf{k} = (x\mathbf{i} + y\mathbf{j} + z\mathbf{k}) - (v_x\mathbf{i} + v_y\mathbf{j} + v_z\mathbf{k})t \\ &\Longrightarrow \bar{x}\mathbf{i} + \bar{y}\mathbf{j} + \bar{z}\mathbf{k} = (x - \mathbf{v}_x t)\mathbf{i} + (y - v_y t)\mathbf{j} + (z - v_z t)\mathbf{k} \end{aligned}$$

and consequently, we must have

$$\bar{x} = x - v_x t, \quad \bar{y} = y - v_y t, \quad \bar{z} = z - v_z t$$

Once again, the relation $\bar{t} = t$ follows from the absolute nature of time in Newtonian mechanics.

3. Show that for a system of particles, equation (1.12) is *invariant* under a general Galilean boost $\mathbf{v}$.

(Solution)

In the barred system, we have

$$\bar{\mathbf{R}}_{CM} = \frac{1}{M}\sum_k m_k \bar{\mathbf{r}}_k = \frac{1}{M}\sum_k m_k (\mathbf{r}_k - \mathbf{v}t) = \left(\frac{1}{M}\sum_k m_k \mathbf{r}_k\right) - \mathbf{v}t$$

and consequently,

$$\bar{\mathbf{R}}_{CM} = \mathbf{R}_{CM} - \mathbf{v}t \tag{1.16}$$

Taking two time derivatives of the last equation, and remembering that $\mathbf{v}$ is constant, we get

$$\bar{\mathbf{A}}_{CM} = \frac{d^2}{dt^2}\bar{\mathbf{R}} = \frac{d^2}{dt^2}(\mathbf{R}_{CM} - \mathbf{v}t) = \mathbf{A}_{CM}$$

so that,

$$\bar{\mathbf{F}} = M\frac{d^2}{dt^2}\bar{\mathbf{R}}_{CM} = \mathbf{F} = M\frac{d^2}{dt^2}\mathbf{R}_{CM} \tag{1.17}$$

4. Show that for a system of particles, the total momentum $\mathbf{P}$ of the system is equal to its total mass M times the velocity of its center of mass $\mathbf{V}_{CM}$:

$$\mathbf{P} = M\mathbf{V}_{CM} \tag{1.18}$$

(Solution)

Differentiating both sides of equation (1.11), we have

$$M\dot{\mathbf{R}}_{CM} = \sum_k m_k \dot{\mathbf{r}}_k = \sum_k \mathbf{p}_k = \mathbf{P} \tag{1.19}$$

as claimed.

5. Show that the law of the convervation of momentum also holds for a system of particles; that is, if the net external force acting on the system is zero ($\mathbf{F} = \mathbf{0}$), then the total momentum of the system is *conserved* (constant).

(Solution)

If the net external force $\mathbf{F}$ acting on the system is zero, then we can write

$$\mathbf{F} = M\,\ddot{\mathbf{R}}_{CM} = \frac{d}{dt}(M\dot{\mathbf{R}}_{CM}) = \frac{d}{dt}\mathbf{P} = \mathbf{0}$$

and consequently, the total momentum $\mathbf{P}$ is constant, or conserved.

6. Show that, for a system of particles, the law of the conservation of momentum for a *covariant* under a general Galilean boost.

(Solution)

Assume that the total momentum **P** of the system is constant relative to the unbarred inertial frame. Then, we have

$$\mathbf{F} = M\frac{d^2}{dt^2}\mathbf{R}_{CM} = \frac{d}{dt}(M\dot{\mathbf{R}}_{CM}) = \frac{d}{dt}(\mathbf{P}) = \mathbf{0}$$

But since $\mathbf{F} = \bar{\mathbf{F}} = \mathbf{0}$, then it follows from equations (1.17) and (1.16) that

$$\begin{aligned}\bar{\mathbf{F}} &= \frac{d}{dt}\bar{\mathbf{P}} = \frac{d}{dt}(M\,\bar{\mathbf{V}}_{CM}) = \frac{d}{dt}\left[M\left(\dot{\mathbf{R}}_{CM} - \mathbf{v}\right)\right] \\ &= \frac{d}{dt}(\mathbf{P} - M\mathbf{v}) = \mathbf{0} - \mathbf{M}\left(\frac{d}{dt}\mathbf{v}\right) = \mathbf{0}\end{aligned}$$

Thus, $\bar{\mathbf{P}}$ is constant, or conserved as well.

Chapter 2

The Structure of Newtonian Spacetime

"The mathematical facts worthy of being studied are those which, by their analogy with other facts, are capable of leading us to the knowledge of a physical law."

Henri Poincare

2.1. Introduction

In this chapter, we will examine the underlying structure of Newtonian spacetime. At the outset, it is apparent that the points (events) of Newtonian spacetime may be put into one-to-one correspondence with $\mathbf{R}^4$, where the latter space is the collection of all ordered 4-tuples of real numbers of the form (x^1, x^2, x^3, x^4). While this comparison is apt, it tells us nothing about the **causal structure** of Newtonian spacetime.

Recall that Newton's notion of absolute time implies that time flows at the same rate for all observers, and is *independent* of the inertial reference frame employed. Consequently, there follows the notion of **absolute simultaneity**; that is, if a given observer attached to the inertial reference frame $S(t,x,y,z)$, notes that a certain event occurs at the instant of time $t = t_0$, then any other observer, equipped with the inertial reference frame $\bar{S}(\bar{t},\bar{x},\bar{y},\bar{z})$, will report that $\bar{t} = t_0$ as well.

Along the same line, suppose an observer in S records two distinct events, $E_1(t_1,x_1,y_1,z_1)$ and $E_2(t_2,x_2,y_2,z_2)$. Assume further that the event E_1 precedes the event E_2, so that $t_1 < t_2$. An observer in $\bar{S}$ will also agree that the event $\bar{E}_1(\bar{t}_1,\bar{x}_1,\bar{y}_1,\bar{z}_1)$ precedes the event $\bar{E}_2(\bar{t}_2,\bar{x}_2,\bar{y}_2,\bar{z}_2)$, since under a Galilean transformation, we must have $\bar{t}_1 = t_1$ and $\bar{t}_2 = t_2$.

From the above considerations, we see that if a cause E_1 precedes an effect E_2 relative to the system S, then the cause $\bar{E}_1$ must also precede the effect $\bar{E}_2$ relative to the system $\bar{S}$. So, the causal structure of Newtonian spacetime is rather simple, and later on, we shall see that this circumstance is a result of the hidden assumption that there is no upper limit on the velocity of signals used for communicating the results of space and time measurements in each inertial frame.

For the sake of simplicity, if we assume that the motion of any particle is confined to two dimensions, say the XY-plane, then Newton's concept of absolute simultaneity implies that at each instant of time, there corresponds a *plane of simultaneity* which contains the *spatial* coordinates of all events that occur at that given instant of time.

Furthermore, there are infinitely many planes of simultaneity, with a unique plane passing through each point of the time axis as shown in Figure 2.1. Each plane, in turn, is a copy of the **Euclidean space**, E^2.

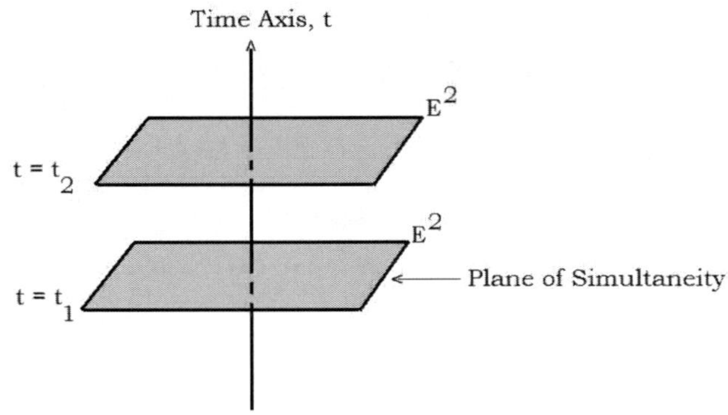

Figure 2.1. The Structure of Newtonian Spacetime.

Now, in general, since we must admit motion in three spatial dimensions, we see that Newtonian spacetime is the Cartesian product of two spaces: $\mathbf{R} \times \mathbf{E}^3$, where we have used the real axis **R** to represent time, and $\mathbf{E}^3$ is Euclidean 3-space.

This type of structure is what mathematicians call a **fibre bundle**. Intuitively speaking, a fibre bundle is a space that, *at least locally*, looks like the Cartesian product of two spaces.[1] A cylinder is a nice example of a fibre bundle, and it is easily seen in Figure 2.2, that the cylinder $T = S^1 \times R_0$, where S^1 is a circle, and R_0 is a segment of the real line.

We call T the **total space**, S^1 the **base space**, and R_0 the **fibre space**, whose elements (line segments) are called **fibres**. It is clear that the cylinder T is also *globally* a product of its base space S^1 and its fibre space R_0, and consequently, it is called a **trivial fibre bundle**.

An example of a *non-trivial* fibre bundle, which is only locally the product of two spaces, is afforded by the Mobius strip.[2]

For Newtonian spacetime, we see that **R** serves as the base space, $\mathbf{E}^3$ serves as a typical fibre, and a copy of $\mathbf{E}^3$ is attached to each point of **R**. It is also clear that Newtonian spacetime, itself, is to be identified with the total space $\mathbf{R} \times \mathbf{E}^3$.

The bundle structure of Newtonian spacetime has some unusual consequences, which we now investigate.

2.2. The Fibre Space, $\mathbf{E}^3$

As Newtonian spacetime has the fibre bundle structure $R \times E^3$, it is important to define the Euclidean space E^3 and to investigate some of its properties. We define the **Euclidean space** E^3 as the vector space R^3, equipped with the standard **Euclidean metric**, or **distance**

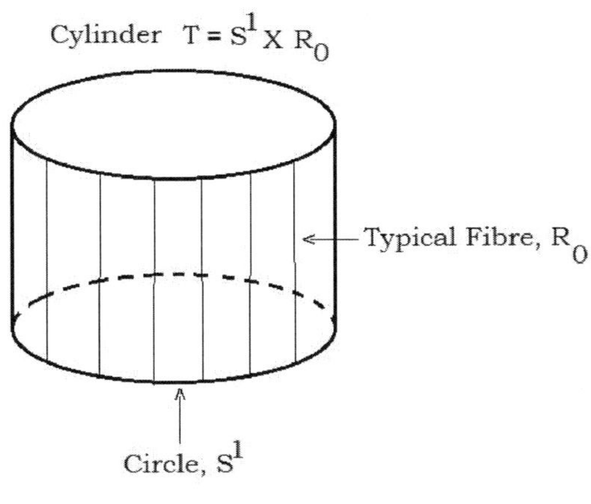

Figure 2.2. A Trivial Fibre Bundle.

function. Before we can define the Euclidean metric, however, we first define the *inner product* of any two vectors in E^3.

If **X** and **Y** are vectors in E^3, we shall agree to represent them as column vectors:

$$\mathbf{X} = \begin{bmatrix} x^1 \\ x^2 \\ x^3 \end{bmatrix} \text{ and } \mathbf{Y} = \begin{bmatrix} y^1 \\ y^2 \\ y^3 \end{bmatrix} \tag{2.1}$$

and we define their **Euclidean inner product** $\langle \mathbf{X}, \mathbf{Y} \rangle$ as

$$\langle \mathbf{X}, \mathbf{Y} \rangle = \mathbf{X}^T \mathbf{Y} = \begin{bmatrix} x^1 & x^2 & x^3 \end{bmatrix} \begin{bmatrix} y^1 \\ y^2 \\ y^3 \end{bmatrix} = x^1 y^1 + x^2 y^2 + x^3 y^3 \tag{2.2}$$

where the row vector $\mathbf{X}^T$ denotes the *transpose* of the column vector **X**. Observe the the Euclidean inner product is just a mapping from $E^3 \times E^3$ into R

$$\langle\ ,\ \rangle : E^3 \times E^3 \to R \tag{2.2a}$$

which assigns a *real number* to each ordered pair of vectors.

The inner product allows us to "do geometry"in E^3 by giving us a concrete way of measuring lengths and angles. Along this line, we can use the inner product to define the length or **norm** $\|\mathbf{X}\|$ of any vector **X** as

$$\|\mathbf{X}\| = \sqrt{\langle \mathbf{X}, \mathbf{X} \rangle} = \sqrt{(x^1)^2 + (x^2)^2 + (x^3)^2} \tag{2.3}$$

From (2.3), it is clear that $\|\mathbf{X}\| \geq 0$ for any vector $\mathbf{X} \in E^3$.

If we now view **X** and **Y** as the *position vectors* of the points P_1 and P_2 respectively, then we define the **distance** $d(P_1, P_2)$ between these two points as

$$s^2 = [d(P_1, P_2)]^2 = \|\mathbf{X} - \mathbf{Y}\|^2 = (x^1 - y^1)^2 + (x^2 - y^2)^2 + (x^3 - y^3)^2 \tag{2.4}$$

Equation (2.4) defines the Euclidean metric in precise way, and it is easy to show that this metric has several useful properties.

Properties of the Euclidean Metric

If $\mathbf{X}, \mathbf{Y}$ and $\mathbf{Z}$ are any vectors in the Euclidean space E^3, then

$$d(\mathbf{X},\mathbf{Y}) \geq 0 \quad \text{(non-negativity)} \tag{2.5a}$$
$$d(\mathbf{X},\mathbf{X}) = 0 \text{ if and only if } \mathbf{X} = \mathbf{0} \quad \text{(non-degeneracy)} \tag{2.5b}$$
$$d(\mathbf{X},\mathbf{Y}) = d(\mathbf{Y},\mathbf{X}) \quad \text{(symmetric)} \tag{2.5c}$$
$$d(\mathbf{X},\mathbf{Z}) \leq d(\mathbf{X},\mathbf{Y}) + d(\mathbf{Y},\mathbf{Z}) \quad \text{(triangle inequality)} \tag{2.5d}$$

Let $\mathbf{X}$ and $\mathbf{X} + d\mathbf{X}$ represent the position vectors of two infinitesimally close points in E^3, then (2.4) tells us that the infinitesimal distance ds^2 between the two points is given by:

$$ds^2 = \|d\mathbf{X}\|^2 = (dx^1)^2 + (dx^2)^2 + (dx^3)^2 = \delta_{ab} dx^a dx^b$$

where δ_{ab} is the **Kronecker delta** defined by

$$\delta_{ab} = \begin{cases} 1, & \text{if } a = b \\ 0, & \text{if } a \neq b \end{cases} \tag{2.6}$$

and the **Einstein summation convention** is employed in (2.6). Under this convention, we agree to *sum* on any index which appears *both* as a lower index and an upper index.

Now, the indices, a and b which appear in (2.4) are sometimes called *dummy indices*, since any other set of indices will also suffice; for example, it is clear that the expressions

$$ds^2 = \delta_{ab} dx^a dx^b = \delta_{cd} dx^c dx^d$$

are equivalent.

We now know how to measure length in E^3, but as of yet, we have said nothing about how we measure angles. We can do this in a roundabout way by first noting that if $\mathbf{X}$ and $\mathbf{Y}$ are vectors in E^3, and the angle between these vectors is θ, then their inner product also may be written in the alternative form:

$$\langle \mathbf{X}, \mathbf{Y} \rangle = \|\mathbf{X}\| \, \|\mathbf{Y}\| \cos\theta \tag{2.7}$$

In order to see this, take a look at Figure 2.3.

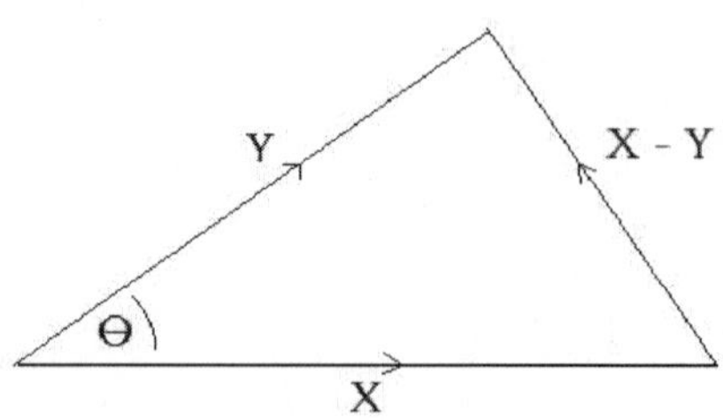

Figure 2.3. A Vector Triangle in E^3.

By the law of cosines, we have

$$\|\mathbf{X}-\mathbf{Y}\|^2 = \|\mathbf{X}\|^2 + \|\mathbf{Y}\|^2 - 2\,\|\mathbf{X}\|\,\|\mathbf{Y}\|\cos\theta$$

and, if we write out both sides of this expression, we get:

$$\begin{aligned}(x^1-y^1)^2+(x^2-y^2)^2+(x^3-y^3)^2 &= \left[(x^1)^2+(x^2)^2+(x^3)^2\right] \\ &\quad + \left[(y^1)^2+(y^2)^2+(y^3)^2\right] - 2\,\|X\|\,\|Y\|\cos\theta \\ &\Longrightarrow -2(x^1y^1+x^2y^2+x^3y^3) = -2\,\|X\|\,\|Y\|\cos\theta \\ &\Longrightarrow \langle\mathbf{X},\mathbf{Y}\rangle = x^1y^1+x^2y^2+x^3y^3 = \|X\|\,\|Y\|\cos\theta\end{aligned}$$

as claimed. Now, the point of all this work is that we can now define the angle θ between two non-zero vectors $\mathbf{X}$ and $\mathbf{Y}$ as

$$\theta = \arccos\left(\frac{\langle\mathbf{X},\mathbf{Y}\rangle}{\|\mathbf{X}\|\,\|\mathbf{Y}\|}\right) \tag{2.7a}$$

2.3. Rotations in the Euclidean Space, E^2

One of the most elementary notions of Euclidean geometry is that the *length* of any given vector, and the *angle* between any two given vectors are unaffected by a rotation. Furthermore, since the laws of Newtonian mechanics are expressed in vector form, then these laws must be *covariant* under such fixed rotations.

In order to explore these ideas, let's first examine rotations in the plane. As shown in Figure 2.4, a *fixed* vector $\mathbf{V}$ is represented in terms of two coordinate systems, where the $\bar{X}$ $\bar{Y}-$ system is rotated thru an angle θ relative to the $XY-$system.

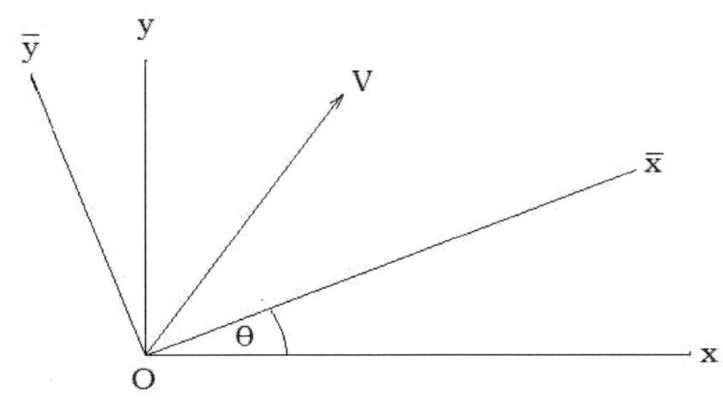

Figure 2.4. Rotation of Axes in E^2.

Let $\mathbf{i}$ and $\boldsymbol{j}$ be the standard *unit basis vectors* in the x and y-directions in the $XY-$system, and let $\bar{\imath}$ and $\bar{\mathbf{j}}$ be the corresponding unit basis vectors in the $\bar{X}\ \bar{Y}-$system. Then, there must exist real numbers c_{ab} such that

$$\begin{aligned}\bar{\imath} &= c_{11}\mathbf{i}+c_{12}\mathbf{j} \\ \bar{\mathbf{j}} &= c_{21}\mathbf{i}+c_{22}\mathbf{j}\end{aligned} \tag{2.8a}$$

Taking successive inner products, we find that

$$\begin{aligned} c_{11} &= \langle \bar{\imath}, \mathbf{i} \rangle = \cos\theta \\ c_{12} &= \langle \bar{\imath}, \mathbf{j} \rangle = \cos\left(\frac{\pi}{2} - \theta\right) = \sin\theta \\ c_{21} &= \langle \bar{\mathbf{j}}, \mathbf{i} \rangle = \cos\left(\frac{\pi}{2} + \theta\right) = -\sin\theta \\ c_{22} &= \langle \bar{\mathbf{j}}, \mathbf{j} \rangle = \cos\theta \end{aligned}$$

so that (2.8a) may be rewritten in matrix form as

$$\begin{bmatrix} \bar{\imath} \\ \bar{\mathbf{j}} \end{bmatrix} = \begin{bmatrix} \cos\theta & \sin\theta \\ -\sin\theta & \cos\theta \end{bmatrix} \begin{bmatrix} \mathbf{i} \\ \mathbf{j} \end{bmatrix} \tag{2.8b}$$

where the matrix

$$C = (c_{ab}) = \begin{bmatrix} \cos\theta & \sin\theta \\ -\sin\theta & \cos\theta \end{bmatrix} \tag{2.8c}$$

is called an **orthogonal matrix**, or **rotation matrix**.

After some simple algebra, equation (2.8b) may be inverted to obtain

$$\begin{bmatrix} \mathbf{i} \\ \mathbf{j} \end{bmatrix} = \begin{bmatrix} \cos\theta & -\sin\theta \\ \sin\theta & \cos\theta \end{bmatrix} \begin{bmatrix} \bar{\imath} \\ \bar{\mathbf{j}} \end{bmatrix} \tag{2.8d}$$

Consequently, comparing (2.8d) and (2.8b), we can conclude that

$$C^{-1} = \begin{bmatrix} \cos\theta & -\sin\theta \\ \sin\theta & \cos\theta \end{bmatrix} = \begin{bmatrix} \cos\theta & \sin\theta \\ -\sin\theta & \cos\theta \end{bmatrix}^T = C^T \tag{2.8e}$$

Clearly, the property $C^{-1} = C^T$ must be satisfied by all rotation matrices. Equation (2.8b) shows us how the basis vectors in each coordinate system transform under a rotation.

A natural question which arises is how do the *components* of the vector $\mathbf{V}$ transform? In order to answer this question, we first observe that the vector $\mathbf{V}$ is a **geometric object** whose existence is *independent* of the coordinate system employed; that is, we can write both

$$\mathbf{V} = \bar{v}^1\bar{\imath} + \bar{v}^2\bar{\mathbf{j}} \ \text{ and } \mathbf{V} = v^1\mathbf{i} + v^2\mathbf{j} \tag{2.9}$$

Consequently, substituting (8d) into the righthand side of (2.9), we obtain

$$\begin{aligned} \mathbf{V} &= \bar{v}^1\bar{\imath} + \bar{v}^2\bar{\mathbf{j}} = v^1\left(\bar{\imath}\cos\theta - \bar{\mathbf{j}}\sin\theta\right)\mathbf{i} + v^2\left(\bar{\imath}\sin\theta + \bar{\mathbf{j}}\cos\theta\right) \\ &\Longrightarrow \bar{v}^1 = v^1\cos\theta + v^2\sin\theta \ \text{ and } \ \bar{v}^2 = -v^1\sin\theta + v^2\cos\theta \end{aligned}$$

from which it follows that

$$\bar{\mathbf{V}} = \begin{bmatrix} \bar{v}^1 \\ \bar{v}^2 \end{bmatrix} = \begin{bmatrix} \cos\theta & \sin\theta \\ -\sin\theta & \cos\theta \end{bmatrix} \begin{bmatrix} v^1 \\ v^2 \end{bmatrix} = C\mathbf{V} \qquad \textit{(Passive Viewpoint)} \tag{2.10}$$

We see that, *under a rotation of coordinate systems, the components of any given vector must transform in exactly the same manner that the corresponding basis vectors of the two coordinate systems transform.*

Now, given almost any type of *transformation*, we have the choice of interpreting that transformation from either a passive viewpoint or an active viewpoint. For example, in the above analysis, we took a **passive viewpoint** of rotation; that is, we assumed that the vector **V** remained *fixed*, while the $\bar{X}\ \bar{Y}-$ coordinate system was rotated thru an angle θ *counterclockwise*, relative to the $XY-$ system.

On the other hand, we could take an **active viewpoint** of this transformation, and view the *vector* as being rotated by an angle θ in the *clockwise* direction, while the coordinate axes remain *fixed*. Both views are equivalent, and in any problem, it is often a matter of convenience as to which viewpoint we take.

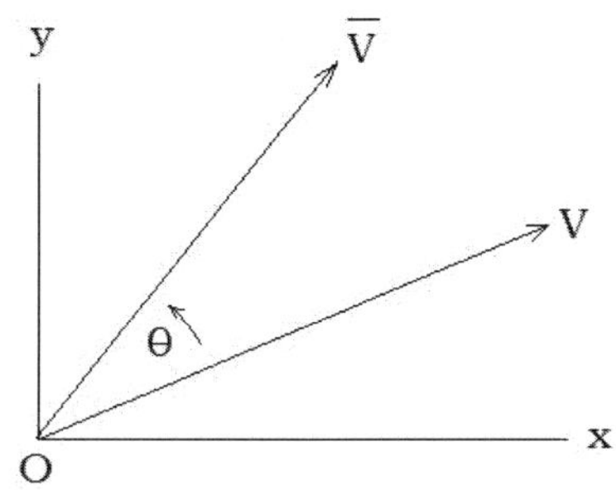

Figure 2.5. Rotation of the vector **V** in counterclockwise direction.

In fact, from an active point of view, if a vector is itself rotated in *counterclockwise* direction, by an angle θ, as shown in Figure 2.5, then we must replace θ by $-\theta$ in (2.10) to obtain

$$\bar{\mathbf{V}} = \begin{bmatrix} \bar{v}^1 \\ \bar{v}^2 \end{bmatrix} = \begin{bmatrix} \cos\theta & -\sin\theta \\ \sin\theta & \cos\theta \end{bmatrix} \begin{bmatrix} v^1 \\ v^2 \end{bmatrix} = R\mathbf{V} \quad \textit{(Active Viewpoint)} \qquad (2.10a)$$

In the last equation, we have denoted the "active rotation"matrix by R to avoid confusion. Now, equation (2.10a) allows us to confirm our intuitive notion that *the length of a vector is invariant under a rotation in the plane,* since

$$\|\bar{\mathbf{V}}\|^2 = \bar{\mathbf{V}}^T\bar{\mathbf{V}} = (R\mathbf{V})^T(R\mathbf{V}) = \mathbf{V}^T\left(R^TR\right)\mathbf{V} = \mathbf{V}^T\mathbf{V} = \|\mathbf{V}\|^2$$

In addition, we have the more general result that *the Euclidean inner product is invariant under rotations in the plane*, since if we are given any two vectors **X** and **Y**, we have

$$\langle R\mathbf{X}, R\mathbf{Y}\rangle = (R\mathbf{X})^T(R\mathbf{Y}) = \mathbf{X}^T\left(R^TR\right)\mathbf{Y} = \mathbf{X}^T\mathbf{Y} = \langle \mathbf{X}, \mathbf{Y}\rangle.$$

Stated differently, the inner product of the image vectors $R\mathbf{X}$ and $R\mathbf{Y}$ is the same as the inner product of the vectors **X** and **Y**.

In a similar fashion, employing equation (2.7a), it is easy to show that the *angle* between the image vectors $R\mathbf{X}$ and $R\mathbf{Y}$ is the also the same as the angle between the vectors **X** and **Y**.

2.4. The Group of Rotations in $\mathbf{E}^2$

It turns out that the set of rotations in E^2 forms a **group**.[3] For the remainder of our discussion, let's take an active viewpoint, and let the matrix

$$R(\theta) = \begin{bmatrix} \cos\theta & -\sin\theta \\ \sin\theta & \cos\theta \end{bmatrix} \tag{2.11}$$

represent the (active) rotation of a vector by an angle θ in the $XY-$plane.

We will now demonstrate that *the set of all such rotations in the plane forms a group under the operation of matrix multiplication.*

First of all, if we have two successive rotations thru angles θ_1 and θ_2, then the product of these two rotations is another rotation thru an angle $(\theta_1 + \theta_2)$, since

$$\begin{aligned} R(\theta_1)R(\theta_2) &= \begin{bmatrix} \cos\theta_1 & -\sin\theta_1 \\ \sin\theta_1 & \cos\theta_1 \end{bmatrix} \begin{bmatrix} \cos\theta_2 & -\sin\theta_2 \\ \sin\theta_2 & \cos\theta_2 \end{bmatrix} \\ &= \begin{bmatrix} \cos(\theta_1+\theta_2) & -\sin(\theta_1+\theta_2) \\ \sin(\theta_1+\theta_2) & \cos(\theta_1+\theta_2) \end{bmatrix} = R(\theta_1+\theta_2) \end{aligned} \tag{2.11a}$$

Consequently, matrix multiplication is a *binary operation* on the set of such matrices. Furthermore, matrix multiplication is *associative*, so the second group property is satisfied. The *identity element* of the group is $R(0)$, since

$$R(0) = \begin{bmatrix} \cos(0) & -\sin(0) \\ \sin(0) & \cos(0) \end{bmatrix} = \begin{bmatrix} 1 & 0 \\ 0 & 1 \end{bmatrix} = I \tag{2.11b}$$

where, I is the 2×2 identity matrix. Finally, to each element $R(\theta)$ there corresponds an inverse, $R^{-1}(\theta)$, such that

$$R^{-1}(\theta) = R(-\theta) \tag{2.12}$$

and clearly, we have both

$$\begin{aligned} R(\theta)R^{-1}(\theta) &= R(\theta)R(-\theta) = R(0) = I \\ R^{-1}(\theta)R(\theta) &= R(-\theta)R(\theta) = R(0) = I \end{aligned}$$

so that the last group property is satisfied.

The group of rotations in E^2 is called the **special orthogonal group**, denoted by $SO(2)$. The term "special" is derived from the fact that such matrices only represent **proper rotations**; that is, rotations such that

$$\det(R) = +1 \tag{2.13}$$

It is easy to show that $SO(2)$ is a **commutative group** in that the *order* of successive rotations in the plane is immaterial.

2.5. The Group of Rotations in E^3

We now turn our attention to rotations in the three dimensional Euclidean space, E^3. A general rotation in E^3 is represented by a transformation of the form

$$\bar{\mathbf{V}} = \begin{bmatrix} \bar{v}^1 \\ \bar{v}^2 \\ \bar{v}^3 \end{bmatrix} = R \begin{bmatrix} v^1 \\ v^2 \\ v^3 \end{bmatrix} = R\mathbf{V} \tag{2.14}$$

Just like rotations in the plane, spatial rotations must preserve the length of an arbitrary vector $\mathbf{V}$; consequently, we must have:

$$\|\bar{\mathbf{V}}\|^2 = \bar{\mathbf{V}}^T\bar{\mathbf{V}} = (R\mathbf{V})^T R\mathbf{V} = \mathbf{V}^T(R^T R)\mathbf{V} = \mathbf{V}^T\mathbf{V} = \|\mathbf{V}\|^2$$

Since $\mathbf{V}$ is arbitrary, we are forced to conclude that

$$R^T R = I \tag{2.15}$$

where I is the 3×3 identity matrix; thus, R must be an orthogonal matrix.

The set of orthogonal matrices, under the operation of matrix multiplication, forms a group called the **orthogonal group**, $O(3)$. To see this, observe that the product of any two orthogonal matrices, say R_1 and R_2, is again an orthogonal matrix, since

$$\begin{aligned} (R_1R_2)^T(R_1R_2) &= R_2^T(R_1^T R_1)R_2 = R_2^T I R_2 \\ &= R_2^T R_2 = I \end{aligned}$$

Also, the matrix I serves as the identity element for the group. Furthermore, any given orthogonal matrix R is invertible, since $R^{-1} = R^T$ always exists.

For any orthogonal matrix, R, we have

$$\det\left(R^T R\right) = (\det R)^2 = 1 \Longrightarrow \det(R) = \pm 1 \tag{2.15a}$$

where *proper rotations* are described by matrices which satisfy $\det(R) = +1$, while *improper rotations* correspond to those orthogonal matrices such that $\det(R) = -1$.

The **special orthogonal group**, $SO(3)$, is a **subgroup**[4] of $O(3)$, and it consists of the set of all *proper spatial rotations* where we restrict $\det(R) = +1$. In order to understand the behaviour of this group, we first note that (passive) rotations around each coordinate axis may be written as

$$R_x(\theta) = \begin{bmatrix} 1 & 0 & 0 \\ 0 & \cos\theta & \sin\theta \\ 0 & -\sin\theta & \cos\theta \end{bmatrix} \tag{2.16a}$$

$$R_y(\phi) = \begin{bmatrix} \cos\phi & 0 & \sin\phi \\ 0 & 1 & 0 \\ -\sin\phi & 0 & \cos\phi \end{bmatrix} \tag{2.16b}$$

$$R_z(\psi) = \begin{bmatrix} \cos\psi & \sin\psi & 0 \\ -\sin\psi & \cos\psi & 0 \\ 0 & 0 & 1 \end{bmatrix} \tag{2.16c}$$

Now, a simple calculation shows that

$$R_x(\theta)R_y(\phi) \neq R_y(\phi)R_x(\theta)$$

so that $SO(3)$, unlike its counterpart $SO(2)$, must be a **non-commutative group**. On the other hand, like its counterpart, $SO(3)$ is an example of a special type of *continuous group*, called a **Lie group**.

A **continuous group** or **Lie group**[5] is one in which each element of the group can be labeled by a set of r *independent*, real parameters, say, $\theta_1, \theta_2, ..., \theta_r$, where each parameter can range in a continuous fashion over some real interval. We shall assume that each element of the group $g(\theta_1, \theta_2, ..., \theta_r)$ corresponds to a *unique point* $P(\theta_1, \theta_2, ..., \theta_r)$ of an r dimensional vector space, S^r, which is equipped with an inner product. The space S^r is called the **group space**, or the **parameter space**.

If the *smallest number* of parameters necessary to characterize a typical element of the group is r, then we say that the **order** of the group is r. For example, $SO(2)$ is a continuous group of order one, since each element may be written as $R(\theta)$, where the parameter θ is allowed to continuously assume values in the interval $[0, 2\pi]$.

Now, let G be a continuous group of order r, so that each element of the group may be characterized by r independent parameters, and a typical element of G may be written as $g(a_1, a_2, ..., a_r)$. As a matter of convenience, we shall choose the parameters such that the identity element of the group is given by $I = g(0, 0, ..., 0)$.

Assume that $g_1 = g(\theta_1, ..., \theta_r)$ and $g_2 = g(\phi_1, ..., \phi_r)$ are elements of G, and $g_3 = g(\psi_1, ...\psi_r) = g_1 g_2$. It is clear that the group parameters corresponding to g_3 can be expressed as *functions* of the parameters corresponding to g_1 and g_2, so that there must exist r functions of the form

$$\psi_k = \psi_k(\theta_1, ...\theta_r; \phi_1, ..., \phi_r) \quad (\text{ where } k = 1, 2, 3, ..., r) \tag{2.17a}$$

Similarly, if $g_1^{-1} = g(\omega_1, ...\omega_r)$ is the inverse of g_1, then the parameters of the element g_1^{-1} can be expressed as functions of the parameters corresponding to g_1:

$$\omega_k = \omega_k(\theta_1, ..., \theta_r) \quad (\text{ where } k = 1, 2, 3, ..., r) \tag{2.17b}$$

We shall say that a continuous group is a **Lie group** if there exists an *open neighborhood* N_ε of the identity element $I = g(0, ..., 0)$, that is,

$$\|P(g) - P(I)\| < \varepsilon$$

for some $\varepsilon > 0$, where $P(g), P(I) \in S^r$, such that the functions in (2.17a) and (2.17b) are continuously differentiable functions of their respective parameters. Stated differently, for infinitesimal elements whose parameters are sufficiently close to the origin $P(0, 0, ..., 0)$ of the parameter space S^r, the operations of group multiplication and inversion possess continuous derivatives.

As we shall see, the behaviour of any element which is a *finite* distance away from the origin can be found by examining the behaviour of that element near the origin. Now, $SO(3)$ is a three-parameter Lie group, and any of its elements may be written in the form $R(\theta, \phi, \psi)$.

If we restrict our attention to those elements of $SO(3)$ near the identity, i.e., elements of the form $R(\delta\theta, \delta\phi, \delta\psi)$, where $\delta\theta$, $\delta\phi$, and $\delta\psi$ are *infinitesimal* parameters, then by Taylor's theorem, we have up to first order terms.

$$\begin{aligned} R(\delta\theta,\delta\phi,\delta\psi) &= R(0,0,0)+\delta\theta\left.\frac{\partial R}{\partial\theta}\right|_{\theta=0}+\delta\phi\left.\frac{\partial R}{\partial\phi}\right|_{\phi=0}+\delta\psi\left.\frac{\partial R}{\partial\psi}\right|_{\psi=0} \\ &= I+iJ_x\,(\delta\theta)+iJ_y\,(\delta\phi)+iJ_z\,(\delta\psi) \end{aligned} \tag{2.18}$$

where the matrices J_x, J_y, and J_z are called the **infinitesimalgenerators** of $SO(3)$, and are given by

$$J_x=\frac{1}{i}\left.\frac{\partial R}{\partial\theta}\right|_{\theta=0}=\begin{bmatrix}0&0&0\\0&0&-i\\0&i&0\end{bmatrix} \tag{2.19a}$$

$$J_y=\frac{1}{i}\left.\frac{\partial R}{\partial\phi}\right|_{\phi=0}=\begin{bmatrix}0&0&i\\0&0&0\\-i&0&0\end{bmatrix} \tag{2.19b}$$

$$J_z=\frac{1}{i}\left.\frac{\partial R}{\partial\psi}\right|_{\psi=0}=\begin{bmatrix}0&-i&0\\i&0&0\\0&0&0\end{bmatrix} \tag{2.19c}$$

The infinitesimal generators are **Hermitean** matrices[6], and it is easy to show that they are related to infinitesimal rotations about the coordinate axes by the relations

$$R_x(\delta\theta)=I+iJ_x(\delta\theta),\quad R_y(\delta\phi)=I+iJ_y(\delta\phi),\quad R_z(\delta\psi)=I+iJ_z(\delta\psi) \tag{2.20}$$

Given any two square matrices of the same order, say A and B, their **commutator** is defined to be

$$[A,B]=AB-BA \tag{2.21}$$

Thus, two matrices *commute*, $AB=BA$, if and only if $[A,B]=0$. It is easy to show that the infinitesimal generators do not commute, and they satisfy the *commutation relations*

$$[J_x,J_y]=iJ_z,\quad [J_y,J_z]=iJ_x,\quad [J_z,J_x]=iJ_y \tag{2.22}$$

Readers familiar with elementary Quantum Mechanics will recognize that, apart from a missing factor of $\hbar=h/2\pi$, the equations in (2.22) are just the commutation relations satisfied by the angular momentum operators. Roughly speaking, we can then say that the infinitesimal generators of $SO(3)$ are none other than the angular momentum operators.

Using the infinitesimal generators, we now derive the rotation matrix corresponding to any finite rotation. To see this, let's take a look at a typical rotation thru an angle ψ about the z-axis. If we assume that $\psi=N(\delta\psi)$ where $N\to\infty$, then

$$\begin{aligned} R_z(\psi) &= \lim_{N\to\infty}\left[I+iJ_z(\delta\psi)\right]^N=\lim_{N\to\infty}\left[I+iJ_z(\frac{\psi}{N})\right]^N \\ &= \exp\left(iJ_z\psi\right) \end{aligned} \tag{2.23}$$

where we have used the well known relation

$$\lim_{N\to\infty}\left(1+\frac{x}{N}\right)^N = e^x$$

Now, the amazing thing about equation (2.23) is that it represents an *exact* equality. To see this, we just define the exponential of any square matrix A by its Taylor series expansion:

$$e^A = \exp(A) = I + A + \frac{A^2}{2!} + \frac{A^3}{3!} + \ldots \tag{2.24}$$

so that (2.23) becomes

$$\exp(iJ_z\psi) = I + iJ_z\psi - J_z^2\frac{\psi^2}{2!} - iJ_z^3\frac{\psi^3}{3!} + \ldots$$

$$= \begin{bmatrix} 1 & 0 & 0 \\ 0 & 1 & 0 \\ 0 & 0 & 1 \end{bmatrix} + \begin{bmatrix} 0 & \psi & 0 \\ -\psi & 0 & 0 \\ 0 & 0 & 0 \end{bmatrix} + \begin{bmatrix} -\frac{1}{2!}\psi^2 & 0 & 0 \\ 0 & -\frac{1}{2!}\psi^2 & 0 \\ 0 & 0 & 0 \end{bmatrix} + \ldots$$

$$= \begin{bmatrix} (1-\frac{1}{2!}\psi^2+\ldots) & (\psi-\frac{1}{3!}\psi^3+\ldots) & 0 \\ -(\psi-\frac{1}{3!}\psi^3+\ldots) & (1-\frac{1}{2!}\psi^2+\ldots) & 0 \\ 0 & 0 & 1 \end{bmatrix} = \begin{bmatrix} \cos\psi & \sin\psi & 0 \\ -\sin\psi & \cos\psi & 0 \\ 0 & 0 & 1 \end{bmatrix}$$

which is in agreement with (2.16c).

In order to generalize (2.23), let's now suppose that $\delta\theta$ is an infinitesimal (active) rotation of a vector $\mathbf{V}$ about a fixed axis, which coincides with the direction of the *unit* vector

$$\mathbf{n} = n_1\mathbf{i} + n_2\mathbf{j} + n_3\mathbf{k}.$$

This situation is shown in Figure 2.6.

From the diagram, it is easy to see, to within terms of first order, that

$$\delta\mathbf{V} = (\mathbf{n}\times\mathbf{V})\,\delta\theta$$

and consequently,

$$\bar{\mathbf{V}} = \mathbf{V} + \delta\mathbf{V} = \mathbf{V} + (\mathbf{n}\times\mathbf{V})\,\delta\theta = \begin{bmatrix} v^1 + (n_2v^3 - n_3v^2)\delta\theta \\ v^2 + (n_3v^1 - n_1v^3)\delta\theta \\ v^3 + (n_1v^2 - n_2v^1)\delta\theta \end{bmatrix}$$

Now, the last equation can written more suggestively as

$$\bar{\mathbf{V}} = R_{\mathbf{n}}(\delta\theta)\mathbf{V} = \begin{bmatrix} 1 & -n_3\delta\theta & n_2\delta\theta \\ n_3\delta\theta & 1 & -n_1\delta\theta \\ -n_2\delta\theta & n_1\delta\theta & 1 \end{bmatrix}\mathbf{V} \tag{2.25}$$

From (2.25), we see that the matrix $R_{\mathbf{n}}(\delta\theta)$, which represents an infinitesimal rotation is *antisymmetric*; furthermore, this matrix corresponds to a counterclockwise rotation about

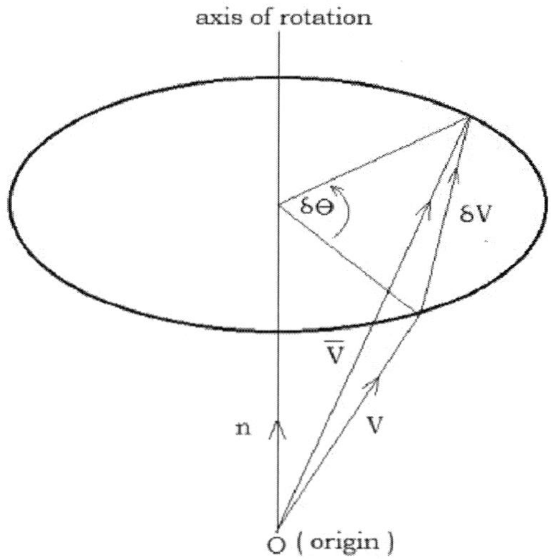

Figure 2.6. Rotation of **V** about the axis **n**.

the axis **n**, through an infinitesimal angle $\delta\theta$. Recalling equations (2.19a) - (2.19c), equation (2.25) may be rewritten as

$$R_{\mathbf{n}}(\delta\theta) = I + i\,(\mathbf{J}\cdot\mathbf{n})\,\delta\theta \tag{2.26}$$

where we have set $\mathbf{J} = J_x\mathbf{i} + J_y\mathbf{j} + J_z\mathbf{k}$.

Once again, we can find the matrix corresponding to a *finite* rotation θ about the axis n by repeated application of (2.26), with $\delta\theta = \theta/N$ as $N \longrightarrow \infty$, to obtain

$$R_{\mathbf{n}}(\theta) = \lim_{N\longrightarrow\infty}\left[I + i\,(\mathbf{J}\cdot\mathbf{n})\,\frac{\theta}{N}\right]^N = \exp\{i\,[(J_x n_1 + J_y n_2 + J_z n_3)\theta]\} \tag{2.27}$$

Finally, we can conclude that if θ is a finite rotation about a fixed axis having the direction of the *unit* vector **n**, then the matrix representing this rotation is

$$R_{\mathbf{n}}(\theta) = \exp\left[i(\mathbf{J}\cdot\mathbf{n})\theta\right] \tag{2.28}$$

In evaluating (2.28), however, we must remember that the infinitesimal generators do not commute.

2.6. The General Galilean Transformation

After our little mathematical detour, we can now discuss Galilean transformations in a manner which is much more general than in the past. Assume there are two inertial frames, $S(t,x,y,z)$ and $\bar{S}(\bar{t},\bar{x},\bar{y},\bar{z})$, where the origin of the second frame $\bar{S}$ moves at a *constant velocity* $\mathbf{v} = v_x\mathbf{i} + v_y\mathbf{j} + v_z\mathbf{k}$ relative to the first frame, as shown in Figure 2.7.

We *do not* assume, however, that the coordinate axes of $\bar{S}$ are parallel to the corresponding axes of S. In addition, we assume that the origin of $\bar{S}$ has an initial displacement $\mathbf{r} = \mathbf{a}$,

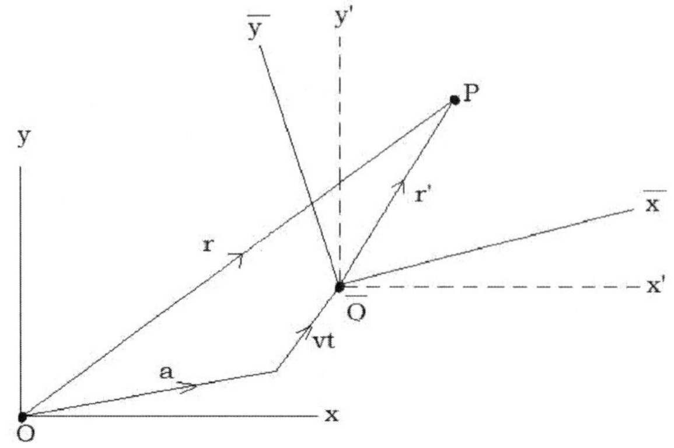

Figure 2.7.

(relative to S), at the time $t = \bar{t} = 0$. Also note in Figure 2.7, for the sake of clarity, we have shown only two space dimensions.

We require the transformation equations between the space and time coordinates in each system. To accomplish this task, by a appropriate rotation R, we can construct a coordinate system (shown by dashed lines) $S'(t', x', y', z')$ whose origin coincides with that of the barred system, but whose axes are parallel to those of $S(t, x, y, z)$, such that

$$\bar{\mathbf{r}} = R\mathbf{r}'$$

But, from the diagram,

$$\mathbf{r}' = \mathbf{r} - \mathbf{v}t - \mathbf{a}$$

so that

$$\bar{\mathbf{r}} = R(\mathbf{r} - \mathbf{v}t - \mathbf{a}) \tag{2.29}$$

We conclude that the most **general Galilean transformation** can be written as

$$\begin{aligned} t &\longrightarrow \bar{t} = t \\ \mathbf{r} &\longrightarrow \bar{\mathbf{r}} = R(\mathbf{r} - \mathbf{v}t - \mathbf{a}) \end{aligned} \tag{2.30}$$

where the matrix $R \in SO(3)$.

The set of transformations defined by (2.30) forms a group, under composition of mappings, called the **general Galilean group**[7]**;** this is the **symmetry group** of Newtonian spacetime in the sense that Newton's laws are *covariant* with respect to any transformation of coordinates affected by a an element of the general Galilean group. This group encodes the Principle of Galilean covariance, and its elements consist of proper rotations, translations, and boosts.

2.7. Exercises

1. Show that the group $SO(2)$ is Abelian.

(Solution)

Let $R(\theta_1)$ and $R(\theta_2)$ be any elements of $SO(2)$, then by (2.11a), we have

$$R(\theta_1)R(\theta_2) = R(\theta_1 + \theta_2) = R(\theta_2 + \theta_1) = R(\theta_2)R(\theta_1)$$

where we have used the commutativity of the addition of real numbers, θ_1 and θ_2.

2. Show that for any vectors $\mathbf{X}$ and $\mathbf{Y}$ in E^3,

$$|\langle \mathbf{X}, \mathbf{Y} \rangle|^2 \leq \|\mathbf{X}\|^2 \cdot \|\mathbf{Y}\|^2 \tag{2.31}$$

This is called the **Cauchy-Schwartz inequality**.

(Solution#1)

By equation (2.7), we can write

$$|\langle \mathbf{X}, \mathbf{Y} \rangle|^2 = \|\mathbf{X}\|^2 \|\mathbf{Y}\|^2 |\cos\theta|^2 \leq \|\mathbf{X}\|^2 \cdot \|\mathbf{Y}\|^2$$

since $|\cos\theta|^2 \leq 1$.

(Solution #2)

Consider the vector $\mathbf{X} + \lambda\mathbf{Y}$ where λ is any real number. We must have

$$\|\mathbf{X} + \lambda\mathbf{Y}\|^2 = \langle \mathbf{X} + \lambda\mathbf{Y}, \mathbf{X} + \lambda\mathbf{Y} \rangle = \langle \mathbf{X}, \mathbf{X} \rangle + 2\langle \mathbf{X}, \mathbf{Y} \rangle \lambda + \langle \mathbf{Y}, \mathbf{Y} \rangle \lambda^2 \geq 0$$

Let $a = \langle \mathbf{Y}, \mathbf{Y} \rangle$, $b = 2\langle \mathbf{X}, \mathbf{Y} \rangle$, and $c = \langle \mathbf{X}, \mathbf{X} \rangle$ so the last inequality may be written as

$$a\lambda^2 + b\lambda + c \geq 0$$

which is a quadratic inequality in λ. This inequality can only be satisfied if the graph of the function $y = a\lambda^2 + b\lambda + c$ is either above the λ-axis, or touchs that axis at just one point. Consequently, the discriminant must satisfy

$$b^2 - 4ac \leq 0$$

or, in other words,

$$[2\langle \mathbf{X}, \mathbf{Y} \rangle]^2 - 4\langle \mathbf{X}, \mathbf{X} \rangle \langle \mathbf{Y}, \mathbf{Y} \rangle \leq 0$$

from which $|\langle \mathbf{X}, \mathbf{Y} \rangle|^2 \leq \|\mathbf{X}\|^2 \cdot \|\mathbf{Y}\|^2$ follows.

3. If $\mathbf{X}$ and $\mathbf{Y}$ are arbitrary vectors in E^3, show that they satisfy the **triangle inequality**:

$$\|\mathbf{X} + \mathbf{Y}\| \leq \|\mathbf{X}\| + \|\mathbf{Y}\| \tag{2.32}$$

(Solution)

We have

$$\begin{aligned}\|\mathbf{X}+\mathbf{Y}\|^2 &= \langle \mathbf{X}+\mathbf{Y},\mathbf{X}+\mathbf{Y}\rangle = \langle \mathbf{X},\mathbf{X}\rangle + 2\langle \mathbf{X},\mathbf{Y}\rangle + \langle \mathbf{Y},\mathbf{Y}\rangle = \|\mathbf{X}\|^2 + 2\langle \mathbf{X},\mathbf{Y}\rangle + \|\mathbf{Y}\|^2 \\ &\leq \|\mathbf{X}\|^2 + 2(\|\mathbf{X}\|\cdot\|\mathbf{Y}\|) + \|\mathbf{Y}\|^2 \quad [\text{ by (2.31) }] \\ &\leq (\|\mathbf{X}\| + \|\mathbf{Y}\|)^2\end{aligned}$$

from which (2.32) follows.

4. Show that for an infinitesimal rotation $\delta\theta$ about the x-axis in E^3:

$$R_x(\delta\theta) = I + iJ_x(\delta\theta)$$

(Solution)

Since $\delta\theta$ is infinitesimal, then up to first-order terms, we have from (2.16a),

$$\begin{aligned}R_x(\delta\theta) &= \begin{bmatrix} 1 & 0 & 0 \\ 0 & \cos(\delta\theta) & \sin(\delta\theta) \\ 0 & -\sin(\delta\theta) & \cos(\delta\theta)\end{bmatrix} \approx \begin{bmatrix} 1 & 0 & 0 \\ 0 & 1 & \delta\theta \\ 0 & -\delta\theta & 1\end{bmatrix} \\ \Longrightarrow \quad & R_x(\delta\theta) = I + i\delta\theta \begin{bmatrix} 0 & 0 & 0 \\ 0 & 0 & -i \\ 0 & i & 0\end{bmatrix} = I + iJ_x(\delta\theta)\end{aligned}$$

5. By direct calculation, use (2.16) to show that

$$R_x(\theta)R_y(\phi) \neq R_y(\phi)R_x(\theta)$$

and hence, $SO(3)$ is a non-commutative group.

(Solution)

$$\begin{aligned}R_x(\theta)R_y(\phi) &= \begin{bmatrix} 1 & 0 & 0 \\ 0 & \cos\theta & \sin\theta \\ 0 & -\sin\theta & \cos\theta\end{bmatrix}\begin{bmatrix} \cos\phi & 0 & \sin\phi \\ 0 & 1 & 0 \\ -\sin\phi & 0 & \cos\phi\end{bmatrix} \\ &= \begin{bmatrix} \cos\phi & 0 & \sin\phi \\ -\sin\theta\sin\phi & \cos\theta & \cos\phi\sin\theta \\ -\cos\theta\sin\phi & -\sin\theta & \cos\theta\cos\phi\end{bmatrix}\end{aligned}$$

while

$$\begin{aligned}R_y(\phi)R_x(\theta) &= \begin{bmatrix} \cos\phi & 0 & \sin\phi \\ 0 & 1 & 0 \\ -\sin\phi & 0 & \cos\phi\end{bmatrix}\begin{bmatrix} 1 & 0 & 0 \\ 0 & \cos\theta & \sin\theta \\ 0 & -\sin\theta & \cos\theta\end{bmatrix} \\ &= \begin{bmatrix} \cos\phi & -\sin\theta\sin\phi & \cos\theta\sin\phi \\ 0 & \cos\theta & \sin\theta \\ -\sin\phi & -\cos\phi\sin\theta & \cos\theta\cos\phi\end{bmatrix}\end{aligned}$$

So, $R_x(\theta)R_y(\phi) \neq R_y(\phi)R_x(\theta)$, as claimed.

6. Verify the commutation relations in (2.22).

(Solution)

We demonstrate only the last relation in (2.22), as the derivation of the remaining relations is similar. Observe that

$$\begin{aligned}[J_x, J_y] &= J_xJ_y - J_yJ_x \\ &= \begin{bmatrix} 0 & 0 & 0 \\ 0 & 0 & -i \\ 0 & i & 0 \end{bmatrix}\begin{bmatrix} 0 & 0 & i \\ 0 & 0 & 0 \\ -i & 0 & 0 \end{bmatrix} - \begin{bmatrix} 0 & 0 & i \\ 0 & 0 & 0 \\ -i & 0 & 0 \end{bmatrix}\begin{bmatrix} 0 & 0 & 0 \\ 0 & 0 & -i \\ 0 & i & 0 \end{bmatrix}\end{aligned}$$

Consequently, after some algebra, we find

$$[J_x, J_y] = \begin{bmatrix} 0 & 1 & 0 \\ -1 & 0 & 0 \\ 0 & 0 & 0 \end{bmatrix} = i\begin{bmatrix} 0 & -i & 0 \\ i & 0 & 0 \\ 0 & 0 & 0 \end{bmatrix} = iJ_z$$

as claimed.

7. Consider the set of all transformations from R^2 into R^2, of the form

$$\bar{\mathbf{x}} = \begin{bmatrix} \bar{x}^1 \\ \bar{x}^2 \end{bmatrix} = \begin{bmatrix} a_{11} & a_{12} \\ a_{21} & a_{22} \end{bmatrix}\begin{bmatrix} x_1 \\ x_2 \end{bmatrix} = A\mathbf{x}$$

where $\det(A) \neq 0$. Show that the set of all such matrices A forms a group, under the operation of matrix multiplication.

[It is a continuous group, known as the **general linear group** in two dimensions, $GL(2)$.]

(Solution)

Let A and B be any two non-singular matrices, then $AB \in GL(2)$ since

$$\det(AB) = \det(A)\det(B) \neq 0$$

Moreover, the group operation is associative since matrix multiplication is associative.

Also, the 2×2 identity matrix I serves as the identity element of $GL(2)$. Finally, the inverse of any element A is given by

$$A^{-1} = \frac{1}{\det(A)}\begin{bmatrix} a_{22} & -a_{12} \\ -a_{21} & a_{11} \end{bmatrix}$$

We are assured that A^{-1} exists, since $\det(A) \neq 0$.

8. *(Galilean Group)* Show that the Galilean transformations of the form:

$$\begin{aligned} t &\longrightarrow \bar{t} = t \\ \mathbf{r} &\longrightarrow \bar{\mathbf{r}} = R(\mathbf{r} - \mathbf{v}t - \mathbf{a}) \end{aligned}$$

where $R \in SO(3)$, comprise a group under the *composition* of such mappings.

(Solution)

Let $\mathbf{G}_1(\mathbf{r}_1) = R_1(\mathbf{r}_1 - \mathbf{v}_1 t - \mathbf{a}_1)$ and $\mathbf{G}_2(\mathbf{r}_2) = R_2(\mathbf{r}_2 - \mathbf{v}_2 t - \mathbf{a}_2)$ then the *composition* of these mappings is *closed*, since

$$\begin{aligned}
\mathbf{G}_3(\mathbf{r}_1) &= (\mathbf{G}_2 \circ \mathbf{G}_1)(\mathbf{r}_1) = \mathbf{G}_2\left[\mathbf{G}_1(\mathbf{r}_1)\right] \\
&= R_2\left[R_1(\mathbf{r}_1 - \mathbf{v}_1 t - \mathbf{a}_1) - \mathbf{v}_2 t - \mathbf{a}_2\right] \\
&= (R_2 R_1)(\mathbf{r}_1 - \mathbf{v}_1 t - \mathbf{a}_1) - R_2(\mathbf{v}_2 t + \mathbf{a}_2) \\
&= (R_2 R_1)(\mathbf{r}_1 - \mathbf{v}_1 t - \mathbf{a}_1) - (R_2 R_1) R_1^{-1}(\mathbf{v}_2 t + \mathbf{a}_2) \\
&= (R_2 R_1)\left[(\mathbf{r}_1 - \left(\mathbf{v}_1 + R_1^{-1}\mathbf{v}_2\right) t - \left(\mathbf{a}_1 + R_1^{-1}\mathbf{a}_2\right)\right] \\
&= R_3(\mathbf{r}_1 - \mathbf{v}_3 t - \mathbf{a}_3)
\end{aligned}$$

where

$$\begin{aligned}
R_3 &= R_2 R_1 \in SO(3) \\
\mathbf{v}_3 &= \left(\mathbf{v}_1 + R_1^{-1}\mathbf{v}_2\right) \\
\mathbf{a}_3 &= \left(\mathbf{a}_1 + R_1^{-1}\mathbf{a}_2\right)
\end{aligned}$$

Clearly, the identity element of the group is the 3×3 identity matrix I, and the inverse of any element $\mathbf{G}(\mathbf{r}) = R(\mathbf{r} - \mathbf{v}t - \mathbf{a})$ is the element

$$\mathbf{G}^{-1}(\mathbf{r}) = R^{-1}(\mathbf{r} + R\mathbf{v}t + R\mathbf{a})$$

since

$$\begin{aligned}
\left(GG^{-1}\right)(\mathbf{r}) &= R\left[R^{-1}(\mathbf{r} + R\mathbf{v}t + R\mathbf{a}) - \mathbf{v}t - \mathbf{a}\right] \\
&= I(\mathbf{r} + R\mathbf{v}t + R\mathbf{a}) - R\mathbf{v}t - R\mathbf{a} \\
&= I\mathbf{r} = \mathbf{r}
\end{aligned}$$

and,

$$\begin{aligned}
\left(G^{-1}G\right)(\mathbf{r}) &= R^{-1}\left[R(\mathbf{r} - \mathbf{v}t - \mathbf{a}) + R\mathbf{v}t + R\mathbf{a}\right] \\
&= I(\mathbf{r} - \mathbf{v}t - \mathbf{a}) + \mathbf{v}t + \mathbf{a} \\
&= I\mathbf{r} = \mathbf{r}
\end{aligned}$$

as well.

Chapter 3

Maxwell's Equations

"From a long view of the history of mankind, the most significant event of the nineteenth century will be judged as Maxwell's discovery of the laws of electrodynamics. "

Richard P. Feynman

3.1. Introduction

At first, Newton's view of space and time, and his mechanistic approach to physics, was spectacularly successful. His theory, along with its application to many related areas of physics, had given physicists an accurate view of physical phenomena. Unfortunately, all of this changed with the advent of Maxwell's theory of electromagnetic waves.

In his *Treatise on Electricity and Magnetism*, first published in 1873, Maxwell unified all electric and magnetic phenomena, and predicted the existence of electromagnetic waves. In just four equations, known as *Maxwell's equations* in his honor, he gave mathematical life to Michael Faraday's intuitive notion of "tubes of force", and created the first classical field theory.

Newton's idea of direct *action at a distance* which had been so successful in explaining gravitational interactions, was replaced by Maxwell's notion of a *field*. The electromagnetic field was to serve as an *intermediary agent* which regulated the interaction of material particles. Maxwell's theory might be called a *classical field theory* since its involves two types of laws.

The first set of laws might be called the *field equations* which impose *geometric* constraints on the nature of the field itself, its time evolution, and its response to the presence of matter, and a second set of laws that provide the *equations of motion* of material particles that move under the influence of the field.[1]

In our discussion of Maxwell's equations, we shall restrict ourselves to his equations for *free space*, and show how this set of equations leads to the prediction of electromagnetic waves. More importantly, we will demonstrate that Maxwell's equations are *not invariant* under the group of Galilean transformations, and consequently, this group must be replaced by another more general symmetry group, called the **Lorentz group**, which can incorporate all electromagnetic phenomena.

Maxwell's equations[2] are:

Maxwell's Equations

$$\nabla \cdot \mathbf{E} = 4\pi\rho \tag{3.1}$$

$$\nabla \cdot \mathbf{B} = 0 \tag{3.2}$$

$$\nabla \times \mathbf{E} = -\frac{1}{c}\frac{\partial \mathbf{B}}{\partial t} \tag{3.3}$$

$$\nabla \times \mathbf{B} = \frac{4\pi}{c}\mathbf{j} + \frac{1}{c}\frac{\partial \mathbf{E}}{\partial t} \tag{3.4}$$

Here, **E** is the **electric field** (dyne/esu), **B** is the **magnetic field**[3] (gauss), ρ is the **charge density** (esu/cm^3), **j** is the **electric current density** (statamp/cm^2), and of course, c is the **speed of light** in a vacuum:

$$c = 2.998 \times 10^{10} \text{ cm/sec} \tag{3.5}$$

Equations (3.1)-(3.4) are Maxwell's **field equations**. Equation (3.1) is *Gauss's law for electric fields*, (3.2) is called *Gauss's law for magnetic fields*, (3.3) is known as *Faraday's law*, and finally, (3.4) is *Ampere's law*.

In addition, the force **F** that a particle of charge q experiences in the presence of an electromagnetic field is given by the **Lorentz force law**:

$$\mathbf{F} = q\mathbf{E} + \frac{q}{c}(\mathbf{v} \times \mathbf{B}) \tag{3.6}$$

where **v** is the velocity of the particle. From Newton's second law, we see that the *equation of motion* of a charged particle of mass m must be

$$m\frac{d\mathbf{v}}{dt} = q\mathbf{E} + \frac{q}{c}(\mathbf{v} \times \mathbf{B}) \tag{3.6a}$$

Equations (3.1) - (3.4), along with (3.6), are sufficient to describe the full range of macroscopic electromagnetic interactions, and to a large extent, all of the laws of electromagnetics may be viewed as mere corollaries to these basic laws.

Before we demonstrate how these equations may be used to predict the existence of electromagnetic waves, we will derive each equation, and conduct a review of the basic laws which govern electricity and magnetism.

3.2. Coulomb's Law

Consider two point charges, q' and q, with position vectors $\mathbf{x}'$ and $\mathbf{x}$, respectively, as shown in Figure 3.1. The electric force **F** that q' exerts on q is given by **Coulomb's law**

$$\mathbf{F} = \frac{qq'}{\|\mathbf{x} - \mathbf{x}'\|^2}\,\hat{\mathbf{e}}_{\mathbf{x}-\mathbf{x}'} \quad \text{where } \hat{\mathbf{e}}_{\mathbf{x}-\mathbf{x}'} = \frac{\mathbf{x} - \mathbf{x}'}{\|\mathbf{x} - \mathbf{x}'\|} \tag{3.7}$$

Observe that $\hat{\mathbf{e}}_{\mathbf{x}-\mathbf{x}'}$ is a *unit* vector in the direction of the vector $\mathbf{x} - \mathbf{x}'$.

From equation (3.7), we see that the force between two point charges varies inversely as the square of the distance between them, and that this force acts along the line connecting the two charges.

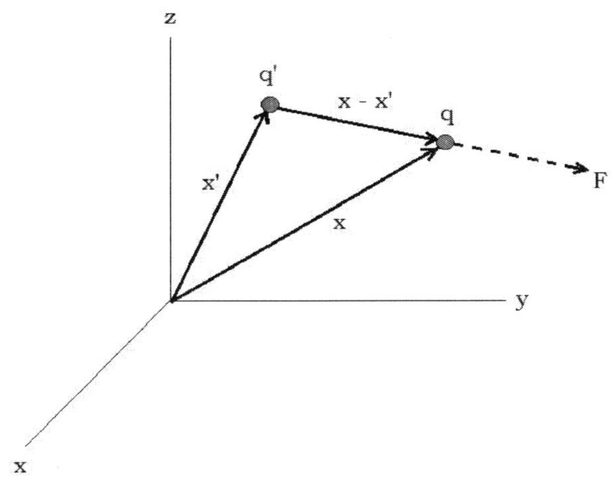

Figure 3.1. Coulomb's Law.

The force **F** may be either *repulsive* or *attractive*, depending upon the signs of the charges q and q'. If these charges have the same sign, then the force is repulsive; if they have opposite signs, the force is attractive.

In equation (3.7), we assume that all of the charges are expressed in the **CGS electrostatic system** of units; that is, in accordance with Coulomb's law, we *define*

$$1 \text{ dyne } = (1 \text{ esu})^2/(1 \text{ cm})^2 \tag{3.7a}$$

Thus, given two identical point charges, placed one centimeter apart, we say that the magnitude of each charge is one **electrostatic unit** (esu) if the force exerted by one charge on the other is one dyne.

As shown in Figure 3.2, we can imagine **lines of force** emerging radially outward from a positive point charge, or directed inward towards a negative point charge. In either case, we assume that these lines of force are *equally spaced* throughout the region surrounding a point charge. Furthermore, *we assume that the number of lines of force is proportional to the magnitude of the charge.*

Once again, let's consider two point charges, q' and q, with position vectors $\mathbf{x}'$ and $\mathbf{x}$, where q is a small test charge. We define the **electric field $\mathbf{E}(\mathbf{x})$** at the point $\mathbf{x}$ as the force per unit test charge.

$$\mathbf{E}(\mathbf{x}) = \frac{\mathbf{F}}{q} \tag{3.8}$$

From Coulomb's law, we can write

$$\mathbf{E}(\mathbf{x}) = q' \frac{\mathbf{x} - \mathbf{x}'}{||\mathbf{x} - \mathbf{x}'||^3} \tag{3.8a}$$

In order to find the electric field associated with a discrete system of point charges, we employ the **principle of linear superposition**; that is, if we are given a system of charges

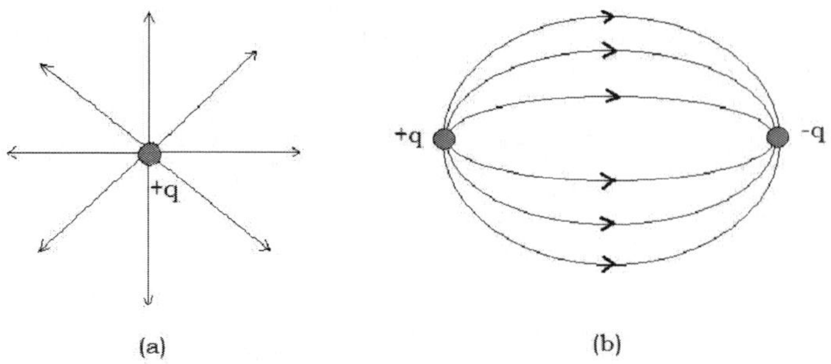

Figure 3.2. (a) Lines of forces leaving a positive charge; (b) Lines of force leaving a positive charge and terminating on a negative charge.

$q_1, q_2, \dots, q_n$ whose corresponding position vectors are $\mathbf{x}'_1, \mathbf{x}'_2, \dots, \mathbf{x}'_n$, then

$$\mathbf{E}(\mathbf{x}) = \sum_{k=1}^{n} q'_k \frac{\mathbf{x} - \mathbf{x}'_k}{\|\mathbf{x} - \mathbf{x}'_k\|^3} \tag{3.8b}$$

so that the electric field $\mathbf{E}(\mathbf{x})$ at the point $\mathbf{x}$ is the *vector sum* of the individual electric fields of each point charge in the collection.

Equation (3.8b) can be easily generalized to handle *continuous* distributions of charge. We first define the **charge density** $\rho(\mathbf{x}')$ as the charge per unit volume. Thus, the charge density $\rho(\mathbf{x}')$ is measured in esu/cm^3, and it is easy to see that the infinitesimal charge dq' contained in a small volume $d^3\mathbf{x}' = dx'dy'dz'$ is simply $dq' = \rho(\mathbf{x}')d^3\mathbf{x}'$.

Consequently, we must replace the summation sign in (3.8b) by an integral to obtain

$$\mathbf{E}(\mathbf{x}) = \int_{\Omega} \rho(\mathbf{x}') \frac{\mathbf{x} - \mathbf{x}'}{\|\mathbf{x} - \mathbf{x}'\|^3} d^3\mathbf{x}' \tag{3.8c}$$

where the integral is a volume integral, taken over the *bounded region* Ω which contains all of the charge.

In the event that charge is continuously distributed over a bounded surface S, with **surface charge density** $\sigma(\mathbf{x}')$ (in esu/cm^2), then we replace the volume integral in (8c) by a surface integral to obtain

$$\mathbf{E}(\mathbf{x}) = \int_{S} \sigma(\mathbf{x}') \frac{\mathbf{x} - \mathbf{x}'}{\|\mathbf{x} - \mathbf{x}'\|^3} dS' \tag{3.8d}$$

Whether we have a discrete or continuous distribution of charge, the electric field $\mathbf{E}(\mathbf{x})$ has an interesting geometric interpretation.

As shown in Figure 3.3, if we look closely at a *small* neighborhood of a fixed point $\mathbf{x}$, then the electric field lines nearly all have the same direction, so we can view $\|\mathbf{E}(\mathbf{x})\|$ as the *number of lines of force per unit area* that pierce an infinitesimal surface element dS whose outward normal $\mathbf{n}$ coincides with the direction of $\mathbf{E}(\mathbf{x})$ at that point.

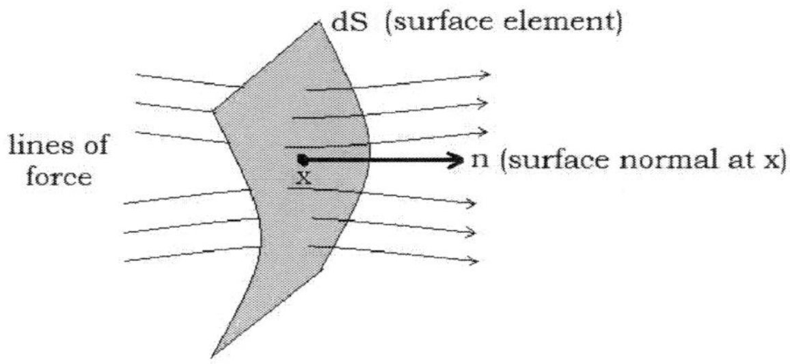

Figure 3.3. Lines of force piercing a small surface element dS.

In the event that the directions of $\mathbf{E}(\mathbf{x})$ and the surface normal $\mathbf{n}$ fail to concide at the point $\mathbf{x}$, then the number of lines of force that pierce dS is simply

$$d\Psi_E = \mathbf{E}(\mathbf{x}) \cdot \mathbf{n}\, dS = \mathbf{E}(\mathbf{x}) \cdot d\mathbf{S} \tag{3.9}$$

We call the quantity $d\Psi_E$ the differential of **electrix flux**. In order to find the total flux through a surface Σ of finite extent, we must sum up the contributions of flux through all of the infinitesimal surface elements comprising Σ; consequently, (3.9) becomes

$$\Psi_E = \int_\Sigma \mathbf{E}(\mathbf{x}) \cdot \mathbf{n} dS \tag{3.9a}$$

Now, assume that ∂R is a *closed* surface which is the boundary of a bounded region R enclosing a charge q'. Then the electrix flux Ψ_E must be interpreted as the *net* number of lines of force that leave the region R. In other words,

$$\Psi_E = \oint_{\partial R} \mathbf{E}(\mathbf{x}) \cdot \mathbf{n} dS \; = \; \text{net number of lines of force leaving } R \tag{3.9b}$$

On the other hand, we know that the number of lines of force emanating from the enclosed charge q' is proportional to the magnitude of q'. So, we suspect there is a relationship between Ψ_E and q'.

Let's investigate this matter further, and assume that the bounded region R contains a point charge q' which is placed at the origin. We want to calculate the flux Ψ_E through the closed surface ∂R. From from (3.8a), the electric field at any point $\mathbf{r}$ can be rewritten as

$$\mathbf{E}(\mathbf{r}) = \frac{q'}{r^2}\,\hat{\mathbf{r}} \tag{3.10}$$

where for convenience, we have set $\mathbf{x} = \mathbf{r}$, $r = \|\mathbf{x}\|$, and $\hat{\mathbf{r}}$ is a unit vector in the direction of $\mathbf{r}$. Since $\mathbf{E}(\mathbf{r})$ has a singularity at $\mathbf{r} = \mathbf{0}$, we construct a small sphere S of radius ε centered at the origin, and we consider the connected region R' which excludes the point $\mathbf{r} = \mathbf{0}$. This state of affairs is shown in Figure 3.4.

Then it is apparent that the flux through the boundary of $\partial R'$ is given by

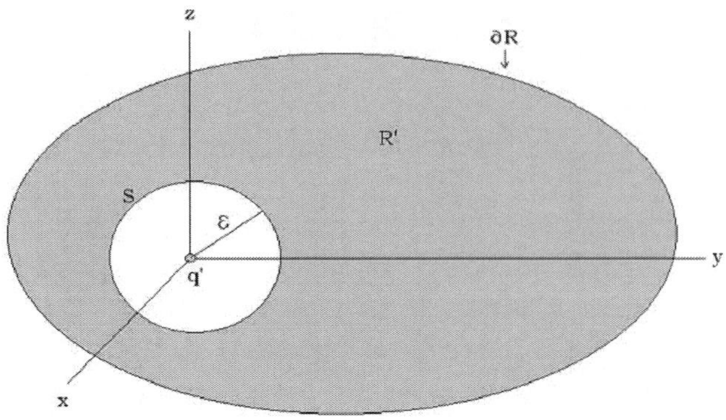

Figure 3.4. The connected region R' does not include the interior of S.

$$\oint_{\partial R'} \mathbf{E}(\mathbf{x}) \cdot \mathbf{n} dS = \oint_{\partial S} \mathbf{E}(\mathbf{x}) \cdot \mathbf{n} dS + \oint_{\partial R} \mathbf{E}(\mathbf{x}) \cdot \mathbf{n} dS$$

or using (3.10), this may be rewritten as

$$\begin{aligned} \oint_{\partial R'} \frac{q'}{r^2} \hat{\mathbf{r}} \cdot \mathbf{n} \, dS &= -\oint_{\partial S} \frac{q'}{r^2} \hat{\mathbf{r}} \cdot \hat{\mathbf{r}} \, dS + \oint_{\partial R} \mathbf{E}(\mathbf{x}) \cdot \mathbf{n} dS \\ &= -\oint_{\partial S} \frac{q'}{\varepsilon^2} \, dS + \oint_{\partial R} \mathbf{E}(\mathbf{x}) \cdot \mathbf{n} dS \\ &= -\frac{q'}{\varepsilon^2}(4\pi\varepsilon^2) + \oint_{\partial R} \mathbf{E}(\mathbf{x}) \cdot \mathbf{n} dS \\ &= -4\pi q' + \oint_{\partial R} \mathbf{E}(\mathbf{x}) \cdot \mathbf{n} dS \qquad (3.11) \end{aligned}$$

Now, it turns out that the lefthand side of (3.11) is zero. To see this, we transform it into a volume integral by the divergence theorem[4] to get

$$\oint_{R'} \left(\nabla \cdot \frac{q'}{r^2} \hat{\mathbf{r}} \right) dV' = \oint_{\partial R'} \frac{q'}{r^2} \hat{\mathbf{r}} \cdot \mathbf{n} \, dS$$

and observe that the integrand on the left vanishes inside R', since

$$\begin{aligned} \nabla \cdot \left(\frac{q'}{r^2} \hat{\mathbf{r}} \right) &= \nabla \cdot \left(\frac{q'}{r^3} \mathbf{r} \right) = \nabla \left(\frac{q'}{r^3} \right) \cdot \mathbf{r} + \frac{q'}{r^3} (\nabla \cdot \mathbf{r}) \\ &= -3 \frac{q'}{r^4} \, (\hat{\mathbf{r}} \cdot \mathbf{r}) + 3 \frac{q'}{r^3} \\ &= -3 \frac{q'}{r^4} \, (r) + 3 \frac{q'}{r^3} = 0 \end{aligned}$$

Consequently, equation (3.11) becomes

$$\oint_{\partial R} \mathbf{E}(\mathbf{x}) \cdot \mathbf{n} dS = 4\pi q' \quad \text{(Gauss's Law)} \tag{3.12}$$

This equation is known as the **integral form of Gauss's law** for electric fields. *It says that the flux of electric field through a closed surface is equal to* 4π *times the total charge it encloses.* Stated differently, 4π lines of force emerge from each unit positive charge.

It should be obvious that if the closed surface ∂R encloses several point charges, or even a continuous distribution of charge, then (3.12) becomes

$$\oint_{\partial R} \mathbf{E}(\mathbf{x}) \cdot \mathbf{n} dS = 4\pi Q \tag{3.12a}$$

where Q is now the *total* charge enclosed by ∂R, and $\mathbf{E}(\mathbf{x})$ is the *resultant* electric field.

Now, Gauss's law also may be written in differential form. To see this, let's assume once again, that ∂R is a *closed* surface which forms the boundary of a bounded region R enclosing a continuous distribution of charge. If the charge density is $\rho(\mathbf{x})$, then the total amount of charge Q, enclosed by this region is

$$Q = \int_R \rho(\mathbf{x}) d^3\mathbf{x} \tag{3.12b}$$

So, combining (3.12a) and (3.12b), we get

$$\oint_{\partial R} \mathbf{E}(\mathbf{x}) \cdot \mathbf{n} dS = 4\pi \int_R \rho(\mathbf{x}) d^3\mathbf{x}$$

Finally, using the divergence theorem, the lefthand side of the last equation can be converted to a volume integral, to give

$$\int_R \nabla \cdot \mathbf{E}(\mathbf{x})\ d^3\mathbf{x} = 4\pi \int_R \rho(\mathbf{x}) d^3\mathbf{x}$$

or, in other words

$$\int_R [\nabla \cdot \mathbf{E}(\mathbf{x}) - 4\pi\rho(\mathbf{x})]\ d^3\mathbf{x} = 0$$

Since the last equation must hold for *any* such region R, then the integrand must vanish. Consequently, we have the differential form of Gauss's law:

$$\nabla \cdot \mathbf{E}(\mathbf{x}) = 4\pi\rho(\mathbf{x}) \tag{3.12c}$$

Observe that (3.12c) is the first of Maxwell's famous equations.

3.3. Electric Potential and Poisson's Equation

It's easy to show that for an *electrostatic* field,

$$\nabla \times \mathbf{E}(\mathbf{x}) = \mathbf{0} \tag{3.13}$$

and this is just a special case of Maxwell's equation (3.3). Since the electrostatic field is curl-free, Stoke's theorem[5] implies that the line integral of $\mathbf{E}(\mathbf{x})$ around any closed path C must be zero; that is,

$$\oint_C \mathbf{E}(\mathbf{x}) \cdot d\mathbf{x} = 0 \tag{3.13a}$$

so the electrostatic field is a *conservative* field. This means that there must exist a *scalar* function $\Phi(\mathbf{x})$, called the **electric potential**, such that

$$\mathbf{E}(\mathbf{x}) = -\nabla\Phi(\mathbf{x}) \tag{3.13b}$$

where the minus sign is taken as a matter of convention. From equation (3.13b), we see that for a small vector displacement $d\mathbf{x}$, the corresponding change in electric potential is simply

$$d\Phi = \nabla\Phi \cdot d\mathbf{x} = -\mathbf{E}(\mathbf{x}) \cdot d\mathbf{x}$$

and so, the electric potential at any point $\mathbf{x}$ may be expressed as a line integral

$$\Phi(\mathbf{x}) = -\int_{\mathbf{x}_0}^{\mathbf{x}} \mathbf{E}(\mathbf{x}') \cdot d\mathbf{x}' \tag{3.14}$$

where $\mathbf{x}_0$ is any convenient reference point at which we set the potential equal to zero, $\Phi(\mathbf{x}_0) = 0$. For a charge distribution that fills a *finite* region, we usually take $\mathbf{x}_0 = \infty$.

We can interpret the potential $\Phi(x)$ at the point $\mathbf{x}$ as the amount of *work* necessary transport a *unit* positive test charge, *against* the field, from the reference point $\mathbf{x}_0$ to the point $\mathbf{x}$. It is important to note that this definition makes sense, since the electrostatic field is conservative; and consequently, we are assured that the line integral is *independent of the path* joining $\mathbf{x}_0$ and $\mathbf{x}$.

To see how this works, let's calculate the potential due to point charge q' which is situated at the origin. Since any path of integration joining $\mathbf{x}_0$ and $\mathbf{x}$ will do, let's use a straight line, so we can take $d\mathbf{x} = \hat{\mathbf{r}}\, dr$, where $\hat{\mathbf{r}}$ is a unit vector in the radial direction, as shown in Figure 3.5. From equations (3.14) and (3.10), we have

$$\begin{aligned}
\Phi(r) &= -\int_{\infty}^{r} \mathbf{E}(\mathbf{r}') \cdot d\mathbf{r}' = -\int_{\infty}^{r} \frac{q'}{r'^2} (\,\hat{\mathbf{r}} \cdot \hat{\mathbf{r}}) \; dr' \\
&= -\int_{\infty}^{r} \frac{q'}{r'^2} \, dr' = \left[\frac{q'}{r'}\right]_{r'=\infty}^{r'=r} = \frac{q'}{r}
\end{aligned}$$

and so, the potential due to a point charge, placed at the origin, is simply

$$\Phi(r) = \frac{q'}{r} \tag{3.14a}$$

If now, we allow a point charge q' to be placed at any point $\mathbf{x}'$, then the resulting potential is

$$\Phi(\mathbf{x}) = \frac{q'}{\|\mathbf{x}-\mathbf{x}'\|} \quad \text{(Potential Due to a Point Charge)} \tag{3.14b}$$

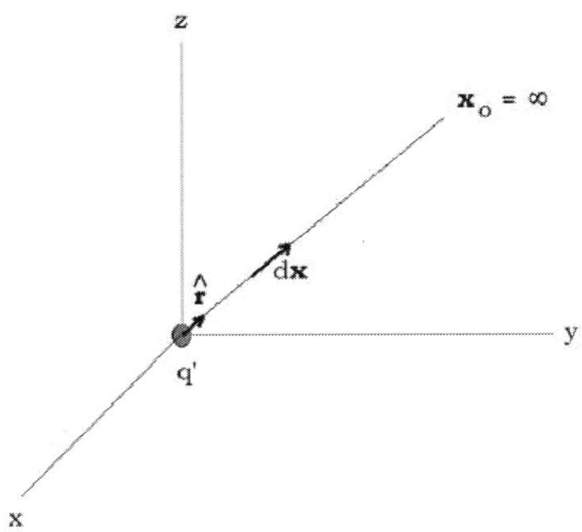

Figure 3.5. Finding the potential due to a point charge q'.

A question which naturally arises is how do we find the potential of a system of point charges? To answer this question, assume that we are given a system of charges $q_1, q_2, \ldots, q_n$ whose corresponding position vectors are $\mathbf{x}'_1, \mathbf{x}'_2, \ldots, \mathbf{x}'_n$. We know from (3.8b) that the electric field is given by

$$\mathbf{E}(\mathbf{x}) = \sum_{k=1}^{n} q'_k \frac{\mathbf{x}-\mathbf{x}'_k}{\|\mathbf{x}-\mathbf{x}'_k\|^3} = \sum_{k=1}^{n} \mathbf{E}_k(\mathbf{x})$$

where for each k, the expression $\mathbf{E}_k(\mathbf{x})$ is the electric field at the point $\mathbf{x}$ which is due to the influence of the charge q'_k alone; this is just the superposition principle at work. At the same time, however, each of the fields $\mathbf{E}_k(\mathbf{x})$ may be written as

$$\mathbf{E}_k(\mathbf{x}) = -\nabla\Phi_k(\mathbf{x}) \qquad (k = 1, 2, 3, \ldots, n) \tag{3.14c}$$

where each potential $\Phi_k(\mathbf{x})$ is due to the presence of the charge q'_k alone. The last two equations may be combined to give

$$\mathbf{E}(\mathbf{x}) = -\sum_{k=1}^{n} \nabla\Phi_k(\mathbf{x}) = -\nabla\left(\sum_{k=1}^{n} \Phi_k(\mathbf{x})\right) \tag{3.14d}$$

We conclude that the potential $\Phi(\mathbf{x})$ is just the sum of all of the individual potentials due to the source charges, taken individually; in other words, *the law of superposition applies to potentials* as well, and so,

$$\Phi(\mathbf{x}) = \sum_{k=1}^{n} \Phi_k(\mathbf{x}) = \Phi_1(\mathbf{x}) + \Phi_2(\mathbf{x}) + \ldots + \Phi_n(\mathbf{x}) \tag{3.14e}$$

Finally, from (3.14b) and (3.14e), we now have an expression for the potential due to a system of discrete charges:

$$\Phi(\mathbf{x}) = \sum_{k=1}^{n} \frac{q'_k}{\|\mathbf{x}-\mathbf{x}'_k\|} \tag{3.15}$$

If we are given a *continuous* distribution of charges in a finite region R, with charge density $\rho(\mathbf{x}')$, then the summation symbol in (3.15) may be replaced by an integral to give

$$\Phi(\mathbf{x}) = \int_R \frac{\rho(\mathbf{x}')}{\|\mathbf{x}-\mathbf{x}'\|} d^3\mathbf{x}' \tag{3.15a}$$

where the integration extends over the *finite* region R.

We now obtain a partial differential equation, known as **Poisson's equation**, which governs the electric potential. If we substitute the expression $\mathbf{E}(\mathbf{x}) = -\nabla\Phi(\mathbf{x})$ into equation (3.12c), we get

$$\nabla\cdot\mathbf{E}(\mathbf{x}) = \nabla\cdot(-\nabla\Phi(\mathbf{x})) = -4\pi\rho(\mathbf{x})$$

or equivalently,

$$\nabla^2\Phi(\mathbf{x}) = -4\pi\rho(\mathbf{x}) \quad \text{(Poisson's Equation)} \tag{3.16}$$

Poisson's equation, along with the appropriate boundary conditions, may be solved to give the potential at any point. For finite distributions of charge, however, we already know that (3.15a) is a *solution* of this equation. On the other hand, if we are interested in solutions of Poisson's equation in charge-free regions, then we must set $\rho(\mathbf{x}) = 0$, and (3.16) reduces to **Laplace's equation**:

$$\nabla^2\Phi(\mathbf{x}) = 0 \quad \text{(Laplace's Equation)} \tag{3.16a}$$

3.4. Exercises

1. Given a scalar field of the form $g = g(R)$ where $R = \mathbf{x}-\mathbf{x}'$, show that

$$\nabla g(R) = g'(R)\,\hat{\mathbf{R}} \tag{3.17}$$

where $\hat{R}$ is a *unit* vector in the direction of R.

(Solution)

We have

$$\begin{aligned} \nabla g(R) &= \frac{\partial g}{\partial x}\mathbf{i} + \frac{\partial g}{\partial y}\mathbf{j} + \frac{\partial g}{\partial z}\mathbf{k} \\ &= g'(R)\left[\frac{\partial R}{\partial x}\mathbf{i} + \frac{\partial R}{\partial y}\mathbf{j} + \frac{\partial R}{\partial z}\mathbf{k}\right] \end{aligned} \tag{3.17a}$$

Let's take a look at the first partial derivative on the righthand side of (3.17a); a simple calculation gives

$$\frac{\partial R}{\partial x} = \frac{\partial}{\partial x}\left[(x-x')^2+(y-y')^2+(z-z')^2\right]^{1/2} = \frac{x-x'}{R}$$

with similar results for the remaining derivatives. Consequently,

$$\nabla g(R) = g'(R)\left[\frac{x-x'}{R}\mathbf{i}+\frac{y-y'}{R}\mathbf{j}+\frac{z-z'}{R}\mathbf{k}\right] = g'(R)\,\hat{\mathbf{R}}$$

2. Show that for any *electrostatic* field $\mathbf{E}(\mathbf{x})$,

$$\nabla\times\mathbf{E}(\mathbf{x}) = \mathbf{0} \tag{3.18}$$

(Solution)

Assume a continuous distribution of charge which fills a region Ω. From (3.8c), we have

$$\mathbf{E}(\mathbf{x}) = \int_{\Omega} \rho(\mathbf{x}')\frac{\mathbf{x}-\mathbf{x}'}{\|\mathbf{x}-\mathbf{x}'\|^3}d^3\mathbf{x}'$$

and consequently,

$$\nabla\times\mathbf{E}(\mathbf{x}) = \int_{\Omega} \rho(\mathbf{x}')\left(\nabla\times\frac{(\mathbf{x}-\mathbf{x}')}{\|\mathbf{x}-\mathbf{x}'\|^3}\right)d^3\mathbf{x}'$$

Recall that for a scalar field g and a vector field $\mathbf{V}$, we have the vector identity

$$\nabla\times(g\mathbf{V}) = g(\nabla\times\mathbf{V})+(\nabla g)\times\mathbf{V}$$

Using this identity, the integrand of the above expression becomes:

$$\left[\nabla\times\left(\frac{1}{\|\mathbf{x}-\mathbf{x}'\|^3}\right)(\mathbf{x}-\mathbf{x}')\right] = \left(\frac{1}{\|\mathbf{x}-\mathbf{x}'\|^3}\right)\nabla\times(\mathbf{x}-\mathbf{x}')+\nabla\left(\frac{1}{\|\mathbf{x}-\mathbf{x}'\|^3}\right)\times(\mathbf{x}-\mathbf{x}')$$

Examining each of the above terms separately, we obtain

$$\nabla\times(\mathbf{x}-\mathbf{x}') = \mathbf{0} \text{ and } \nabla\left(\frac{1}{\|\mathbf{x}-\mathbf{x}'\|^3}\right) = -\left(\frac{3}{\|\mathbf{x}-\mathbf{x}'\|^4}\right)\hat{\mathbf{e}}_{x-x'}$$

where $\hat{\mathbf{e}}_{x-x'}$ is a unit vector in the direction of $(\mathbf{x}-\mathbf{x}')$. Consequently, the second term vanishes as well, since $\hat{\mathbf{e}}_{x-x'}\times(\mathbf{x}-\mathbf{x}') = \mathbf{0}$; we conclude that, $\nabla\times\mathbf{E}(\mathbf{x}) = \mathbf{0}$.

3. *(Field Due to an Infinite Line Charge)* Assume that charge is uniformly distributed with density λ along an infinite straight line which coincides with the z-axis, as shown in Figure 3.6.

Show that the electric field $\mathbf{E}(r)$, at a perpendicular distance r from the z-axis, is given by

$$\mathbf{E}(r) = \frac{2\lambda}{r}\,\hat{\mathbf{r}} \tag{3.19}$$

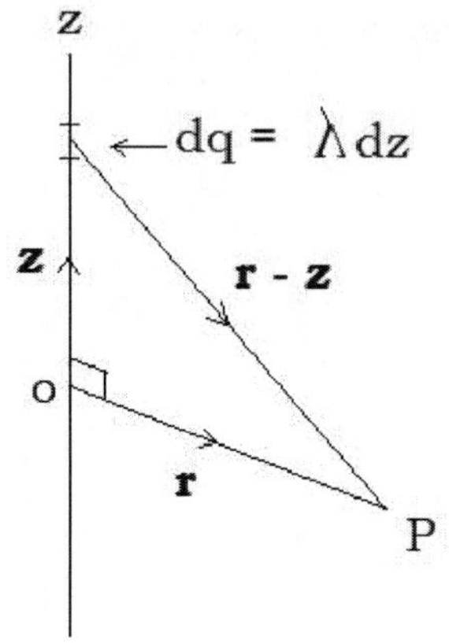

Figure 3.6. An infinite line charge along the z-axis.

(Solution)

It's convenient to assume that the point P lies in the xy-plane. We then obtain

$$d\mathbf{E} = \frac{\lambda dz}{\|\mathbf{r}-\mathbf{z}\|^2}\hat{e}_{\mathbf{r}-\mathbf{z}} = \frac{\lambda dz}{r^2+z^2}\left(\frac{r\hat{\mathbf{r}}-z\hat{\mathbf{z}}}{\sqrt{r^2+z^2}}\right) = \frac{\lambda r dz}{(r^2+z^2)^{3/2}}\hat{\mathbf{r}} - \frac{\lambda z dz}{(r^2+z^2)^{3/2}}\hat{\mathbf{z}}$$

Consequently,

$$\mathbf{E} = \left(\int_{-\infty}^{\infty}\frac{\lambda r dz}{(r^2+z^2)^{3/2}}\right)\hat{\mathbf{r}} - \left(\int_{-\infty}^{\infty}\frac{\lambda z dz}{(r^2+z^2)^{3/2}}\right)\hat{\mathbf{z}}$$

The second integral vanishes, since the integrand is an odd function of z, so we obtain

$$\mathbf{E} = \left(\int_{-\infty}^{\infty}\frac{\lambda r}{(r^2+z^2)^{3/2}}dz\right)\hat{\mathbf{r}}$$

If we use the substitution $z = r\tan\theta$, then $dz = r\sec^2\theta$, and the integral becomes:

$$\mathbf{E} = \frac{\lambda}{r}\left(\int_{-\pi/2}^{\pi/2}\cos\theta\, d\theta\right)\hat{\mathbf{r}} = \frac{2\lambda}{r}\hat{\mathbf{r}}$$

4. *(Field Due to a Charged Disk)* As shown in Figure–3.7, a uniformly charged disk of radius a, with surface charge density σ, sits in the xy-plane. Find the electric field $\mathbf{E}$ at a point which is h units above the xy-plane.

(Solution)

Using cylindrical coordinates (r,ϕ,z), we see that at the point P,

$$d\mathbf{E} = \frac{dq}{R^2}\hat{\mathbf{R}} = \frac{\sigma dS}{(r^2+h^2)^{3/2}}(r\hat{\mathbf{r}}-h\hat{\mathbf{z}})$$

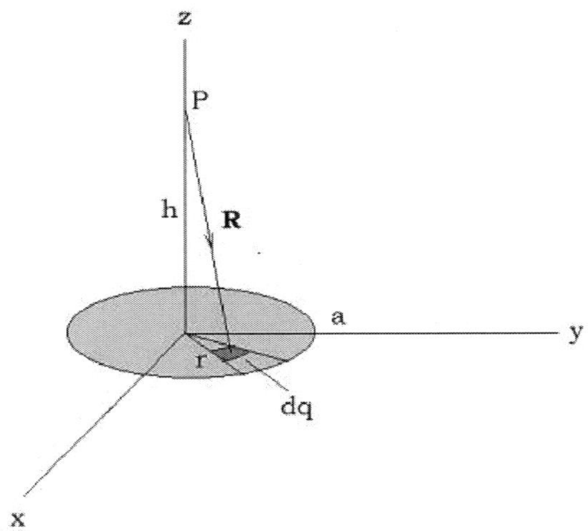

Figure 3.7. A uniformly charged disk.

But, in cylindrical coordinates, the element of area is $dS = rdrd\phi$, so the expression for $d\mathbf{E}$ becomes

$$d\mathbf{E} = \left(\frac{\sigma r^2}{(r^2+h^2)^{3/2}} drd\phi\right)\hat{\mathbf{r}} - \left(\frac{\sigma rh}{(r^2+h^2)^{3/2}} drd\phi\right)\hat{\mathbf{z}}$$

and

$$\mathbf{E} = \int_{r=0}^{r=a}\int_{\phi=0}^{\phi=2\pi}\left(\frac{\sigma r^2}{(r^2+h^2)^{3/2}} drd\phi\right)\hat{\mathbf{r}} - \int_{r=0}^{r=a}\int_{\phi=0}^{\phi=2\pi}\left(\frac{\sigma rh}{(r^2+h^2)^{3/2}} drd\phi\right)\hat{\mathbf{z}}$$

The first integral vanishes, since by symmetry, the radial components must cancel. So we now obtain

$$\mathbf{E} = -2\pi\sigma h\int_{r=0}^{r=a}\left(\frac{rdr}{(r^2+h^2)^{3/2}} dr\right)\hat{\mathbf{z}} = -2\pi\sigma h\left[-\frac{1}{\sqrt{r^2+h^2}}\right]_{r=0}^{r=a}\hat{\mathbf{z}}$$

so that the field at P is given by

$$\mathbf{E} = 2\pi\sigma h\left(\frac{1}{h} - \frac{1}{\sqrt{a^2+h^2}}\right)\hat{\mathbf{z}} \tag{3.20}$$

5. *(Field Due to an Infinite Charged Plane)* A uniformly charged plane, with surface charge density σ, coincides with the xy-plane. Find the electric field $\mathbf{E}$ at a point P which is h units above the plane, as shown in Figure 3.8.

(Solution)

Using cylindrical coordinates (r, ϕ, z), we have at the point P,

$$d\mathbf{E} = \frac{dq}{R^2}\hat{\mathbf{R}} = \frac{\sigma rdrd\phi}{(r^2+h^2)^{3/2}}(r\hat{\mathbf{r}} - h\hat{\mathbf{z}})$$

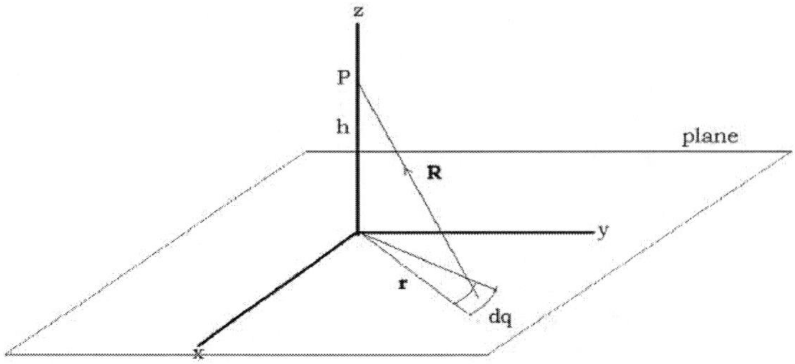

Figure 3.8. An infinite plane of charge.

so that

$$\mathbf{E} = \int_{r=0}^{r=a}\int_{\phi=0}^{\phi=2\pi}\left(\frac{\sigma r^2}{(r^2+h^2)^{3/2}}drd\phi\right)\hat{\mathbf{r}} - \int_{r=0}^{r=a}\int_{\phi=0}^{\phi=2\pi}\left(\frac{\sigma rh}{(r^2+h^2)^{3/2}}drd\phi\right)\hat{\mathbf{z}}$$

As in the previous example, the radial component of **E** vanishes due to symmetry, and so, we obtain

$$\mathbf{E} = -2\pi\sigma h\int_{r=0}^{\infty}\left(\frac{rdr}{(r^2+h^2)^{3/2}}dr\right)\hat{\mathbf{z}} = -2\pi\sigma h\left[-\frac{1}{\sqrt{r^2+h^2}}\right]_{r=0}^{r=\infty}\hat{\mathbf{z}} = 2\pi\sigma\hat{\mathbf{z}}$$

and

$$\mathbf{E} = (2\pi\sigma)\,\hat{\mathbf{z}} \tag{3.21}$$

Observe that (3.21) could have been obtained by simply letting $a \to \infty$ in equation (3.20).

6. Use Gauss's law to solve Example 5, above.

(Solution)

As shown in Figure 3.9, we construct a small Gaussian cylinder S, with cross-sectional area A. We assume that the electric field **E** is normal to the plane. Consequently, above the plane, $\mathbf{E}_{above} = E\mathbf{n}$, while below the plane $\mathbf{E}_{below} = -E\mathbf{n}$, where **n** is the unit normal to the *upper* surface.

By Gauss's law, we can write

$$\oint_S \mathbf{E}\cdot\mathbf{n}dS = 4\pi Q = 4\pi\sigma S$$

But

$$\begin{aligned}\oint_S \mathbf{E}\cdot\mathbf{n}\,dS &= \int_{top}\mathbf{E}\cdot\mathbf{n}\,dS + \int_{bottom}\mathbf{E}\cdot\mathbf{n}\,dS + \int_{side}\mathbf{E}\cdot\mathbf{n}\,dS \\ &= \int_{top}E\mathbf{n}\cdot\mathbf{n}dS + \int_{bottom}-E\mathbf{n}\cdot(-\mathbf{n})\,dS = 2ES\end{aligned}$$

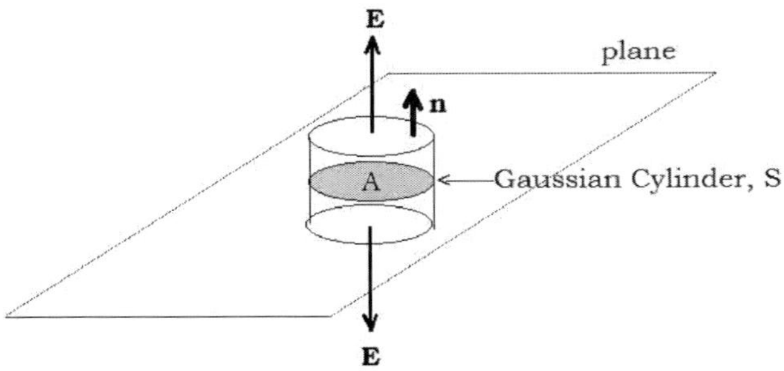

Figure 3.9.

where the surface integral over the "side" vanishes since E is perpendicular to the normal to the curved surface. Combining the last two equations gives

$$2ES = 4\pi\sigma S \Longrightarrow E = 2\pi\sigma$$

and consequently, we again obtain

$$\mathbf{E} = (2\pi\sigma)\,\mathbf{n}$$

7. *(Field Due to a Uniformly Charged Sphere)* A sphere of radius a carries a uniform charge throughout its interior. If the charge density is ρ, then find:

(a) the electric field $\mathbf{E}$ at any interior point, a distance r from its center, where $r \leq a$.
(b) the electric field $\mathbf{E}$ at any exterior point, a distance r from its center, where $r \geq a$.

(Solution)
(a) As shown in Figure 3.10, we let the center of the sphere S coincide with the origin, and we construct a Gaussian sphere S_1 of radius r inside S.

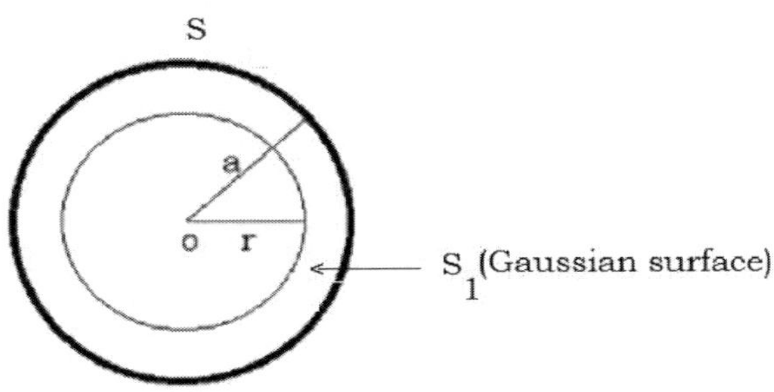

Figure 3.10. Constructing a Gaussian surface S_1 inside S.

From Gauss's law, we have

$$\oint_{S_1} \mathbf{E} \cdot \mathbf{n} dS = 4\pi Q$$

By symmetry, however, $\mathbf{E} = E\,\hat{\mathbf{r}}$ where $\hat{\mathbf{r}}$ is a unit vector in the radial direction, and the magnitude of $\mathbf{E}$ is constant on S_1. So, the last equation becomes

$$\oint_{S_1} \mathbf{E}\cdot\mathbf{n}\,dS = \oint_{S_1} E\,(\hat{\mathbf{r}}\cdot\hat{\mathbf{r}})\,dS = E\oint_{S_1} dS = E(4\pi r^2) = 4\pi Q \Longrightarrow E = \frac{Q}{r^2}$$

But the total charge enclosed is $Q = \rho\left(\frac{4}{3}\pi r^3\right)$, so that

$$\boldsymbol{E} = \left(\frac{4}{3}\pi\rho r\right)\hat{\mathbf{r}} \qquad (\text{ for } r \le a) \tag{3.22a}$$

(b) As shown in Figure 11, we now construct a Gaussian sphere S_2 of radius r which lies outside S.

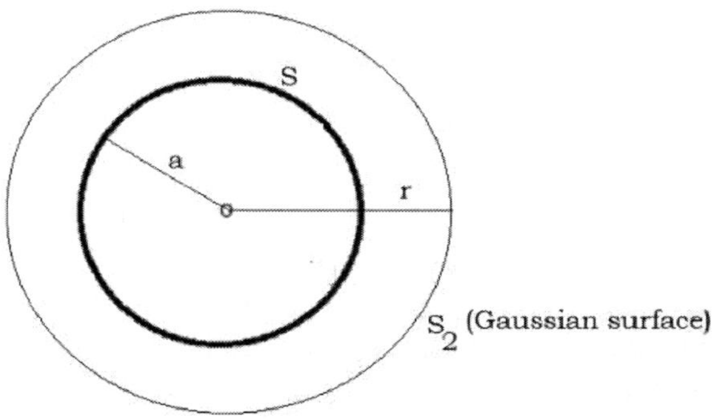

Figure 3.11. Constructing a Gaussian surface S_2 outside S.

Once again, from Gauss's law, we have

$$\oint_{S_2} \mathbf{E}\cdot\mathbf{n}dS = 4\pi Q$$

and, proceeding as before, we can write

$$\oint_{S_2} \mathbf{E}\cdot\mathbf{n}\,dS = \oint_{S_2} E\,(\hat{\mathbf{r}}\cdot\hat{\mathbf{r}})\,dS = E\oint_{S_2} dS = E(4\pi r^2) = 4\pi Q \Longrightarrow E = \frac{Q}{r^2}$$

But now, $Q = \rho\left(\frac{4}{3}\pi a^3\right)$, so the last equation becomes

$$\boldsymbol{E} = \left(\frac{4}{3}\pi\rho a^3\right)\frac{\hat{\mathbf{r}}}{r^2} \qquad (\text{ for } r \ge a) \tag{3.22b}$$

We see from (3.22b), that the sphere behaves like a *point charge* with all of its charge concentrated at origin.

8. Show that (3.22b) satisfies the differential form of Gauss's law.

(Solution)

We must show that $\nabla \cdot \mathbf{E} = 0$, since $\rho = 0$ outside S. Given a vector field $\mathbf{V}$, its divergence in spherical coordinates is given by

$$\nabla \cdot \mathbf{V} = \frac{1}{r^2}\frac{\partial}{\partial r}\left(r^2 V_r\right) + \frac{1}{r\sin\theta}\frac{\partial}{\partial\theta}\left(A_\theta \sin\theta\right) + \frac{1}{r\sin\theta}\frac{\partial}{\partial\phi}\left(A_\phi\right)$$

From (3.22b), we have $E_r = \frac{4}{3}\pi\rho a^3/r^2$, while $E_\theta = E_\phi = 0$; so we get

$$\nabla \cdot \mathbf{E} = \frac{1}{r^2}\frac{\partial}{\partial r}\left(\frac{4}{3}\pi\rho a^3\right) = 0$$

9. *(Potential Due to Finite Line Charge)* A line of charge of length $2a$ sits on the x-axis, as shown in Figure 3.12.

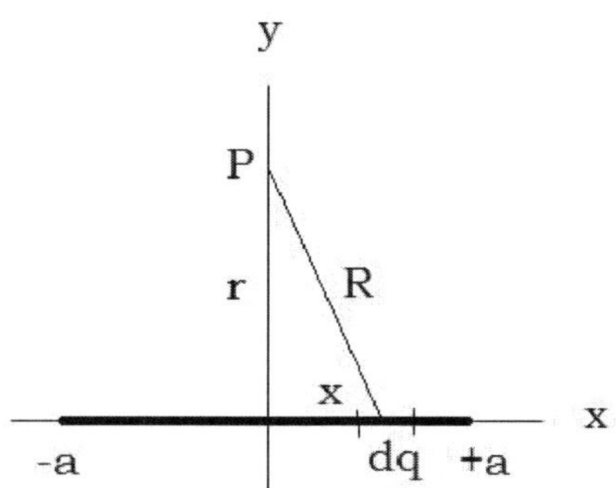

Figure 3.12. Potential due to a finite line-charge.

If the charge density λ is uniform, find the electric potential at the point $P(0,r)$, on the y-axis.

(Solution)

We have $d\Phi(r) = dq/R = \lambda dx/R$, so

$$\Phi(r) = \int_{-a}^{+a} \frac{1}{\sqrt{r^2+x^2}}\lambda dx = 2\lambda \int_0^{+a} \frac{dx}{\sqrt{r^2+x^2}}$$

Use the substitution $x = r\tan\theta \Longrightarrow dx = r\sec^2\theta d\theta$, so the last integral becomes

$$\Phi(r) = 2\lambda \int_0^{\tan^{-1}(a/r)} \sec\theta \, d\theta = 2\lambda[\,\ln|\sec\theta + \tan\theta|]_{\theta=0}^{\theta=\tan^{-1}(a/r)}$$

or,

$$\Phi(r) = 2\lambda \ln\left(\frac{a+\sqrt{r^2+a^2}}{r}\right) \qquad (3.23)$$

10. Find the electric field $\mathbf{E}(r)$ which corresponds to the potential given in (3.23).

(Solution)

We have

$$\mathbf{E}(r) = -\nabla\Phi(r) = -\frac{\partial}{\partial r}\left[2\lambda\ln\left(\frac{a+\sqrt{r^2+a^2}}{r}\right)\right]\hat{\mathbf{r}} = \left(\frac{2\lambda a}{r\sqrt{a^2+r^2}}\right)\hat{\mathbf{r}}$$

Observe that as $a \longrightarrow \infty$, we have $\mathbf{E}(r) \to (2\lambda/r)\hat{\mathbf{r}}$ which is the field due an *infinite* line charge, that we already derived in (3.19).

11. *(Potential Due to a Hollow Sphere)* A uniformly charged hollow sphere of radius a, and with surface charge density σ, sits with its center at the origin. Find the electric potential $\Phi(r)$ at any exterior point P which is a distance r from the center. Refer to Figure 3.13.

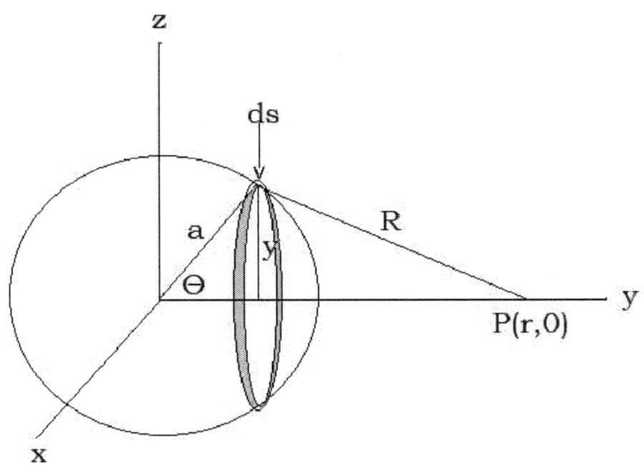

Figure 3.13. A uniformly charged hollow sphere.

(Solution)

Without loss of generality, we assume that P sits on the x-axis, with coordinates $(r,0)$, where $r > a$. We construct a ring of thickness ds and radius y as shown. Evidently, the amount of charge dq residing on this ring is

$$dq = (2\pi y ds)\sigma = 2\pi a^2\sigma\sin\theta d\theta$$

and consequently, the potential at P must be

$$\Phi(r) = \oint_{sphere}\frac{dq}{R} = \int_0^{\pi}\frac{2\pi a^2\sigma\sin\theta d\theta}{\sqrt{a^2+r^2-2ar\cos\theta}} = \frac{2\pi a\sigma}{r}\left[\sqrt{a^2+r^2-2ar\cos\theta}\right]_{\theta=0}^{\theta=\pi}$$

so that

$$\Phi_{interior}(r) = \frac{4\pi a^2\sigma}{r} \qquad \text{(provided } r > a\text{)} \tag{3.24}$$

Observe that since $4\pi a^2\sigma$ is the total charge on the sphere, (3.24) tells us that the hollow sphere behaves like a *point charge* of magnitude $Q = 4\pi a^2\sigma$, situated at the origin, when it is viewed from an exterior point *P*.

12. Assuming the same information as in Example 11, find the potential at any *interior* point *P*.

(Solution)

Let *P* be an interior point which is a distance r from the center of the sphere. If we imagine constructing a Gaussian sphere of radius r which contains the point *P*, it is obvious that this sphere encloses zero charge. Consequently, we must have $\mathbf{E}(r) = \mathbf{0}$ inside the sphere. But we know that $\mathbf{E}(r) = -\nabla\Phi(r)$, so the potential must be constant throughout the interior. Since $\Phi(r)$ must be a *continuous* function of r, we can use (3.24) as a guide, and for convenience, set

$$\Phi_{interior} = \Phi_{exterior}(a) = 4\pi a\sigma = \text{constant}$$

3.5. Current Density and Charge Conservation

If the terminals of a battery are connected to the two ends of a conductor, an electric field $\mathbf{E}$ is set up in the interior of a conductor. We know that this electric field must exert a force $q\mathbf{E}$ on each charge within the conductor. In most metals, one or two electrons in the outermost shell of each atom are free to move; consequently, the applied electric field drives the flow of free electrons, resulting in an *electric current*.

We define the magnitude of the **current** in a wire as the amount of charge, per unit time, that passes a fixed point in the wire. If the linear charge density of *moving* charges is λ, then the **current** is given by

$$\mathbf{I} = \lambda\mathbf{v} \tag{3.25}$$

Equation (3.25) tells us that current is a *vector*. This equation makes sense, since the amount of charge Δq that passes a fixed point in time Δt must be $\Delta q = \lambda(v\Delta t)$, and therefore, the magnitude of the current is $I = \Delta q/\Delta t = \lambda v$. In the CGS electrostatic system of units, current is measured in *statamperes* where

$$1 \text{ statamp} = 1 \frac{\text{esu}}{\text{sec}} \tag{3.25a}$$

In the event that the flow of charge fills a three dimensional region of space, it is advantageous to describe it in terms of the **volume current density j**. We define the current density as

$$\mathbf{j} = \rho\boldsymbol{v} \tag{3.25b}$$

where ρ is the volume charge density (esu/cm^3), and $\mathbf{v}$ is the velocity.

In order to understand the meaning of volume current density, consider a three dimensional flow of charge along a "tube", as shown in Figure 3.14.

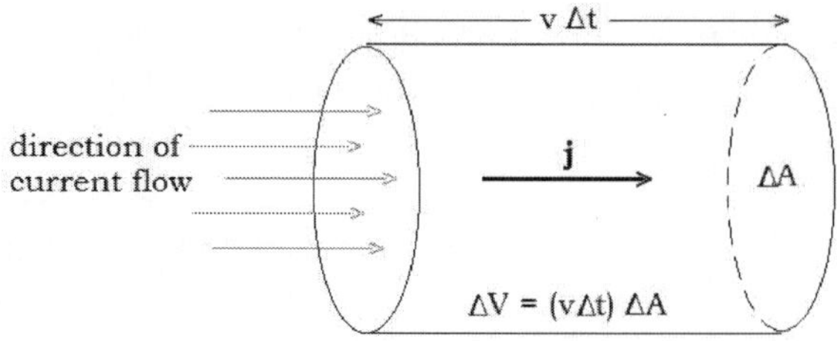

Figure 3.14. A "tube" of volume current.

If the infinitesimal cross-sectional area of the tube is ΔA, and its unit normal has the *same direction* as the flow, then the magnitude of the current density is given by

$$j = \rho v = \frac{\Delta q}{\Delta V} v = \frac{\Delta q}{(v\Delta t)\Delta A} v = \frac{\Delta q}{(\Delta A)(\Delta t)}$$

So, the magnitude of the current density represents the amount of charge that crosses a surface which is perpendicular to the direction of flow, per unit area, per unit time. Stated differently, in terms of current, the magnitude of **j** is simply the current per unit cross-sectional area; so, the current density may be expressed in the CGS electrostatic system in the unit *statamp/cm*^{2}.

In general, the direction of **j** may not be parallel to the unit surface normal **n**. So, we define the electric current passing through a surface S as the surface integral

$$I = \int_S \mathbf{j} \cdot \mathbf{n} \, dS \tag{3.26}$$

Now, numerous experiments have shown that *charge is conserved*. This may sound like a trivial fact, but it is not; it is a *fundamental law* of physics. Charge conservation has been verified even in the interaction of subatomic particles; the total charge before each interaction is exactly equal to the total charge after the interaction.[6]

For example, in a process known as *beta-decay*, a neutron n decays into a proton p^+, an electron e^-, and antineutrino $\bar{\nu}$:

$$n \longrightarrow p^+ + e^- + \bar{\nu}$$

Since an antineutrino carries no charge, and a neutron is electrically neutral, we see that the total amount of charge before and after the decay process is the same.

Consequently, if we are given a *closed* volume V, the amount of charge which is contained in V cannot change unless charge enters or exits this volume through its boundary ∂V. This is shown in Figure 3.15.

From (3.26), the amount of charge leaving V per unit time is just the integral of the current density taken over the closed surface; in other words,

$$I = \left(\begin{array}{c} \text{Charge leaving volume } V \\ \text{per unit time} \end{array} \right) = \oint_{\partial V} \mathbf{j} \cdot \mathbf{n} \, dS \tag{3.27}$$

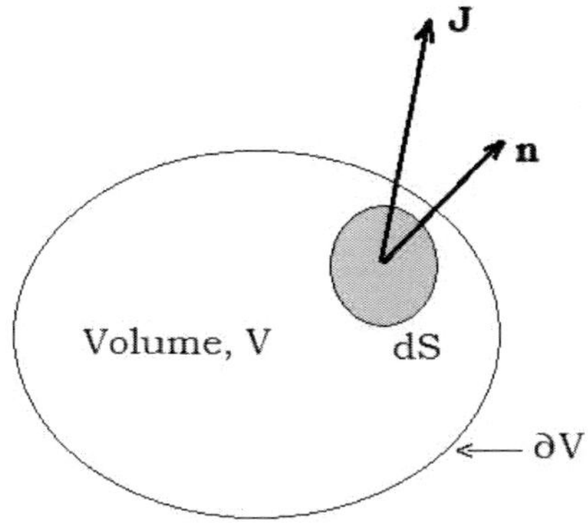

Figure 3.15. Charge leaving a closed volume V.

On the other hand, the time rate of *decrease* of the total charge in V is given by

$$-\frac{dQ}{dt} = -\frac{d}{dt}\int_V \rho \, d^3\mathbf{x} = -\int_V \frac{\partial \rho}{\partial t}\, d^3\mathbf{x} \tag{3.28}$$

Equating (3.27) and (3.28), we have

$$-\int_V \frac{\partial \rho}{\partial t}\, d^3\mathbf{x} = \oint_{\partial V} \mathbf{j}\cdot\mathbf{n}\, dS\,, \tag{3.29}$$

and employing the divergence theorem, we now convert the righthand side to a volume integral to obtain

$$-\int_V \frac{\partial \rho}{\partial t}\, d^3\mathbf{x} = \int_V (\nabla\cdot\mathbf{j})\; d^3\mathbf{x} \Longrightarrow \int_V \left[\frac{\partial \rho}{\partial t} + \nabla\cdot\mathbf{j}\right] d^3\mathbf{x} = 0$$

Since the volume V is arbitrary, we must have

Continuity Equation

$$\frac{\partial \rho}{\partial t} + \nabla\cdot\mathbf{j} = 0 \tag{3.30}$$

and this equation is known as the **continuity equation**; it enforces charge conservation.

3.6. Biot-Savart Law and Ampere's Law

We now begin a discussion of **magnetostatics**, or the study of the magnetic fields accompanying *steady* currents. We assume that the charge density is independent of the time, so that $\partial\rho\partial t = 0$, and consequently, the continuity equation reduces to

$$\nabla\cdot\mathbf{j} = 0 \tag{3.31}$$

The field of magnetostatics was initiated by Oersted, who in 1819, during a public demonstration, discovered that the current in wire deflected a compass needle at right angles to the direction of current flow. Later on, in 1820, Biot and Savart derived a result for the elemental magnetic field $d\mathbf{B}$ which is due to a current element in a wire.

As shown in Figure 3.16, a segment of wire carries a steady current I.

If the current element $I\,\mathbf{dl}'$ has position vector $\mathbf{x}'$ and the point P has position vector $\mathbf{x}$, then the elemental magnetic field $d\mathbf{B}$ at the point P is given by

$$d\mathbf{B} = \frac{I}{c}\,\frac{d\mathbf{l}' \times (\mathbf{x}-\mathbf{x}')}{\|\mathbf{x}-\mathbf{x}'\|^3} \tag{3.32}$$

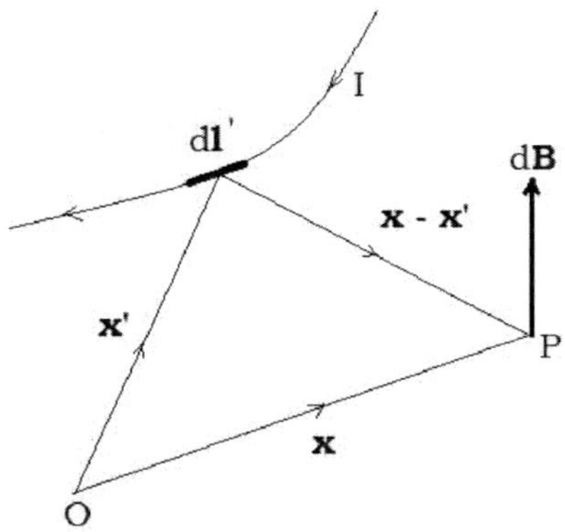

Figure 3.16. Magnetic field due to a current element.

Observe that due to the presence of the vector product in (3.32), the elemental magnetic field $d\mathbf{B}$ is perpendicular to the plane determined by the vectors $d\mathbf{l}'$ and $(\mathbf{x}-\mathbf{x}')$. Also, we see that the magnetic field $\mathbf{B}(\mathbf{x})$ at the point P is given by the line integral.

Biot-Savart Law

$$\mathbf{B}(\mathbf{x}) = \frac{I}{c}\int_C \frac{d\mathbf{l}' \times (\mathbf{x}-\mathbf{x}')}{\|\mathbf{x}-\mathbf{x}'\|^3} \tag{3.33}$$

where integration is performed along a current path C.

Equation (3.33) may be expressed as a volume integral. To see this, let dV represent the elemental volume of the wire segment $d\mathbf{l}'$, and let $\mathbf{v}$ denote the charge velocity. Now, observe that

$$Id\mathbf{l}' = \frac{dq}{dt}d\mathbf{l}' = (dq)\frac{d\mathbf{l}'}{dt} = (dq)\mathbf{v} = \rho\mathbf{v}dV = \mathbf{j}dV$$

and consequently, (3.33) may be replaced by the volume integral

$$\mathbf{B}(\mathbf{x}) = \frac{1}{c}\int_V \frac{\mathbf{j}(\mathbf{x}') \times (\mathbf{x}-\mathbf{x}')}{\|\mathbf{x}-\mathbf{x}'\|^3}\,d^3\mathbf{x}' \tag{3.34}$$

Now, let's take a closer look at the integrand in (3.34). Recalling the vector identity

$$\nabla \times (g\mathbf{V}) = g(\nabla \times \mathbf{V}) + (\nabla g) \times \mathbf{V}$$

and setting $\mathbf{V} = \mathbf{j}(\mathbf{x}')$ and $g = \|\mathbf{x}-\mathbf{x}'\|^{-1}$, we obtain

$$\nabla \times \left(\frac{1}{\|\mathbf{x}-\mathbf{x}'\|}\mathbf{j}(\mathbf{x}')\right) = \frac{1}{\|\mathbf{x}-\mathbf{x}'\|}\nabla \times \mathbf{j}(\mathbf{x}') + \nabla\left(\frac{1}{\|\mathbf{x}-\mathbf{x}'\|}\right) \times \mathbf{j}(\mathbf{x}') \tag{3.34a}$$

But $\nabla \times \mathbf{j}(\mathbf{x}') = \mathbf{0}$, since the vector operator involves derivatives with respect to the unprimed (field) variables. Also, a simple calculation gives

$$\nabla\left(\frac{1}{\|\mathbf{x}-\mathbf{x}'\|}\right) = -\frac{(\mathbf{x}-\mathbf{x}')}{\|\mathbf{x}-\mathbf{x}'\|^3} \tag{3.34b}$$

Thus, (3.34a) simplifies to:

$$\nabla \times \left(\frac{1}{\|\mathbf{x}-\mathbf{x}'\|}\mathbf{j}(\mathbf{x}')\right) = \frac{\mathbf{j}(\mathbf{x}') \times (\mathbf{x}-\mathbf{x}')}{\|\mathbf{x}-\mathbf{x}'\|^3} \tag{3.34c}$$

which is precisely the integrand in (3.34). Employing (3.32) and (3.34c), we can now write:

$$\mathbf{B}(\mathbf{x}) = \nabla \times \mathbf{A}(\mathbf{x}) \tag{3.35}$$

where the vector field $\mathbf{A}(\mathbf{x})$ is given by

$$\mathbf{A}(\mathbf{x}) = \frac{1}{c}\int_V \frac{\mathbf{j}(\mathbf{x}')}{\|\mathbf{x}-\mathbf{x}'\|}\, d^3\mathbf{x}' \tag{3.36}$$

and is called the **vector potential**.

Equations (3.35) and (3.36) have some interesting consequences. First, from (3.35), we see that the vector potential $\mathbf{A}(\mathbf{x})$ plays a role in magnetostatics which is somewhat analogous to the role played by the electric (scalar) potential $\Phi(\mathbf{x})$ in electrostatics. Unfortunately, the former is not as easy to work with, due to its vector character.

Secondly, since the divergence of the curl of a vector is always zero, we have:

$$\nabla \cdot (\nabla \times \mathbf{A}(\mathbf{x})) = 0 \Longrightarrow \nabla \cdot \mathbf{B}(\mathbf{x}) = 0 \tag{3.37}$$

which is just equation (3.2), the second equation in the set of Maxwell's equations.

Equation (3.37) tells us, at least classically, that point sources of magnetism, i.e., **magnetic monopoles**, cannot exist.[7] *Magnetic field lines must always form closed loops so that the net flux of a magnetic field through any closed surface always must be zero.*

Although we know the curl of the vector potential, since $\nabla \times \mathbf{A}(\mathbf{x}) = \mathbf{B}(\mathbf{x})$, we have left its divergence unspecified, so that the vector potential is not yet *uniquely determined*. We will now show that under *magneto-static conditions*, we can always set $\nabla \cdot \mathbf{B}(\mathbf{x}) = 0$.

For the sake of brevity, set $\mathbf{R} = \mathbf{x} - \mathbf{x}'$ so that $R = \|\mathbf{x} - \mathbf{x}'\|$. By direct calculation, one can show that

$$\nabla\left(\frac{1}{R}\right) = -\nabla'\left(\frac{1}{R}\right) \tag{3.38}$$

where, in rectangular coordinates, the operator ∇' involves derivatives with respect to the primed (source) variables, i.e.,

$$\nabla' = \mathbf{i}\frac{\partial}{\partial x'} + \mathbf{j}\frac{\partial}{\partial y'} + \mathbf{k}\frac{\partial}{\partial z'} \tag{3.39}$$

Consequently,

$$\begin{aligned} \nabla' \cdot \left(\frac{1}{R}\mathbf{j}(\mathbf{x}')\right) &= \left(\nabla'\frac{1}{R}\right) \cdot \mathbf{j}(\mathbf{x}') + \frac{1}{R}\left(\nabla' \cdot \mathbf{j}(\mathbf{x}')\right) \\ &= -\left(\nabla\frac{1}{R}\right) \cdot \mathbf{j}(\mathbf{x}') + \frac{1}{R}\left(\nabla' \cdot \mathbf{j}(\mathbf{x}')\right) \end{aligned}$$

so that

$$\left(\nabla\frac{1}{R}\right) \cdot \mathbf{j}(\mathbf{x}') = \frac{1}{R}\left(\nabla' \cdot \mathbf{j}(\mathbf{x}')\right) - \nabla' \cdot \left(\frac{1}{R}\mathbf{j}(\mathbf{x}')\right) \tag{3.40}$$

We now take the divergence of the vector potential in (3.36), and use (3.40) to obtain

$$\begin{aligned} \nabla \cdot \mathbf{A}(\mathbf{x}) &= \frac{1}{c}\int_V \left(\nabla \cdot \frac{1}{R}\right) \mathbf{j}(\mathbf{x}')\, d^3\mathbf{x}' \\ &= \frac{1}{c}\int_V \left[\frac{1}{R}\left(\nabla' \cdot \mathbf{j}(\mathbf{x}')\right) - \nabla' \cdot \left(\frac{1}{R}\mathbf{j}(\mathbf{x}')\right)\right] d^3\mathbf{x}' \\ &= (0) - \frac{1}{c}\int_V \nabla' \cdot \left(\frac{1}{R}\mathbf{j}(\mathbf{x}')\right) d^3\mathbf{x}' \\ &= -\frac{1}{c}\oint_{\partial V} \frac{1}{R}\mathbf{j}(\mathbf{x}') \cdot \mathbf{n}\, dS \end{aligned}$$

where we used the fact that under magnetostatic conditions, we must have $\nabla' \cdot \mathbf{j}(\mathbf{x}') = 0$, and we transformed the second integral into a surface integral via the divergence theorem. If we now choose the volume V sufficiently large so that $\mathbf{j}(\mathbf{x}') = \mathbf{0}$ for all points $\mathbf{x}' \in \partial V$, then the surface integral vanishes as well, and so, we may take

$$\nabla \cdot \mathbf{A}(\mathbf{x}) = 0 \tag{3.41}$$

Equation (3.41) leads to some interesting consequences. First of all, we have

$$\begin{aligned} \nabla \times \mathbf{B}(\mathbf{x}) &= \nabla \times (\nabla \times \mathbf{A}(\mathbf{x})) \\ &= -\nabla^2\mathbf{A}(\mathbf{x}) - \nabla(\nabla \cdot \mathbf{A}(\mathbf{x})) \\ &= -\nabla^2\mathbf{A}(\mathbf{x}) \quad (\text{ since } \nabla \cdot \mathbf{A}(\mathbf{x}) = 0\,) \end{aligned} \tag{3.42}$$

Employing (3.35) and (3.41), we can write

$$\nabla \times \mathbf{B}(\mathbf{x}) = -\nabla^2 \mathbf{A}(\mathbf{x}) = -\frac{1}{c} \int_V \mathbf{j}(\mathbf{x}')\, \nabla^2 \left(\frac{1}{\|\mathbf{x}-\mathbf{x}'\|} \right) d^3\mathbf{x}'$$

But it may be shown [See the Exercises that follow.] that

$$\nabla^2 \left(\frac{1}{\|\mathbf{x}-\mathbf{x}'\|} \right) = -4\pi\delta^3 \left(\mathbf{x}-\mathbf{x}'\right) \tag{3.43}$$

where $\delta^3 (\mathbf{x}-\mathbf{x}')$ is the three-dimensional delta function, so the last equation becomes

Ampere's Law

$$\nabla \times \mathbf{B}(\mathbf{x}) = \frac{4\pi}{c}\, \mathbf{j}(\mathbf{x}) \tag{3.44}$$

which is known as **Ampere's law**. Observe that this is special case of Maxwell's equation (3.4) and holds under magnetostatic conditions. Now, we can rewrite Ampere's law in integral form. Integrating both sides of (3.44), we obtain

$$\int_S \nabla \times \mathbf{B}(\mathbf{x}) \cdot \mathbf{n}\, dS = \frac{4\pi}{c} \int_S \mathbf{j}(\mathbf{x}) \cdot \mathbf{n}\, dS \tag{3.45}$$

Finally, if we apply Stoke's theorem to the integral on the left, and recall the definition of current, we obtain

(Integral Form)

$$\oint_C \mathbf{B}(\mathbf{x}) \cdot d\mathbf{l} = \frac{4\pi}{c}\, I_{enclosed} \tag{3.46}$$

where $I_{enclosed}$ is the current which is enclosed by the curve C that forms the boundary of the surface S.

3.7. Exercises

1. Current flows in a cylindrical wire of radius a. If the current density $j = \alpha r$ where α is a constant, and r is the distance from the cylinder's axis, show that the current is given by

$$I = \left(\frac{2\pi\alpha}{3} \right) a^3 \tag{3.47}$$

(Solution)

We employ cylindrical coordinates. As shown in Figure 3.17, we assume that current flows in the positive z-direction so that $\mathbf{j} = (\alpha r)\hat{\mathbf{z}}$. Also, the unit normal to the surface S is $\mathbf{n} = \hat{\mathbf{z}}$. From equation (3.26), we obtain

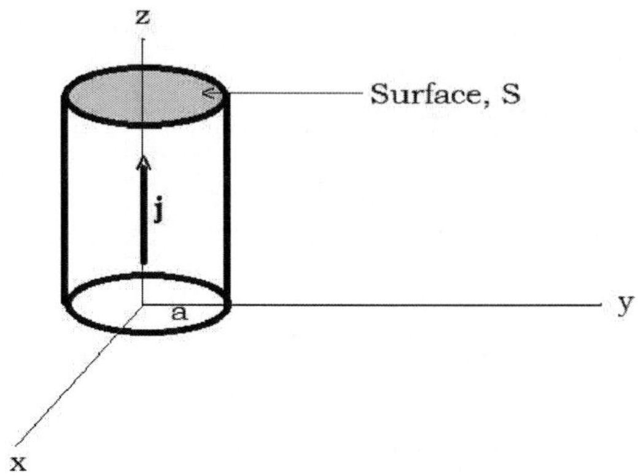

Figure 3.17. Current flow in a cylindrical wire.

$$\begin{aligned} I &= \int_S \mathbf{j}\cdot\mathbf{n}\, dS = \int_S (\alpha r)\,(\hat{\mathbf{z}}\cdot\hat{\mathbf{z}})\, dS \\ &= \int_0^a \int_0^{2\pi} (\alpha r)\,(r dr d\phi) = 2\pi\alpha \int_0^a r^2 dr \\ &= \frac{2\pi\alpha}{3} a^3 \end{aligned}$$

2. *(Magnetic Field Due to an Infinite Straight Wire)* An infinitely long, straight wire carries a steady current I as shown in Figure 3.18. Show that in cylindrical coordinates, the magnetic field $\mathbf{B}$ is given by

$$\mathbf{B} = \frac{2I}{rc}\,\hat{\phi} \tag{3.48}$$

(Solution)

Using cylindrical coordinates, we set $Id\mathbf{l}' = (Idz)\,\hat{\mathbf{z}}$. Employing (3.33), and setting $\mathbf{R} = \mathbf{x} - \mathbf{x}'$, we have:

$$\begin{aligned} d\mathbf{B} &= \frac{I}{c}\,\frac{d\mathbf{l}'\times\mathbf{R}}{R^3} = \frac{I}{c}\,\frac{(dz\,\hat{\mathbf{z}})\times(r\hat{\mathbf{r}} - z\hat{\mathbf{z}})}{\|\mathbf{x}-\mathbf{x}'\|^3} \\ &= \frac{I}{c}\left[\frac{rdz}{(r^2+z^2)^{3/2}}\right]\hat{\phi} \end{aligned}$$

Integrating, we obtain

$$\mathbf{B} = \frac{I}{c}\left[\int_{-\infty}^{+\infty} \frac{rdz}{(r^2+z^2)^{3/2}}\right]\hat{\phi}$$

Let $z = r\tan\theta$, so $dz = r\sec^2\theta d\theta$, and the integral becomes

$$\mathbf{B} = \frac{I}{rc}\left[\int_{-\pi/2}^{+\pi/2} \cos\theta d\theta\right]\hat{\phi} = \left(\frac{2I}{rc}\right)\hat{\phi}$$

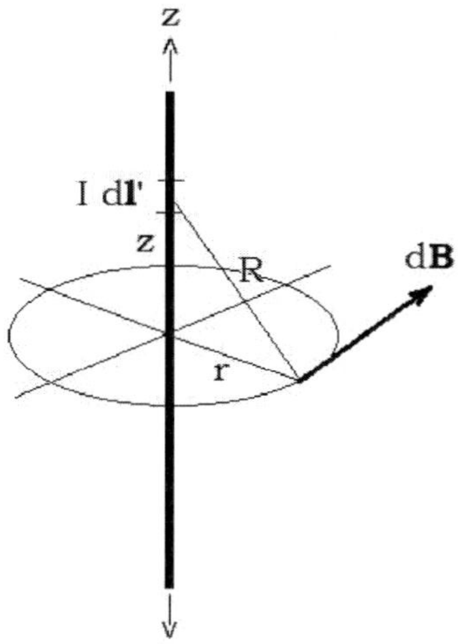

Figure 3.18. Magnetic field surrounding an infinitely long wire.

3. *(Magnetic Field Due to a Finite Straight Wire)* A thin wire of length L lies on the z-axis as shown in Figure 3.19. Show that, in cylindrical coordinates, the magnetic field at any point P which is a radial distance of r from the z-axis is given by

$$\mathbf{B} = \frac{I}{rc}\frac{L}{\sqrt{L^2+r^2}}\hat{\phi} \tag{3.49}$$

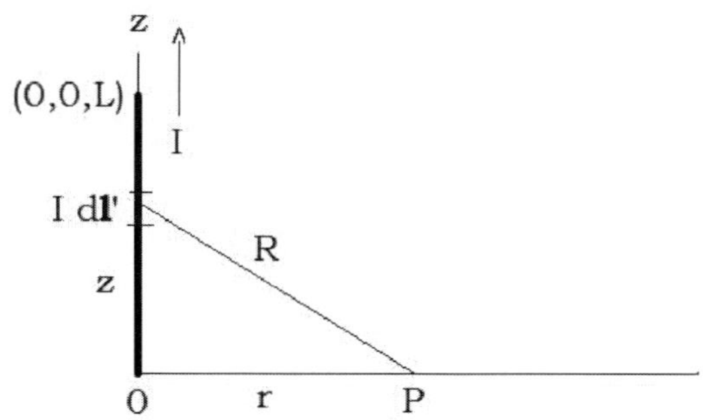

Figure 3.19. A thin wire of length L.

(Solution)

We employ cylindrical coordinates, and we set $Id\mathbf{l}' = (Idz)\,\hat{\mathbf{z}}$. Setting $\mathbf{R} = \mathbf{x} - \mathbf{x}'$, we have by (3.33),

$$\begin{aligned} d\mathbf{B} &= \frac{I}{c}\frac{d\mathbf{l}' \times \mathbf{R}}{R^3} = \frac{I}{c}\frac{(dz\,\hat{\mathbf{z}}) \times (r\hat{\mathbf{r}} - z\hat{\mathbf{z}})}{\|\mathbf{x}-\mathbf{x}'\|^3} \\ &= \frac{I}{c}\left[\frac{rdz}{(r^2+z^2)^{3/2}}\right]\hat{\phi} \end{aligned}$$

so that

$$\mathbf{B} = \frac{I}{c}\left[\int_{-\infty}^{+\infty} \frac{r dz}{(r^2+z^2)^{3/2}}\right]\hat{\phi}$$

As in the previous problem, we let $z = r\tan\theta$, so $dz = r\sec^2\theta d\theta$, and we obtain

$$\mathbf{B} = \frac{I}{rc}\left[\int_0^{\theta = \tan^{-1}(L/r)} \cos\theta d\theta\right]\hat{\phi} = \frac{I}{rc}\frac{L}{\sqrt{L^2+r^2}}\hat{\phi}$$

4. *(Field Due to Circular Current Loop)* As shown in Figure 3.20, a steady current I circulates in the counterclockwise direction in a circular loop of radius a, sitting in the xy-plane. Show that, in cylindrical coordinates, the magnetic field at the point $P(0,0,h)$ is given by

$$\mathbf{B} = \frac{2\pi I}{c}\frac{a^2}{(a^2+h^2)^{3/2}}\hat{\mathbf{z}} \tag{3.50}$$

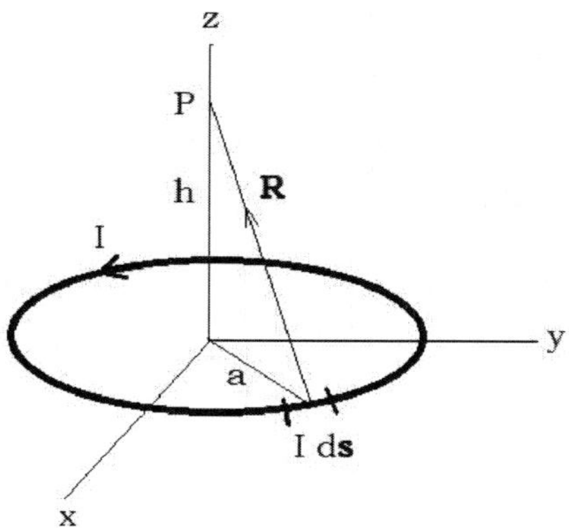

Figure 3.20. A circular current loop of radius a.

(Solution)

Employing cylindrical coordinates, and we set $Id\mathbf{l}' = (Ids)\hat{\phi}$. Setting $\mathbf{R} = \mathbf{x} - \mathbf{x}'$, we have by (3.33),

$$\begin{aligned} d\mathbf{B} &= \frac{I}{c}\frac{d\mathbf{l}' \times \mathbf{R}}{R^3} = \frac{I}{c}\frac{(ds\,\hat{\phi}) \times (a\hat{\mathbf{r}} - h\hat{\mathbf{z}})}{R^3} \\ &= \frac{I}{c}\left[\frac{a^2 d\phi}{(a^2+h^2)^{3/2}}\hat{\mathbf{z}} - \frac{h d\phi}{(a^2+h^2)^{3/2}}\hat{\mathbf{r}}\right] \end{aligned}$$

By symmetry, the components of $\mathbf{B}$ in the $\hat{\mathbf{r}}$-direction must cancel, so upon integration, we obtain

$$\mathbf{B} = \frac{I}{c}\int_0^{2\pi} \frac{a^2 d\phi}{(a^2+h^2)^{3/2}}\hat{\mathbf{z}} = \frac{2\pi I}{c}\frac{a^2}{(a^2+h^2)^{3/2}}\hat{\mathbf{z}}$$

5. *(Dirac Delta Function)* Show that

$$\nabla^2 \left(\frac{1}{\|\mathbf{x}-\mathbf{x}'\|} \right) = -4\pi\delta^3(\mathbf{x}-\mathbf{x}') \tag{3.51}$$

where $\delta^3(\mathbf{x}-\mathbf{x}')$ is the three-dimensional Dirac delta function.

(Solution)

Consider a region V with boundary ∂V, as shown in Figure 3.21.

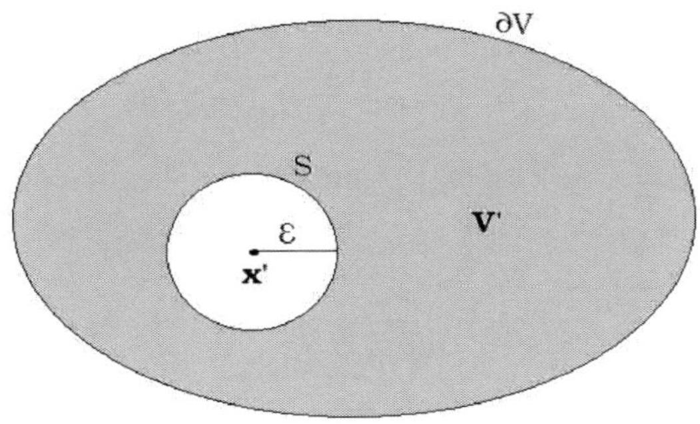

Figure 3.21. Region V with boundary ∂V.

We construct a sphere S of radius ε, and with center at $\mathbf{x}'$. Let the region V' be that portion of V which *excludes* the interior of the sphere. Direct calculation shows that

$$\nabla \cdot \left(\frac{\mathbf{x}-\mathbf{x}'}{\|\mathbf{x}-\mathbf{x}'\|^3} \right) = 0 \;\text{ for all } \mathbf{x} \in V'$$

On the other hand, by the divergence theorem,

$$\begin{aligned} \int_V \nabla \cdot \left(\frac{\mathbf{x}-\mathbf{x}'}{\|\mathbf{x}-\mathbf{x}'\|^3} \right) d^3\mathbf{x} &= \oint_{\partial V} \frac{(\mathbf{x}-\mathbf{x}')}{\|\mathbf{x}-\mathbf{x}'\|^3} \cdot d\mathbf{S} \\ &= \oint_S \frac{\mathbf{x}-\mathbf{x}'}{\|\mathbf{x}-\mathbf{x}'\|^3} \cdot d\mathbf{S} \\ &= \oint_S \frac{1}{\varepsilon^2} dS = 4\pi \end{aligned}$$

since $(\mathbf{x}-\mathbf{x}')/\|\mathbf{x}-\mathbf{x}'\|^3$ is just the electric field due to the *unit positive charge* situated at $\mathbf{x}'$, and consequently the electric flux through ∂V must be the same as the flux through S. Note that the entire contribution to the volume integral is coming from the single point $\mathbf{x}'$, and the result is independent of ε. Thus, we have shown that

$$\int_V \nabla \cdot \left(\frac{\mathbf{x}-\mathbf{x}'}{\|\mathbf{x}-\mathbf{x}'\|^3} \right) d^3\mathbf{x} = \begin{cases} 4\pi, \text{ if } \mathbf{x}' \in V \\ 0, \text{ if } \mathbf{x}' \notin V \end{cases}$$

Evidently,

$$\nabla \cdot \left(\frac{\mathbf{x}-\mathbf{x}'}{\|\mathbf{x}-\mathbf{x}'\|^3} \right) = 4\pi\delta^3(\mathbf{x}-\mathbf{x}') \tag{3.52}$$

Now, let us recall that

$$\nabla \left(\frac{1}{\|\mathbf{x}-\mathbf{x}'\|} \right) = -\frac{\mathbf{x}-\mathbf{x}'}{\|\mathbf{x}-\mathbf{x}'\|^3}$$

so

$$\nabla^2 \left(\frac{1}{\|\mathbf{x}-\mathbf{x}'\|} \right) = \nabla \cdot \nabla \left(\frac{1}{\|\mathbf{x}-\mathbf{x}'\|} \right) = -\nabla \cdot \left(\frac{\mathbf{x}-\mathbf{x}'}{\|\mathbf{x}-\mathbf{x}'\|^3} \right) = -4\pi\delta^3(\mathbf{x}-\mathbf{x}')$$

as required.

6. *(Green's Symmetrical Theorem)* If $f(\mathbf{x})$ and $g(\mathbf{x})$ are sufficiently smooth scalar fields, show that:

$$\int_V \left(f\, \nabla^2 g - g\, \nabla^2 f\right) dV = \oint_{\partial V} (f\, \nabla g - g\, \nabla f) \cdot d\mathbf{S} \tag{3.53}$$

where the region V is bounded by the closed surface ∂V.

(Solution)

First, observe that

$$\nabla \cdot (f\, \nabla g) = f\, \nabla^2 g + \nabla f \cdot \nabla g \quad \text{and} \quad \nabla \cdot (g\, \nabla f) = g\nabla^2 f + \nabla f \cdot \nabla g$$

Subtracting the last two equations, we get

$$\nabla \cdot (f\, \nabla g - g\, \nabla f) = f\, \nabla^2 g - g\nabla^2 f$$

Employing the last equation, and the divergence theorem, we have

$$\begin{aligned} \int_V \left(f\, \nabla^2 g - g\nabla^2 f\right) dV &= \int_V \nabla \cdot (f\, \nabla g - g\, \nabla f)\, dV \\ &= \oint_{\partial V} (f\, \nabla g - g\, \nabla f) \cdot d\mathbf{S} \end{aligned}$$

7. *(Uniqueness Theorem)* Let $\mathbf{V}(\mathbf{x})$ be a vector field such that $\|\mathbf{V}(\mathbf{x})\| \sim 1/\|\mathbf{x}\|^2$ as $\|\mathbf{x}\| \to \infty$. Show that if both $\nabla\times\, \mathbf{V}(\mathbf{x})$ and $\nabla \cdot \mathbf{V}(\mathbf{x})$ are known at all points $\mathbf{x} \in R^3$, then $\mathbf{V}(\mathbf{x})$ is unique.

(Solution)

Assume there exists another vector field $\mathbf{V}'(\mathbf{x})$ such that $\|\mathbf{V}'(\mathbf{x})\| \sim 1/\|\mathbf{x}\|^2$ as $\|\mathbf{x}\| \to \infty$, while at the same time,

$$\nabla \times \mathbf{V}'(\mathbf{x}) = \nabla \times \mathbf{V}(\mathbf{x}) \quad \text{and} \quad \nabla \cdot \mathbf{V}'(\mathbf{x}) = \nabla \cdot \mathbf{V}(\mathbf{x})$$

for all $\mathbf{x} \in R^3$. Let $\varepsilon(\mathbf{x}) = \mathbf{V}'(\mathbf{x}) - \mathbf{V}(\mathbf{x})$ so that

$$\nabla \times \varepsilon(\mathbf{x}) = \nabla \times \left(\mathbf{V}'(\mathbf{x}) - \mathbf{V}(\mathbf{x})\right) = \mathbf{0}$$

Now, $\nabla \times \varepsilon(\mathbf{x}) = \mathbf{0}$ implies there exists a potential function $\Phi(\mathbf{x})$ such that $\varepsilon(\mathbf{x}) = \nabla\Phi(\mathbf{x})$. Consequently,

$$\nabla \cdot \varepsilon(\mathbf{x}) = \nabla^2\Phi(\mathbf{x}) = 0$$

On the other hand,

$$\begin{aligned}
\oint_{\partial V} (\Phi\nabla\Phi) \cdot d\mathbf{S} &= \int_V \nabla \cdot (\Phi\nabla\Phi)\, dV = \int_V \Phi\nabla^2\Phi\, dV + \int_V (\nabla\Phi)^2\, dV \\
&\Longrightarrow \oint_{\partial V} (\Phi\nabla\Phi) \cdot d\mathbf{S} = \int_V \|\nabla\Phi\|^2\, dV \quad (\text{ since } \nabla^2\Phi = 0\ )
\end{aligned}$$

But $\|\Phi(\mathbf{x})\| \sim 1/\|\mathbf{x}\|$ and $\|\nabla\Phi(\mathbf{x})\| \sim 1/\|\mathbf{x}\|^2$ as $\|\mathbf{x}\| \to \infty$. Thus, if we let $\|\mathbf{x}\| \to \infty$, then the surface integral on the left vanishes

$$\left|\oint_{\partial V} (\Phi\nabla\Phi) \cdot d\mathbf{S}\right| \to 0 \text{ as } \|\mathbf{x}\| \to \infty$$

and we conclude that $\nabla\Phi = \varepsilon(\mathbf{x}) = \mathbf{0}$, and so, $\mathbf{V}'(\mathbf{x}) = \mathbf{V}(\mathbf{x})$.

8. *(Helmholtz Theorem)* Let $\mathbf{V}(\mathbf{x})$ be a vector field such that $\|\mathbf{V}(\mathbf{x})\| \sim 1/\|\mathbf{x}\|^2$ as $\|\mathbf{x}\| \to \infty$. Show that $\mathbf{V}(\mathbf{x})$ may be written *uniquely* as

$$\mathbf{V}(\mathbf{x}) = \nabla \times \mathbf{A}(\mathbf{x}) + \nabla\Phi(\mathbf{x}) \tag{3.54}$$

where

$$\mathbf{A}(\mathbf{x}) = \frac{1}{4\pi}[\nabla \times \mathbf{W}(\mathbf{x})] \tag{3.54a}$$

$$\Phi(\mathbf{x}) = -\frac{1}{4\pi}[\nabla \cdot \mathbf{W}(\mathbf{x})] \tag{3.54b}$$

and where $\mathbf{W}(\mathbf{x})$ is given by

$$\mathbf{W}(\mathbf{x}) = \int_V \frac{\mathbf{V}(\mathbf{x}')}{\|\mathbf{x} - \mathbf{x}'\|} d^3\mathbf{x}' \tag{3.54c}$$

and the volume V may be taken arbitrarily large.

(Solution)

We first show that $\nabla^2\mathbf{W}(\mathbf{x}) = -4\pi\mathbf{V}(\mathbf{x})$. To this end, let

$$\begin{aligned}
\mathbf{W}(\mathbf{x}) &= W_1\mathbf{i} + W_2\mathbf{j} + W_3\mathbf{k} \\
\mathbf{V}(\mathbf{x}) &= V_1\mathbf{i} + V_2\mathbf{j} + V_3\mathbf{k}
\end{aligned}$$

so that by (3.54c), we have

$$W_1(\mathbf{x}) = \int\limits_V \frac{V_1(\mathbf{x}')}{\|\mathbf{x}-\mathbf{x}'\|} d^3\mathbf{x}'$$

and consequently,

$$\begin{aligned} \nabla^2 W_1(\mathbf{x}) &= \int\limits_V V_1(\mathbf{x}') \left[\nabla^2 \frac{1}{\|\mathbf{x}-\mathbf{x}'\|}\right] d^3\mathbf{x}' = -4\pi \int\limits_V V_1(\mathbf{x}')\delta^3(\mathbf{x}-\mathbf{x}')d^3\mathbf{x}' \\ &= 4\pi V_1(\mathbf{x}) \end{aligned}$$

Similarly, we can show that $\nabla^2 W_2(\mathbf{x}) = 4\pi V_2(\mathbf{x})$ and $\nabla^2 W_3(\mathbf{x}) = 4\pi V_3(\mathbf{x})$; so that

$$\nabla^2 \mathbf{W}(\mathbf{x}) = -4\pi \mathbf{V}(\mathbf{x}) \tag{3.54d}$$

as claimed. Now, recall the vector identity

$$\nabla^2 \mathbf{W} = -\nabla \times (\nabla \times \mathbf{W}) + \nabla (\nabla \cdot \mathbf{W})$$

so that by (3.54d), (3.54a), and (3.54b) we can write

$$-4\pi \mathbf{V}(\mathbf{x}) = -4\pi [\nabla \times \mathbf{W}(\mathbf{x})] - 4\pi [\nabla \Phi(x)]$$

from which (3.54) follows immediately.

The uniqueness of the decomposition follows from the Uniqueness Theorem above. Observe the interesting similarity between equation (3.36) for vector potential, and equation (3.52c).

9. Use the integral form of Ampere's law, to determine the magnetic field **B** in the interior of a cylindrical conductor of radius a, as shown in Figure 3.22. Assume the current I is both steady and uniformly distributed.

(Solution)

We construct a circular Amperian loop C with radius r. The direction of **B** may be obtained from the *righthand rule*, which states that if we grasp the wire with our right hand, and place our thumb in the direction of current flow, then our fingers will point in the direction of the magnetic field lines. Consequently, using cylindrical coordinates, we must have $\mathbf{B} = B\,\hat{\phi}$ where B is constant.

By Ampere's law, we have

$$\oint\limits_C \mathbf{B}(\mathbf{x}) \cdot d\mathbf{l} = \frac{4\pi}{c} I_{enclosed}$$

where $I_{enclosed}$ is the current enclosed by the loop C. Since $d\mathbf{l} = (ds)\hat{\phi}$, we have

$$\oint\limits_C \mathbf{B}(\mathbf{x}) \cdot d\mathbf{l} = \int_0^{2\pi} B\hat{\phi} \cdot \hat{\phi}\,(rd\phi) = B(2\pi r)$$

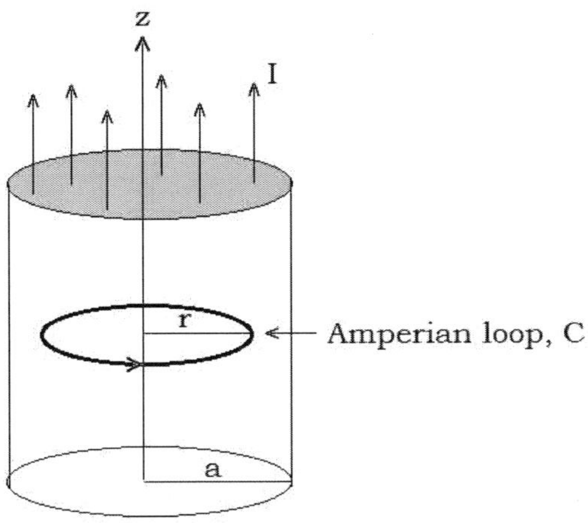

Figure 3.22. Steady current I flowing through a cylindrical conductor.

On the other hand, since the current I is uniform,

$$I_{enclosed} = I\left(\frac{\pi r^2}{\pi a^2}\right) = I\,\frac{r^2}{a^2}$$

Combining the last two results, we obtain

$$B(2\pi r) = \frac{4\pi}{c}\cdot I\,\frac{r^2}{a^2} \Longrightarrow B = \frac{2rI}{a^2 c}$$

so that magnetic field inside the conductor is

$$\mathbf{B}_{interior} = \left(\frac{2rI}{a^2 c}\right)\hat{\phi} \tag{3.55}$$

Incidently, if we apply Ampere's law to determine the external magnetic field, we find that

$$\mathbf{B}_{external} = \left(\frac{2I}{rc}\right)\hat{\phi} \tag{3.55a}$$

a result which is in agreement with equation (3.48).

3.8. Faraday's Law

In a series of beautiful experiments, Faraday discovered that an electric field is induced in a conducting loop by the *relative motion* of a magnet and the conducting loop. Whether the magnet moves, and the loop is stationary, or vice versa, the result is the same; an electric field is induced in the loop.

Additionally, even if the conducting loop and the source of the magnetic field are *both stationary*, but the magnetic field varies with time, again an electric field is induced in the conducting loop. Such a varying magnetic field may be created, for example, by repeatedly turning on and off a simple electromagnet connected to a storage battery.

Unlike a *Coulomb field*, or an electrostatic field, which arises from a stationary distribution of charges, the *induced electric field* is *non-conservative*; and consequently, its integral around a closed path is, in general, not zero.

We define the **electromotive force** or **emf**, ε as the line integral of the induced field $\mathbf{E}$ around a closed path C, or

$$\varepsilon = \oint_C \mathbf{E} \cdot d\mathbf{l} \tag{3.56}$$

Examining this equation, we see that the emf is the amount of work done (by the induced field) in moving a *unit* charge around the contour C. The units of ε are ergs/statcoul, or more simply, statvolts.

Faraday's law states that *the emf around a closed loop is proportional to the time rate of change of the magnetic flux through the loop.* Simply stated, the Faraday's law is

$$\varepsilon = -\frac{1}{c}\frac{d}{dt}(\Psi_B) \tag{3.57}$$

where the **magnetic flux**, Ψ_B is given by the surface integral

$$\Psi_B = \int_S \mathbf{B} \cdot d\mathbf{S} \tag{3.58}$$

Clearly, Ψ_B is proportional to the number of magnetic field lines which pass through the surface S. Note that for the special case of a closed surface S, we must have $\Psi_B = 0$, owing to the fact that the magnetic field is divergence-free.

Combining the above equations, Faraday's law can be written in the form

$$\oint_C \mathbf{E} \cdot d\mathbf{l} = -\frac{1}{c}\frac{d}{dt}\int_S \mathbf{B} \cdot d\mathbf{S} \tag{3.59}$$

Let's further investigate the meaning of equation (3.59). As shown in Figure 3.23, a magnetic field $\mathbf{B}(t)$, which is time-dependent, interacts with a filimentary conducting loop, C. The magnetic lines of force pierce through the surface S creating magnetic flux.

Let's assume that $B(t)$ is decreasing, so that $dB/dt < 0$, and the righthand side of (3.59) is positive. As a result, the induced electric field $\mathbf{E}$ must have a direction similar to the direction of $d\mathbf{l}$, and an induced electric current must flow in the counterclockwise direction of the loop, C. Finally, the induced current I will, in turn, create a magnetic field of its own, whose direction (according to the right- hand rule) will oppose the direction of $\mathbf{B}(t)$.

The above analyis is just an instance of **Lenz's law** which states that *the direction of the induced emf is such that the magnetic field (which it creates) tends to oppose the change of the original, inducing field* $\mathbf{B}(t)$.

Finally, we can write Faraday's law in differential form, by applying Stoke's theorem to the lefthand side of (3.59), to obtain

$$\int_S (\nabla \times \mathbf{E}) \cdot d\mathbf{S} = \int_S -\frac{1}{c}\frac{\partial \mathbf{B}}{\partial t} \cdot d\mathbf{S}$$

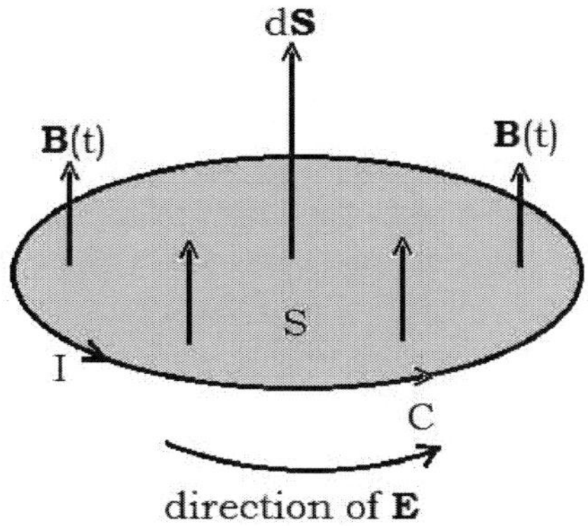

Figure 3.23. A magnetic field $\mathbf{B}(t)$ within the circuit loop, C.

so that

$$\int_S \left[(\nabla \times \mathbf{E}) + \frac{1}{c}\frac{\partial \mathbf{B}}{\partial t} \right] \cdot d\mathbf{S} = \mathbf{0}$$

Since the last equation must hold for *any* surface S whose boundary is C, we must have

Faraday's Law

$$\nabla \times \mathbf{E} = -\frac{1}{c}\frac{\partial \mathbf{B}}{\partial t} \tag{3.60}$$

We see that (3.60) is precisely Maxwell's third equation.

3.9. Maxwell's Displacement Current

In Section 3.7, we derived Ampere's law; i.e.,

$$\nabla \times \mathbf{B} = \frac{4\pi}{c}\mathbf{j}$$

which was shown to hold under *magnetostatic* conditions. A question which naturally arises is whether this law holds for the more general situation in which we allow time-varying fields. A simple analysis shows that it does not.

To see this, recall that by the continuity equation, we have

$$\nabla \cdot \mathbf{j} = -\frac{\partial \rho}{\partial t}$$

At the same time, if we take the divergence of both sides of Ampere's equation, recalling that the divergence of the curl is zero, we get

$$\nabla \cdot [\nabla \times \mathbf{B}] = \frac{4\pi}{c}[\nabla \cdot \mathbf{j}] = 0$$

This implies that $\nabla \cdot \mathbf{j} = 0$, which violates the equation of charge conservation. So, we conclude that Ampere's law, as it stands, doesn't hold good for time-varying fields.

In order to correct this problem, let's assume that some unknown vector field, say $\chi(\mathbf{x})$, must be added to the righthand side of Ampere's law to help us preserve charge conservation. Thus, our modified form of Ampere's law now reads

$$\nabla \times \mathbf{B} = \frac{4\pi}{c}(\mathbf{j} + \chi) \tag{3.61}$$

Taking the divergence of both sides of (3.61), and again enforcing charge conservation, we obtain

$$0 = \frac{4\pi}{c}(\nabla \cdot \mathbf{j} + \nabla \cdot \chi) = \frac{4\pi}{c}\left(-\frac{\partial \rho}{\partial t} + \nabla \cdot \chi\right)$$

and we conclude that

$$-\frac{\partial \rho}{\partial t} + \nabla \cdot \chi(\mathbf{x}) = 0$$

But recall, that by Gauss's law, $\nabla \cdot \mathbf{E} = 4\pi\rho$; so the last equation becomes

$$\nabla \cdot \chi = \frac{\partial \rho}{\partial t} = \frac{1}{4\pi}\frac{\partial}{\partial t}(\nabla \cdot \mathbf{E}) = \frac{1}{4\pi}\left(\nabla \cdot \frac{\partial \mathbf{E}}{\partial t}\right)$$

Now, this equation, in turn, will hold if we simply set

$$\chi = \frac{1}{4\pi}\frac{\partial \mathbf{E}}{\partial t}$$

Thus, Maxwell postulated that the general form of Ampere's law must be

$$\nabla \times \mathbf{B} = \frac{4\pi}{c}\mathbf{j} + \frac{1}{c}\frac{\partial \mathbf{E}}{\partial t}$$

and this is identical to equation (3.4), Maxwell's fourth equation. Also, Maxwell called the new term

$$\mathbf{j}_d = \frac{1}{c}\frac{\partial \mathbf{E}}{\partial t} \tag{3.62}$$

the **displacement current density** to distinguish it from the conduct- ion current density $\mathbf{j}$.

3.10. Electromagnetic Waves

As we promised in the beginning of this chapter, we will show how Maxwell's equations leads to the prediction of electromagnetic waves, and we will demonstrate that Maxwell's equations are *not invariant* under the group of Galilean transformations,

For convenience, we first list Maxwell's equations:

$$\nabla \cdot \mathbf{E} = 4\pi\rho \tag{3.1}$$

$$\nabla \cdot \mathbf{B} = 0 \tag{3.2}$$

$$\nabla \times \mathbf{E} = -\frac{1}{c}\frac{\partial \mathbf{B}}{\partial t} \tag{3.3}$$

$$\nabla \times \mathbf{B} = \frac{4\pi}{c}\mathbf{j} + \frac{1}{c}\frac{\partial \mathbf{E}}{\partial t} \tag{3.4}$$

Under *free space* conditions, we assume that there are no charges present ($\rho = 0$) and no *conduction* currents ($\mathbf{j} = \mathbf{0}$). So, Maxwell's equations assume the form

Maxwell's Equations *(Free Space)*

$$\nabla \cdot \mathbf{E} = 0 \tag{3.63}$$

$$\nabla \cdot \mathbf{B} = 0 \tag{3.64}$$

$$\nabla \times \mathbf{E} = -\frac{1}{c}\frac{\partial \mathbf{B}}{\partial t} \tag{3.65}$$

$$\nabla \times \mathbf{B} = \frac{1}{c}\frac{\partial \mathbf{E}}{\partial t} \tag{3.66}$$

We now show how these equations lead to the prediction of electromagnetic waves. We start by taking the curl of both sides of equation (3.65), to obtain

$$\nabla \times (\nabla \times \mathbf{E}) = -\frac{1}{c}\frac{\partial}{\partial t}(\nabla \times \mathbf{B}) \tag{3.67}$$

where, we have used the fact that space and time operators commute. Substituting (3.66) into the righthand side of (3.67), and using the vector identity

$$\nabla \times (\nabla \times \mathbf{E}) = \nabla (\nabla \cdot \mathbf{E}) - \nabla^2 \mathbf{E}$$

we get

$$\nabla (\nabla \cdot \mathbf{E}) - \nabla^2 \mathbf{E} = -\frac{1}{c}\frac{\partial}{\partial t}\left(\frac{1}{c}\frac{\partial \mathbf{E}}{\partial t}\right) = -\frac{1}{c^2}\frac{\partial^2 \mathbf{E}}{\partial t^2}$$

But by (3.63), $\nabla \cdot \mathbf{E} = 0$, so the last equation becomes

$$\nabla^2 \mathbf{E} = \frac{1}{c^2}\frac{\partial^2 \mathbf{E}}{\partial t^2} \tag{3.68}$$

Equation (3.68) is the *vector form* of the famous **wave equation**. If we use rectangular coordinates, and write $\mathbf{E}$ in terms of its components as

$$\mathbf{E} = E_x\,\hat{\mathbf{x}} + E_y\hat{\mathbf{y}} + E_z\hat{\mathbf{z}}\ ,$$

then

$$\nabla^2 \mathbf{E} = \left(\nabla^2 E_x\right)\hat{\mathbf{x}} + \left(\nabla^2 E_y\right)\hat{\mathbf{y}} + \left(\nabla^2 E_z\right)\hat{\mathbf{z}}$$

and the vector wave equation for $\mathbf{E}$ becomes three separate *scalar* wave equations involving its components.

$$\nabla^2 E_x = \frac{1}{c^2}\frac{\partial^2 E_x}{\partial t^2}\,,\quad \nabla^2 E_y = \frac{1}{c^2}\frac{\partial^2 E_y}{\partial t^2}\,,\quad \nabla^2 E_z = \frac{1}{c^2}\frac{\partial^2 E_z}{\partial t^2} \tag{3.68a}$$

On the other hand, if we take the curl of both sides of equation (3.66), and then use (3.64), we also obtain the same vector wave equation for $\mathbf{B}$:

$$\nabla^2 \mathbf{B} = \frac{1}{c^2}\frac{\partial^2 \mathbf{B}}{\partial t^2} \tag{3.69}$$

Similarly, if we resolve **B** into its rectangular components, we again obtain three scalar wave equations involving its components:

$$\nabla^2 B_x = \frac{1}{c^2}\frac{\partial^2 B_x}{\partial t^2}\,, \quad \nabla^2 B_y = \frac{1}{c^2}\frac{\partial^2 B_y}{\partial t^2}\,, \quad \nabla^2 B_z = \frac{1}{c^2}\frac{\partial^2 B_z}{\partial t^2} \tag{3.69a}$$

In order to understand the significance of (3.68) and (3.69), let's take a look at a one-dimensional wave equation of the form

$$\frac{\partial^2 u}{\partial x^2} = \frac{1}{c^2}\frac{\partial^2 u}{\partial t^2} \tag{3.70}$$

where $u = u(x,t)$. As shown in the Exercises that follow, equation (3.70) has solutions of the form $u(x,t) = F(x-ct)$ and $u(x,t) = G(x+ct)$, where F and G are *arbitrary* functions. If we look at Figure 3.24, we see that the solution $u = F(x-ct)$ represents a *wave* traveling in the positive x-direction, while maintaining its initial profile (or shape) with speed c. Similarly, the solution $u = G(x+ct)$ represents a wave moving in the negative x-direction, with speed c.

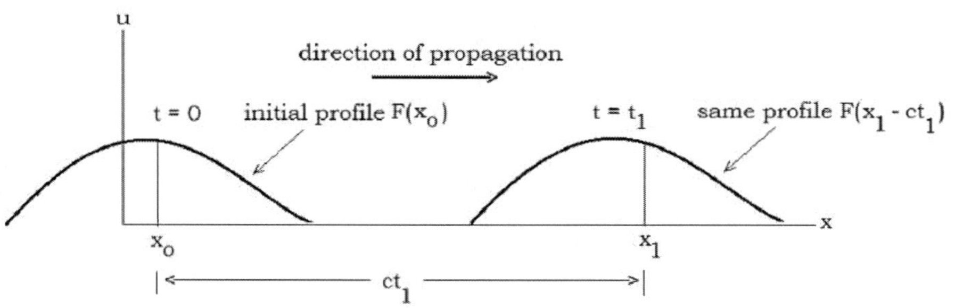

Figure 3.24. A traveling wave moving to the right, with speed c.

Now, we seek solutions of (3.68) and (3.69) form

$$\mathbf{E}(\mathbf{x},t) = \mathbf{E}_0 \exp\left[i(\mathbf{k}\cdot\mathbf{x} - \omega t + \delta)\right] \tag{3.71a}$$

$$\mathbf{B}(\mathbf{x},t) = \mathbf{B}_0 \exp\left[i(\mathbf{k}\cdot\mathbf{x} - \omega t + \delta)\right] \tag{3.71b}$$

where the *amplitudes* $\mathbf{E}_0, \mathbf{B}_0$, and *propagation vector* **k** are constant vectors, the *angular frequency* ω, and the *phase lag* δ are real constants, and **x** is the usual *position vector*, which is given in rectangular coordinates by

$$\mathbf{x} = x\hat{\mathbf{x}} + y\hat{\mathbf{y}} + z\hat{\mathbf{z}}$$

It is understood that the physical fields are the real parts of the complex wave functions in (3.71).

In addition, if we introduce *complex amplitudes*

$$\tilde{\mathbf{E}}_0 = \mathbf{E}_0 \exp(i\delta) = (E_{01}\hat{\mathbf{x}} + E_{02}\hat{\mathbf{y}} + E_{03}\hat{\mathbf{z}})e^{i\delta} \tag{3.72a}$$

$$\tilde{\mathbf{B}}_0 = \mathbf{B}_0 \exp(i\delta) = (B_{01}\hat{\mathbf{x}} + B_{02}\hat{\mathbf{y}} + B_{03}\hat{\mathbf{z}})e^{i\delta} \tag{3.72b}$$

then we can "absorb" the phase lag into the complex amplitudes $\tilde{\mathbf{E}}_0$ and $\tilde{\mathbf{B}}_0$, and equations (3.71) simplify to

$$\mathbf{E}(\mathbf{x},t) = \tilde{\mathbf{E}}_0 \exp\left[i(\mathbf{k}\cdot\mathbf{x}-\omega t)\right] \quad (3.73a)$$
$$\mathbf{B}(\mathbf{x},t) = \tilde{\mathbf{B}}_0 \exp\left[i(\mathbf{k}\cdot\mathbf{x}-\omega t)\right] \quad (3.73b)$$

Let's take a look at the **E**-wave in (3.73a). The real quantity $\mathbf{k}\cdot\mathbf{x}-\omega t$ is called the *phase* of the wave. It is clear that, at any given instant t_0 of time, this quantity will be *constant* along each point of the surface

$$\mathbf{k}\cdot\mathbf{x}-\ \omega t_0 = k_1 x + k_2 y + k_3 z - \omega t_0 = C = \text{constant} \quad (3.74)$$

This surface is called the *wave front*; by definition, it is a surface of *constant* phase. From (3.74), we see that the wave front must be a *plane* with normal **k**. Consequently, the **E**-wave is called a plane wave, and the **B**-wave is a plane wave as well.

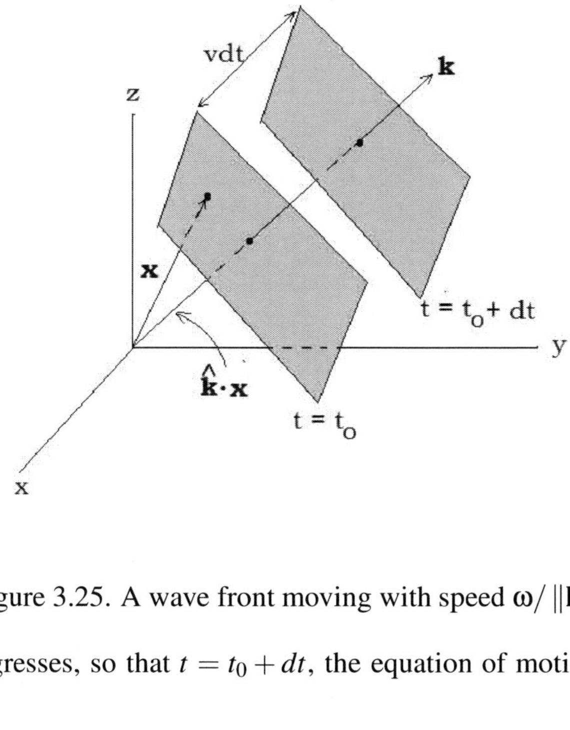

Figure 3.25. A wave front moving with speed $\omega/\left\|\mathbf{k}\right\|$.

As the time progresses, so that $t = t_0 + dt$, the equation of motion of the *same* wave front must be

$$\mathbf{k}\cdot(\mathbf{x}+d\mathbf{x}) - \omega\,(t+dt) = C$$

so, subtracting the last two equations, and simplifying, we see that the propagation *velocity* of the wave must be

$$\mathbf{v} = \frac{d\mathbf{x}}{dt} = \frac{\omega}{\left\|\mathbf{k}\right\|}\,\hat{\mathbf{k}}$$

and its propagation *speed* is $\omega/\left\|\mathbf{k}\right\|$. The movement of the wave front is shown in Figure 3.25. Also, it's easy to show that its spatial period, or *wavelength* λ, and its *frequency* f are respectively given by

$$\lambda = \frac{2\pi}{\left\|\mathbf{k}\right\|} \text{ and } f = \frac{\omega}{2\pi}$$

so that the familiar formula, $v = \lambda f$, still holds true.

Now, if we substitute (3.73a) into (3.68), we find that (3.68) will be satisfied as long as

$$\|\mathbf{k}\|^2 = \frac{\omega^2}{c^2} \tag{3.75}$$

so that the *wave propagation speed* is $c = \omega / \|\mathbf{k}\|$. Substituting (3.73a) into (3.63), and (3.73b) into (3.64), yields

$$\mathbf{k} \cdot \tilde{\mathbf{E}}_0 = 0 \text{ and } \mathbf{k} \cdot \tilde{\mathbf{B}}_0 = 0 \tag{3.76}$$

so that the electric and magnetic fields must oscillate in planes that are perpendicular to the direction of propagation, and we conclude that electromagnetic waves are *transverse waves*.

For the sake of brevity, let $\xi(\mathbf{x},t) = \exp[i(\mathbf{k}\cdot\mathbf{x} - \omega t + \delta)]$, so that

$$\mathbf{E}(\mathbf{x},t) = (E_{01}\hat{\mathbf{x}} + E_{02}\hat{\mathbf{y}} + E_{03}\hat{\mathbf{z}})\xi(\mathbf{x},t) \tag{3.77a}$$

$$\mathbf{B}(\mathbf{x},t) = (B_{01}\hat{\mathbf{x}} + B_{02}\hat{\mathbf{y}} + B_{03}\hat{\mathbf{z}})\xi(\mathbf{x},t) \tag{3.77b}$$

If we substitute equations (3.77) into (3.65), we obtain

$$\begin{aligned} \nabla \times \mathbf{E} &= \begin{vmatrix} \hat{\mathbf{x}} & \hat{\mathbf{y}} & \hat{\mathbf{z}} \\ \frac{\partial}{\partial x} & \frac{\partial}{\partial y} & \frac{\partial}{\partial z} \\ E_{01}\xi & E_{02}\xi & E_{03}\xi \end{vmatrix} \\ &= i\xi\left[(k_2E_{03} - k_3E_{02})\hat{\mathbf{x}} + (k_3E_{01} - k_1E_{03})\hat{\mathbf{y}} + (k_1E_{02} - k_2E_{01})\hat{\mathbf{z}}\right] \end{aligned}$$

and

$$-\frac{1}{c}\frac{\partial \mathbf{B}}{\partial t} = i\xi\frac{\omega}{c}(B_{01}\hat{\mathbf{x}} + B_{02}\hat{\mathbf{y}} + B_{03}\hat{\mathbf{z}})$$

Equating components of the last two expressions, we obtain:

$$k_2E_{03} - k_3E_{02} = \frac{\omega}{c}B_{01}, \quad k_3E_{01} - k_1E_{03} = \frac{\omega}{c}B_{02}, \quad k_1E_{02} - k_2E_{01} = \frac{\omega}{c}B_{03}$$

or, in other words,

$$\mathbf{k} \times \mathbf{E}_0 = \frac{\omega}{c}\mathbf{B}_0 \tag{3.78}$$

so the amplitudes of the **E**-wave and the **B**-wave are related.

Now, (3.78) can be cast into a more illuminating form. We take $\mathbf{E}_0\times$ of both sides of (3.78), to obtain

$$\mathbf{E}_0 \times (\mathbf{k} \times \mathbf{E}_0) = \frac{\omega}{c}\mathbf{E}_0 \times \mathbf{B}_0$$

But,

$$\begin{aligned} \mathbf{E}_0 \times (\mathbf{k} \times \mathbf{E}_0) &= \mathbf{k}(\mathbf{E}_0 \cdot \mathbf{E}_0) - \mathbf{E}_0(\mathbf{E}_0 \cdot \mathbf{k}) \\ &= \|\mathbf{E}_0\|^2 \mathbf{k} \end{aligned}$$

since $\mathbf{E}_0 \cdot \mathbf{k} = 0$ by virtue of (3.76). Combining last two equations, gives

$$\mathbf{E}_0 \times \mathbf{B}_0 = \frac{c}{\omega}\|\mathbf{E}_0\|^2\,\mathbf{k} \tag{3.79}$$

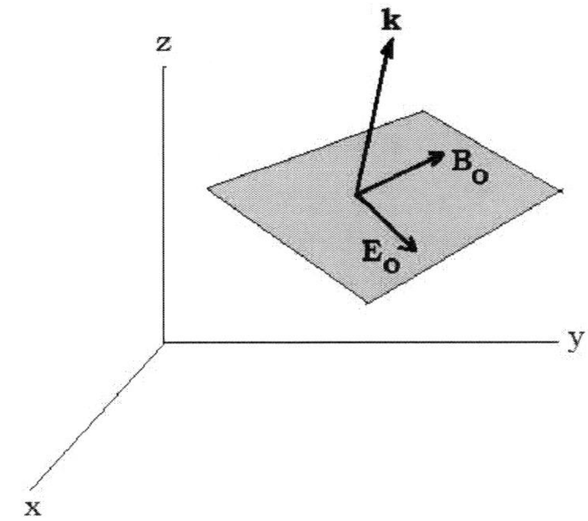

Figure 3.26. The propagation vector **k**, and the vectors $\mathbf{E}_0$ and $\mathbf{B}_0$.

We conclude that $\mathbf{E}_0$, $\mathbf{B}_0$, and **k** form a righthanded triad, so the planes of oscillation of the **E**-wave and the **B**-wave are mutually perpendicular; this state of affairs is shown in Figure 3.26.

Incidentally, if we take the vector norm of both sides of (3.78), and recall (3.75), we obtain

$$||\mathbf{E}_0|| = ||\mathbf{B}_0|| \tag{3.80}$$

so, the amplitudes of the **E**-wave and the **B**-wave are equal as well.

In conclusion, we have demonstrated that electromagnetic waves are transverse waves which propagate through space at the speed of light. This led Maxwell to conclude that light waves are nothing more than electromagnetic waves.

Now, the problem with the Postulate of Galilean Relativity, and hence, Newtonian physics, is that Maxwell's equations are *not covariant* under a Galilean transformation. We will show this indirectly, by showing that the electromagnetic wave equation, which is a direct result of Maxwell's theory, is not covariant under such a transfor- mation.

To see this, let's consider an **E**-wave of the form $\mathbf{E}(x,t) = E(x,t)\,\hat{\mathbf{z}}$

$$E(x,t) = E_0 \exp\left[i(kx - \omega t)\right] \tag{3.81}$$

where E_0 is a real constant. This is a special case of (3.71a), where for convenience, we have set $\delta = 0$. This wave moves in the positive x-direction, and its wave fronts are always parallel to the xz-plane Furthermore, it satisfies the scalar wave equation

$$\frac{\partial^2 E}{\partial x^2} = \frac{1}{c^2}\frac{\partial^2 E}{\partial t^2} \tag{3.82}$$

as long as we take $c = \omega/k$. This equation holds true, relative to an observer situated in an inertial reference frame, say $S(t,x)$.

Now, let's assume that there is another observer, situated in the inertial frame $\bar{S}(\bar{t},\bar{x})$, where the second frame $\bar{S}$ moves in the direction of the positive x$-$axis, at a *constant velocity* **u** relative to the first frame. Under the Galilean transformation, (1.6), we know that

$$\bar{t} = t \quad (3.83a)$$

$$\bar{x} = x - ut \quad (3.83b)$$

Recall that this set of equations represents a *Galilean boost* $\mathbf{u} = u\hat{\mathbf{x}}$ in the positive x-direction of S. We now use equations (3.83) to see what the wave equation looks like in the barred reference frame. Now, in terms of the coordinates in the frame $\bar{S}$, the electric field is given by

$$\bar{E}(\bar{x},\bar{t}) = \bar{E}(x-ut,t) = E(x,t)$$

By the chain rule,

$$\frac{\partial E}{\partial t} = \frac{\partial \bar{E}}{\partial \bar{t}}\frac{\partial \bar{t}}{\partial t} + \frac{\partial \bar{E}}{\partial \bar{x}}\frac{\partial \bar{x}}{\partial t} = \frac{\partial \bar{E}}{\partial \bar{t}} - u\frac{\partial \bar{E}}{\partial \bar{x}} \quad (3.84)$$

Similarly, we find that

$$\frac{\partial^2 E}{\partial t^2} = \frac{\partial^2 \bar{E}}{\partial \bar{t}^2} - 2u\frac{\partial^2 \bar{E}}{\partial \bar{x}\partial \bar{t}} + u^2\frac{\partial^2 \bar{E}}{\partial \bar{x}'^2} \quad (3.85a)$$

$$\frac{\partial^2 E}{\partial x^2} = \frac{\partial^2 \xi}{\partial \bar{x}^2} \quad (3.85b)$$

If we now substitute equations (3.85) into (3.82), we obtain

$$\frac{\partial^2 E}{\partial x^2} - \frac{1}{c^2}\frac{\partial^2 E}{\partial t^2} = \left(\frac{\partial^2 \bar{E}}{\partial \bar{x}^2} - \frac{1}{c^2}\frac{\partial^2 \bar{E}}{\partial \bar{t}^2}\right) + \frac{1}{c^2}\left(2v\frac{\partial^2 \bar{E}}{\partial \bar{x}\partial \bar{t}} - v^2\frac{\partial^2 \bar{E}}{\partial \bar{x}^2}\right) = 0 \quad (3.86)$$

Due to the presence of the second term on the righthand side of (3.86), we see that the wave equation is not *covariant* under a Galilean transformation.

This violates the Postulate of Galilean Covariance, where we assumed that the laws of physics are *covariant* with respect to Galilean trans- formations. This is a serious problem, since when one writes down any fundamental law of physics, all inertial reference frames must be equivalent.

Either the philosophy of Galilean covariance is no longer valid in the electromagnetic realm, or we must make some alteration to the Galilean transformation law that connects two inertial observers. In the next chapter, we'll see how Einstein, remedied this situation.

3.11. Exercises

1. A circular conducting loop of variable radius $r(t)$ sits in the xy-plane with its center at the origin. The radius of the loop at any time t, is given by

$$r(t) = a(1+\alpha t)$$

where a and α are positive constants. If the loop is placed in a constant magnetic field $\mathbf{B} = -B_0\hat{\mathbf{z}}$, then find the emf induced in the loop.

(Solution)

By Faraday's law, the induced emf is given by

$$\begin{aligned}\varepsilon &= -\frac{1}{c}\frac{d}{dt}\left(\int_S (-B_0\hat{\mathbf{z}})\cdot\hat{\mathbf{z}}\,dS\right) = \frac{B_0}{c}\frac{d}{dt}\left(\pi r^2\right) = \frac{aB_0\pi}{c}\frac{d}{dt}\left\{(1+\alpha t)^2\right\} \\ &= \frac{2\pi a\alpha B_0}{c}(1+\alpha t)\end{aligned}$$

2. As shown in Figure–3.27, a vertical segment of wire slides in the positive x-direction with speed v, while maintaining contact with two parallel conductors, sitting in the xy-plane. Assuming the presence of a uniform magnetic field $\mathbf{B} = -B_0\hat{\mathbf{z}}$, use Faraday's law to calculate the induced emf and the induced current. What is the direction of the induced current?

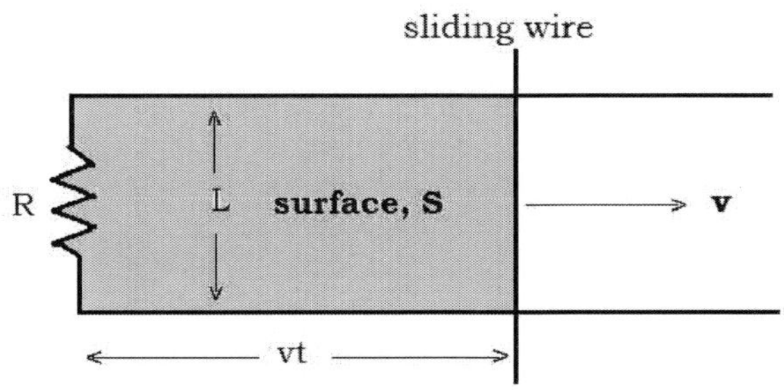

Figure 3.27.

(Solution)

The magnetic flux through S is given by

$$\Psi_B = \int_S \mathbf{B}\cdot d\mathbf{S} = \int_S (-B_0\hat{\mathbf{z}})\cdot(\hat{\mathbf{z}}dS) = -B_0 S = -B_0 Lvt$$

By Faraday's law, the induced emf is

$$\varepsilon = -\frac{1}{c}\frac{d\Psi_B}{dt} = -\frac{1}{c}(-B_0 Lv) = \frac{B_0 Lv}{c}$$

and the induced current is $I = \varepsilon/R = B_0 Lv/Rc$. From Lenz's law, the induced current moves counterclockwise in the circuit.

3. As shown in Figure 3.28, a circular conducting coil of radius a rotates about the x-axis, with a constant angular speed ω. Assuming the presence of a steady magnetic field $\mathbf{B} = B_0\hat{\mathbf{z}}$, find the induced emf in the loop.

(Solution)

From the Figure 3.28, we see that the outward unit normal $\hat{\mathbf{n}}$ to the surface of the loop is

$$\hat{\mathbf{n}} = \cos\left(\omega t + \frac{\pi}{2}\right)\hat{\mathbf{y}} + \sin\left(\omega t + \frac{\pi}{2}\right)\hat{\mathbf{z}} = (-\sin\omega t)\,\hat{\mathbf{y}} + (\cos\omega t)\,\hat{\mathbf{z}}$$

So, the magnetic flux through the surface bounded by the loop is

$$\begin{aligned}\Psi_B &= \int_S \mathbf{B}\cdot d\mathbf{S} = \int_S (B_0\hat{\mathbf{z}})\cdot[(-\sin\omega t)\,\hat{\mathbf{y}} + (\cos\omega t)\,\hat{\mathbf{z}}]\,dS \\ &= B_0\cos\omega t\int_S dS = B_0\pi a^2\cos\omega t\end{aligned}$$

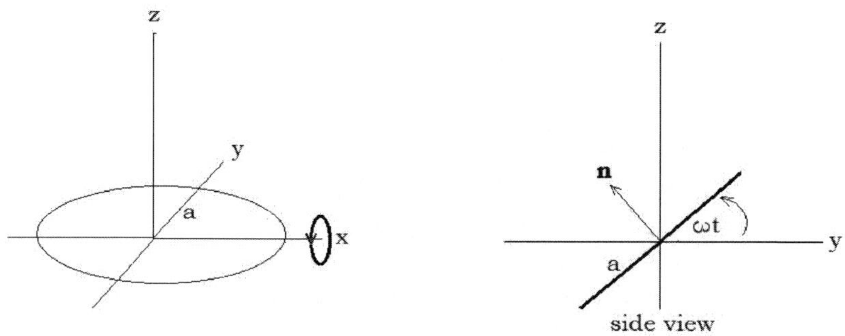

Figure 3.28. A conducting loop rotating around the x-axis.

According to Faraday's law, the induced emf is

$$\varepsilon = -\frac{1}{c}\frac{d\Psi_B}{dt} = -\frac{B_0\pi a^2}{c}\frac{d}{dt}(\cos\omega t) = \frac{B_0\pi a^2}{c}\sin\omega t$$

This gives rises to an alternating current in the loop, which changes direction $\omega/2\pi$ cycles per second.

4. *(Faraday's Homopolar Generator)* Assume that we are given a *closed* system, which contains moveable charges, such as a conductor. Assume further, that $\mathbf{F}$ is *any* force which is capable of moving such charges. We can generalize our previous definition of emf to read

$$\varepsilon = \int_{\mathbf{x}_0}^{\mathbf{x}} \frac{\mathbf{F}}{q}\cdot d\mathbf{l} \qquad (3.87)$$

Thus, the emf is the work done, per unit charge, in moving a charge from the initial position $\mathbf{x}_0$ to the final position $\mathbf{x}$. As a result, a potential difference (voltage) is created between the points $\mathbf{x}$ and $\mathbf{x}_0$. Use this general definition of emf to find the voltage reading on the meter of *Faraday's homopolar generator* in Figure 3.29.

(Solution)

We assume that the magnetic field is $\mathbf{B} = B_0\hat{\mathbf{z}}$, and that the disk rotates in the counter-clockwise direction, with angular speed ω.

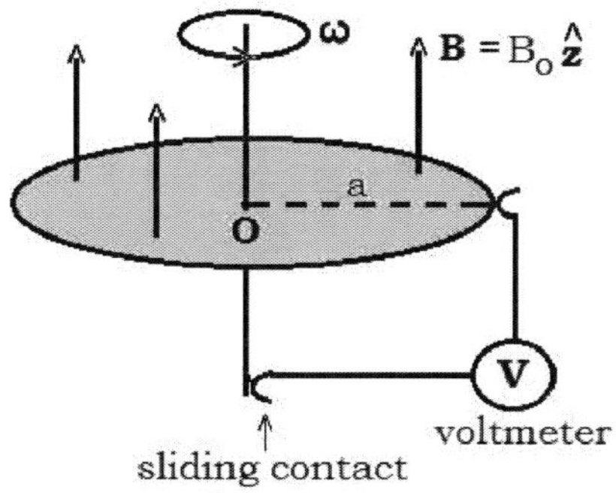

Figure 3.29. Faraday's homopolar generator.

According to the Lorentz force law, the magnetic field $\mathbf{B}$ must exert a force per unit charge of

$$\mathbf{F}_m = \frac{1}{c}(\mathbf{v}\times\mathbf{B})$$

This force is exerted on free charges in the conducting disk. Now, a point on the disk which is a distance r from the center must have a velocity $\mathbf{v} = (r\omega)\,\hat{\phi}$, so that

$$\mathbf{F}_m = \frac{1}{c}\,(r\omega)\,\hat{\phi}\times B_0\hat{\mathbf{z}} = \left(\frac{r\omega B_0}{c}\right)\hat{\mathbf{r}}$$

Let C be any radial path extending from the center O to the rim of the disk. The emf must be

$$\varepsilon = \int_C \mathbf{F}_m\cdot d\mathbf{l} = \int_0^a \left(\frac{r\omega B_0}{c}\right)\hat{\mathbf{r}}\cdot(\hat{\mathbf{r}}dr) = \frac{\omega B_0}{c}\int_0^a r\,dr = \frac{\omega B_0 a^2}{2c}$$

and this is the reading on the voltmeter.

5. *(General Solution of the Wave Equation)* Show that the general solution of the scalar wave equation

$$\frac{\partial^2 u}{\partial x^2} = \frac{1}{c^2}\frac{\partial^2 u}{\partial t^2}$$

is given by

$$u(x,t) = F(x-ct) + G(x+ct) \tag{3.88}$$

where F and G are *arbitrary* functions.

(Solution)

We introduce a change of a variables

$$\xi = x+ct \quad \text{and} \quad \chi = x-ct$$

and express the original wave equation in terms of the new variables. By the chain rule,

$$\begin{aligned}
\frac{\partial u}{\partial t} &= \frac{\partial u}{\partial \xi}\frac{\partial \xi}{\partial t} + \frac{\partial u}{\partial \chi}\frac{\partial \chi}{\partial t} = c\frac{\partial u}{\partial \xi} - c\frac{\partial u}{\partial \chi} \Longrightarrow \frac{\partial}{\partial t} = c\frac{\partial}{\partial \xi} - c\frac{\partial}{\partial \chi} \\
\frac{\partial u}{\partial x} &= \frac{\partial u}{\partial \xi}\frac{\partial \xi}{\partial x} + \frac{\partial u}{\partial \chi}\frac{\partial \chi}{\partial x} = \frac{\partial u}{\partial \xi} + \frac{\partial u}{\partial \chi} \Longrightarrow \frac{\partial}{\partial x} = \frac{\partial}{\partial \xi} + \frac{\partial}{\partial \chi}
\end{aligned}$$

After some operator algebra, we obtain

$$\begin{aligned}
\frac{\partial^2}{\partial t^2} &= c^2\left(\frac{\partial^2}{\partial \xi^2} - 2\frac{\partial}{\partial \xi \partial \chi} + \frac{\partial^2}{\partial \chi^2}\right) \\
\frac{\partial^2}{\partial x^2} &= \frac{\partial^2}{\partial \xi^2} + 2\frac{\partial^2}{\partial \xi \partial \chi} + \frac{\partial^2}{\partial \chi^2}
\end{aligned}$$

So, in terms of the new variables, the wave equation becomes

$$\frac{\partial^2 u}{\partial \xi^2} + 2\frac{\partial^2 u}{\partial \xi \partial \chi} + \frac{\partial^2 u}{\partial \chi^2} = \frac{\partial^2 u}{\partial \xi^2} - 2\frac{\partial u}{\partial \xi \partial \chi} + \frac{\partial^2 u}{\partial \chi^2}$$

This equation, in turn, simplifies to

$$\frac{\partial^2 u}{\partial \xi \partial \chi} = \frac{\partial}{\partial \xi}\left(\frac{\partial u}{\partial \chi}\right) = 0$$

Integration with respect to ξ gives

$$\frac{\partial u}{\partial \chi} = g(\chi)$$

where $g(\chi)$ is an arbitrary function of χ. Integration with respect to χ now gives

$$u = \int g(\chi)\partial \chi = F(\xi) + G(\chi)$$

where $G(\chi)$ is the antiderivative of $g(\chi)$, and $F(\xi)$ is an arbitrary function of ξ. In other words, returning to the original variables, we have shown that

$$u(x,t) = F(x - ct) + G(x + ct)$$

6. *(D'Alembert's Solution)* Show that the solution of the *initial value problem*

$$\begin{aligned}
\frac{\partial^2 u}{\partial x^2} &= \frac{1}{c^2}\frac{\partial^2 u}{\partial t^2} && \text{(3.89a)} \\
u(x,0) &= f(x) \quad \text{(initial profile)} && \text{(3.89b)} \\
u_t(x,0) &= g(x) \quad \text{(initial speed)} && \text{(3.89c)}
\end{aligned}$$

is given by

$$u(x,t) = \frac{f(x-ct) + f(x+ct)}{2} + \frac{1}{2c}\int_{x-ct}^{x+ct} g(\xi)\,d\xi \tag{3.90}$$

(Solution)

We know that the general solution of (3.89a) may be written

$$u(x,t) = F(x-ct) + G(x+ct)$$

Substituting this solution into the initial conditions gives

$$F(x) + G(x) = f(x) \text{ and } -cF'(x) + cG'(x) = g(x)$$

Integrating the second equation yields

$$-cF(x) + cG(x) = \int_{x_0}^{x} g(\xi)d\xi + C_0$$

where C_0 is a constant. We now can solve the following system of equations for $F(x)$ and $G(x)$

$$\begin{aligned} F(x) + G(x) &= f(x) \\ -cF(x) + cG(x) &= \int_{x_0}^{x} g(\xi)d\xi + C_0 \end{aligned}$$

to obtain:

$$\begin{aligned} F(x) &= \frac{1}{2}f(x) - \frac{1}{2c}\int_{x_0}^{x} g(\xi)d\xi - \frac{C_0}{2c} \\ G(x) &= \frac{1}{2}f(x) + \frac{1}{2c}\int_{x_0}^{x} g(\xi)d\xi + \frac{C_0}{2c} \end{aligned}$$

so that the solution of the original problem is

$$\begin{aligned} u(x,t) &= F(x-ct) + G(x+ct) \\ &= \left(\frac{1}{2}f(x-ct) - \frac{1}{2}\int_{x_0}^{x-ct} g(\xi)d\xi\right) + \left(\frac{1}{2}f(x+ct) + \frac{1}{2c}\int_{x_0}^{x+ct} g(\xi)d\xi\right) \\ &= \frac{f(x-ct) + f(x+ct)}{2} + \frac{1}{2c}\int_{x-ct}^{x+ct} g(\xi)\,d\xi \end{aligned}$$

Chapter 4

Introduction to Special Relativity

I sometimes ask myself how it came about that I was the one to develop the theory of relativity. The reason, I think, is that a normal adult never stops to think about problems of space and time. These are things which he has thought about as a child.

Albert Einstein

4.1. The Michelson-Morley Experiment

We have seen that Maxwell's theory provided strong evidence that light was electromagnetic in nature, and that light waves are *transverse waves* where the component electric and magnetic fields which comprise such waves, oscillate in planes that are perpendicular to the direction of propagation.

At the same time, we have shown that the wave equation, and hence, indirectly, the equations of Maxwell's theory, are not covariant under a Galilean transformation. This violates the fundamental philosophy that all inertial frames of reference should be equivalent when one formulates the laws of nature.

Such was the state of affairs in physics near the close of the nineteenth century.[1] The intellectual debate resulting from this situation was to be radically increased with the negative result of the Michelson-Morley experiment, which we will examine.

As inheritors of Newton's mechanistic legacy, physicists at that time were very familiar with *mechanical waves* of all sorts. It was well known that a typical mechanical wave required a medium for its propagation, and that the properties of the medium had a direct effect on the speed of any such wave.

For example, the speed of sound waves in air is dependent upon the density of the air, and its temperature. The speed of sound is reckoned with respect to some inertial reference frame, which is usually taken to be a frame which is at rest relative to the un- disturbed air.

Thus, many physicists assumed that electromagnetic waves required a medium for their propagation, called the *ether*, and that the speed of light waves - which appears so prominently in the wave equation - must be reckoned relative to the ether.

Some scientists believed that the ether must be at rest relative to the system of fixed stars, and consequently, any laboratory frame which is at rest in the ether should serve as an ideal platform from which to measure the *absolute motion* of the Earth through space.

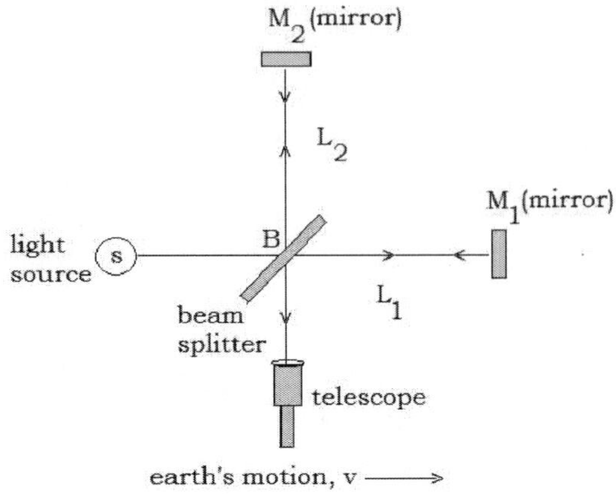

Figure 4.1. Michelson-Morley experiment.

In 1881, Albert Michelson devised a clever experiment which was designed to measure the absolute motion of the earth through the ether. In 1887, Michelson then teamed up with E.A. Morley to repeat essentially the same experiment, but with a more sensitive experimental setup.

The basic idea behind the Michelson interferometer is shown in Figure 1. Light emerges from a light source S and strikes a beam splitter B which separates the light into two perpendicular beams. The beam traveling along arm L_1 moves parallel to the direction of the earth's motion through the ether, while the other beam moves along the arm L_2 which is perpendicular to this motion. The beams are then reflected from two mirrors, M_1 and M_2, and eventually arrive within the field of view of an optical telescope.

Now, owing to the wave nature of light, the two beams which emerge from the beam splitter are initially in phase. So, if the beams are out of phase when they arrive at the telescope, then an interference pattern will be observed. As we will see, the existence of interference fringes would serve as evidence of the earth's motion through the ether.

Since the arm L_1 is moving parallel to the earth's motion, the total roundtrip time for this beam to arrive at the telescope is simply

$$T_1 = \frac{L_1}{c-v} + \frac{L_1}{c+v} = \frac{2L_1}{c}\frac{1}{(1-v^2/c^2)} \qquad (4.1a)$$

where we have assumed that the speed of light relative to the ether is c, and the Galilean law of velocity addition is valid. Similarly, using elementary vector addition, the total roundtrip time for the second beam is given by

$$T_2 = 2\cdot\frac{L_2}{\sqrt{c^2-v^2}} = \frac{2L_2}{c}\frac{1}{\sqrt{1-v^2/c^2}} \qquad (4.1b)$$

Any interference pattern will depend upon the difference in total travel times, $T_1 - T_2$. Michelson and Morley then rotated the arms of the apparatus by 90^0. In this manner, the roles of each beam were interchanged. After such a rotation, the travel times were

$$T_1' = \frac{2L_1}{c}\frac{1}{\sqrt{1-v^2/c^2}} \tag{4.2a}$$

$$T_2' = \frac{2L_2}{c}\frac{1}{(1-v^2/c^2)} \tag{4.2b}$$

and consequently, the total time difference ΔT is given by

$$\Delta T = (T_1' - T_2') - (T_1 - T_2) = \frac{2(L_1+L_2)}{c}\frac{1}{\sqrt{1-v^2/c^2}}\left[1 - \frac{1}{(1-v^2/c^2)}\right] \tag{4.3}$$

Furthermore, if $v^2/c^2 \ll 1$, then we can expand the righthand side of (4.3) to obtain the approximation

$$|\Delta T| = \frac{(L_1+L_2)}{c}\cdot\frac{v^2}{c^2} \qquad (\text{for } v^2/c^2 \ll 1) \tag{4.4}$$

Although at the outset of the experiment, an effort was made to make the arms L_1 and L_2 as equal in length as possible, but we can see from (4.3) and (4.4) that the total time difference is dependent upon the sum of L_1 and L_2, rather then their difference.

From (4.4), we see that the phase difference between the two beams is given by

$$\Delta\phi = 2\pi\cdot\frac{c\,|\Delta T|}{\lambda} = \frac{2\pi(L_1+L_2)}{\lambda}\left(\frac{v^2}{c^2}\right) \qquad (\text{for } v^2/c^2 \ll 1) \tag{4.5}$$

and the total number of fringes shifted should be

$$\Delta N = \frac{\Delta\phi}{2\pi} \approx \frac{(L_1+L_2)}{\lambda}\left(\frac{v^2}{c^2}\right) \qquad (\text{for } v^2/c^2 \ll 1) \tag{4.6}$$

Now, the fringe shift expressed in (4.6) should have been detectable.

In order to see this, we recall that the earth's orbital speed is about $v \approx 3 \times 10^4$ m/sec. In addition, in 1887, the experimenters employed an effective path length of $L_1 \approx L_2 \approx 11$ meters which was obtained using a system of multiple reflections, and they used monochromatic yellow light with a wavelength $\lambda \approx 5.8 \times 10^{-7}$m. From (4.6), the fringe shift should have been

$$\Delta N \approx \frac{(22\text{ m})}{(5.79\times 10^{-7}\text{ m})}\left(\frac{3\times 10^4\text{ m/sec}}{3\times 10^8\text{ m/sec}}\right)^2 \approx 0.38$$

or about, 0.4 fringes. But to the contrary, no fringe shift was ever observed. Although virtually the same experiment (in various guises) was repeated many times in the future, the same null result was always obtained! This created even more controversy in the world of physics.

4.2. The Advent of Special Relativity

The Theory of Special Relativity officially sprang to life in 1905, when Albert Einstein, published his paper entitled *On the Electrodynamics of Moving Bodies*.[2] Needless to say, it is difficult to reconstruct the actual chain of thoughts which led Einstein to his theory; however, Einstein gave us two hints of his thought process at the very outset of his paper.

First, Einstein was struck with some of the "asymmetries" he found in Maxwell's theory. In order to understand his thought process, let us consider the interaction between a square conducting loop of side b and a permanent magnet with square poles of side a, where $a >> b$, as shown in Figure 4.2a. Here, we assume that the poles of the two magnets are close enough in order to prevent "fringing" and the magnetic field is given by $\mathbf{B} = -B_0\hat{\mathbf{z}}$, where B_0 is a positive constant.

If, as shown in Figure 4.2b, the conducting loop moves to the right with a constant velocity $\mathbf{v} = v\hat{\mathbf{x}}$, while the magnets are stationary, then a *motional emf* is induced in the loop by the action of the Lorentz force on free charges in the loop. Here, the induced emf is given by

$$\text{emf} = \frac{1}{c}\int_Q^P (\mathbf{v}\times\mathbf{B})\cdot d\mathbf{l} = \frac{1}{c}\int_{-b/2}^{b/2}(vB_0\hat{\mathbf{y}})\cdot(\hat{\mathbf{y}}dy) = \frac{bvB_0}{c} \tag{4.7}$$

Since the potential at the point Q is greater than that at the point P, a counterclockwise current is set up in the loop. This may also be seen by Lenz's law, since the direction of the induced magnetic field must oppose the direction of the original magnetic field.

Let us now suppose that conducting loop is held stationary, while the set of permanent magnets approach the square loop, with a velocity of $\mathbf{v} = -v\hat{\mathbf{x}}$. In this case, Faraday's law tells us that the resulting change in magnetic flux will give rise to an electromotive force in the loop, thus, producing an induced current. Here, we have

$$\text{emf} = -\frac{1}{c}\frac{d}{dt}\int_S \mathbf{B}\cdot d\mathbf{S} = -\frac{1}{c}\frac{d}{dt}(-B_0avt) = \frac{bvB_0}{c} \tag{4.8}$$

a result which agrees with (4.7).

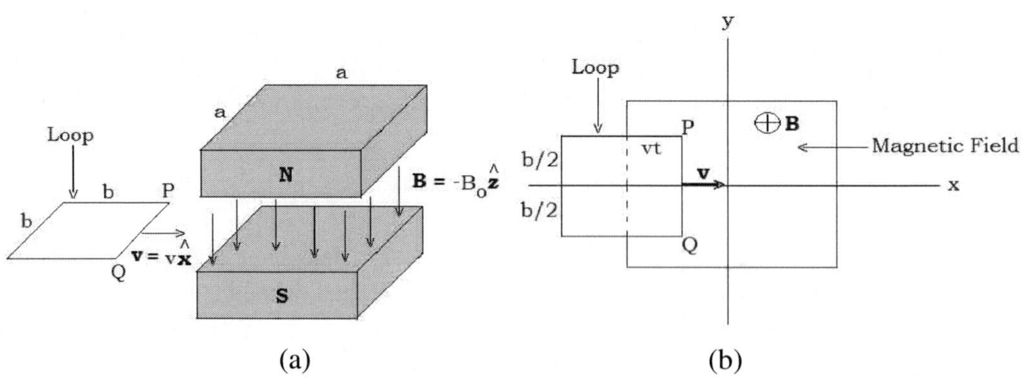

Figure 4.2. (a) Interaction of a square loop with a magnetic field; (b) A birds eye view of the interaction.

Thus, in both cases, the result is the same, even though *different physical principles* were used to obtain them. Einstein saw this as more than a mere coincidence; he observed that it is the *relative motion* of the set of permanent magnets and the loop that really matters. In other words, *the principle of relativity must apply to electromagnetic phenomena as well.*

Secondly, the null result of the Michelson-Morley experiment led him to believe that the *absolute motion* of any body cannot be detected by any means, and light must move past all inertial observers at the same speed, regardless of the state of their relative motion.[3]

Einstein thus formulated two basic postulates at the beginning of his paper. We have taken the liberty of restating them as follows:

Postulate 1 *(Special Principle of Relativity)*

The laws of physics (including the laws of mechanics and electromagnetism) must assume the same mathematical form for all *inertial* observers.

Postulate 2 *(Constancy of Light Velocity)*

The speed of light in free space, is the same in all *inertial* reference frames, and hence, it is a universal constant.

The first postulate emphasizes that not only the laws of mechanics, but all the laws of physics must be assume the same mathematical form for all inertial observers. That is, the laws of physics must be *covariant* with respect to the set of all inertial observers.

Note that the first postulate does not restrict us to a particular type of coordinate transformation, but hints that *the correct coordinate transformation between any two inertial frames must satisfy this principle.* In this regard, the first postulate is much more general than the Principle of Galilean Covariance, but it is, admittedly, in the same spirit.

The second postulate is merely a statement of experimental fact. Instead of trying to "explain away" the negative result of the Michelson-Morley experiment, Einstein simply took the constancy of the speed of light as a "given" in his theory. The ether concept was superfluous to Einstein, since the ether couldn't be detected in the first place.

In the next section, we will see how these two postulates can be employed to obtain the correct set of transformation equations that relate the space and time measurements of two inertial observers.

4.3. The Lorentz Transformation

Einstein's second postulate clearly implies that the Galilean trans- formation, previously used to relate the space and time coordinates of two inertial reference frames, is not correct. There must exist some other set of transformation equations which preserves the covariance of physical laws, and at the same time, the constancy of the velocity of light. We now proceed to find this fundamental set of transformation equations.

We first assume that we have two inertial frames, $K(x,y,z,t)$ and $\bar{K}(\bar{x},\bar{y},\bar{z},\bar{t})$, whose origins coincide at the time $t = \bar{t} = 0$, such that the second frame moves in the positive x-direction at a constant speed v relative to the first frame. as shown in Figure 4.3.

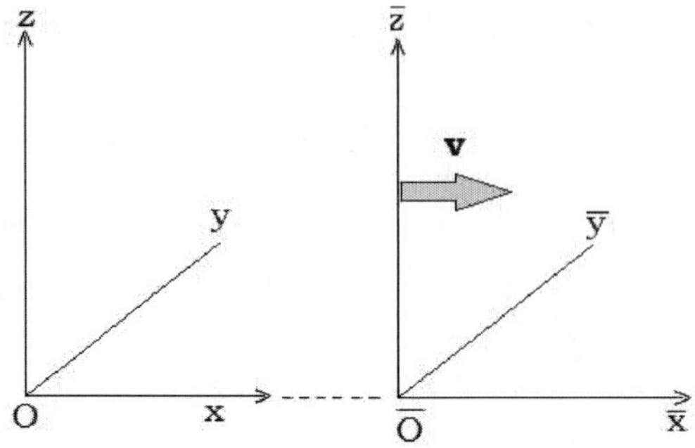

Figure 4.3. The inertial frame $\bar{K}$ moves at a constant speed v (in the positive x-direction) relative to the inertial frame K.

At the time $t = \bar{t} = 0$, suppose that a light pulse is emitted from the origins of both inertial frames. When viewed from the first frame $K(x, y, z, t)$, the resulting wave front will lie on the surface of the sphere:

$$x^2 + y^2 + z^2 - c^2t^2 = 0 \tag{4.9a}$$

Similarly, owing to the *invariance of the velocity of light*, an observer in the frame $\bar{K}(\bar{x}, \bar{y}, \bar{z}, \bar{t})$, will describe the *same* wavefront by the equation:

$$\bar{x}^2 + \bar{y}^2 + \bar{z}^2 - c^2\bar{t}^2 = 0 \tag{4.9b}$$

We conclude that the principle of the invariance of the velocity of light requires that:

$$x^2 + y^2 + z^2 - c^2t^2 = 0 \Longleftrightarrow \bar{x}^2 + \bar{y}^2 + \bar{z}^2 - c^2\bar{t}^2 = 0 \tag{4.10}$$

We now inquire as to what type of transformation will relate the space and time coordinates of both inertial frames, while enforcing (4.10). We first define the column vectors

$$\mathbf{x} = \begin{bmatrix} x^0 \\ x^1 \\ x^2 \\ x^3 \end{bmatrix} = \begin{bmatrix} ct \\ x \\ y \\ z \end{bmatrix} \text{ and } \bar{\mathbf{x}} = \begin{bmatrix} \bar{x}^0 \\ \bar{x}^1 \\ \bar{x}^2 \\ \bar{x}^3 \end{bmatrix} = \begin{bmatrix} c\bar{t} \\ \bar{x} \\ \bar{y} \\ \bar{z} \end{bmatrix} \tag{4.11}$$

where we number the components of each vector from 0 to 3, in keeping with the modern convention, and in addition, we have set $x^0 = ct$, and $\bar{x}^0 = c\bar{t}$.

We recall that, since the *law of inertia* must hold in both inertial frames, any object which travels in a straight line relative to the frame $K(x, y, z, t)$, must also be viewed as moving along a straight line by an observer in the frame $\bar{K}(\bar{x}, \bar{y}, \bar{z}, \bar{t})$.

Since a *linear* transformation has the property of mapping straight lines into straight lines, we begin by assuming that there exists an invertible linear transformation of the form:

$$\bar{\mathbf{x}} = \Lambda \mathbf{x} \tag{4.12}$$

which relates the space and time coordinates of both frames, where Λ is a 4×4 matrix whose elements we have yet to determine. When written out in explicit form, (4.12) becomes:

$$\begin{aligned}\bar{x}^0 &= \Lambda^0_0 x^0 + \Lambda^0_1 x^1 + \Lambda^0_2 x^2 + \Lambda^0_3 x^3 \\ \bar{x}^1 &= \Lambda^1_0 x^0 + \Lambda^1_1 x^1 + \Lambda^1_2 x^2 + \Lambda^1_3 x^3 \\ \bar{x}^2 &= \Lambda^2_0 x^0 + \Lambda^2_1 x^1 + \Lambda^2_2 x^2 + \Lambda^2_3 x^3 \\ \bar{x}^3 &= \Lambda^3_0 x^0 + \Lambda^3_1 x^1 + \Lambda^3_2 x^2 + \Lambda^3_3 x^3\end{aligned} \tag{4.12a}$$

Since the $\bar{x}^2$ and $\bar{x}^3$ axes of the frame $\bar{K}$ are perpendicular to its direction of motion, then by symmetry, we must have

$$\bar{x}^2 = x^2 \text{ and } \bar{x}^3 = x^3 \tag{4.13}$$

and consequently,

$$\begin{aligned}\Lambda^2_0 &= 0,\ \Lambda^2_1 = 0,\ \Lambda^2_2 = 1,\ \Lambda^2_3 = 0 \\ \Lambda^3_0 &= 0,\ \Lambda^3_1 = 0,\ \Lambda^3_2 = 0,\ \Lambda^3_3 = 1\end{aligned} \tag{4.13a}$$

When written in matrix form, (4.12a) now becomes

$$\begin{bmatrix}\bar{x}^0 \\ \bar{x}^1 \\ \bar{x}^2 \\ \bar{x}^3\end{bmatrix} = \begin{bmatrix}\Lambda^0_0 & \Lambda^0_1 & \Lambda^0_2 & \Lambda^0_3 \\ \Lambda^1_0 & \Lambda^1_1 & \Lambda^1_2 & \Lambda^1_3 \\ 0 & 0 & 1 & 0 \\ 0 & 0 & 0 & 1\end{bmatrix}\begin{bmatrix}x^0 \\ x^1 \\ x^2 \\ x^3\end{bmatrix} \tag{4.14}$$

In order to enforce condition (4.10), we shall *require* that:

$$\left(x^1\right)^2 + \left(x^2\right)^2 + \left(x^3\right)^2 - (x^0)^2 = \left(\bar{x}^1\right)^2 + \left(\bar{x}^2\right)^2 + \left(\bar{x}^3\right)^2 - (\bar{x}^0)^2 \tag{4.14a}$$

If, for the time being, we introduce the imaginary time coordinates

$$T = ix^0 = ict \quad \text{and } \bar{T} = i\bar{x}^0 = ic\bar{t}$$

and use the fact that $\bar{x}^2 = x^2$ and $\bar{x}^3 = x^3$, then the last equation simplifies to:

$$(x^1)^2 + T^2 = (\bar{x}^1)^2 + \bar{T}^2$$

which we see is just an *orthogonal transformation*, or rotation by a fixed but imaginary angle θ in the x^0x^1-plane. We can exploit this idea to rewrite the transformation (4.14) as

$$\begin{bmatrix}ic\bar{t} \\ \bar{x}^1 \\ \bar{x}^2 \\ \bar{x}^3\end{bmatrix} = \begin{bmatrix}\cos\theta & \sin\theta & 0 & 0 \\ -\sin\theta & \cos\theta & 0 & 0 \\ 0 & 0 & 1 & 0 \\ 0 & 0 & 0 & 1\end{bmatrix}\begin{bmatrix}ict \\ x^1 \\ x^2 \\ x^3\end{bmatrix} \tag{4.15}$$

Taking differentials of both sides of the last equation, and recalling that θ is constant, we obtain:

$$\begin{aligned} icd\bar{t} &= \cos\theta\,(icdt) + \sin\theta\, dx^1 \\ d\bar{x}^1 &= -\sin\theta\,(icdt) + \cos\theta\, dx^1 \\ d\bar{x}^2 &= dx^2 \\ d\bar{x}^3 &= dx^3 \end{aligned}$$

Now, a point which is *fixed* relative to the $\bar{K}$ frame must satisfy the condition

$$d\bar{x}^1 = d\bar{x}^2 = d\bar{x}^3 = 0$$

while the *same* point must have the velocity v relative to the K frame. So we obtain, in particular,

$$0 = -\sin\theta\ (icdt) + \cos\theta\ dx^1$$

or in other words,

$$v = \frac{dx^1}{dt} = ic\tan\theta \Longrightarrow \tan\theta = -i\frac{v}{c} \tag{4.16}$$

Consequently, we have

$$\cos\theta = \frac{1}{\sec\theta} = \frac{1}{\sqrt{1+\tan^2\theta}} = \frac{1}{\sqrt{1-v^2/c^2}} \tag{4.17a}$$

$$\sin\theta = \tan\theta\cos\theta = \frac{-iv/c}{\sqrt{1-v^2/c^2}} \tag{4.17b}$$

If, following the standard convention, we set

$$\gamma = \frac{1}{\sqrt{1-v^2/c^2}} \quad \text{and} \quad \beta = \frac{v}{c} \tag{4.18}$$

and then substitute (4.17a) and (4.17b) into (4.15), we get

$$\begin{bmatrix} ic\bar{t} \\ \bar{x}^1 \\ \bar{x}^2 \\ \bar{x}^3 \end{bmatrix} = \begin{bmatrix} \gamma & -i\beta\gamma & 0 & 0 \\ i\beta\gamma & \gamma & 0 & 0 \\ 0 & 0 & 1 & 0 \\ 0 & 0 & 0 & 1 \end{bmatrix} \begin{bmatrix} ict \\ x^1 \\ x^2 \\ x^3 \end{bmatrix}$$

and, after removing the appearance of the imaginary unit, this simplifies to

$$\begin{aligned} \bar{x}^0 &= \gamma x^0 - \beta\gamma x^1 \\ \bar{x}^1 &= -\beta\gamma\, x^0 + \gamma x^1 \\ \bar{x}^2 &= x^2 \\ \bar{x}^3 &= x^3 \end{aligned}$$

Thus, we obtain the **Lorentz transformation** in matrix form as

$$\begin{bmatrix} \bar{x}^0 \\ \bar{x}^1 \\ \bar{x}^2 \\ \bar{x}^3 \end{bmatrix} = \begin{bmatrix} \gamma & -\beta\gamma & 0 & 0 \\ -\beta\gamma & \gamma & 0 & 0 \\ 0 & 0 & 1 & 0 \\ 0 & 0 & 0 & 1 \end{bmatrix} \begin{bmatrix} x^0 \\ x^1 \\ x^2 \\ x^3 \end{bmatrix} \tag{4.19}$$

We shall call the 4×4 transformation matrix in (4.19), the *Lorentz matrix.* When written out fully, the Lorentz transformation is

Lorentz Transformation

$$\begin{aligned} \bar{t} &= \gamma\left(t - vx/c^2\right) \\ \bar{x} &= \gamma(x - vt) \\ \bar{y} &= y \\ \bar{z} &= z \end{aligned} \tag{4.20a}$$

Several remarks are in order. First, in order to obtain the inverse transformation, we observe that since a fixed point in the inertial frame K must be seen as moving with a speed $-v$ along the $\bar{x}$-axis by an observer in the $\bar{K}$ frame, we can invert the transformation (4.20a) by simply replacing v with $-v$,and interchanging the barred and unbarred coordinates. In doing so, we obtain the *inverse transformation*

$$\begin{aligned} t &= \gamma\left(\bar{t} + v\bar{x}/c^2\right) \\ x &= \gamma(\bar{x} + v\bar{t}) \\ y &= \bar{y} \\ z &= \bar{z} \end{aligned} \tag{4.20b}$$

Secondly, if we take the differential of both sides of equation (4.20a), then square and rearrange the terms, we find that

$$c^2dt^2 - dx^2 - dy^2 - dz^2 = c^2d\bar{t}^2 - d\bar{x}^2 - d\bar{y}^2 - d\bar{z}^2 \ ,$$

and consequently, the quantity

Spacetime Interval

$$ds^2 = c^2dt^2 - dx^2 - dy^2 - dz^2 \tag{4.21}$$

must be *invariant* under a Lorentz transformation. This important quantity is called the *spacetime interval*, and as we shall see later on, it forms the foundation of spacetime geometry associated with the Theory of Special Relativity.

Finally, for velocities small compared to the velocity of light, or for $(v/c) \ll 1$, we should expect the Lorentz transformation to reduce to the Galilean transformation. Now, it may be seen from (4.18) that

$$\gamma \approx 1 \ \text{ for } (v/c) \ll 1$$

so (4.20a) becomes

$$\begin{aligned}\bar{t} &= t \\ \bar{x} &= x - vt \\ \bar{y} &= y \\ \bar{z} &= z\end{aligned}$$

This is just a special case of the *Galilean transformation*, as we predicted.

4.4. The Lorentz Contraction

We now begin to investigate some of consequences of the Lorentz transformation. Assume that the inertial frames $K(x,y,z,t)$ and $\bar{K}(\bar{x},\bar{y},\bar{z},\bar{t})$ are in the standard configuration shown in Figure 4.4, and that a rigid rod lies at rest along the $\bar{x}$-axis of the inertial frame $\bar{K}$, with its left endpoint at the position $\bar{x} = \bar{x}_1$, and its right endpoint at $\bar{x} = \bar{x}_2$.

Since the rod is at rest in this frame, the length L_0 of the rod is simply

$$L_0 = \bar{x}_2 - \bar{x}_1 \tag{4.22}$$

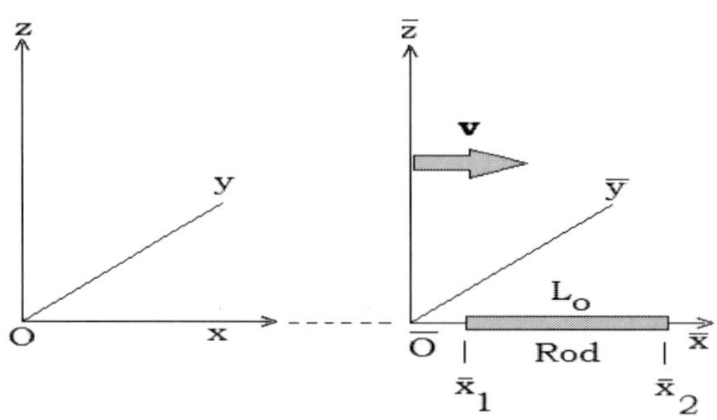

Figure 4.4. A rod of length L_0, at rest in the frame $\bar{K}$.

The length L_0 is called the *proper length* of the rod, as it is the length of the rod as determined by an observer who is *at rest* relative to the rod.

Let us now determine the length L of the rod as determined by an observer in the inertial frame K. From the Lorentz transformation, we have

$$\bar{x}_2 = \gamma(x_2 - vt_2) \quad \text{and} \quad \bar{x}_1 = \gamma(x_1 - vt_1)$$

Subtracting these two equations, we obtain:

$$\bar{x}_2 - \bar{x}_1 = \gamma[(x_2 - x_1) - v(t_2 - t_1)] \tag{4.23}$$

Now, in order to make an accurate length measurement, an observer in the frame K must record the positions of both ends of the rod at the *same time*, so $t_2 = t_1$; equation (4.21) then simplifies to

$$\bar{x}_2 - \bar{x}_1 = \gamma(x_2 - x_1) \Longrightarrow L_0 = \gamma L$$

and, the length of the rod, as measured in the frame K, is simply

Length Contraction Formula

$$L = \gamma^{-1} L_0 = \sqrt{1 - v^2/c^2}\, L_0 \tag{4.24}$$

Since the coefficient $\gamma^{-1} \leq 1$, we see that relative to the frame K, in which the rod is moving at a constant speed $v > 0$, the length of the rod will be reduced or *contracted* along its direction of motion by a factor of γ^{-1}. We call this quantity the *Lorentz contraction factor*. Thus, we see that *the length of a rigid rod is always the longest when measured relative to its proper or rest frame*.

4.5. Time Dilation and Proper Time

Assume that a particle moves in the positive x-direction, at a *constant* speed v, relative to the inertial frame $K(x,t)$. Let $\bar{K}(\bar{x},\bar{t})$ be the rest frame of the particle, so that the two frames are in the standard configuration. Furthermore, let us imagine that a clock attached to this particle records successive values of the time $\bar{t}$, while our laboratory clock, which is identical in nature with its counterpart, records successive values of the laboratory time t.

Now, according to the Lorentz transformation, we have

$$t = \gamma\left(\bar{t} - \frac{v}{c^2}\bar{x}\right)$$

and upon taking differentials, remembering that v is constant, we obtain:

$$dt = \gamma\left(d\bar{t} - \frac{v}{c^2}d\bar{x}\right) \tag{4.25}$$

Since the particle is at rest relative to the frame $\bar{K}(\bar{x},\bar{t})$, then $d\bar{x} = 0$, so the last equation becomes:

$$dt = \gamma\, d\bar{t} \tag{4.26}$$

Since $d\bar{t}$ is the time interval recorded by a clock attached to the particle, and hence, this clock is at rest relative to the frame $\bar{K}(\bar{x},\bar{t})$, it is called the *proper time interval*, and is traditionally denoted by $d\tau$. In terms of our new symbolism, (4.26) becomes:

$$d\tau = \frac{1}{\gamma}dt = \sqrt{1 - \frac{v^2}{c^2}}dt \tag{4.27}$$

Later on, we will show that for a *material particle* (unlike a photon), we must have $v < c$. Consequently, the Lorentz factor γ^{-1} which appears in (4.27), is always less than unity. *This means that the proper time interval which is recorded by a clock attached to a material particle will record a smaller interval of time than that of a laboratory clock recording the same motion.* Stated differently, moving clocks will always go slow by a factor of γ^{-1}, and we call this the phenomenon the **time dilation effect**.

The proper time interval $d\tau$ has an important property that we shall find to be extremely useful later on: *the proper time interval $d\tau$ is invariant under a Lorentz transformation.* To see this, first recall that the spacetime interval (4.21) is invariant under a Lorentz transformation, so we have

$$c^2dt^2 - dx^2 - dy^2 - dz^2 = c^2d\bar{t}^2 - d\bar{x}^2 - d\bar{y}^2 - d\bar{z}^2 \tag{4.28}$$

Rearranging terms in (4.28), we obtain

$$c^2dt^2 - \left(\frac{dx^2+dy^2+dz^2}{dt^2}\right)dt^2 = c^2d\bar{t}^2 - \left(\frac{d\bar{x}^2+d\bar{y}^2+d\bar{z}^2}{d\bar{t}^2}\right)d\bar{t}^2 \,. \tag{4.28a}$$

But the quantities which appear in parentheses are simply the squares of the particle's velocity relative to each inertial frame:

$$v^2 = \left(\frac{dx^2+dy^2+dz^2}{dt^2}\right) \text{ and } \bar{v}^2 = \left(\frac{d\bar{x}^2+d\bar{y}^2+d\bar{z}^2}{d\bar{t}^2}\right) \tag{4.28b}$$

Substituting (4.28b) into (4.28a), and simplifying gives

$$d\tau^2 = c^2\left(1-\frac{v^2}{c^2}\right)dt^2 = c^2\left(1-\frac{\bar{v}^2}{c^2}\right)d\bar{t}^2 = d\bar{\tau}^2$$

and consequently,

$$d\tau = d\bar{\tau} \,. \tag{4.29}$$

We must conclude that *the proper time interval is invariant under a Lorentz transformation.*

Let us now consider the case of *accelerated* motion. Here, we assume that a material particle undergoing an acceleration a, moves in the positive x-direction, relative to the inertial frame $K(x,t)$. Once again, we assume that a clock is attached to the particle, and we inquire whether the proper time interval recorded by this clock will still be given by (4.27).

We are led to make the assumption that if the clock attached to the particle is an *ideal clock*; that is, one whose operating characteristics are not affected by the acceleration of the particle, then equation (4.27) should still hold good. This assumption, which has been the source of much debate, is traditionally called the **clock hypothesis**. It was not included in Einstein's original formulation of Special Relativity, as his initial work was confined to the study of inertial, and hence, unaccelerated reference frames.

Assuming this hypothesis to be true, if relative to the laboratory frame $K(x,t)$, the particle moves from an initial position $x = x_1$ at time $t = t_1$ to a terminal position $x = x_2$ at time $t = t_2$, then we must calculate the proper time interval $\Delta\tau$ to be:

Clock Hypothesis

$$\Delta\tau = \tau_2 - \tau_1 = \int_{t_1}^{t_2} \sqrt{1 - \frac{v^2}{c^2}}\, dt \tag{4.30}$$

In (4.30), we must remember that the particle's velocity v need not remain constant, but may be a function $v = v(t)$ of the time recorded in the laboratory frame.

The veracity of this assumption has been demonstrated experimentally by the use of modern particle accelerators which, under the right conditions, can impart enormous *accelerations* to certain species of subatomic particles, thereby affecting their observed lifetimes.

4.6. Minkowski Diagrams

We now turn our attention to the geometric representation of the Lorentz transformation by constructing what is commonly called a **Minkowski diagram**. This is just a graphical depiction of the spacetime relationship between the inertial frames $K(x,t)$ and $\bar{K}(\bar{x},\bar{t})$ under a Lorentz transformation. Once again, we assume that these two frames are in the standard configuration.

As shown in Figure 4.5, we shall represent the laboratory frame $K(x,t)$ by using an *orthogonal* set of axes, x and ct. The mutual orthogonality of these axes is not a mandatory requirement, but merely a matter of convenience.

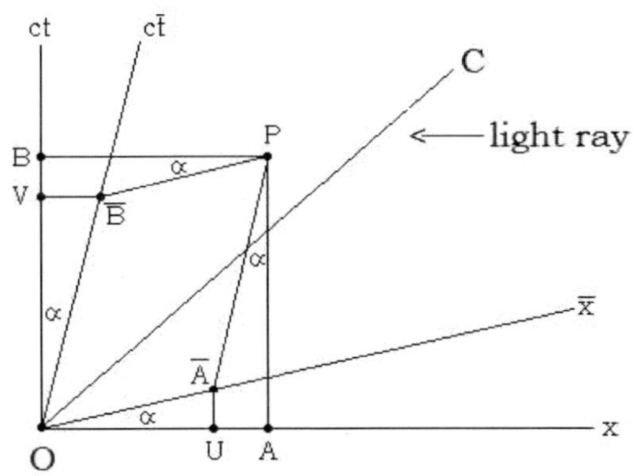

Figure 4.5. A simple Minkowski diagram.

We first consider the motion (relative to $K(x,t)$) of a *fixed* point on the $\bar{x}$-axis of the frame $\bar{K}(\bar{x},\bar{t})$. Since this point must have a constant velocity v relative to $K(x,t)$, then we must have:

$$\frac{dx}{dt} = v \Longrightarrow \frac{dx}{d(ct)} = \frac{v}{c} \tag{4.31}$$

Consequently, the $\bar{t}$-axis of $\bar{K}$ must be inclined at an angle α with respect to the t-axis of K, where

$$\tan\alpha = \frac{v}{c} = \beta \tag{4.32}$$

Also, from (4.32), we easily derive the useful relations:

$$\cos\alpha = \frac{1}{\sqrt{1+\beta^2}}, \quad \text{and} \quad \sin\alpha = \frac{\beta}{\sqrt{1+\beta^2}}. \tag{4.33}$$

As shown in Figure 4.5, it turns out that the $\bar{x}$-axis of the frame $\bar{K}$ must make the same angle α with respect to the x-axis of the frame K. To see this, let's consider the motion of the light ray OC which leaves the origin at the time $t = \bar{t} = 0$. Since the equation of this ray, relative to the frame K, must be $x = ct$, then we see that the slope of the line OC must be unity.

On the other hand, the same light ray, when viewed from the frame $\bar{K}$, must also have velocity c, and its equation must be $\bar{x} = c\bar{t}$. This means that the line OC must also have a slope of unity relative to the frame $\bar{K}$, and this can only happen if the $\bar{x}$-axis of $\bar{K}$, is also inclined at an angle α relative to the x-axis of K . Thus, the axes of the frame $\bar{K}$ are *oblique* and *symmetric* with respect to the path of the light ray OC.

Now, it is customary to refer to any point in spacetime as an *event*. From Figure 4.6, we see that any event P has two sets of spacetime coordinates. For example, the event P has, relative to the frame K, the coordinates $(x, ct) = (OA, OB)$; while relative to $\bar{K}$, the same point has the coordinates $(\bar{x}, c\bar{t}) = (O\bar{A}, O\bar{B})$. It is the *relationship* existing between these two pairs of coordinates that is, of course, governed by the Lorentz transformation.

One must be careful to note, however, the the length-scale which represents a *unit distance* in the frame $\bar{K}$ is not the same as that in the frame $\bar{K}$. Perhaps the best way of seeing this is to first recall that the *spacetime interval*

$$s^2 = c^2t^2 - x^2 = c^2\bar{t}^2 - \bar{x}^2 \tag{4.34}$$

in *invariant* under a Lorentz transformation. This means that the quantity, s^2, *whatever its value may be*, must be the same for all inertial observers. In particular, if we set $s^2 = -1$, then both of the equations

$$x^2 - c^2t^2 = 1 \tag{4.35a}$$

$$\bar{x}^2 - c^2\bar{t}^2 = 1 \tag{4.35b}$$

must be satisfied.

The first of these two equations represents a rectangular hyperbola in the frame K, while the latter equation defines the same hyperbola in the frame $\bar{K}$. We will now use this fact to *calibrate* the axes in each frame.

We begin by noting that the $\bar{x}$-axis of the frame $\bar{K}$ corresponds to the line $c\bar{t} = 0$. So from (4.35b), we conclude that the right branch of the hyperbola $\bar{x}^2 - c^2\bar{t}^2 = 1$ must intersect the $\bar{x}$-axis at $\bar{x} = 1$, and so, the line segment $O\bar{A}$ has unit length relative to $\bar{K}$. Once again, we take the length scale along the $c\bar{t}$-axis of $\bar{K}$ to be the same, so that $O\bar{B} = O\bar{A}$.

Similarly, the x-axis of the frame K corresponds to the line $ct = 0$. So from (4.35a), we see that the right branch of the hyperbola $x^2 - c^2t^2 = 1$ must intersect the x-axis at $x = 1$.

Thus, the line segment *OA* has unit length in this frame. As a matter of convention, we take the length scale along the *ct*-axis to be the same, so that $OB = OA$.

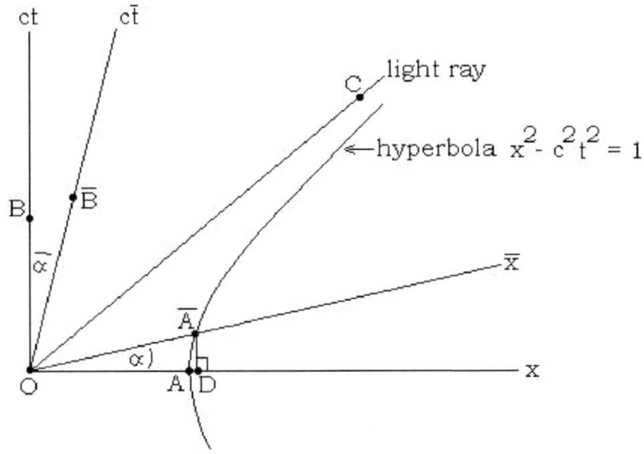

Figure 4.6. Scheme for calibration of the axes.

But, in general, it can be seen that *the unit-length scales in each frame are not the same*; that is, $OA \neq O\bar{A}$. Relative to K, the equation of the $\bar{x}$-axis is given by

$$ct = (\tan\alpha)\, x \tag{4.36}$$

and substituting this expression into (4.35a), we can obtain the x-coordinate of the point $\bar{A}$ where the branch of the hyperbola intersects the $\bar{x}$-axis. In doing so, we find that

$$x^2\left(1 - \tan^2\alpha\right) = 1 \Longrightarrow x = \frac{1}{\sqrt{1 - \tan^2\alpha}} = \frac{1}{\sqrt{1 - \beta^2}}$$

In other words, the length of the segment *OD* is given by

$$OD = \frac{1}{\sqrt{1 - \beta^2}} \tag{4.37}$$

and consequently, the length of the segment $O\bar{A}$ must be

$$O\bar{A} = \frac{OD}{\cos\alpha} \Longrightarrow O\bar{A} = \sqrt{\frac{1 + \beta^2}{1 - \beta^2}} \tag{4.38}$$

and, it is the length $O\bar{A}$ which is the unit-length in the frame $\bar{K}$. In other words, this is the *scale factor* introduced naturally by the invariance of the spacetime interval in (4.34), using the convention $s^2 = -1$. *This means we must multiply any length reckoned relative to the frame $\bar{K}$, by this scale factor when converting to the corresponding length in the frame K.*

Finally, we now examine how Minkowski diagrams can be used to derive the Lorentz transformation itself. To accomplish this task, we turn our attention to Figure 4.7, in which

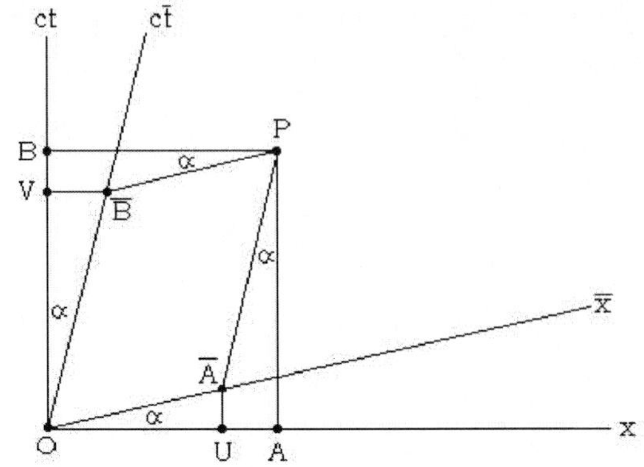

Figure 4.7. Minkowski diagram for the Lorentz transformation.

two inertial frames, K and $\bar{K}$, are drawn in the standard configuration. Here, the event P has, relative to the frame K, the coordinates $(x, ct) = (OA, OB)$. While relative to $\bar{K}$, the same event has the coordinates $(\bar{x}, c\bar{t}) = (O\bar{A}, O\bar{B})$. From the geometry of Figure 4.7, and the scale factor (4.38), we can write

$$\begin{aligned} x &= OU + UA \\ &= (O\bar{A})\cos\alpha + (P\bar{A})\sin\alpha \\ &= \left(\bar{x}\cdot\sqrt{\frac{1+\beta^2}{1-\beta^2}}\right)\frac{1}{\sqrt{1+\beta^2}} + \left(c\bar{t}\cdot\sqrt{\frac{1+\beta^2}{1-\beta^2}}\right)\frac{\beta}{\sqrt{1+\beta^2}} \\ &= \frac{1}{\sqrt{1-\beta^2}}(\bar{x}+v\bar{t}) \\ \Longrightarrow &\quad x = \gamma(\bar{x}+v\bar{t}) \end{aligned} \tag{4.39}$$

Similarly,

$$\begin{aligned} ct &= OV + VB \\ &= (O\bar{B})\cos\alpha + (P\bar{B})\sin\alpha \\ &= \left(c\bar{t}\cdot\sqrt{\frac{1+\beta^2}{1-\beta^2}}\right)\frac{1}{\sqrt{1+\beta^2}} + \left(\bar{x}\cdot\sqrt{\frac{1+\beta^2}{1-\beta^2}}\right)\frac{\beta}{\sqrt{1+\beta^2}} \\ &= \frac{1}{\sqrt{1-\beta^2}}(c\bar{t}+\beta\bar{x}) = c\gamma\left(\bar{t}+v\bar{x}/c^2\right) \\ \Longrightarrow &\quad t = \gamma\left(\bar{t}+v\bar{x}/c^2\right) \end{aligned} \tag{4.40}$$

Equations (4.39) and (4.40) are just the equations of the "inverse"Lorentz transformation that we derived previously.

4.7. Exercises

1. By direct calculation, show that the *spacetime interval*

$$ds^2 = c^2dt^2 - dx^2$$

is invariant under a special Lorentz transformation.

(Solution)

Taking differentials of the Lorentz transformation equations given by (4.20a), and recalling that v is constant, we obtain

$$\begin{aligned} d\bar{x} &= \gamma(dx - vdt) \\ d\bar{t} &= \gamma\left(dt - vdx/c^2\right) \end{aligned}$$

Consequently,

$$d\bar{s}^2 = c^2 d\bar{t}^2 - d\bar{x}^2 = \gamma^2\left[c^2\left(dt - vdx/c^2\right)^2 - (dx - vdt)^2\right]$$

Recalling the definition of γ, we can write

$$\begin{aligned} d\bar{s}^2 &= \left(1 - \frac{v^2}{c^2}\right)^{-1} \cdot \left[c^2\left(1 - \frac{v^2}{c^2}\right)dt^2 - \left(1 - \frac{v^2}{c^2}\right)dx^2\right] \\ &= c^2dt^2 - dx^2 \\ &= ds^2 \end{aligned}$$

2. Assume that two inertial frames, $K(x,t)$ and $\bar{K}(\bar{x},\bar{t})$, are in the standard configuration, while a rigid rod lies at rest, at an angle ϕ with respect to the $\bar{x}$-axis of the second frame. If the proper length of the rod (relative to $\bar{K}$) is L_0, then what is its length as relative to the frame $K(x,t)$?

(Solution)

Without loss of generality, we assume that the left endpoint of the rod is at the common origin of both reference systems at time $t = \bar{t} = 0$. At that time, the length L of the rod, relative to $K(x,t)$, is:

$$L^2 = x^2 + y^2$$

But owing to the Lorentz contraction, only the *longitudinal component* of the rod's length will be affected, so from (4.24), we have

$$x = \gamma^{-1}\bar{x} = \gamma^{-1}\left(L_0 \cos\phi\right)$$

while the *tranverse component* of its length remains the same; that is,

$$y = \bar{y} = L_0 \sin\phi$$

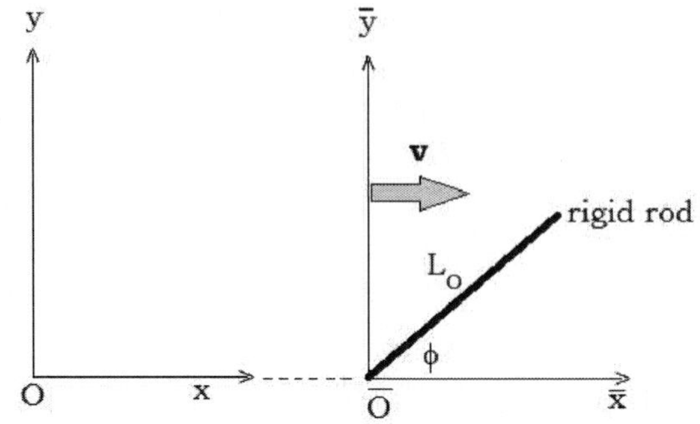

Figure 4.8. A rigid rod inclined at an angle ϕ.

Substituting the last two equations into the first one, we get:

$$\begin{aligned} L^2 &= x^2+y^2 = \left[\gamma^{-1}\ (L_0\cos\phi)\right]^2 + (L_0\sin\phi)^2 \\ &= L_0^2\,\gamma^{-2}\left(\cos^2\phi+\gamma^2\sin^2\phi\right) \\ &= L_0^2\,\gamma^{-2}\left[1+(\gamma^2-1)\sin^2\phi\right] \end{aligned}$$

So, the length of the rod relative to an observer in the frame $K(x,t)$ is

$$L=\gamma^{-1}L_0\sqrt{1+(\gamma^2-1)\sin^2\phi}$$

Observe that as $\phi \longrightarrow 0$, $L \longrightarrow \gamma^{-1}L_0$ as expected.

3. In the inertial frame $\bar{K}(\bar{x},\bar{t})$, at time $\bar{t}=0$, a pulse of light is emitted from the origin O at an angle ϕ relative to the $\bar{x}$-axis, and is transmited to the point P. If the two inertial frames $K(x,t)$ and $\bar{K}(\bar{x},\bar{t})$ are in the standard configuration, with their origins coinciding at time $t=\bar{t}=0$, then how much time elapses during the light ray's journey from O to P, as measured by an observer in the laboratory frame $K(x,t)$?

(Solution)

From the Lorentz transformation, we have

$$\Delta t = \gamma(\Delta\bar{t}+\frac{v}{c^2}\Delta\bar{x})$$

Since the speed of light is *invariant*, we have

$$\Delta\bar{t} = \bar{L}/c\,,$$

and since $\Delta\bar{x}=\bar{L}\cos\phi$, we see that during the light ray's journey from O to P, the amount of time that elapses in the laboratory frame $K(x,t)$ is simply

$$\begin{aligned} \Delta t &= \gamma\left[\left(\frac{\bar{L}}{c}\right)+\frac{v}{c^2}\left(\bar{L}\cos\phi\right)\right] \\ &= \frac{\gamma\bar{L}}{c}\left(1+\beta\cos\phi\right) \end{aligned}$$

4. As shown in Figure 4.9, a rocket, with a proper length of L_0 moves in the positive x-direction at a constant speed v. At the instant of time, $t = 0$, when the rear of the rocket (point R) coincides with the origin of the laboratory frame, an astronaut at the front (point F) of the rocket, sends a radio signal directly towards the rear. Relative to an observer in the laboratory frame, at what time does the radio signal reach the point R?

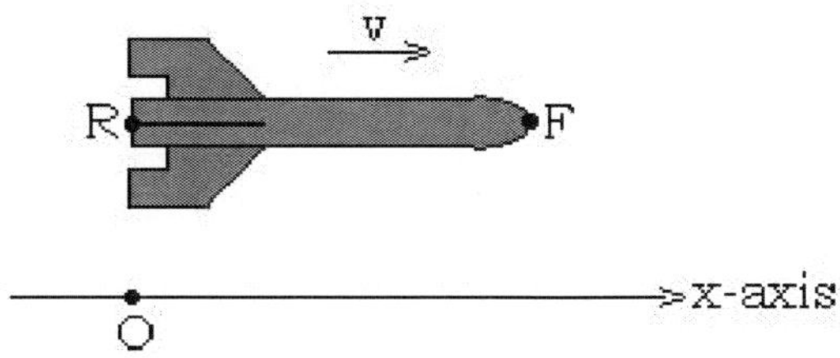

Figure 4.9. A rocket moving at a constant speed v.

(Solution)

Let the frame $\bar{K}(\bar{x},\bar{t})$ be the rest frame of the rocket, and let $K(x,t)$ be the laboratory frame. From the Lorentz transformation,

$$\Delta t = \gamma(\Delta\bar{t} + \frac{v}{c^2}\Delta\bar{x})$$

Since the proper length of the rocket is L_0 relative to the frame $\bar{K}(\bar{x},\bar{t})$, then the amount of time $\Delta\bar{t}$ is takes the radio signal to travel from F to R must be:

$$\Delta\bar{t} = \frac{L_0}{c}$$

Here, $\Delta\bar{t}$ is the amount of time recorded by a clock onboard the rocket. Also, the distance $\Delta\bar{x}$ is simply the proper length of the rocket

$$\Delta\bar{x} = L_0$$

Substituting these values into the first equation gives

$$\begin{aligned}\Delta t &= \gamma\frac{L_0}{c}\left[1+\frac{v}{c}\right] = \frac{L_0}{c}\frac{1}{\sqrt{1-\beta^2}}(1+\beta) \\ &= \frac{L_0}{c}\sqrt{\frac{1+\beta}{1-\beta}}\end{aligned}$$

5. Given that the frames $K(x,t)$ and $\bar{K}(\bar{x},\bar{t})$ are in the standard configuration, let $\bar{\mathbf{u}}$ and $\bar{\mathbf{v}}$ be basis vectors in the frame $\bar{K}$ such that, relative to this frame,

$$\bar{\mathbf{u}} = [1,0]_{\bar{K}} \text{ and } \bar{\mathbf{v}} = [0,1]_{\bar{K}}$$

If $\mathbf{u}$ and $\mathbf{v}$ are basis vectors in the frame $K(x,t)$ such that $\mathbf{u} = [1,0]_K$ and $\mathbf{v} = [0,1]_K$, then show that

$$\bar{\mathbf{u}} = \gamma\mathbf{u} + \gamma\beta\mathbf{v} \text{ and } \bar{\mathbf{v}} = \gamma\beta\mathbf{u} + \gamma\mathbf{v}$$

where the Euclidean lengths of $\bar{\mathbf{u}}$ and $\bar{\mathbf{v}}$ satisfy

$$\|\bar{\mathbf{u}}\| = \|\bar{\mathbf{v}}\| = \sqrt{\frac{1+\beta^2}{1-\beta^2}}$$

(Solution)

Beginning with the Lorentz transformation equations, we have

$$\begin{aligned} x &= \gamma(\bar{x} + \beta c\bar{t}) \\ ct &= \gamma(\bar{t} + \beta\bar{x}) \end{aligned}$$

Since the basis vector $\bar{\mathbf{u}} = [1,0]_{\bar{K}}$, we substitute the values $\bar{x} = 1$ and $c\bar{t} = 0$ into the first transformation equation to obtain

$$\begin{aligned} \bar{\mathbf{u}} &= [1,0]_{\bar{K}} = [\gamma, \beta\gamma]_K = \gamma[1,0]_K + \beta\gamma[0,1]_K \\ &\implies \bar{\mathbf{u}} = \gamma\mathbf{u} + \gamma\beta\mathbf{v} \end{aligned}$$

as claimed. Similarly, since $\bar{\mathbf{v}} = [0,1]_{\bar{K}}$, we put $\bar{x} = 0$ and $c\bar{t} = 1$ in the second transformation equation to obtain

$$\begin{aligned} \bar{\mathbf{v}} &= [0,1]_{\bar{K}} = [\beta\gamma, \gamma]_K = \beta\gamma[1,0]_K + \gamma[0,1]_K \\ &\implies \bar{\mathbf{v}} = \beta\gamma\mathbf{u} + \gamma\mathbf{v} \end{aligned}$$

Finally, observe that

$$\begin{aligned} \|\bar{\mathbf{u}}\|^2 &= \bar{\mathbf{u}} \cdot \bar{\mathbf{u}} = (\gamma\mathbf{u} + \gamma\beta\mathbf{v}) \cdot (\gamma\mathbf{u} + \gamma\beta\mathbf{v}) \\ &= \gamma^2(\mathbf{u} \cdot \mathbf{u}) + 2\gamma^2\beta(\mathbf{u} \cdot \mathbf{v}) + \gamma^2\beta^2(\mathbf{v} \cdot \mathbf{v}) \\ &= \gamma^2(1) + 2\gamma^2\beta(0) + \gamma^2\beta^2(1) \\ &= \gamma^2\left(1+\beta^2\right) = \frac{1+\beta^2}{1-\beta^2} \end{aligned}$$

and consequently,

$$\|\bar{\mathbf{u}}\| = \sqrt{\frac{1+\beta^2}{1-\beta^2}}$$

The expression for $\|\bar{\mathbf{v}}\|$ may be similarly obtained. We observe that the last equation is merely a restatement of equation (4.38), where we found by *geometric* means that the length $O\bar{A}$, which is the unit-length in the frame $\bar{K}$, is given by

$$O\bar{A} = \sqrt{\frac{1+\beta^2}{1-\beta^2}}$$

6. A rocket moves away from a spacestation at a constant velocity v. If identical clocks in the both the rocket and the spacestation record the times $t = \bar{t} = 0$ initially, then show, using a Minkowski diagram, that the proper time duration $\Delta\bar{T}$,indicated by a clock onboard the rocket, is related to the time elapsed ΔT by a clock in the spacestation by

$$\Delta\bar{T} = \Delta T\sqrt{1-\frac{v^2}{c^2}}$$

(Solution)

Let the space station be situated at the origin of the laboratory frame K. Also, let $\bar{K}$ be the *rest frame* of the rocket so that at time $\bar{t} = 0$, the rocket is at the origin of this system, and remains at the point $\bar{x} = 0$. Relative to an observer in K, however, the rocket traces out a path in spacetime which coincides with the $c\bar{t}$-axis, as shown in Figure 4.10.

If $\bar{T}_1$ and $\bar{T}_2$ are two successive readings on the rocket's clock, then from the geometry of the diagram, we see that

$$c(T_2 - T_1) = \left[c(\bar{T}_2 - \bar{T}_1)\sqrt{\frac{1+\beta^2}{1-\beta^2}}\right]\cos\alpha$$

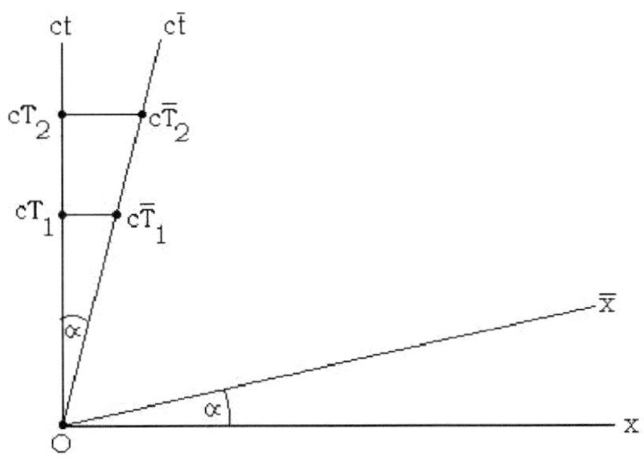

Figure 4.10. A rocket leaving the space station at O.

where the scale factor (4.38) has been used. Consequently, by (4.33), the last expression becomes:

$$c\Delta T = c\Delta\bar{T}\cdot\left(\sqrt{\frac{1+\beta^2}{1-\beta^2}}\frac{1}{\sqrt{1+\beta^2}}\right) = \frac{c\Delta\bar{T}}{\sqrt{1-\beta^2}}$$

from which we find

$$\Delta\bar{T} = \Delta T\sqrt{1-\beta^2} = \Delta T\sqrt{1-\frac{v^2}{c^2}}$$

7. Show that the scalar wave equation

$$\frac{1}{c^2}\frac{\partial^2 u}{\partial t^2} - \frac{\partial^2 u}{\partial x^2} = 0$$

is *covariant* with respect to the special Lorentz transformation given by (4.20a).

(Solution)

From (4.20a), we have

$$\bar{t} = \gamma\left(t - vx/c^2\right) \quad \text{and} \quad \bar{x} = \gamma(x - vt)$$

Employing the chain rule, we find that

$$\begin{aligned}
\frac{\partial}{\partial t} &= \frac{\partial \bar{t}}{\partial t}\cdot\frac{\partial}{\partial \bar{t}} + \frac{\partial \bar{x}}{\partial t}\cdot\frac{\partial}{\partial \bar{x}} = \gamma\frac{\partial}{\partial \bar{t}} - \gamma v\frac{\partial}{\partial \bar{x}} \\
\frac{\partial}{\partial x} &= \frac{\partial \bar{t}}{\partial x}\cdot\frac{\partial}{\partial \bar{t}} + \frac{\partial \bar{x}}{\partial x}\cdot\frac{\partial}{\partial \bar{x}} = -\frac{\gamma v}{c^2}\frac{\partial}{\partial \bar{t}} + \gamma\frac{\partial}{\partial \bar{x}}
\end{aligned}$$

After some operator algebra, we obtain

$$\begin{aligned}
\frac{\partial^2}{\partial t^2} &= \gamma^2\frac{\partial^2}{\partial \bar{t}^2} - 2\gamma^2 v\frac{\partial^2}{\partial \bar{t}\partial \bar{x}} + \gamma^2 v^2\frac{\partial^2}{\partial \bar{x}^2} \\
\frac{\partial^2}{\partial x^2} &= \frac{\gamma^2 v^2}{c^4}\frac{\partial^2}{\partial \bar{t}^2} - 2\frac{\gamma^2 v}{c^2}\frac{\partial^2}{\partial \bar{t}\partial \bar{x}} + \gamma^2\frac{\partial^2}{\partial \bar{x}^2}
\end{aligned}$$

After more algebra, we obtain

$$\begin{aligned}
\frac{1}{c^2}\frac{\partial^2 u}{\partial t^2} - \frac{\partial^2 u}{\partial x^2} &= \left(\frac{1}{c^2}\frac{\partial^2 u}{\partial \bar{t}^2} - \frac{\partial^2 u}{\partial \bar{x}^2}\right)\left[\gamma^2\left(1 - \frac{v^2}{c^2}\right)\right] \\
&= \frac{1}{c^2}\frac{\partial^2 u}{\partial \bar{t}^2} - \frac{\partial^2 u}{\partial \bar{x}^2} = 0
\end{aligned}$$

as required.

4.8. General Lorentz Boosts

We now examine a somewhat more general transformation known as a *general Lorentz boost* in the direction of the vector $\mathbf{v} = [v_1, v_2, v_3]$. As shown in Figure 4.10, the origin $\bar{O}$ of the barred frame has a constant velocity $\mathbf{v}$ relative to the origin O of the laboratory frame.

We also assume that the origins of the frames K and $\bar{K}$ coincide at $t = \bar{t} = 0$, and that the position vectors

$$\mathbf{x} = [x^1, x^2, x^3] \text{ and } \bar{\mathbf{x}} = [\bar{x}^1, \bar{x}^2, \bar{x}^3], \tag{4.41}$$

represent the position of the *same* point P in each reference frame.

Although $K(\mathbf{x},t)$ and $\bar{K}(\bar{\mathbf{x}},\bar{t})$ are *not* in the *standard position* (as in Figure 4.3), we assume that their corresponding axes are *parallel* in the sense that the frames can be brought into standard position by applying the *same* proper rotation to each frame.

Let $\hat{\mathbf{v}}$ be a unit vector in the direction of the velocity vector $\mathbf{v}$. As shown in Figure 4.12, we can use this vector to resolve $\mathbf{x}$ into the component vectors $\mathbf{x}_{\parallel}$ and $\mathbf{x}_{\perp}$, that are parallel and perpendicular, respectively, to the direction of $\mathbf{v}$.

We have

$$\hat{\mathbf{v}} = \frac{1}{v}\mathbf{v} \tag{4.42}$$

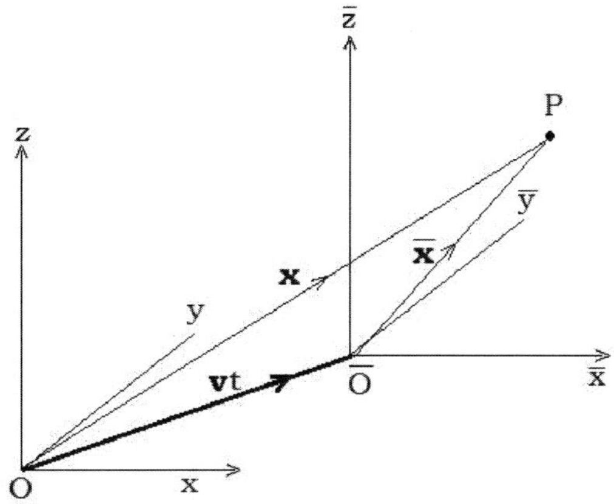

Figure 4.11. A general Lorentz boost in the direction of the vector **v**.

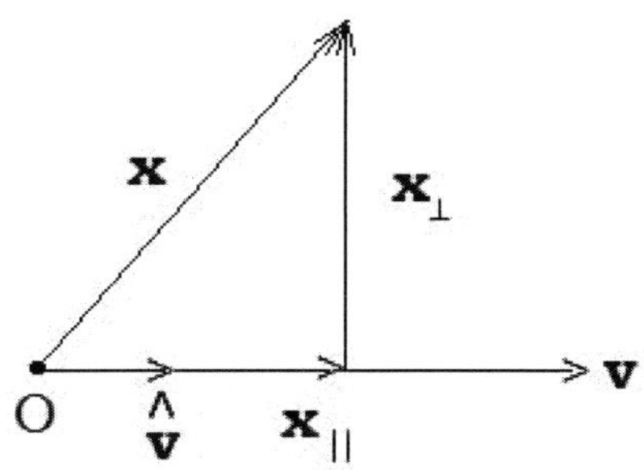

Figure 4.12.

so that

$$\mathbf{x}_{\parallel} = (\mathbf{x}\cdot\hat{\mathbf{v}})\,\hat{\mathbf{v}} = \frac{(\mathbf{x}\cdot\mathbf{v})}{v^2}\mathbf{v} \tag{4.43}$$

$$\mathbf{x}_{\perp} = \mathbf{x}-\mathbf{x}_{\parallel} = \mathbf{x}-\frac{(\mathbf{x}\cdot\mathbf{v})}{v^2}\mathbf{v} \tag{4.44}$$

Since the vector projections $\mathbf{x}_{\perp}$ and $\bar{\mathbf{x}}_{\perp}$, are each perpendicular to $\hat{\mathbf{v}}$, we assume that $\mathbf{x}_{\perp} = \bar{\mathbf{x}}_{\perp}$, and consequently,

$$\begin{aligned}\bar{\mathbf{x}} &= \bar{\mathbf{x}}_{\perp}+\bar{\mathbf{x}}_{\parallel} = \mathbf{x}_{\perp}+\bar{\mathbf{x}}_{\parallel} = \left[\mathbf{x}-\frac{(\mathbf{x}\cdot\mathbf{v})}{v^2}\mathbf{v}\right]+\gamma\left(\mathbf{x}_{\parallel}-vt\right)\\ &= \left[\mathbf{x}-\frac{(\mathbf{x}\cdot\mathbf{v})}{v^2}\mathbf{v}\right]+\gamma\left[\left(\frac{(\mathbf{x}\cdot\mathbf{v})}{v^2}\mathbf{v}\right)-vt\right]\end{aligned}$$

Upon rearranging terms, the last expression becomes

$$\bar{\mathbf{x}} = \mathbf{x}+\frac{(\mathbf{x}\cdot\mathbf{v})}{v^2}(\gamma-1)\,\mathbf{v}-\gamma\mathbf{v}t \tag{4.45}$$

Similarly, we find the transformation of the time t, as

$$\begin{aligned} \bar{t} &= \gamma\left[t - \left(\frac{\mathbf{v}\cdot\mathbf{x}_{\parallel}}{c^2}\right)\right] = \gamma\left[t - \frac{1}{c^2}\mathbf{v}\cdot\left(\frac{(\mathbf{x}\cdot\mathbf{v})}{v^2}\mathbf{v}\right)\right] \\ &= \gamma\left[t - \frac{(\mathbf{x}\cdot\mathbf{v})}{c^2}\right] \end{aligned} \tag{4.46}$$

Assembling (4.45) and (4.46), we now have the vector form of the equations for a general boost

General Lorentz Boost

$$\bar{\mathbf{x}} = \mathbf{x} + \frac{(\mathbf{x}\cdot\mathbf{v})}{v^2}(\gamma-1)\mathbf{v} - \gamma\mathbf{v}t \tag{4.45}$$

$$\bar{t} = \gamma\left[t - \frac{(\mathbf{x}\cdot\mathbf{v})}{c^2}\right] \tag{4.46}$$

We now derive the transformation matrix which corresponds to a general boost. We first observe (4.45) may be written in component form as

$$\begin{aligned} \bar{x}^i &= x^i + \frac{1}{v^2}\left(x^k v^k\right)(\gamma-1)v^i - \gamma v^i t = \delta^{ik}x^k + \frac{1}{v^2}\left(x^k v^k\right)(\gamma-1)v^i - \gamma v^i t \\ &= \left[\delta^{ik} + \frac{(\gamma-1)}{v^2}v^i v^k\right]x^k - \gamma v^i t \end{aligned}$$

where the indices $i,k = 1,2,3$. If we define the quantities ω^{ik} as

$$\omega^{ik} = \delta^{ik} + \frac{(\gamma-1)}{v^2}v^i v^k, \tag{4.47}$$

then the last equation can be neatly written as

$$\bar{x}^i = \omega^{ik}x^k - \gamma v^i t \tag{4.48}$$

Similarly, it is easy to see that the component form of (4.46) is simply

$$\bar{t} = \gamma\left[t - \frac{x^k v^k}{c^2}\right] \tag{4.49}$$

Upon setting $x^0 = ct$, $\bar{x}^0 = c\bar{t}$, and $\beta^k = v^k/c$, our general boost may be written in component form as

$$\bar{x}^0 = \gamma\left[x^0 - \beta^k x^k\right] \tag{4.50}$$

$$\bar{x}^i = \omega^{ik}x^k - \gamma\beta^i x^0 \quad (\text{for } i,k = 1,2,3) \tag{4.51}$$

Clearly, (4.50) and (4.51) may now be written in matrix form as

$$\begin{bmatrix} \bar{x}^0 \\ \bar{x}^1 \\ \bar{x}^2 \\ \bar{x}^3 \end{bmatrix} = \begin{bmatrix} \gamma & -\beta^1\gamma & -\beta^2\gamma & -\beta^3\gamma \\ -\beta^1\gamma & \omega^{11} & \omega^{12} & \omega^{13} \\ -\beta^2\gamma & \omega^{21} & \omega^{22} & \omega^{23} \\ -\beta^3\gamma & \omega^{31} & \omega^{32} & \omega^{33} \end{bmatrix} \begin{bmatrix} x^0 \\ x^1 \\ x^2 \\ x^3 \end{bmatrix} \tag{4.52}$$

Thus, the matrix Λ_B, corresponding to a general boost is

$$\Lambda_B = \begin{bmatrix} \gamma & -\beta^1\gamma & -\beta^2\gamma & -\beta^3\gamma \\ -\beta^1\gamma & \omega^{11} & \omega^{12} & \omega^{13} \\ -\beta^2\gamma & \omega^{21} & \omega^{22} & \omega^{23} \\ -\beta^3\gamma & \omega^{31} & \omega^{32} & \omega^{33} \end{bmatrix} \tag{4.53}$$

Now, the since quantities ω^{ik} are easily seen to be symmetric in (4.47), we see from (4.52), that Λ_B is a *symmetric matrix*; that is,

$$\Lambda_B^T = \Lambda_B\ . \tag{4.54}$$

Also, for the special case, $\mathbf{v} = [v,0,0]$, equation (4.53) must reduce to the matrix corresponding to the special Lorentz transformation found in (4.19).

One must keep in mind, however, that the validity of (4.53) depends upon the crucial assumption that we made at the beginning of our analysis; namely, the inertial frames $K(\mathbf{x},t)$, and $\bar{K}(\bar{\mathbf{x}},\bar{t})$ can be brought into the standard position by the applying the *same* proper rotation matrix to each frame.

4.9. Transformation of Velocities

If a material particle has velocity $\bar{\mathbf{u}}$ relative to the inertial frame $\bar{K}$, then it is sensible to inquire what this particle's velocity would be relative to the laboratory frame K. In order to answer this question, we first assume that these frames are in the standard position, so that the special Lorentz transformation (4.20a) may be invoked.

Recall that, under this set of circumstances, we have

$$\begin{aligned} \bar{x}^1 &= \gamma(x^1 - vt) \\ \bar{x}^2 &= x^2 \\ \bar{x}^3 &= x^3 \\ \bar{t} &= \gamma\left(t - vx^1/c^2\right) \end{aligned} \tag{4.55}$$

Upon taking differentials of the quantities in (4.55), while remembering that v is constant, we obtain

$$\begin{aligned} d\bar{x}^1 &= \gamma(dx^1 - vdt) \\ d\bar{x}^2 &= dx^2 \\ d\bar{x}^3 &= dx^3 \\ d\bar{t} &= \gamma\left(dt - \frac{v}{c^2}dx^1\right) \end{aligned} \tag{4.56}$$

Now, the velocities of the particle relative to each frame are given by

$$\mathbf{u}=\left[\frac{dx^1}{dt},\frac{dx^2}{dt},\frac{dx^3}{dt}\right] \text{ and } \bar{\mathbf{u}}=\left[\frac{d\bar{x}^1}{d\bar{t}},\frac{d\bar{x}^2}{d\bar{t}},\frac{d\bar{x}^3}{d\bar{t}}\right] \tag{4.57}$$

respectively; so that combining (4.56) and (4.57), we find that

$$\bar{u}^1=\frac{\gamma(dx^1-vdt)}{\gamma\left(dt-\frac{v}{c^2}dx^1\right)},\ \bar{u}^2=\frac{dx^2}{\gamma\left(dt-\frac{v}{c^2}dx^1\right)},\ \bar{u}^3=\frac{dx^3}{\gamma\left(dt-\frac{v}{c^2}dx^1\right)}$$

Upon simplifying the right hand sides of each equation, we see that under a special Lorentz transformation, the velocity components transform according to:

$$\bar{u}^1=\frac{u^1-v}{[1-(u^1v/c^2)]},\ \bar{u}^2=\frac{u^2}{\gamma[1-(u^1v/c^2)]},\ \bar{u}^3=\frac{u^3}{\gamma[1-(u^1v/c^2)]} \tag{4.58}$$

In terms of one-dimensional motion, (4.58) has an interesting geometric interpretation which is due to Synge [56][4]. For this special case, we have $\bar{\mathbf{u}}=\left[\bar{u}^1,0,0\right]$, while $\mathbf{u}=\left[u^1,0,0\right]$. In the corresponding Minkowski diagram shown in Figure 4.13, we assume that the particle moves along the path OP in spacetime, and the line PQ has been drawn parallel to the $c\bar{t}$-axis of the barred frame.

From Figure 4.13, it is easy to see that

$$\frac{u^1}{c}=\frac{OR}{RP}=\tan\delta \text{ and } \frac{v}{c}=\tan\alpha$$

Using the law of sines on triangle OPR, we obtain

$$\frac{\bar{u}^1}{c}=\frac{OQ}{PQ}=\frac{\sin(\delta-\alpha)}{\sin\left[\frac{\pi}{2}-(\delta+\alpha)\right]}=\frac{\sin(\delta-\alpha)}{\cos(\delta+\alpha)}$$

and consequently,

$$\frac{\bar{u}^1}{c}=\frac{\tan\delta-\tan\alpha}{1+\tan\delta\cdot\tan\alpha}=\frac{u^1/c-v/c}{1+u^1v/c^2}$$

Finally, simplifying the last equation, we obtain

$$\bar{u}^1=\frac{u^1-v}{1+u^1v/c^2}$$

which is just the one-dimensional version of (4.58).

Under a general Lorentz boost, or a Lorentz transformation without rotation, the velocity transformation formulas become slightly more complicated. To see this, we begin with the equations for a general boost

$$\begin{aligned}\bar{\mathbf{x}} &= \mathbf{x}+\frac{(\mathbf{x}\cdot\mathbf{v})}{v^2}(\gamma-1)\,\mathbf{v}-\gamma\mathbf{v}t\\ \bar{t} &= \gamma\left[t-\frac{(\mathbf{x}\cdot\mathbf{v})}{c^2}\right]\end{aligned}$$

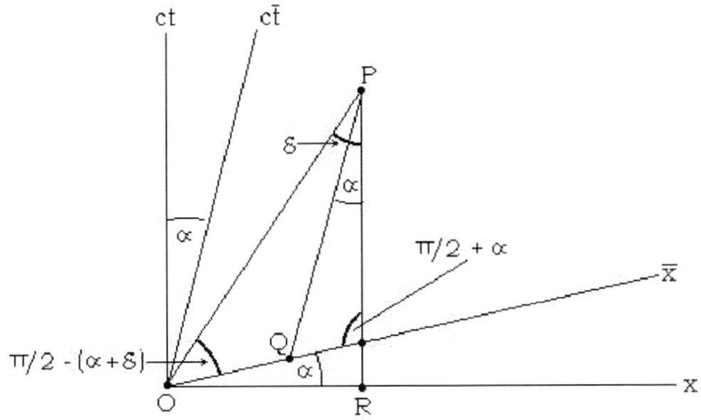

Figure 4.13. The geometry of one dimensional motion.

and we take differentials, remembering that **v** is constant, to obtain

$$\begin{aligned} d\bar{\mathbf{x}} &= d\mathbf{x}+\frac{(d\mathbf{x}\cdot\mathbf{v})}{v^2}(\gamma-1)\mathbf{v}-\gamma\mathbf{v}dt \\ d\bar{t} &= \gamma\left[dt-\frac{(d\mathbf{x}\cdot\mathbf{v})}{c^2}\right] \end{aligned}$$

and consequently,

$$\begin{aligned} \frac{d\bar{\mathbf{x}}}{d\bar{t}} &= \frac{(d\mathbf{x}/dt)+\left[(\mathbf{u}\cdot\mathbf{v})(\gamma-1)/v^2\right]\mathbf{v}-\gamma\mathbf{v}}{\gamma[1-(\mathbf{u}\cdot\mathbf{v})/c^2]} \\ &= \frac{\gamma^{-1}(d\mathbf{x}/dt)+\left[(\mathbf{u}\cdot\mathbf{v})\left(1-\gamma^{-1}\right)/v^2\right]\mathbf{v}-\mathbf{v}}{[1-(\mathbf{u}\cdot\mathbf{v})/c^2]} \end{aligned}$$

Upon using the definitions

$$\bar{\mathbf{u}}=\frac{d\bar{\mathbf{x}}}{d\bar{t}},\quad \text{and } \mathbf{u}=\frac{d\mathbf{x}}{dt}, \tag{4.59}$$

the last equation becomes

$$\bar{\mathbf{u}}=\frac{\left(\sqrt{1-v^2/c^2}\right)\mathbf{u}+\left[(\mathbf{u}\cdot\mathbf{v})/v^2\right]\left(1-\sqrt{1-v^2/c^2}\right)\mathbf{v}-\mathbf{v}}{[1-(\mathbf{u}\cdot\mathbf{v})/c^2]} \tag{4.60}$$

Now equation (4.60) has some interesting consequences. First, this equation implies, that if both $\|\mathbf{u}\| << c$ and $\|\mathbf{v}\| << c$, then

$$\bar{\mathbf{u}}\approx\mathbf{u}-\mathbf{v}$$

which is just the *Galilean velocity-addition formula*, as we would expect.

Secondly, as shown in the Exercises that follow, if we use (4.60) to relate the *speeds* $\bar{u}^2=\bar{\mathbf{u}}\cdot\bar{\mathbf{u}}$ and $u^2=\mathbf{u}\cdot\mathbf{u}$, then we find that

$$\left(1-\bar{u}^2/c^2\right)=\frac{\left(1-u^2/c^2\right)\left(1-v^2/c^2\right)}{[1-(\mathbf{u}\cdot\mathbf{v})/c^2]^2} \tag{4.61a}$$

and

$$\left(1-u^2/c^2\right)=\frac{\left(1-\bar{u}^2/c^2\right)\left(1-v^2/c^2\right)}{\left[1+\left(\bar{\mathbf{u}}\cdot\mathbf{v}\right)/c^2\right]^2} \tag{4.61b}$$

Furthermore, if in keeping with our previous convention, we now define the function $\gamma(\xi)$ by

$$\gamma(\xi)=\left(1-\xi^2/c^2\right)^{-1/2}, \tag{4.62}$$

then (4.61a) and (4.61b) may be neatly written as

$$\begin{aligned}\gamma(\bar{u}) &= \left[1-\left(\mathbf{u}\cdot\mathbf{v}\right)/c^2\right]\gamma(u)\gamma(v) && (4.63\text{a})\\ \gamma(u) &= \left[1+\left(\bar{\mathbf{u}}\cdot\mathbf{v}\right)/c^2\right]\gamma(\bar{u})\gamma(v) && (4.63\text{b})\end{aligned}$$

It is interesting to note that the second postulate of special relativity; that is, the invariance of the velocity of light is encoded in equations (4.61). To see this, let's consider a light ray moving with the speed $u=c$ relative to the inertial frame K. In this case, the righthand side of (4.61a) becomes zero, so we obtain $\bar{u}=c$, as one would expect.

Conversely, if a light ray has speed $\bar{u}=c$ relative to the frame $\bar{K}$, then (4.61b) the righthand side of this equation vanishes, which implies that $u=c$ as well. Also, it is worth noting that the speed of the light ray relative to each frame is independent of its *direction*.

4.10. Transformation of Acceleration

We are now in a position to derive the laws governing the transformation of *acceleration* between two inertial frames. Contrary to popular misconceptions, we can, in fact, discuss acceleration in Special Relativity as long as we confine ourselves to accelerations measured relative to *inertial* frames.

We imagine that a particle undergoes an acceleration relative to the inertial frame K, and inquire how the components of that acceleration appear to an observer at rest in the frame $\bar{K}$. Once again, we assume that the frames K and $\bar{K}$ are in the standard position, so that the special Lorentz transformation applies. From (4.58), we know that, under this special set of circumstances, the velocity components transform as

$$\bar{u}^1=\frac{u^1-v}{\left[1-\left(u^1v/c^2\right)\right]},\quad \bar{u}^2=\frac{u^2}{\gamma\left[1-\left(u^1v/c^2\right)\right]},\quad \bar{u}^3=\frac{u^3}{\gamma\left[1-\left(u^1v/c^2\right)\right]} \tag{4.58}$$

Taking differentials of both sides of each equation in (4.58), and remembering that v is constant, we obtain:

$$\begin{aligned}d\bar{u}^1 &= \frac{du^1}{\gamma^2\left(1-u^1v/c^2\right)^2}\\ d\bar{u}^2 &= \frac{\left(1-u^1v/c^2\right)du^2+u^2vdu^1/c^2}{\gamma\left(1-u^1v/c^2\right)^2}\\ d\bar{u}^3 &= \frac{\left(1-u^1v/c^2\right)du^3+u^3vdu^1/c^2}{\gamma\left(1-u^1v/c^2\right)^2}\end{aligned} \tag{4.64}$$

Now, the time-differential in the barred system is given by

$$d\bar{t} = \gamma\left(dt - \frac{v}{c^2}dx^1\right) = \gamma dt\left(1 - \frac{v}{c^2}\frac{dx^1}{dt}\right) = \gamma dt\left(1 - \frac{u^1 v}{c^2}\right) \tag{4.65}$$

where, we have used the fact that $u^1 = dx^1/dt$. So, dividing both sides of of each equation in (4.64) by (4.65), we obtain the components of the acceleration in the frame $\bar{K}$:

$$\begin{aligned}
\bar{a}^1 &= \frac{d\bar{u}^1}{d\bar{t}} = \frac{a^1}{\gamma^3\left(1-u^1 v/c^2\right)^2} \\
\bar{a}^2 &= \frac{d\bar{u}^2}{d\bar{t}} = \frac{a^2}{\gamma^2\left(1-u^1 v/c^2\right)^2} + \frac{\left(u^2 v/c^2\right)a^1}{\gamma^2\left(1-u^1 v/c^2\right)^3} \\
\bar{a}^3 &= \frac{d\bar{u}^3}{d\bar{t}} = \frac{a^3}{\gamma^2\left(1-u^1 v/c^2\right)^2} + \frac{\left(u^3 v/c^2\right)a^1}{\gamma^2\left(1-u^1 v/c^2\right)^3}
\end{aligned} \tag{4.66}$$

The inverse transformation may be obtained by interchanging the barred and unbarred quantities, and replacing v by $-v$; in so doing, we obtain

$$\begin{aligned}
a^1 &= \frac{du^1}{dt} = \frac{\bar{a}^1}{\gamma^3\left(1+\bar{u}^1 v/c^2\right)^2} \\
a^2 &= \frac{du^2}{dt} = \frac{\bar{a}^2}{\gamma^2\left(1+\bar{u}^1 v/c^2\right)^2} - \frac{\left(\bar{u}^2 v/c^2\right)\bar{a}^1}{\gamma^2\left(1+\bar{u}^1 v/c^2\right)^3} \\
a^3 &= \frac{du^3}{dt} = \frac{\bar{a}^3}{\gamma^2\left(1+\bar{u}^1 v/c^2\right)^2} + \frac{\left(\bar{u}^3 v/c^2\right)\bar{a}^1}{\gamma^2\left(1+\bar{u}^1 v/c^2\right)^3}
\end{aligned} \tag{4.67}$$

Now these equations are quite messy, but if we choose the frame $\bar{K}$ to be the *instantaneous rest frame* of the particle, so that at *each* instant, $\bar{u}^1 = \bar{u}^2 = \bar{u}^3 = 0$, then (4.67) simplifies to

$$\mathbf{a} = \left[\frac{1}{\gamma^3}\bar{a}^1, \frac{1}{\gamma^2}\bar{a}^2, \frac{1}{\gamma^2}\bar{a}^3\right] \tag{4.68}$$

4.11. Exercises

1. As shown in Figure 4.14, assume that the frame K, and the frame $\bar{K}$ both lie in the same *plane*, and these frames can be brought into the standard position by performing a counter-clockwise rotation through the *same* angle θ in each frame. Assume further that the frames coincide at $t = \bar{t} = 0$, and the boost has the special form $\mathbf{v} = \left[v^1, v^2, 0\right]$. Use this arrangement to derive equations (4.45) and (4.46).[5]

(Solution)

Assuming a passive, counterclockwise rotation θ of the xy-axes, we obtain for the left-hand frame,

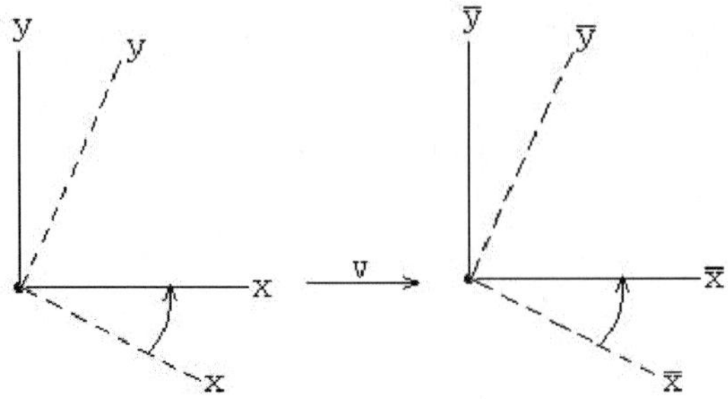

Figure 4.14. Geometry of the boost $v = \left[v^1, v^2, 0\right]$.

$$\begin{bmatrix} X \\ Y \end{bmatrix} = \begin{bmatrix} \cos\theta & \sin\theta \\ -\sin\theta & \cos\theta \end{bmatrix} \begin{bmatrix} x \\ y \end{bmatrix} = \begin{bmatrix} x\cos\theta + y\sin\theta \\ -x\sin\theta + y\cos\theta \end{bmatrix}$$

and similarly, performing the same rotation of axes in the righthand frame, we obtain

$$\begin{bmatrix} \bar{X} \\ \bar{Y} \end{bmatrix} = \begin{bmatrix} \cos\theta & \sin\theta \\ -\sin\theta & \cos\theta \end{bmatrix} \begin{bmatrix} \bar{x} \\ \bar{y} \end{bmatrix} = \begin{bmatrix} \bar{x}\cos\theta + \bar{y}\sin\theta \\ -\bar{x}\sin\theta + \bar{y}\cos\theta \end{bmatrix}$$

Since the frames K and $\bar{K}$ are now in the standard position, then the special Lorentz transformation may be used to give

$$\begin{aligned} \bar{x}\cos\theta + \bar{y}\sin\theta &= \gamma[(x\cos\theta + y\sin\theta) - vt] \\ -\bar{x}\sin\theta + \bar{y}\cos\theta &= -x\sin\theta + y\cos\theta \\ \bar{z} &= z \\ \bar{t} &= \gamma t - \frac{\gamma v}{c^2}(x\cos\theta + y\sin\theta) \end{aligned}$$

Upon eliminating $\cos\theta$ and $\sin\theta$ from the lefthand sides of the first two equations, and rearranging terms, we find that

$$\begin{aligned} \bar{x} &= x + (\gamma - 1)(x\cos\theta + y\sin\theta)\cos\theta - \gamma vt\cos\theta \\ \bar{y} &= y + (\gamma - 1)(x\cos\theta + y\sin\theta)\sin\theta - \gamma vt\sin\theta \\ \bar{z} &= z \\ \bar{t} &= \gamma t - \frac{\gamma v}{c^2}(x\cos\theta + y\sin\theta) \end{aligned}$$

Next, we define the position vectors $\mathbf{x} = [x, y, z]$ and $\bar{\mathbf{x}} = [\bar{x}, \bar{y}, \bar{z}]$, and let $\mathbf{v} = \left[v^1, v^2, 0\right]$. then

$$\frac{\mathbf{x} \cdot \mathbf{v}}{v} = (x\cos\theta + y\sin\theta), \quad \cos\theta = \frac{v^1}{\|\mathbf{v}\|}, \quad \sin\theta = \frac{v^2}{\|\mathbf{v}\|}$$

Finally, substituting these relations into the previous set of equations, we obtain

$$\begin{aligned} \bar{\mathbf{x}} &= \mathbf{x} + \frac{(\mathbf{x} \cdot \mathbf{v})}{v^2}(\gamma - 1)\mathbf{v} - \gamma \mathbf{v}t \\ \bar{t} &= \gamma\left[t - \frac{(\mathbf{x} \cdot \mathbf{v})}{c^2}\right] \end{aligned}$$

as required.

2. Find the inverse transformation corresponding to the Lorentz boost given in (4.45) and (4.46). Also, find the matrix Λ_B^{-1} which corresponds to the inverse transformation.

(Solution)

The inverse transformation may be obtained from (4.45) and (4.46) by simply making the substitutions $\mathbf{x} \longrightarrow \bar{\mathbf{x}}$, $t \longrightarrow \bar{t}$, and $\mathbf{v} \longrightarrow -\mathbf{v}$; in doing so, we obtain

$$\mathbf{x} = \bar{\mathbf{x}} + \frac{(\bar{\mathbf{x}} \cdot \mathbf{v})}{v^2}(\gamma - 1)\mathbf{v} + \gamma\mathbf{v}\bar{t} \tag{4.69}$$

$$t = \gamma\left[\bar{t} + \frac{(\bar{\mathbf{x}} \cdot \mathbf{v})}{c^2}\right] \tag{4.70}$$

We now write these equations in component form. First, observe that (4.69) may be written as

$$x^i = \left[\delta^{ik} + \frac{(\gamma-1)}{v^2}v^i v^k\right]\bar{x}^k + \gamma v^i t \tag{4.71}$$

$$x^0 = \gamma\left[\bar{x}^0 + \beta^k \bar{x}^k\right] \quad (\text{for } i,k = 1,2,3)$$

If, as before, we set

$$\omega^{ik} = \delta^{ik} + \frac{(\gamma-1)}{v^2}v^i v^k,$$

then we find that

$$x^i = \omega^{ik}\bar{x}^k + \gamma v^i\,\bar{t} \quad (\text{for } i,k = 1,2,3)$$

$$x^0 = \gamma\left[\bar{x}^0 + \beta^k \bar{x}^k\right]$$

where $\beta^k = v^k/c$ and $i,k = 1,2,3$. Consequently,

$$\begin{bmatrix} x^0 \\ x^1 \\ x^2 \\ x^3 \end{bmatrix} = \begin{bmatrix} \gamma & \beta^1\gamma & \beta^2\gamma & \beta^3\gamma \\ \beta^1\gamma & \omega^{11} & \omega^{12} & \omega^{13} \\ \beta^2\gamma & \omega^{21} & \omega^{22} & \omega^{23} \\ \beta^3\gamma & \omega^{31} & \omega^{32} & \omega^{33} \end{bmatrix} \begin{bmatrix} \bar{x}^0 \\ \bar{x}^1 \\ \bar{x}^2 \\ \bar{x}^3 \end{bmatrix} \tag{4.72}$$

and we conclude that matrix of the inverse transformation is

$$\Lambda_B^{-1} = \begin{bmatrix} \gamma & \beta^1\gamma & \beta^2\gamma & \beta^3\gamma \\ \beta^1\gamma & \omega^{11} & \omega^{12} & \omega^{13} \\ \beta^2\gamma & \omega^{21} & \omega^{22} & \omega^{23} \\ \beta^3\gamma & \omega^{31} & \omega^{32} & \omega^{33} \end{bmatrix} \tag{4.73}$$

as required.

It is important to note, in passing, that (4.73) could have been immediately obtained from (4.53) by making the substitutions $\beta_k \rightarrow -\beta_k$, for $k = 1,2,3$.

3. Use equation (4.60) to show that

$$\left(1-\bar{u}^2/c^2\right)=\frac{\left(1-u^2/c^2\right)\left(1-v^2/c^2\right)}{\left[1-(\mathbf{u}\cdot\mathbf{v})/c^2\right]^2}$$

(Solution)

Starting with (4.60), and performing some vector algebra, we find

$$\begin{aligned}\bar{u}^2 &= \|\bar{u}\|^2\\ &= \bar{\mathbf{u}}\cdot\bar{\mathbf{u}}\\ &= \frac{[\gamma(v)]^{-2}u^2-2(\mathbf{u}\cdot\mathbf{v})+(\mathbf{u}\cdot\mathbf{v})^2/v^2\left(1-[\gamma(v)]^{-2}\right)+v^2}{\left[1-(\mathbf{u}\cdot\mathbf{v})/c^2\right]^2}\end{aligned}$$

Consequently,

$$\begin{aligned}\left(1-\frac{\bar{u}^2}{c^2}\right) &= \frac{\left[\left(1-v^2/c^2\right)-\left(1-v^2/c^2\right)\left(u^2/c^2\right)\right]}{\left[1-(\mathbf{u}\cdot\mathbf{v})/c^2\right]^2}\\ &= \frac{\left(1-u^2/c^2\right)\left(1-v^2/c^2\right)}{\left[1-(\mathbf{u}\cdot\mathbf{v})/c^2\right]^2}\end{aligned}$$

as required.

4. Derive equations (4.63a) and (4.63b):

$$\gamma(\bar{u}) = \left[1-(\mathbf{u}\cdot\mathbf{v})/c^2\right]\gamma(u)\gamma(v) \qquad (4.63a)$$

$$\gamma(u) = \left[1+(\bar{\mathbf{u}}\cdot\mathbf{v})/c^2\right]\gamma(\bar{u})\gamma(v) \qquad (4.63b)$$

(Solution)

Equation (4.60) may be rewritten as

$$\left(1-\bar{u}^2/c^2\right)^{-1}=\left[1-(\mathbf{u}\cdot\mathbf{v})/c^2\right]^2\left(1-u^2/c^2\right)^{-1}\left(1-v^2/c^2\right)^{-1}$$

Upon taking the square root of both sides, and recalling (4.62), we obtain

$$\gamma(\bar{u})=\left[1-(\mathbf{u}\cdot\mathbf{v})/c^2\right]\gamma(u)\gamma(v)$$

which is (4.63a). To obtain (4.63b), we start with the last equation, and simply make the substitutions $u\longrightarrow\bar{u}$, $\mathbf{u}\longrightarrow\bar{\mathbf{u}}$, and $\mathbf{v}\longrightarrow-\bar{\mathbf{v}}$, to obtain

$$\begin{aligned}\gamma(u) &= \left[1-(\bar{\mathbf{u}}\cdot(-\bar{\mathbf{v}}))/c^2\right]\gamma(\bar{u})\gamma(v)\\ &= \left[1+(\bar{\mathbf{u}}\cdot\mathbf{v})/c^2\right]\gamma(\bar{u})\gamma(v)\end{aligned}$$

as required.

5. Show that

$$\left(1-\frac{v^2}{c^2}\right)=\left[1-(\mathbf{u}\cdot\mathbf{v})/c^2\right]\left[1+(\bar{\mathbf{u}}\cdot\mathbf{v})/c^2\right] \tag{4.74}$$

(Solution)

Employing (4.63a) and (4.63b), we can write

$$\begin{aligned}
\gamma(\bar{u}) &= \left[1-(\mathbf{u}\cdot\mathbf{v})/c^2\right]\gamma(u)\gamma(v)\\
&= \left[1-(\mathbf{u}\cdot\mathbf{v})/c^2\right]\left[1+(\bar{\mathbf{u}}\cdot\mathbf{v})/c^2\right]\gamma(\bar{u})\left[\gamma(v)\right]^2\\
&\Longrightarrow 1=\left[1-(\mathbf{u}\cdot\mathbf{v})/c^2\right]\left[1+(\bar{\mathbf{u}}\cdot\mathbf{v})/c^2\right]\left[\gamma(v)\right]^2\\
&\Longrightarrow \left(1-\frac{v^2}{c^2}\right)=\left[1-(\mathbf{u}\cdot\mathbf{v})/c^2\right]\left[1+(\bar{\mathbf{u}}\cdot\mathbf{v})/c^2\right]
\end{aligned}$$

as claimed.

6. An observer in the laboratory frame sees two particles moving parallel to the x-axis, approaching each other with the same speed u. What is their *relative* speed?

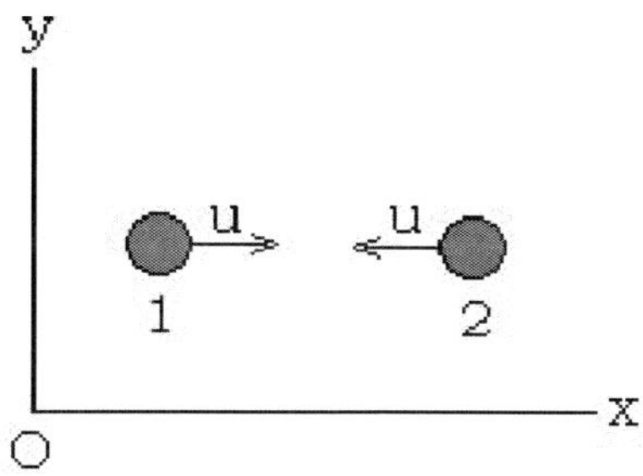

Figure 4.15. Two particles approaching each other.

Refering to Figure 4.15, let the frame $\bar{K}$ be attached to the particle #1, so that $v=u$. Also, relative to the laboratory frame K, the velocity components of the particle #2 are $u^1=-u$, and $u^2=u^3=0$. Consequently, from (4.58), the velocity of particle #2 relative to particle #1 is given by

$$\bar{u}^1=\frac{u^1-v}{\left[1-(u^1v/c^2)\right]}=\frac{(-u)-(u)}{1+u^2/c^2}=-\frac{2u}{1+u^2/c^2}$$

Observe that for $u << c$, we have $\bar{u}^1\approx -2u$, which is the *non-relativistic* relative velocity that one would expect.

Chapter 5

Introduction to Differential Geometry

There is geometry in the humming of the strings; there is music in the spacing of the spheres.

Pythagoras

5.1. Introduction

Thus far, we have used the term *spacetime* in a rather cavalier fashion, without fully discussing what this term really means. At the very foundation of all spacetime theories - including Special Relativity - is the assumption that *at least locally*, spacetime is a special type of mathematical artifice which is called a *differentiable manifold.*

In most relativistic theories, spacetime is assumed to be a four dimensional differentiable manifold, and it is on this canvas that we shall paint the mathematical relationships between various physical events, and to describe the evolution of physical processes.

Consequently, in this chapter and the next, we shall take a mathematical detour in order to fix our concepts, and to build a firm foundation for the discussion of the spacetimes of both Special Relativity, and General Relativity.

5.2. Topological Spaces

Now, the concept of a space as a mere set of points is insufficient as a platform for physics. This primitive notion of space is too amorphous, and it will not suit our needs. We must have an some *additional structure* which enables us to meaningfully assign coordinates to various events, write down transformations of coordonates, and to do calculus.

We begin adding some additional structure by first requiring that a manifold must be a topological space. Let X be any set of points, and let $T = \{U_\alpha : \alpha \in I\}$ be a certain collection of some or all of the subsets of X, where the index set I may be either finite or infinite, then we say that the pair (X, T) is a *topological space* if the set T satisfies the following axioms:

Topological Space Axioms

1. The empty set $\varnothing$, and the original set X are both elements of T; i.e.,

$$\varnothing \in T \text{ and } X \in T. \tag{5.1}$$

2. The union of an *arbitrary number* of sets in T is again an element of T. That is, if $\{U_\alpha : \alpha \in A\}$ where the index set A is either a finite or infinite subset of the index set I, we have

$$\bigcup_{\alpha \in A} U_\alpha \in T \tag{5.2}$$

3. The intersection of a *finite number* of sets in T is again an element of T. In other words, if $\{U_\alpha : \alpha \in B\}$ where the index set B is a finite subset of the index set I, then

$$\bigcap_{\alpha \in B} U_\alpha \in T \tag{5.3}$$

We call the members of T *open sets*, and the pair (X,T) is called a **topological space**. We say that the set T defines a topology on X, since we may associate many different topologies with the same set. In order to fix this new concept, let's look at some simple examples.

Example 1. If X is any set of points, let $T = \{\varnothing, X\}$. It is apparent that T satisfies all of the above axioms, so the pair (X,T) forms a topology called the *trivial topology* on X.

Example 2. For any set X, let T be the collection of *all* possible subsets of X. Clearly, axioms (1)-(3) are satisfied and we call the pair (X,T) the *discrete topology* on X.

Example 3. Let $X = R$, the real line, and let T denote the collection of all open intervals (a,b), along with all of their possible unions, where $a,b \in R$. Then T defines a topology on X which is called the *standard topology* on R.

Example 4. Let $X = R^2$, and let T denote the collection of all Cartesian products of the form $(a_1,b_1) \times (a_2,b_2)$ along with all of their possible unions. Here, non-empty members of T consist open rectangles, and their unions. This defines a topology on R^2, known as the *standard topology* on R^2. In a similar fashion, we can define the *standard topology* on R^n by defining T as the collection of all Cartesian products of the form

$$(a_1,b_1) \times (a_2,b_2) \times \ \dots \ \times (a_n,b_n)$$

along with all of their possible unions.

There are many ways of constructing new topological spaces. For example, if T_1 and T_2 are any two topologies defined on a the same set X, then it may be shown that their intersection $T_1 \cap T_2$ is also a topology on X; that is, the pair $(X, T_1 \cap T_2)$ is again a topological space. One particularly useful way of constructing new topological spaces is by defining what are called *relative topologies*.

If (X,T) is any topological space, and if X_0 is any subset of X, then we can construct a new collection of sets T' where

$$T' = \{X_0 \cap U_\alpha : U_\alpha \in T\} \tag{5.4}$$

It may be shown [See the Exercises] that the pair (X_0, T') is again a topological space, and we say that the set T induces a *relative topology* on X_0.

Now, given any set of points X, a *metric* on X is any real-valued function $d : X \times X \longrightarrow R$ which satisfies the metric axioms previously given in (2.5a) - (2.5d). In other words, the metric d provides us with the *notion of distance* by assigning a *non-negative* real number $d(x,y)$ to each *ordered pair* of points (x,y).

Now, if the set X is equipped with a metric d, and $\varepsilon > 0$ is any positive real number, then we can make X into a topological space by defining a topology T on X, such that the elements of T consist of all sets of the form:

$$U_\varepsilon(x) = \{y \in X : d(x,y) < \varepsilon\} \tag{5.4a}$$

along with all of their possible unions. We call T the *metric topology* on X, and we call the triple (X,T,d) a **metric space**. It is clear that the open sets of T are the 'open balls' given by (4), along with all of their possible unions.

Example 5. Let $X = R^2$ and let S^1 be the unit circle, given by

$$S^1 = \left\{(x,y) \in R^2 : x^2 + y^2 = 1\right\}$$

Now, S^1 can be given the relative topology which is induced by the metric topology on R^2. That is, we can define a topology T' on S^1 by simply taking

$$T' = \left\{S^1 \cap U_\alpha : U_\alpha \in T\right\}$$

where the set T consists of all open disks in the plane:

$$U_\varepsilon(x) = \left\{y \in R^2 : d(x,y) < \varepsilon\right\}$$

and all of their possible unions. Consequently, the open sets in T' will consist of the empty set $\varnothing$, along with all 'open arcs'.

Most of the types of topological spaces that physicists regularly encounter are examples of what are called *Hausdorf spaces*. Before we can define such spaces, we need the notion of a neighborhood. Let (X,T) be a topological space, and let $x \in X$. If N_x is a subset of X such that $x \in N_x \subseteq U$ where $U \in T$, then we say that N_x is a **neighborhood** of the point x. Now, if it so happens that in addition, $N_x = U$ for some open set $U \in T$, then we call N_x an **open neighborhood** of the point x.

We say that a topological space (X,T) is a **Hausdorf space**, or has the **Hausdorf property**, if given any two distinct points $x, y \in X$ there exist neighborhoods N_x and N_y containing x and y respectively, such that $N_x \cap N_y = \varnothing$. This state of affairs is shown in Figure 5.1.

It is easy to show that *any metric space has the Hausdorf property* and we assume that any of the spaces we shall consider will enjoy this property as well.

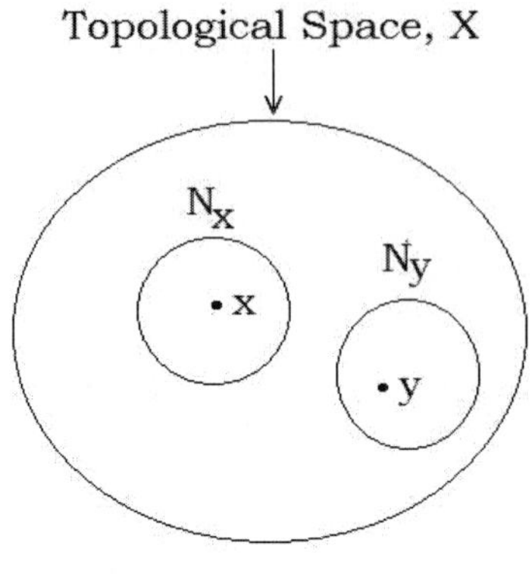

Figure 5.1. A space X with the Hausdorf property.

Before we can proceed to our discussion of manifolds, we need the concept of a continuous mapping. From elementary calculus, recall that a real-valued function $f : R \longrightarrow R$ is said to be **continuous** at the point x_0 if for every $\varepsilon > 0$ one can find a $\delta > 0$ such that

$$|x - x_0| < \delta \Longrightarrow |f(x) - f(x_0)| < \varepsilon \tag{5.5}$$

Assuming that R is equipped with the standard topology, then statement (5.5) is equivalent to

$$x \in (x_0 - \delta,\ x_0 + \delta) \Longrightarrow f(x) \in (f(x_0) - \varepsilon,\ f(x_0) + \varepsilon) \tag{5.5a}$$

or in other words, given any *open set* $(f(x_0) - \varepsilon,\ f(x_0) + \varepsilon)$ which contains $f(x_0)$, we can always find an *open set* $(x_0 - \delta,\ x_0 + \delta)$ which contains the the point x_0.

This means that the function $f : R \longrightarrow R$ is continuous at the point x_0 if when we are given any open set, say V, which contains the point $f(x_0)$, there exists an open set U containing x_0 such that the image $f(U) \subseteq V$, as shown in Figure 5.2.

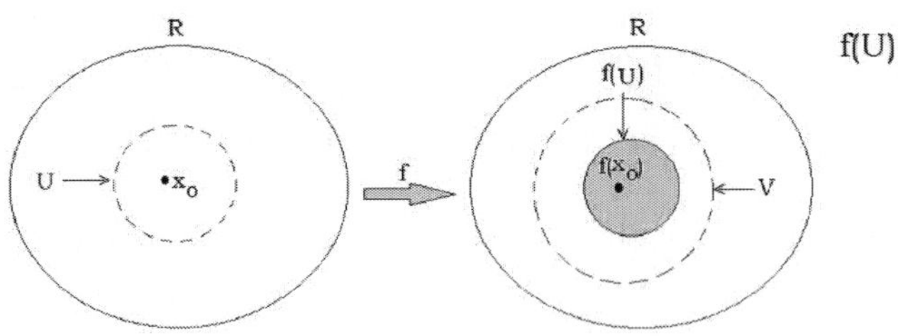

Figure 5.2. A mapping $f : R \longrightarrow R$ which is continuous at the point x_0.

It's easy to show that such *a function is continuous at each point of its domain if the inverse image of every open set in the range is again an open set in its domain.* Now, we can easily extend this concept to handle topological spaces.

If (X_1, T_1) and (X_2, T_2) are topological spaces, then any mapping of the form $f : X_1 \longrightarrow X_2$ is **continuous** if the inverse image of any open set in T_2 is an open set in T_1. Furthermore, if both f and its inverse f^{-1} are continuous, then we call the mapping f a **homeomorphism**, and we say that X_1 is **homeomorphic** to X_2, and conversely.

Homeomorphisms are nothing more than **bicontinuous** mappings between topological spaces, and at an intuitive level, two spaces are homeomorphic if any one space may be continuously deformed into the other space, without tearing or pasting. From the point of view of a trained topologist, homeomorphic spaces are equivalent.

Example 6. Consider the topological space (R,T) where T is the standard topology on R; that is, T is the collection of all open intervals (a,b), along with all of their possible unions, where $a,b \in R$. Now, consider the function $f : R \longrightarrow R$ defined by

$$f(x) = \begin{cases} x, \text{ if } x \leq 1 \\ x+2, \text{ if } x > 1 \end{cases}$$

It is easy to see that this function is not continuous, since the inverse image of the open interval $(1/2, 3/2)$ (on the y-axis) is the half-open interval $(1/2, 1]$ (on the x-axis), and the latter interval is *not* open, since $(1/2, 1] \notin T$.

5.3. Differentiable Manifolds

We now turn our attention to manifolds. We define an n-dimensional differentiable manifold M, as a set which satisfies the following properties:

Definition of a Differentiable Manifold

(a) There exists a topology T defined on M, such that (M,T) is a topological space with the Hausdorf property.

(b) The topology T contains a family of open sets $F = \{U_\alpha \in T : \alpha \in I\}$ which **covers** M; that is,

$$M = \bigcup_{\alpha \in I} U_\alpha \tag{5.6}$$

(c) Corresponding to each open set U_α, there is a mapping ϕ_α, such that ϕ_α is a homeomorphism from U_α onto an open subset V_α of R^n.

(d) If $U_\alpha, U_\beta \in F$ where $U_\alpha \bigcap U_\beta \neq \varnothing$, then the composite mapping $\psi_{\alpha\beta} = \phi_\alpha \phi_\beta^{-1}$ given by

$$\psi_{\alpha\beta} : \phi_\beta(U_\alpha \bigcap U_\beta) \longrightarrow \phi_\alpha(U_\alpha \bigcap U_\beta) \tag{5.7}$$

is **smooth**; i.e., infinitely differentiable.

This definition may appear a bit too abstract at first glance, so let's take a closer look at each of its key elements. First, by requiring M to be a topological space, we are assured that we can incorporate a crude notion of the *nearness* of any two points in M, along with the concept of *continuity*.

Secondly, since M is provided with a family of open sets F that cover M, we are guaranteed that for any point $p \in M$, there exists at least one open set $U_\alpha \in F$ such that $p \in U$.

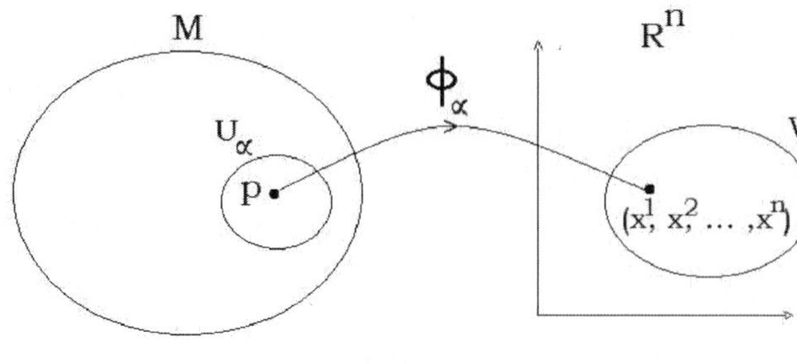

Figure 5.3. *M* is *locally homeomorphic* to R^n.

In other words, given any point $p \in M$, we can always find an *open neighborhood* which contains p.

As shown in Figure 5.3, axiom (c) guarantees that M is *locally homeomorphic* to R^n. To see this, note that if we are given any $p \in U_\alpha \subseteq M$, then there exists a homeomorphism ϕ_α from U_α onto an open subset $V_\alpha \in R^n$. Now, if we explicitly write

$$\phi_a(p) = \left(x^1(p), x^2(p), \ldots, x^n(p)\right) \tag{5.8}$$

for each $p \in U_\alpha$, we see that $\left(x^1(p), x^2(p), \ldots, x^n(p)\right) \in R^n$ are merely the coordinates of p which *correspond to* the mapping ϕ_α. Thus, we conclude that property (c) provides us with a means of *coordinatizing* the space M.

We call each pair (U_α, ϕ_α) a **chart** or **local coordinate system**, while the collection $\{(U_\alpha, \phi_\alpha) : \alpha \in I\}$ is called an **atlas**. Also, we call the open set U_α a **coordinate neighborhood** of p, while ϕ_α is called a **coordinate function**.

Observe that we *do not* require that a manifold must be covered by a single coordinate system, but we allow for the fact that this may not be possible in most realistic situations, and therefore, a collection of such charts, or an atlas, is required to coordinatize the entire manifold. In other words, we have not required that a manifold should be *globally homeomorphic* to R^n, but only *locally homeomorphic* to R^n.

Now in the event that $U_\alpha, U_\beta \in F$ where $U_\alpha \bigcap U_\beta \neq \varnothing$, then there must be some point $p \in U_\alpha \bigcap U_\beta$; this means that *two sets of coordinates* may be assigned to this point. This may appear to be rather strange, but one must remember that a point $p \in M$ is a *geometrical object*, and as such, its existence is *independent* of the manner in which coordinates are assigned to it.

Axiom (d) provides us with **compatibility conditions** on any two charts. It requires that the transition from one coordinate system to another must be *smooth* (C^∞). As shown in Figure 5.4, we see that the coordinate function ψ_β assigns the coordinates (x^i) to the point $p \in M$, while the coordinate function ψ_α assigns the coordinates $(\bar{x}^i)$ to the *same* point.

Observe that the map $\psi_{\alpha\beta} = \phi_\alpha \phi_\beta^{-1}$ maps the open set $\phi_\beta(U_\alpha \bigcap U_\beta)$ of R^n onto the open set $\phi_\alpha(U_\alpha \bigcap U_\beta)$ of R^n; in particular, it maps the coordinates $(x^i) \equiv (x^1, x^2, \ldots, x^n)$ of p in the first coordinate system into the coordinates $(\bar{x}^i) \equiv (\bar{x}^1, \bar{x}^2, \ldots, \bar{x}^n)$ of p in the second coordinate system. In other words, the map $\psi_{\alpha\beta}$ provides us with a change of coordinates from one system to the another in a manner such that the transformation

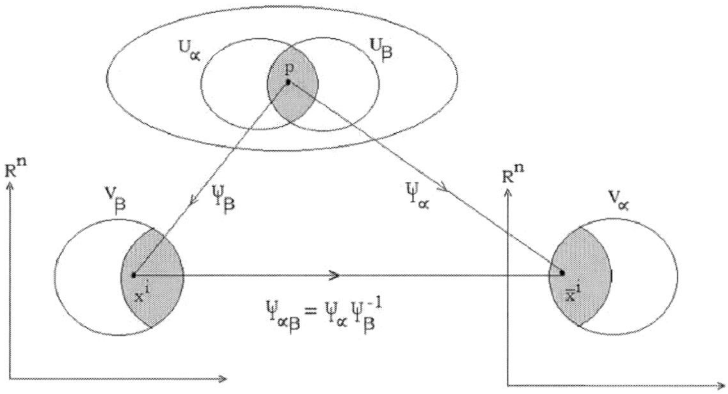

Figure 5.4. A smooth change of coordinates $\bar{x}^i = \psi_{\alpha\beta}(x^i)$.

$$\bar{x}^i = \psi_{\alpha\beta}(x^i) \tag{5.9}$$

is smooth.

The map $\psi_{\alpha\beta} = \phi_\alpha \phi_\beta^{-1}$ is called a **transition map**, and since it maps an open set in R^n into another open set in R^n, we can now exploit the usual concept of smoothness in R^n to say that *the map $\psi_{\alpha\beta}$ is* **smooth** *(or C^∞) provided that the partial derivatives (of all orders) of the functions $\bar{x}^i$ with respect to the x^i, exist and are continuous.*

Now, a **differentiable structure** (or C^∞ structure) on M is a *maximal* C^∞ atlas; that is, an atlas which contains *all* possible charts for M, such that each pair of charts satisfies the compatibility conditions given in Axiom (d). We assume that such a structure exists for any of the manifolds we shall deal with in the future.

Perhaps, the simplest example of a manifold is an open set U in R^n. It's easy to see that the lone chart (U, I_U), where I_U denotes the identity map on U, will serve as an atlas for U. Also, it may be shown that the unit circle S^1 and the unit sphere S^2 are one and two dimensional manifolds, respectively. The reader is referred to the Exercises for additional details.

5.4. Exercises

1. *(Open Sets)* Let (X,T) be a topological space, and let A be a subset of X with the property that given any point $p \in A$, there exists an open set $U_p \in T$ such that $p \in U_p \subseteq A$. Show that A is an open set.

(Solution)

Given any point $p \in A$, we have $p \in U_p \subseteq A$. Consequently,

$$\bigcup_{p \in A} U_p = A$$

Since A is the union of open sets, then by (5.2), $A \in T$ and therefore, A is open.

2. *(Relative Topology)* Let (X,T) be a topological space, and let $X_0 \subseteq X$. Define the collection of sets T' where

$$T' = \{X_0 \cap U_\alpha : U_\alpha \in T\}$$

Show that the pair (X_0, T') is again a topological space.

(Solution)

Clearly, $\varnothing \in T'$ and $X_0 \in T'$. Consider the sets of the form $U'_\alpha = X_0 \cap U_\alpha$ where each $U_\alpha \in T$. We have

$$\bigcup_\alpha U'_\alpha = \bigcup_\alpha (X_0 \cap U_\alpha) = X_0 \bigcap \left(\bigcup_\alpha U_\alpha \right) \in T'$$

since by (5.2), we know that $\bigcup_\alpha U_\alpha \in T$. Thus, T' is "closed" under arbitrary unions. Also, if the index $\alpha \in I$ where I is a *finite* index set, then

$$\bigcap_\alpha U'_\alpha = \bigcap_\alpha (X_0 \cap U_\alpha) = X_0 \bigcap \left(\bigcap_\alpha U_\alpha \right) \in T'$$

since by (5.3), we must have $\bigcap_\alpha U_\alpha \in T$. So, we have shown that T' is closed upon finite intersections. We conclude that (X_0, T') is a topological space, as claimed.

3. *(Composite Mappings)* Let $(X,T_1), (Y,T_2)$, and (Z,T_3) be topological spaces, and assume that the mappings $g : X \longrightarrow Y$ and $f : Y \longrightarrow Z$ are both continous. Show that the composite mapping $fg : X \longrightarrow Z$ is also continuous.

(Solution)

Let U and V be open sets, where $U \in T_1$ and $V \in T_3$. We must show that $(fg)^{-1}(V) \in T_1$. Since f is continuous, we have $f^{-1}(V) \in T_2$. Similarly, since g is continuous, we have $g^{-1}\left[f^{-1}(V)\right] \in T_1$ as well. Observe that

$$g^{-1}\left[f^{-1}(W)\right] = g^{-1}f^{-1}(W) = (fg)^{-1}(W) \in T_1$$

so fg is a continuous mapping from X to Z; in other words, *the composition of continuous mappings is continuous.*

4. *(Intersection of Topologies)* Let (X,T_1) and (X,T_2) both be topological spaces. Show that $(X, T_1 \bigcap T_2)$ is again a topological space.

(Solution)

Since $X \in T_1$ and $X \in T_2$, then $X \in T_1 \bigcap T_2$. Similarly, since $\varnothing \in T_1$ and $\varnothing \in T_2$, then $\varnothing \in T_1 \bigcap T_2$ as well.

We first show that axiom (2) is satisfied. Now, assume that

$$\begin{aligned} T_1 &= \{U_\alpha : \alpha \in I\} \\ T_2 &= \{V_\beta : \beta \in J\}. \end{aligned}$$

Let $W_{\alpha\beta} = U_\alpha \bigcap V_\beta$ for $\alpha \in I$, and for $\beta \in J$. We have

$$\begin{aligned}\bigcup_{\alpha,\beta} W_{\alpha\beta} &= \bigcup_{\alpha,\beta} (U_\alpha \bigcap V_\beta) = \bigcup_{\alpha}\left[\bigcup_{\beta}(U_\alpha \bigcap V_\beta)\right] = \bigcup_{\alpha}\left[U_\alpha \bigcap \left(\bigcup_{\beta} V_\beta\right)\right] \\ &= \left(\bigcup_{\alpha} U_\alpha\right) \bigcap \left(\bigcup_{\beta} V_\beta\right) \in T_1 \bigcap T_2\end{aligned}$$

since by (5.2), arbitrary unions are "closed" in both T_1 and T_2.

Now assume that $U, V \in T_1 \bigcap T_2$. Then clearly, $U, V \in T_1$ and $U, V \in T_2$. But since T_1 and T_2 are topologies, then by (3), we must have $U \bigcap V \in T_1$ and $U \bigcap V \in T_2$. Hence,

$$U \bigcap V \in T_1 \bigcap T_2$$

and we see that $T_1 \bigcap T_2$ satisfies axiom (3) as well.

5. *(Metric Space)* Let (X, T, d) be a metric space where T is the *metric topology* on X, as defined by (5.4a). Show that this metric space has the Hausdorf property.

(Solution)

Let $x_1, x_2 \in X$ where $x_1 \neq x_2$. Consider the sets

$$U_{\varepsilon_1}(x_1) = \{x \in X : d(x, x_1) < \varepsilon_1\} \text{ and } U_{\varepsilon_2}(x_2) = \{x \in X : d(x, x_2) < \varepsilon_2\}$$

By definition, $U_{\varepsilon_1}(x_1) \in T$ and $U_{\varepsilon_2}(x_2) \in T$, for all $\varepsilon_1, \varepsilon_2 > 0$. It suffices to show that we can choose ε_1, and ε_2 sufficiently small so that $U_{\varepsilon_1}(x_1) \bigcap U_{\varepsilon_2}(x_2) = \varnothing$. Since the points x_1 and x_2 are distinct, then $d(x_1, x_2) > 0$.

So, we let's take $\varepsilon_1, \varepsilon_2 < \frac{1}{2} d(x_1, x_2)$ and assume to the contrary, that there exists a point $x \in X$ such that $x \in U_{\varepsilon_1}(x_1) \bigcap U_{\varepsilon_2}(x_2)$. Then we must have both $d(x, x_1) < \frac{1}{2} d(x_1, x_2)$ and $d(x, x_2) < \frac{1}{2} d(x_1, x_2)$ so that by the triangle inequality, we can write

$$d(x_1, x_2) \leq d(x, x_1) + d(x, x_2) < \frac{1}{2} d(x_1, x_2) + \frac{1}{2} d(x_1, x_2) = d(x_1, x_2).$$

In other words, we are forced to conclude that $d(x_1, x_2) < d(x_1, x_2)$, which is an obvious contradiction. Thus, $U_{\varepsilon_1}(x_1)$ and $U_{\varepsilon_2}(x_2)$ must be disjoint open sets, and the metric space (X, T, d) has the Hausdorf property.

6. *(The Unit Circle S^1)* Show that the unit circle S^1 given by

$$S^1 = \left\{(x, y) \in R^2 : x^2 + y^2 = 1\right\}$$

is a one-dimensional, differentiable manifold.

(Solution)

We set up polar coordinates (r, θ) on S^1 where we set $r = 1$, and define the sets

$$U_1 = S^1 - \{(1, 0)\} \text{ and } U_2 = S^1 - \{(-1, 0)\}$$

Both of these sets are open in the relative topology which is induced by the metric topology on R^2, and furthermore,

$$S^1 = U_1 \bigcup U_2.$$

Define the mappings ϕ_1, ϕ_2 according to

$$\begin{aligned} \phi_1 &: U_1 \longrightarrow R \text{ such that } \phi_1((x,y)) = \theta_1 \text{ where } \theta_1 \in (0, 2\pi) \subseteq R. \\ \phi_2 &: U_2 \longrightarrow R \text{ such that } \phi_2((x,y)) = \theta_2 \text{ where } \theta_2 \in (-\pi, \pi) \subseteq R. \end{aligned}$$

Clearly, each of these mappings is a *homeomorphism* from U_1 and U_2 respectively, onto open subsets of R. In other words, (ϕ_1, U_1) and (ϕ_2, U_2) are *charts*, while $\{(\phi_1, U_1), (\phi_2, U_2)\}$ is an *atlas* for S^1.

Now, consider the point $(x,y) \in U_1 \bigcap U_2$, where $\phi_1((x,y)) = \theta_1$ and $\phi_2((x,y)) = \theta_2$. We find the transition function $\psi_{12} = \phi_1 \phi_2^{-1}$:

$$\theta_1 = \psi_{12}(\theta_2) = \phi_1 \phi_2^{-1}(\theta_2) = \phi_1(x,y) = \begin{cases} \theta_2, & \text{if } y > 0 \\ \theta_2 + 2\pi, & \text{if } y < 0 \end{cases}$$

So, the transition function $\psi_{12} = \phi_1 \phi_2^{-1}$ is certainly *smooth* since in either case, $\partial\theta_1 / \partial\theta_2 = 1$, while all of the other partial derivatives vanish.

7. *(The Sphere, S^2)* Show that the two-dimensional sphere S^2,

$$S^2 = \left\{ (x,y,z) \in R^3 : x^2 + y^2 + z^2 = 1 \right\}$$

is a two-dimensional manifold.

(Solution)

We begin by defining the sets

$$\begin{aligned} U_1 &= S^2 - \{(0,0,1)\} = S^2 - \{\text{North Pole}\} \\ U_2 &= S^2 - \{(0,0,-1)\} = S^2 - \{\text{South Pole}\} \end{aligned}$$

It's easy to see that both of these sets are open in the relative topology which is induced by the metric topology on R^3, and

$$S^1 = U_1 \bigcup U_2.$$

We define the mappings ϕ_1 and ϕ_2, given by

$$\phi_1 : U_1 \longrightarrow R^2 \text{ such that } \phi_1(x,y,z) = \left(\frac{x}{1-z}, \frac{y}{1-z} \right) \equiv (\bar{u}, \bar{v}) \text{ where } z \neq 1$$

$$\phi_2 : U_2 \longrightarrow R^2 \text{ such that } \phi_2(x,y,z) = \left(\frac{x}{1+z}, \frac{y}{1+z} \right) \equiv (u, v) \text{ where } z \neq -1$$

As shown in Figure 5.5, ϕ_1 and ϕ_2 correspond to the *stereographic projections* of the sphere from the north pole, and the south pole, respectively, onto the xy-plane. Each of

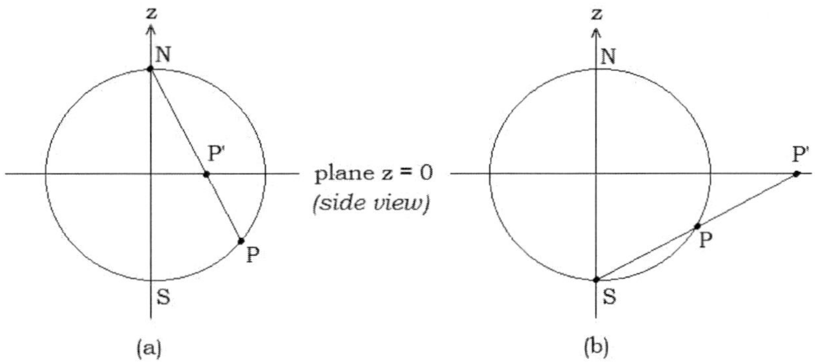

Figure 5.5. Stereographic projections of the unit sphere S^2 from the north pole (a) and from the south pole (b), onto the xy-plane.

these mappings is a *homeo-morphism* from U_1 and U_2 respectively, onto open subsets of the plane, R^2. Thus, (ϕ_1, U_1) and (ϕ_2, U_2) are *charts* and $\{(\phi_1, U_1), (\phi_2, U_2)\}$ is an *atlas* for S^1.

Now, after some algebra, we find that

$$\phi_2^{-1}(u,v) = \left(\frac{2u}{1+u^2+v^2}, \frac{2v}{1+u^2+v^2}, \frac{1-u^2-v^2}{1+u^2+v^2}\right) \equiv (x,y,z)$$

So, if we are given any point $(x,y,z) \in U_1 \bigcap U_2$, the transition function $\psi_{12} = \phi_1\phi_2^{-1}$ becomes

$$\begin{aligned}
(\bar{u},\bar{v}) &= \psi_{12}(u,v) = \phi_1\phi_2^{-1}(u,v) = \phi_1\left(\frac{2u}{1+u^2+v^2}, \frac{2v}{1+u^2+v^2}, \frac{1-u^2-v^2}{1+u^2+v^2}\right) \\
&= \left(\frac{u}{u^2+v^2}, \frac{v}{u^2+v^2}\right) \\
\Longrightarrow & \quad \bar{u} = \frac{u}{u^2+v^2} \text{ and } \bar{v} = \frac{v}{u^2+v^2}
\end{aligned}$$

Since $u^2+v^2 \neq 0$ whenever $(x,y,z) \in U_1 \bigcap U_2$, we see that all of the partial derivatives of $\bar{u}$ and $\bar{v}$ with respect to the unbarred variables exist and are continuous, so that $\psi_{12} = \phi_1\phi_2^{-1}$ is *smooth*.

5.5. Differentiable Maps

Suppose that $f : M \longrightarrow N$ is a map from an M-dimensional manifold M to an n-dimensional manifold N.[1] As shown in Figure 5.6, this map assigns a *unique* point $f(p) \in N$ to each point $p \in M$. Let (U, ϕ) be a coordinate chart which contains the point p, and let (V, ψ) be a chart which contains the point $f(p)$, and define the vectors,

$$\mathbf{x} \equiv (x^b) = \phi(p) \in R^m \text{ and } \mathbf{y} \equiv (y^a) = \psi(f(p)) \in R^n . \tag{5.10}$$

From the diagram, the map f must have the *coordinate representation* given by

$$\mathbf{y} = \psi f \phi^{-1}(\mathbf{x}) \tag{5.11}$$

or in other words, **y** is just a *vector-valued function* of M variables; it maps the vector $\mathbf{x} \in R^m$ to the vector $\mathbf{y} \in R^n$. Alternatively, (5.11) may be written as

$$y^a = f^a(x^b) \tag{5.11a}$$

where $a = 1, 2, ..., m$ and $b = 1, 2, ..., n$.

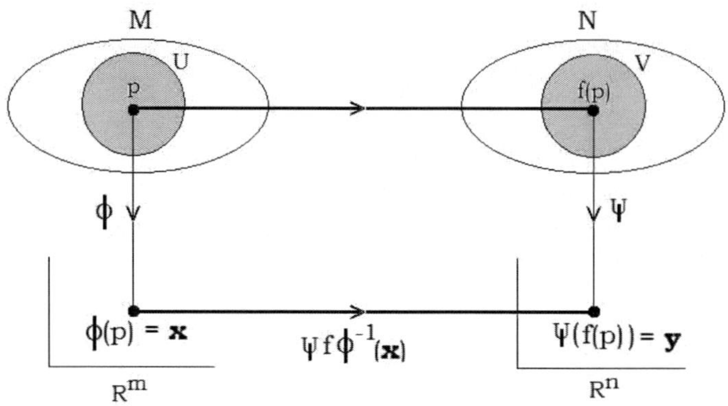

Figure 5.6. Coordinate representation of the map $\psi f \phi^{-1} : R^m \longrightarrow R^n$.

Since the coordinate representation of f is just a mapping from R^m into R^n, we can exploit the notion of differentiability such spaces to define the differentiability of the given map. That is, if for *each* y^a, we can compute the partial derivatives (of all orders) with respect to each x^b, and if all these partial derivatives exist and are continuous, then we say that the mapping f is **smooth,** or **differentiable**.

If, in addition, the smooth map $\psi f \phi^{-1}$ has an inverse $\phi f^{-1} \psi^{-1}$, which is also smooth, then the map f is said to be **diffeomorphism**, and we say that the manifolds M and N are **diffeomorphic** (to each other). If M and N are diffeomorphic, then we shall write $M \equiv N$. It should be clear that the notion of a diffeomorphism is an *equivalence relation*, and furthermore,

$$M \equiv N \Longrightarrow \dim(M) = \dim(N) \tag{5.12}$$

The notion of diffeomorphism bears the same relation to manifolds, as the notion of homeomorphism does to topological spaces. In other words, *any two diffeomorphic manifolds are considered to be equivalent; they are one and the same manifold for all practical purposes.*

5.6. Vectors

We now direct our attention towards defining geometric objects, called *vectors* or *tangent vectors*, on a general manifold M. We want to do this in an *intrinsic manner*, or in a way which does not rely on the assumption that M is embedded in R^n. We can do this by first defining a *curve* on a manifold. A **curve** α on an n-dimensional manifold M is a *smooth* map

$$\alpha : (-\varepsilon, \varepsilon) \longrightarrow M \tag{5.13}$$

where $(-\varepsilon, \varepsilon) \subseteq R$. As shown in Figure 5.7, the map α assigns to each real number $t \in (-\varepsilon, \varepsilon)$, a *unique* point $\alpha(t) \in M$. We must remember that the curve is to be identified with the *map* α itself, and not the collection of image points $\alpha(t) \in M$.

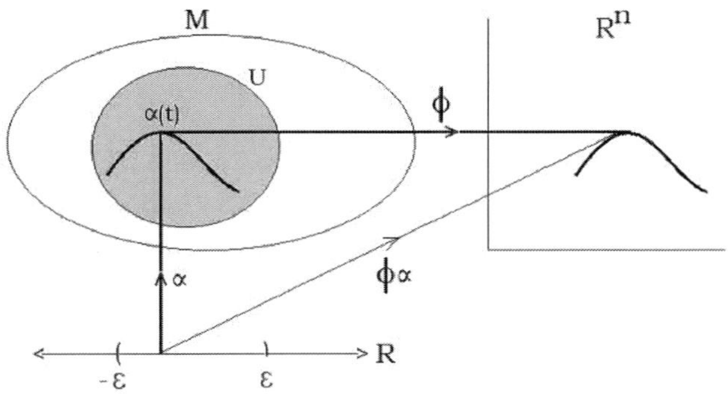

Figure 5.7. A curve α on the manifold M.

Let (U, ϕ) be a coordinate chart on M, and without loss of generality, assume that $\alpha(t) \in U$ for all $t \in (-\varepsilon, \varepsilon)$. From Figure 5.7, we see that the curve α has the *coordinate representation*

$$x^k = \phi^k(\alpha(t)) \qquad (k = 1, 2, 3, ..., n) \tag{5.14}$$

where ϕ^k denotes the k-th coordinate[2] of the point $\phi(\alpha(t)) \in R^n$.

When we say that the map α is **smooth**, we mean that *all* of the derivatives of the x^k with respect to t exist and are continuous. In particular, (real-valued) **function**[3] on M is a *smooth* map $f : M \longrightarrow R$, and we shall denote the set of smooth functions on M by $C^\infty(M)$.

The function f assigns a *unique* real number $f(p)$ to each point $p \in M$. If we employ the same chart as above, it's easy to see that the coordinate representation of f is given by $f\phi^{-1} : R^n \longrightarrow R$, or by,

$$y = f\phi^{-1}\left(x^1, x^2, ..., x^n\right) \equiv f\phi^{-1}(\mathbf{x}),$$

so that y is just a real-valued function of n-variables.

For a fixed $t_0 \in (-\varepsilon, \varepsilon)$, we can define the *tangent vector* at the point $\alpha(t_0)$ with the aid of the *directional derivative* of the real-valued function $f(\alpha(t))$ at $t = t_0$. From elementary analysis, this is given in terms of local coordinates by

$$\left.\frac{d}{dt} f(\alpha(t))\right|_{t=t_0} = \left.\frac{\partial f}{\partial x^k} \cdot \frac{dx^k}{dt}\right|_{t=t_0} \tag{5.15}$$

where in (5.15), we have taken the liberty of using the shorthand notation

$$\frac{\partial f}{\partial x^k} \equiv \left.\frac{\partial}{\partial x^k}\left(f\phi^{-1}(\mathbf{x})\right)\right|_{\phi(p)} \quad \text{and} \quad \left.\frac{dx^k}{dt}\right|_{t=t_0} \equiv \left.\frac{d}{dt}\left(\phi^k(\alpha(t))\right)\right|_{t=t_0}. \tag{5.16}$$

If we now define the *differential operator* **v** given by

$$\mathbf{v} = v^k \frac{\partial}{\partial x^k} \quad \text{where } v^k = \frac{d}{dt}\left(\phi^k(\alpha(t))\right)\Big|_{t=t_0} \tag{5.17}$$

then (5.15) may be rewritten as

$$\frac{d}{dt} f(\alpha(t))\Big|_{t=t_0} = v^k \frac{\partial f}{\partial x^k} \equiv \mathbf{v}[f] \tag{5.18}$$

We define the **tangent vector** to M at the point $p = \alpha(t_0)$ as the *differential operator* **v**, given in (5.17). *We see that a tangent vector acts on a smooth function f to give a real number $\mathbf{v}[f]$ which represents the rate of change of the function along the curve α at the point $\alpha(t_0)$.*

The set of *all* tangent vectors at $p \in M$ is called the **tangent space** of M at p, which we denote by $T_p(M)$. We can easily visualize $T_p(M)$ by imagining all of the possible smooth curves on M that pass through the point p, and then drawing their respective tangent vectors at p. For example, as shown in Figure 5.8, if $M = S^2$, then all of the vectors $\mathbf{v} \in T_p(M)$ lie in the plane which is tangent to the sphere at p.

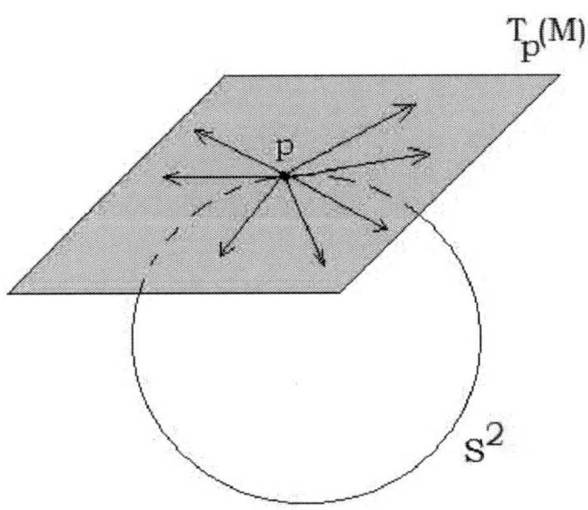

Figure 5.8. The tangent space $T_p(M)$ to the sphere $M = S^2$ at p.

The tangent space at any point $p \in M$ has a the structure of a **vector space**. Since, if $\mathbf{v}, \mathbf{w} \in T_p(M)$, and f is any smooth function on M, then we can define **vector addition** and **scalar multiplication** by

$$(\mathbf{v}+\mathbf{w})[f] = \mathbf{v}[f] + \mathbf{w}[f] \tag{5.19a}$$

$$(\lambda\mathbf{v})[f] = \lambda\mathbf{v}[f] \quad (\text{for all } \lambda \in R) \tag{5.19b}$$

Furthermore, it turns out that $T_p(M)$ has the *same dimension* as the given manifold M:

$$\dim T_p(M) = \dim M = n \tag{5.20}$$

In order to understand (5.20), let (U, ϕ) again be a coordinate chart containing the point $p \in M$. Using (5.16), we can then define the operators $\mathbf{e}_k$ at p by their action on any smooth function f:

$$\mathbf{e}_k\,[f] \equiv \left.\frac{\partial}{\partial x^k}\right|_p [f] = \frac{\partial}{\partial x^k}\left(f\phi^{-1}(\mathbf{x})\right)\Big|_{\phi(p)} \quad ; \quad (k = 1,2,3,...,n) \tag{5.21}$$

From (5.18), we see that for any vector $\mathbf{v} \in T_p(M)$, we can write

$$\mathbf{v} = v^k \mathbf{e}_k \tag{5.22}$$

and consequently, each vector $\mathbf{v}$ at p may be written as a *linear combination* of the $\mathbf{e}_k$. The quantities v^k are called the **components** of $\mathbf{v}$, *relative* to this basis. These quantities are *linearly independent*, since, if we assume that there exist $c^k \in R$ such that $\mathbf{v} = c^k\mathbf{e}_k = \mathbf{0}$, and we apply $\mathbf{v}$ to the m-th coordinate function $x^m(t) = \phi^m(\alpha(t))$, then we obtain

$$\mathbf{v}\,[x^m] = c^k\mathbf{e}_k\,[x^m] = c^k\frac{\partial}{\partial x^k}(x^m) = c^k\delta^m_k = c^k = 0 \tag{5.23}$$

Thus, the set of quantities $\{\mathbf{e}_1, \mathbf{e}_2, ..., \mathbf{e}_n\}$ forms a basis for $T_p(M)$ which is called a **coordinate basis** or **natural basis**.

A tangent vector $\mathbf{v} \in T_p(M)$ is a geometrical object, and therefore, it has an existence which is *independent* of the coordinate chart used to describe it. This simple observation permits us to determine how the components of $\mathbf{v}$ transform under a change of coordinates.

To see this, let (U,ϕ) and $(V,\bar{\phi})$ be overlapping charts on M, where $p \in U \bigcap V$. Furthermore, assume that $x^k = \phi^k(p)$ and $\bar{x}^k = \bar{\phi}^k(p)$. Then, we find that there are two equivalent expressions for $\mathbf{v} \in T_p(M)$,

$$\mathbf{v} = v^k\frac{\partial}{\partial x^k} \text{ and } \mathbf{v} = \bar{v}^m\frac{\partial}{\partial \bar{x}^m}\,. \tag{5.24}$$

Evidently, by the chain rule, the components of $\mathbf{v}$ must transform according to

$$\bar{v}^m = \frac{\partial \bar{x}^m}{\partial x^k}v^k \tag{5.25}$$

and we can even use (5.25) to *define* vectors, by saying that a **vector** (or **contravariant vector**) is any object whose components transform according to this rule. While such a definition is certainly useful, it is not coordinate free.

5.7. Dual Vector Spaces and One-forms

Let V be an n-dimensional vector space over the field of real numbers, and let $f : V \longrightarrow R$ be a *linear* function defined on V. Thus, f assigns a *unique* real number $f(\mathbf{v})$ to each vector $\mathbf{v} \in V$, and it does this in a linear fashion; that is

$$\begin{aligned} f(\mathbf{v}+\mathbf{w}) &= f(\mathbf{v}) + f(\mathbf{w}) \\ f(\lambda\mathbf{v}) &= \lambda f(\mathbf{v}) \end{aligned} \tag{5.26}$$

for all $\mathbf{v}, \mathbf{w} \in V$, and for all $\lambda \in R$. If the collection $\{\mathbf{e}_1, \mathbf{e}_2, ..., \mathbf{e}_n\}$ is a basis for V, then by (5.26)

$$f(\mathbf{v}) = f(v^k \mathbf{e}_k) = v^k f(\mathbf{e}_k), \tag{5.27}$$

and consequently, the action of f on any vector $\mathbf{v} \in V$ is completely determined by the collection of n real numbers $f(\mathbf{e}_k)$.

Now, let V^* denote the set of all such real-valued linear functions on V. It's easy to see that V^* is also a *vector space* in its own right, where the operations of addition, and scalar multiplication are given by

$$\begin{aligned} (f_1 + f_2)(\mathbf{v}) &= f_1(\mathbf{v}) + f_2(\mathbf{v}) \\ (\lambda f)(\mathbf{v}) &= \lambda f(\mathbf{v}) \end{aligned} \tag{5.28}$$

for all $f, f_1, f_2 \in V^*$, $\mathbf{v} \in V$, and $\lambda \in R$. We call this vector space the **dual vector space** V^* to V, and its elements are called **dual vectors**, **covectors**, **covariant vectors**, or **one-forms**.

Motivated by (5.27), we can define a basis on the dual space V^* by simply specifying how the elements of this basis shall act on the original basis vectors $\{\mathbf{e}_k\}$ of V. In particular, we can *define* a special type of basis $\{\omega^1, \omega^2, ..., \omega^n\}$ on V^*, called the **dual basis**, by requiring that

$$\omega^k(\mathbf{e}_m) = \delta^k_m \tag{5.29}$$

Consequently, for an arbitrary $f \in V^*$, there must exist real numbers $f_1, f_2, ..., f_n$ such that

$$f = f_i \omega^i \,, \tag{5.30}$$

and so, for any $\mathbf{v} \in V$ we have

$$f(\mathbf{v}) = f_i \omega^i \left(v^k \mathbf{e}_k\right) = f_i v^k \omega^i(\mathbf{e}_k) = f_i v^k \delta^i_k = f_i v^i \tag{5.31}$$

From (5.31), we see that the action of f on any vector $\mathbf{v}$ may be viewed as an **inner product** $\langle \,, \rangle : V^* \times V \longrightarrow R$, where

$$f(\mathbf{v}) = \langle f, \mathbf{v} \rangle \equiv f_i v^i \tag{5.32}$$

Observe that this inner product is defined only for a one-form and a vector, and not for two vectors.

After this small digression, let us again consider a differentiable manifold M. Given any point $p \in M$, we know that $T_p(M)$ is a vector space, and consequently, there must exist a vector space which is dual to $T_p(M)$, called the **cotangent space** at p. We denote this new space by $T^*_p(M)$, and each element $\omega \in T^*_p(M)$ is a linear function which sends each vector $\mathbf{v} \in T_p(M)$ to a unique real number. A typical element of the dual space $T^*_p(M)$ is called a **covector**, **covariant vector**, or a **one-form**.

Given a smooth function $f \in C^\infty(M)$, its *differential* df is again a smooth function, which turns out to be one-form. In fact, we define the action of the differential df on any vector $\mathbf{v} \in T_p(M)$ by

$$df(\mathbf{v}) \equiv \langle df, \mathbf{v} \rangle \equiv \mathbf{v}[f] = v^k \frac{\partial f}{\partial x^k} \in R \tag{5.33}$$

Thus, df assigns the real number $df(\mathbf{v})$ to each vector $\mathbf{v} \in T_p(M)$. Also, from (5.19), we conclude that df is *linear* as well, so $df \in T_p^*(M)$, as claimed.

On the other hand, if (U, ϕ) is a coordinate chart containing the point $p \in M$, so that $x^k = \phi^k(p)$, we know from elementary analysis that df must have the coordinate representation

$$df = \frac{\partial f}{\partial x^k} dx^k \tag{5.34}$$

Since df can be written as a *linear combination* of the elements in the set $\{dx^k\}$, we suspect that this set might provide a basis for $T_p^*(M)$.

In fact, these elements form a *dual basis* for the cotangent space.

To see this, let $\mathbf{v} = \delta_p^m \mathbf{e}_m$ where m is *fixed*, so that the p-th component of $\mathbf{v}$ is unity, and all of its other components vanish. Then, by (5.33), we can write

$$dx^k(\mathbf{v}) \equiv \mathbf{v}\left[x^k\right] = \delta_p^m \frac{\partial x^k}{\partial x^m} = \delta_p^m \delta_m^k = \delta_p^k, \tag{5.35}$$

or in other words,

$$\left\langle dx^k, \frac{\partial}{\partial x^p} \right\rangle = \delta_p^k \tag{5.36}$$

Thus, the set of coordinate differentials $\{dx^1, dx^2, ..., dx^n\}$ comprises a dual basis for the cotangent space $T_p^*(M)$. As a result of our construction, we should note that *each coordinate differential is a one-form in its own right.*

Since any one-form may be written as $\omega = \omega_k dx^k$, then as in the case of vector spaces, we can define the inner product

$$\langle\ ,\rangle : T_p^*(M) \times T_p(M) \longrightarrow R \tag{5.37}$$

by

$$\begin{aligned} \langle \omega, \mathbf{v} \rangle &= \left\langle \omega_k dx^k, v^m \frac{\partial}{\partial x^m} \right\rangle = \omega_k v^m \left\langle dx^k, \frac{\partial}{\partial x^m} \right\rangle \\ &= \omega_k v^m \delta_m^k = \omega_k v^k \end{aligned} \tag{5.38}$$

We must remember that this inner product is only defined on *ordered* pairs of one-forms and vectors.

Given an arbitrary one-form $\omega = \omega_k dx^k$, let's see how its **components** ω^k transform under a change of coordinates. Let (U, ϕ) and $(V, \bar{\phi})$ be overlapping charts on M, where $p \in U \bigcap V$. Furthermore, let $x^k = \phi^k(p)$ and $\bar{x}^k = \bar{\phi}^k(p)$. Then, we have both

$$\omega = \omega_k dx^k \text{ and } \omega = \bar{\omega}_m d\bar{x}^m. \tag{5.39}$$

But, under the resulting coordinate transformation $\bar{x}^m = \bar{x}^m(x^k)$, we also can write

$$d\bar{x}^m = \frac{\partial \bar{x}^m}{\partial x^k} dx^k \tag{5.40}$$

Inserting (40) into (39), and simplifying, we find

$$\left(\omega^k - \frac{\partial \bar{x}^m}{\partial x^k} \bar{\omega}_m\right) dx^k = 0. \tag{5.41}$$

Recalling that the quantities in the set $\{dx^k\}$ are *linearly independent*, we have

$$\omega_k = \frac{\partial \bar{x}^m}{\partial x^k} \bar{\omega}_m \tag{5.42}$$

Finally, if we interchange barred and unbarred quantities, we see that under the coordinate transformation $\bar{x}^m = \bar{x}^m(x^k)$, the components of any one-form ω must transform according to the rule

$$\bar{\omega}_k = \frac{\partial x^m}{\partial \bar{x}^k} \omega_m \tag{5.43}$$

5.8. Exercises

1. Show that the one-forms $\{dx^k\}$ comprise a linearly independent set.

(Solution)

Assume that there exist constants $c_1, c_2, ..., c_n$ such that

$$c_k dx^k = c_1 dx^1 + c_2 dx^2 + ... + c_n dx^n = 0$$

Then by (5.36), we have

$$c_k dx^k \left(\frac{\partial}{\partial x^m} \right) = c_k \delta^k_m = c_m = 0$$

2. *(Derivation)* A **derivation** at a point $p \in M$ is a map

$$\mathbf{v}_p : C^\infty(M) \longrightarrow R$$

such that $\mathbf{v}_p$ is *linear*, i.e.,

$$\mathbf{v}_p(f+g) = \mathbf{v}_p(f) + \mathbf{v}_p(g) \quad \text{for all } f, g \in C^\infty(M) \tag{5.44a}$$

$$\mathbf{v}_p(\lambda f) = \lambda \mathbf{v}_p(f) \quad \text{for all } f \in C^\infty(M), \text{ and } \lambda \in R \tag{5.44b}$$

and which satisfies the **Leibnitz rule**:

$$\mathbf{v}_p(fg) = f\mathbf{v}_p(g) + g\mathbf{v}_p(f) \tag{5.45}$$

Show that any vector $\mathbf{v} \in T_p(M)$ is a *derivation* at the point $p \in M$.

(Solution)

Assume that $\mathbf{v} \in T_p(M)$ and simply define $\mathbf{v}(fg) \equiv \mathbf{v}[fg]$. Now, by (5.17) and (5.18), we have

$$\mathbf{v}[f+g] = v^k \frac{\partial}{\partial x^k}(f+g) = v^k \frac{\partial f}{\partial x^k} + v^k \frac{\partial g}{\partial x^k} = \mathbf{v}[f] + \mathbf{v}[g]$$

$$\mathbf{v}[\lambda f] = v^k \frac{\partial}{\partial x^k}(\lambda f) = \lambda \left(v^k \frac{\partial f}{\partial x^k} \right) = \lambda \mathbf{v}[f]$$

so $\mathbf{v}$ is linear. Similarly, we can write

$$\begin{aligned}\mathbf{v}[fg] &= v^k\frac{\partial}{\partial x^k}(fg) = v^k\left(f\frac{\partial g}{\partial x^k} + g\frac{\partial f}{\partial x^k}\right) = f\left(v^k\frac{\partial g}{\partial x^k}\right) + g\left(v^k\frac{\partial f}{\partial x^k}\right) \\ &= f\mathbf{v}[g] + g\mathbf{v}[f]\end{aligned}$$

so that $\mathbf{v}$ satisfies the Leibnitz rule as well.

3. Assume that the vectors $\{\mathbf{e}_1, \mathbf{e}_2, ..., \mathbf{e}_n\}$ form a basis for $T_p(M)$. If $A = (A^p_q)$ is any non-singular $n \times n$ matrix, then show that the n-vectors $\mathbf{b}_k = A^p_k\mathbf{e}_p$ also form a basis for $T_p(M)$.

(Solution)

It suffices to show that the vectors $\{\mathbf{b}_k\}$ are linearly independent. So, assume there exist constants c^k such that

$$c^k\mathbf{b}_k = c^1\mathbf{b}_1 + c^2\mathbf{b}_2 + ... + c^n\mathbf{b}_n = \mathbf{0}\,.$$

This implies that

$$c^k\mathbf{b}_k = \left(c^kA^p_k\right)\mathbf{e}_p = \mathbf{0} \Longrightarrow c^kA^p_k = 0$$

where the last implication follows from the assumed linear independence of the vectors $\{\mathbf{e}_1, \mathbf{e}_2, ..., \mathbf{e}_n\}$. Since A is nonsingular, let $B = A^{-1}$, and multiply both sides of the last equation by B^m_p to obtain

$$c^k\left(B^m_pA^p_k\right) = c^k\delta^m_k = c^m = \mathbf{0}$$

as required.

5.9. Change of Basis

From the previous worked example, it is apparent that the coordinate basis $\{\mathbf{e}_1, \mathbf{e}_2, ..., \mathbf{e}_n\}$ for $T_p(M)$ is not unique, and consequently, it is important to consider how the components of any tangent vector $\mathbf{v} \in T_p(M)$ transform under an arbitrary change of basis.

To accomplish this task, assume that $\{\bar{\mathbf{b}}_k\}$ and $\{\mathbf{b}_k\}$ are *any* two bases for the tangent space $T_p(M)$ at the point p. Since by hypothesis, each $\bar{\mathbf{b}}_k \in T_p(M)$,then by the very definition of basis, there must exist n^2-constants A^m_p such that

$$\bar{\mathbf{b}}_p = A^m_p\mathbf{b}_m \tag{5.46}$$

Conversely, there must exist n^2-constants B^q_m such that

$$\mathbf{b}_m = B^q_m\bar{\mathbf{b}}_q \tag{5.47}$$

Substituting (5.47) into (5.46) gives

$$\bar{\mathbf{b}}_p = \left(B^q_mA^m_p\right)\bar{\mathbf{b}}_q \Longrightarrow (\delta^q_p - B^q_mA^m_p)\bar{\mathbf{b}}_q = \mathbf{0} \tag{5.48}$$

But since $\{\bar{\mathbf{b}}_q\}$ is a linearly independent set of vectors, (5.48) implies that

$$B^q_m A^m_p = \delta^q_p \tag{5.49}$$

or, in the language of square matrices, $B = A^{-1}$. *Thus, we see that general basis transformations can only be effected by non-singular matrices.*

Now, consider an arbitrary vector $\mathbf{v} \in T_p(M)$. Under the above change of basis, we must have

$$\mathbf{v} = v^a \mathbf{e}_a = \bar{v}^b \bar{\mathbf{e}}_b \tag{5.50}$$

Substituting (5.46) into (5.50), we find that

$$v^a \left(B^b_a \bar{\mathbf{e}}_b \right) = \bar{v}^b \bar{\mathbf{e}}_b \Longrightarrow \left(B^b_a v^a - \bar{v}^b \right) \bar{\mathbf{e}}_b = \mathbf{0} \tag{5.51}$$

Recalling that $B = A^{-1}$, (5.51) shows that under the change of basis which is performed by (5.46), the components of v must transform according to the rule

$$\bar{v}^b = (A^{-1})^b_a v^a \tag{5.52}$$

We say that the components of $\mathbf{v}$ transform *contravariantly*, and this is why tangent vectors are also called **contravariant vectors**.

The transformation rule given in (5.52) is a generalization of the special rule given in (5.25) for *coordinate bases*. To see this, let (U, ϕ) and $(V, \bar{\phi})$ be overlapping charts on M, where $p \in U \bigcap V$, and assume that the corresponding coordinates of p are given by $x^k = \phi^k(p)$ and $\bar{x}^k = \bar{\phi}^k(p)$, respectively. Also, assume that $\mathbf{b}_k = \partial/\partial x^k$ and $\bar{\mathbf{b}}_k = \partial/\partial \bar{x}^k$ are both coordinate bases for $T_p(M)$.

Given any function $f \in C^\infty(M)$, the chain rule tells us that

$$\frac{\partial f}{\partial x^k} = \frac{\partial \bar{x}^m}{\partial x^k} \frac{\partial f}{\partial \bar{x}^m} \tag{5.53}$$

Since f is arbitrary, the last equation implies that

$$\frac{\partial}{\partial x^k} = \frac{\partial \bar{x}^m}{\partial x^k} \frac{\partial}{\partial \bar{x}^m} \tag{5.54}$$

If we compare (5.54) with (5.47), we can make the identifications

$$B^m_k \equiv \left(A^{-1}\right)^m_k = \frac{\partial \bar{x}^m}{\partial x^k} \tag{5.55}$$

so (5.52) and (5.55) yield

$$\bar{v}^m = \frac{\partial \bar{x}^m}{\partial x^k} v^k \tag{5.56}$$

which is the rule we previously obtained for the transformation of vector components under a change of *coordinate* bases.

Let us again assume a change of basis in $T_p(M)$ which is given by (5.46), and we now determine how the corresponding *dual bases* in the cotangent space $T^*_p(M)$ are related. We shall denote these dual bases by the sets of *covectors* $\left\{\beta^k\right\}$ and $\left\{\bar{\beta}^k\right\}$ respectively, and in keeping with (5.29), we shall require that

$$\beta^k(\mathbf{b}_m) = \delta^k_m \ \text{ and } \ \bar{\beta}^k(\bar{\mathbf{b}}_m) = \delta^k_m \ . \tag{5.57}$$

From (5.46) and (5.57), we can write

$$\begin{aligned} \delta^k_p &= \bar{\beta}^k(\bar{\mathbf{b}}_p) = A^m_p\, \bar{\beta}^k(\mathbf{b}_m) \\ &\Longrightarrow \bar{\beta}^k(\mathbf{b}_m) = (A^{-1})^k_p \end{aligned} \tag{5.58}$$

But, (5.58) may be rewritten as,

$$\begin{aligned} \bar{\beta}^k(\mathbf{b}_m) &= (A^{-1})^k_r\, \delta^r_m = (A^{-1})^k_r\, \beta^r(\mathbf{b}_m) \\ &\Longrightarrow \left(\bar{\beta}^k - (A^{-1})^k_r\, \beta^r\right)(\mathbf{b}_m) = 0 \end{aligned} \tag{5.59}$$

We conclude that under a general change of basis as given in (5.46), in the tangent space $T_p(M)$, there corresponds a change of basis in the cotangent space $T^*_p(M)$ which is given by

$$\bar{\beta}^k = (A^{-1})^k_r \beta^r \tag{5.60}$$

For an arbitrary covector (or one-form) $\omega \in T^*_p(M)$ we must have both

$$\omega = \omega_k \beta^k \ \text{ and } \ \omega = \bar{\omega}_k \bar{\beta}^k \ , \tag{5.61}$$

so that by (5.60), we have

$$\begin{aligned} \bar{\omega}_k\left((A^{-1})^k_r \beta^r\right) &= \omega_k \beta^k = \omega_k \delta^r_k \beta^r \\ &\Longrightarrow \left(\bar{\omega}_k (A^{-1})^k_r - \omega_r\right)\beta^r = 0 \end{aligned} \tag{5.62}$$

and consequently,

$$(A^{-1})^k_r\, \bar{\omega}_k = \omega_r \tag{5.63}$$

Multiplying both sides of the last equation by A^r_p gives

$$\begin{aligned} (A^{-1})^k_r A^r_p\, \bar{\omega}_k &= \delta^k_p \bar{\omega}_k = A^r_p \omega_r \\ &\Longrightarrow \bar{\omega}_p = A^r_p \omega_r \end{aligned} \tag{5.64}$$

We see that (5.64) is the rule which governs the transformation of the components of any covector ω under a general change of basis. We say that the components of a covector (or one-form) transform *covariantly* under a change of basis. Consequently, covectors (and one-forms) are sometimes called **covariant vectors**.

Once again, let's now assume that (U,ϕ) and $(V,\bar{\phi})$ are overlapping charts containing the point $p \in U \bigcap V$, and that the coordinates of p are given by $x^k = \phi^k(p)$ and $\bar{x}^k = \bar{\phi}^k(p)$, respectively. Also, suppose that $\beta^k = dx^k$ and $\bar{\beta}^k = d\bar{x}^k$ are both *coordinate* bases for $T^*_p(M)$.

For an arbitrary covector $\omega \in T_p^*(M)$, we must have both

$$\omega = \omega_k dx^k \text{ and } \bar{\omega} = \bar{\omega}_k d\bar{x}^k \tag{5.65}$$

At the same time, under the corresponding coordinate transformation $\bar{x}^p = \bar{x}^p(x^k)$, the definition of differential implies that

$$d\bar{x}^p = \frac{\partial \bar{x}^p}{\partial x^k} dx^k \,. \tag{5.66}$$

Comparing (5.66) with (5.60), we see that

$$(A^{-1})^k_{\ r} = \frac{\partial \bar{x}^k}{\partial x^r} \,, \tag{5.67}$$

which implies

$$A^r_{\ p} = \frac{\partial x^r}{\partial \bar{x}^p} \tag{5.68}$$

Finally, inserting (5.68) into (5.64) gives the familiar result

$$\bar{\omega}_p = \frac{\partial x^r}{\partial \bar{x}^p} \omega_r \tag{5.69}$$

which is the same as (5.43). It is important to note that (5.69) holds only under a change of *coordinate* bases, and is a special case of (5.64).

5.10. Maps of Manifolds

Let M and N be manifolds, and let $F : M \longrightarrow N$ be a smooth map from M into N. It turns out that the smooth map F naturally induces a a map, called the **tangent map**, between the tangent spaces in each manifold.

In order to see this, let $\alpha : (-\varepsilon, \varepsilon) \longrightarrow M$ be a smooth curve on M, such that $p_0 = \alpha(0)$ is a fixed point on α. Under the map F, the curve α is mapped to another curve $\bar{\alpha}$ which passes through the point $F(p_0) \in N$. In fact, this image curve is given by

$$\bar{\alpha}(t) = (F\alpha)(t) \tag{5.70}$$

where $t \in (-\varepsilon, \varepsilon)$.

At the same time, as shown in Figure 5.9, the vector $\mathbf{v} \in T_{p_0}(M)$ is mapped to the tangent vector $\bar{\mathbf{v}} \in T_{F(p_0)}(M)$. We say that the tangent vector $\bar{\mathbf{v}}$ is the **push-forward** of the vector $\mathbf{v}$ under F, and write

$$\bar{\mathbf{v}} = F_*(\mathbf{v}) \tag{5.71}$$

We call F_* the **tangent map**, and we see that for *any* point $p \in M$, the *induced map* F_* maps the tangent space $T_p(M)$ into the corresponding tangent space $T_{F(p)}(N)$.

Now, if $g \in C^\infty(N)$ is any smooth function on N, observe that

$$\bar{\mathbf{v}}[g] = \frac{d}{dt}(g\bar{\alpha})(t)\bigg|_{t=0} = \frac{d}{dt}(gF\alpha)(t)\bigg|_{t=0} = \mathbf{v}[gF] \tag{5.72}$$

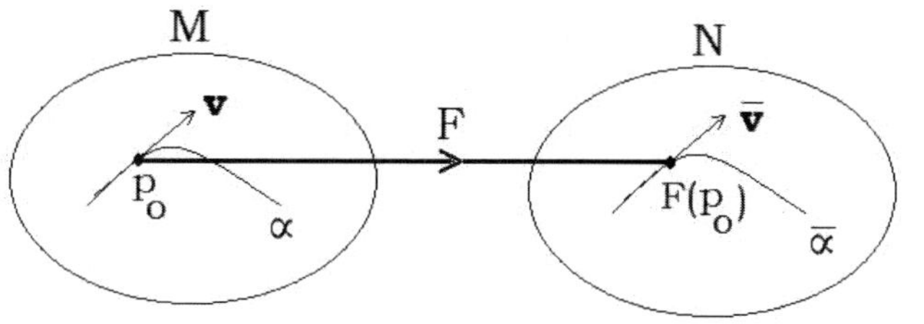

Figure 5.9. F induces a map between $T_p(M)$ and $T_{F(p)}(N)$.

Motivated by (5.72), we are led to define the tangent map F_* by

$$(F_*\mathbf{v})\,[g] = \mathbf{v}\,[gF] \tag{5.73}$$

for any smooth function $g \in C^\infty(N)$, and for any vector $\mathbf{v} \in T_p(M)$. Once again, we must emphasize that while $F : M \to N$, the induced mapping $F_* : T_p(M) \longrightarrow T_{F(p)}(N)$.

Let us now determine the coordinate representation of the tangent map in (5.73). To accomplish this task, let (U, ϕ) be a coordinate neighborhood of the point $p \in M$, and let $(\bar{U}, \bar{\phi})$ be a coordinate neighborhood of the corresponding point $F(p) \in N$. Also, let $x^k = \phi^k(p)$ and $\bar{x}^k = (\bar{\phi}F)^k(p)$ represent the coordinates of the points p and $F(p)$, respectively. Observe that

$$\begin{aligned}
(F_* v)^k\Big|_{F(p)} &= \bar{v}^k\big|_{F(p)} = \frac{d\bar{x}^k}{dt}\Big|_{F(p)} \\
&= \frac{d}{dt}\,(\bar{\phi}F)^k\,(p)\Big|_{F(p)} = \frac{d}{dt}\,(\bar{\phi}F\phi^{-1})\,(x^k)\Big|_p \\
&= \frac{\partial \bar{x}^k}{\partial x^l}\cdot\frac{dx^l}{dt}\Big|_p \\
&= \frac{\partial \bar{x}^k}{\partial x^l}\cdot v^l\big|_p
\end{aligned}$$

Thus, we conclude that the coordinate representation of the tangent mapping is given by

$$(F_*\mathbf{v})^k\big|_{F(p)} = \frac{\partial \bar{x}^k}{\partial x^l}\,v^l\Big|_p \tag{5.74}$$

Observe that the coefficient $\partial\bar{x}^k/\partial x^l$ which appears in (5.74) is just the *Jacobian matrix* of the mapping $\bar{\mathbf{x}} = F(\mathbf{x})$. In particular, if $\dim(M) = m$ and $\dim(N) = n$, then the coefficients $\partial\bar{x}^k/\partial x^l$ comprise the elements of an $n \times m$ Jacobian matrix.

For the special case $M = N$, if we take F equal to the identity map, then the Jacobian matrix becomes a *square* matrix, and (5.74) reduces to the usual transformation rule for the components of a tangent (or contravariant) vector under a change of coordinates.

Now, the smooth map $F : M \longrightarrow N$ also induces another type of map, called the **dual tangent map**, between the *cotangent* spaces:

$$F^* : T^*_{F(p)}(N) \longrightarrow T^*_p(M)$$

such that

$$(F^*\omega)(\mathbf{v}) = \omega(F_*\mathbf{v}) \tag{5.75}$$

for any one-form $\omega \in T^*_{F(p)}(N)$, and for all tangent vectors $\mathbf{v} \in T_p(M)$. Observe that the dual tangent mapping *works backwards*; it takes each one-form $\omega \in T^*_{F(p)}(N)$, which lives in the cotangent space at the point $F(p) \in N$, and maps it to a one-form $F^*\omega \in T^*_p(M)$, which lives at the point $p \in M$. For this reason, $F^*\omega$ is often called the **pullback** of ω under F.

We already know that for any smooth function $f \in C^\infty(N)$, that its differential df is a one-form on N. So it makes sense to look at the pullback of df under F^*. For any $\mathbf{v} \in T_p(M)$, by (5.75), we can write

$$\begin{aligned}
(F^*df)(\mathbf{v}) &= df(F_*\mathbf{v}) \\
&= (F_*\mathbf{v})[f] \quad [\text{ by } (5.33)\ ] \\
&= \mathbf{v}[Ff] \quad [\text{ by } (5.73)\ ] \\
&= d(Ff)(\mathbf{v})
\end{aligned}$$

Thus, we obtain the useful formula

$$(F^*df)(\mathbf{v}) = d(Ff)(\mathbf{v}) \tag{5.76}$$

We now determine the coordinate representation of the pullback operation. So, let (U, ϕ) be a coordinate neighborhood of the point $p \in M$, and assume $(\bar{U}, \bar{\phi})$ is a coordinate neighborhood of $F(p) \in N$. Also, let $x^k = \phi^k(p)$ and $\bar{x}^k = (\bar{\phi}F)^k(p)$ be the coordinates of the points p and $F(p)$, respectively. Assume further that

$$F^*\omega = \omega_m dx^m \in T^*_p(M) \ \text{ and } \bar{\omega} = \bar{\omega}_k d\bar{x}^k \in T^*_p(N)$$

Consequently, for any vector $\mathbf{v} \in T_p(M)$, we can write

$$\begin{aligned}
\omega_m v^m &= (F^*\omega)(\mathbf{v}) = \omega(F_*\mathbf{v}) = \langle \omega, F_*\mathbf{v} \rangle \\
&= \bar{\omega}_k \ (F_*\mathbf{v})^k \\
&= \bar{\omega}_k \frac{\partial \bar{x}^k}{\partial x^m} v^m \ [\text{ by } (5.74)\ ]
\end{aligned}$$

Since $\mathbf{v}$ is arbitrary, we must have

$$\omega_m = \frac{\partial \bar{x}^k}{\partial x^m} \bar{\omega}_k \tag{5.77}$$

and so, in the language of pullbacks, we have

$$(F^*\omega)_m \mid_p = \frac{\partial \bar{x}^k}{\partial x^m} \bar{\omega}_k \mid_{F(p)} \tag{5.78}$$

which is the relation we originally sought.

5.11. Tensors

We now turn our attention to consider important types of multi- linear mappings that can be defined at each point of a manifold. These special types of multilinear mappings are called *tensors*.

Consider a differentiable manifold M, where $\dim M = n$, and a point $p \in M$. We shall say that $\mathbf{V}$ is a **tensor of type** $(1,0)$ at p, if $\mathbf{V}$ is a *linear mapping* which assigns a *unique* real number to each one- form $\omega \in T_p^*(M)$. More specifically, we require that $\mathbf{V}$ is linear in the sense that

$$\begin{aligned} \mathbf{V}(\omega+\sigma) &= \mathbf{V}(\omega)+\mathbf{V}(\sigma) \\ \mathbf{V}(c\omega) &= c\mathbf{V}(\omega) \end{aligned} \tag{5.79}$$

for all one-forms $\omega, \sigma \in T_p^*(M)$, and for any real number c. It should be clear that any contravariant vector (or tangent vector) $\mathbf{V} \in T_p(M)$, is a tensor of type $(1,0)$ at p.

Next, we shall say that $\mathbf{T}$ is a **tensor of type** $(2,0)$ at p, if $\mathbf{T}$ assigns a *unique* real number $\mathbf{T}(\omega,\sigma)$ to each *ordered* pair of one-forms $\omega, \sigma \in T_p^*(M)$, and it does so in a *bilinear* fashion; that is, $\mathbf{T}$ is linear in each of its arguments:

$$\begin{aligned} \mathbf{T}(c_1\omega_1+c_2\omega_2,\sigma) &= c_1\mathbf{T}(\omega_1,\sigma)+c_2\mathbf{T}(\omega_2,\sigma) \\ \mathbf{T}(\omega,c_1\sigma_1+c_2\sigma_2) &= c_1\mathbf{T}(\omega,\sigma_1)+c_2\mathbf{T}(\omega,\sigma_2) \end{aligned} \tag{5.80}$$

Now, the collection of all such bilinear maps $\mathbf{T} : T_p^*(M) \times T_p^*(M) \longrightarrow R$ is called the **tensor product space** $\otimes^2 T_p(M)$; it is simply the collection of all tensors of type $(2,0)$ at p. The tensor product space is a *vector space* in its own right, with the operations of addition and scalar multiplication defined by

$$\begin{aligned} (\mathbf{T}_1+\mathbf{T}_2)(\omega,\sigma) &= \mathbf{T}_1(\omega,\sigma)+\mathbf{T}_2(\omega,\sigma) \\ (c\mathbf{T})(\omega,\sigma) &= c(\mathbf{T}(\omega,\sigma)) \quad (\text{for all } c \in R) \end{aligned} \tag{5.81}$$

In particular, if we are given two tangent vectors $\mathbf{u}, \mathbf{v} \in T_p(M)$, we can define a special type of bilinear mapping $\mathbf{u}\otimes\mathbf{v} : T_p^*(M) \times T_p^*(M) \longrightarrow R$ by

$$(\mathbf{u}\otimes\mathbf{v})(\omega,\sigma) = \omega(\mathbf{u})\,\sigma(\mathbf{v}) \tag{5.82}$$

for all one-forms $\omega, \sigma \in T_p^*(M)$. We call $\mathbf{u}\otimes\mathbf{v}$ the **tensor product** of $\mathbf{u}$ and $\mathbf{v}$, and it should be clear that $\mathbf{u}\otimes\mathbf{v} \in \otimes^2 T_p(M)$.

Now, if the set of one-forms $\{\beta_k\}$ is a basis for $T_p^*(M)$, and if we expand $\omega, \sigma \in T_p^*(M)$ in terms of this basis as $\omega = \omega_a\beta^a$ and $\sigma = \sigma_b\beta^b$ respectively, then, owing to (5.80), we can write

$$\mathbf{T}(\omega,\sigma) = \mathbf{T}\left(\omega_a\beta^a, \sigma_b\beta^b\right) = \mathbf{T}\left(\beta^a,\beta^b\right)\,\omega_a\sigma_b \equiv T^{ab}\,\omega_a\sigma_b \tag{5.83}$$

where $T^{ab} = \mathbf{T}\left(\beta^a, \beta^b\right)$ are called the **components** of **T** relative to this basis. On the other hand, if the set $\{\mathbf{b}_k\}$ forms a basis for $T_p(M)$, then we also have

$$\omega(\mathbf{b}_a) = \omega_b \beta^b(\mathbf{b}_a) = \omega_b \delta^b_a = \omega_a \quad (5.84a)$$

$$\sigma(\mathbf{b}_b) = \sigma_a \beta^a(\mathbf{e}_b) = \sigma_a \delta^b_a = \sigma_b \quad (5.84b)$$

If we substitute equations (5.84) into (5.83), we obtain

$$\mathbf{T}(\omega, \sigma) = T^{ab}\,\omega_a \sigma_b = T^{ab} \omega(\mathbf{b}_a)\sigma(\mathbf{b}_b) = T^{ab}(\mathbf{b}_a \otimes \mathbf{b}_b)(\omega, \sigma)$$

or, in other words,

$$\mathbf{T} = T^{ab}\,\mathbf{b}_a \otimes \mathbf{b}_b \quad (5.85)$$

Thus, any tensor $\mathbf{T} \in \otimes^2 T_p(M)$ can be written as a linear combination of the tensor products of basis vectors, $\mathbf{b}_a \otimes \mathbf{b}_b$. It is easy to see that such products are *linearly independent*, since if there exist n^2 constants k^{ab} such that $k^{ab}\mathbf{b}_a \otimes \mathbf{b}_b = \mathbf{0}$, then, in particular, we must have

$$k^{ab}\mathbf{b}_a \otimes \mathbf{b}_b\left(\beta^c, \beta^d\right) = k^{ab}\beta^c(\mathbf{b}_a)\beta^d(\mathbf{b}_b) = k^{ab}\delta^c_a\delta^d_b = k^{cd} = 0$$

So, we see that the set of tensor products of basis vectors $\{\mathbf{b}_a \otimes \mathbf{b}_b\}$ forms a *basis* for the vector space $\otimes^2 T_p(M)$, and $\dim \otimes^2 T_p(M) = n^2$.

Now, let's determine how the components T^{ab} of **T** transform under a change of coordinate basis. Assume that (U, ϕ) and $(V, \bar{\phi})$ are overlapping charts containing the point $p \in U \bigcap V$, and the coordinates of p are given by $x^k = \phi^k(p)$ and $\bar{x}^k = \bar{\phi}^k(p)$, respectively. Also, suppose that $\{dx^k\}$ and $\{d\bar{x}^k\}$ are *coordinate* bases for $T^*_p(M)$.

By the definition of tensor components and equation (5.66), we have:

$$\bar{T}^{ab} = \mathbf{T}(d\bar{x}^a, d\bar{x}^b) = \mathbf{T}\left(\frac{\partial \bar{x}^a}{\partial x^k}dx^k, \frac{\partial \bar{x}^b}{\partial x^l}dx^l\right) = \frac{\partial \bar{x}^a}{\partial x^k}\frac{\partial \bar{x}^b}{\partial x^l}\,\mathbf{T}(dx^k, dx^l)$$

$$= \frac{\partial \bar{x}^a}{\partial x^k}\frac{\partial \bar{x}^b}{\partial x^l}\,T^{kl}$$

We conclude that, under the corresponding coordinate transformation $\bar{x}^p = \bar{x}^p(x^k)$, the components of **T** must transform according to

$$\bar{T}^{ab} = \frac{\partial \bar{x}^a}{\partial x^k}\frac{\partial \bar{x}^b}{\partial x^l}\,T^{kl}, \quad (5.86)$$

and furthermore, this law must apply to any tensor of type $(2,0)$.

We can easily generalize the above results by defining a **tensor T of type** $(r,0)$ at $p \in M$ as a *multilinear mapping*

$$\mathbf{T} : T^*_p(M) \times T^*_p(M) \times \ \ldots \times \ T^*_p(M) \longrightarrow R$$

which maps each ordered r-tuple of one-forms $(\omega_1, \omega_2, ..., \omega_r)$ into a unique real number. When we say that T is *multilinear*, we mean that it is *linear in each one of its arguments*.

The tensor product space $\otimes^r T_p(M)$ is now the tensor product of $T_p(M)$ with itself, taking r factors:

$$\otimes^r T_p(M) = T_p(M) \otimes T_p(M) \otimes \ ... \ \otimes T_p(M) \quad (r \text{ factors}) \tag{5.87}$$

Each element $\mathbf{T} \in \otimes^r T_p(M)$ is a tensor of type $(r,0)$, and $\otimes^r T_p(M)$ is a vector space with the operations of addition and scalar multiplication given by

$$\begin{aligned} (\mathbf{T}_1 + \mathbf{T}_2)(\omega_1, \omega_2, ..., \omega_r) &= \mathbf{T}_1(\omega_1, \omega_2, ..., \omega_r) + \mathbf{T}_2(\omega_1, \omega_2, ..., \omega_r) \\ (c\mathbf{T})(\omega_1, \omega_2, ..., \omega_r) &= c\mathbf{T}(\omega_1, \omega_2, ..., \omega_r)) \quad (\text{ for all } c \in R\ ) \end{aligned} \tag{5.88}$$

Following the same argument as above, it's easy to show that each $\mathbf{T} \in \otimes^r T_p(M)$ may be written in the form

$$\mathbf{T} = T^{k_1 k_2 \ldots k_r} \mathbf{b}_{k_1} \otimes \mathbf{b}_{k_2} \otimes \ldots \otimes \mathbf{b}_{k_r} \tag{5.89}$$

so that the set of tensor products of basis vectors $\{\mathbf{b}_{k_1} \otimes \mathbf{b}_{k_2} \otimes \ldots \otimes \ \mathbf{b}_{k_r}\}$ forms a *basis* for the vector space $\otimes^r T_p(M)$. Furthermore, since $\dim(M) = n$, then we must have $\dim \otimes^r T_p(M) = n^r$, as well.

Similarly, assuming a change of coordinate basis, it's easy to show that under the corresponding coordinate transformation $\bar{x}^p = \bar{x}^p(x^k)$, the components of $\mathbf{T}$ must transform according to

$$\bar{T}^{k_1 k_2 \ldots k_r} = \frac{\partial \bar{x}^{k_1}}{\partial x^{j_1}} \frac{\partial \bar{x}^{k_2}}{\partial x^{j_2}} \cdots \frac{\partial \bar{x}^{k_r}}{\partial x^{j_r}} T^{j_1 j_2 \ldots j_r}, \tag{5.90}$$

and this law must apply to any tensor of type $(r,0)$. Any such tensor $\mathbf{T}$ is said to be r-times contravariant.

Now, let's reverse the process, and see how we can map *tangent vectors*
into real numbers. We shall say that ω is a **tensor of type** $(0,1)$ at p, if ω is a *linear mapping* which assigns a *unique* real number to each tangent vector $\mathbf{u} \in T_p(M)$. That is, we require that

$$\begin{aligned} \omega(\mathbf{u} + \mathbf{v}) &= \omega(\mathbf{u}) + \omega(\mathbf{v}) \\ \omega(c\mathbf{u}) &= c\omega(\mathbf{u}) \end{aligned} \tag{5.91}$$

for all vectors $\mathbf{u}, \mathbf{v} \in T_p(M)$, and for any real number c. Obviously, any one-form (or covector) $\omega \in T_p^*(M)$ is a tensor of type $(0,1)$ at p. *From now on, for consistency of notation, we shall write all one-forms in boldface print in order to emphasize their tensor character.*

Next, we shall say that $\mathbf{S}$ is a **tensor of type** $(0,2)$ at p, if $\mathbf{S}$ assigns a *unique* real number $\mathbf{S}(\mathbf{u}, \mathbf{v})$ to each *ordered* pair of tangent vectors $\mathbf{u}, \mathbf{v} \in T_p(M)$, in a *bilinear* fashion. That is, $\mathbf{S}$ is linear in each of its arguments:

$$\begin{aligned} \mathbf{S}(c_1\mathbf{u}_1 + c_2\mathbf{u}_2, \mathbf{v}) &= c_1\mathbf{S}(\mathbf{u}_1, \mathbf{v}) + c_2\mathbf{S}(\mathbf{u}_2, \mathbf{v}) \\ \mathbf{S}(\mathbf{u}, c_1\mathbf{v}_1 + c_2\mathbf{v}_2) &= c_1\mathbf{S}(\mathbf{u}, \mathbf{v}_1) + c_2\mathbf{S}(\mathbf{u}, \mathbf{v}_2) \end{aligned} \tag{5.92}$$

In a similar construction, we call the collection of all such bilinear maps $\mathbf{S} : T_p(M) \times T_p(M) \longrightarrow R$ the **tensor product space** $\otimes^2 T_p^*(M)$; it is the collection of all tensors of type

$(0,2)$ at p. Furthermore, this new tensor product space is a *vector space*, with the operations of addition and scalar multiplication defined by

$$\begin{aligned}(\mathbf{S}_1+\mathbf{S}_2)(\mathbf{u},\mathbf{v}) &= \mathbf{S}_1(\mathbf{u},\mathbf{v})+\mathbf{S}_2(\mathbf{u},\mathbf{v}) \\ (c\mathbf{S})(\mathbf{u},\mathbf{v}) &= c(\mathbf{S}(\mathbf{u},\mathbf{v})) \quad (\text{for all } c\in R)\end{aligned} \tag{5.93}$$

For the particular case of two one-forms $\omega,\sigma\in T_p^*(M)$, we can define a special type of bilinear mapping $\omega\otimes\sigma : T_p(M)\times T_p(M)\longrightarrow R$ by

$$(\omega\otimes\sigma)(\mathbf{u},\mathbf{v}) = \omega(\mathbf{u})\sigma(\mathbf{v}) \tag{5.93a}$$

for all vectors $\mathbf{u},\mathbf{v}\in T_p(M)$. In a similar fashion, we call $\omega\otimes\sigma$ the **tensor product** of ω and σ, and we see that $\omega\otimes\sigma\in\otimes^2 T_p^*(M)$.

If the set of tangent vectors $\{\mathbf{b}_k\}$ is a basis for $T_p(M)$, then we can expand each vector in terms of this basis as $\mathbf{u}=u^a\mathbf{b}_a$ and $\mathbf{v}=v^b\mathbf{b}_b$ respectively, and write

$$\mathbf{S}(\mathbf{u},\mathbf{v}) = \mathbf{S}\left(u^a\mathbf{b}_a, v^b\mathbf{b}_b\right) = \mathbf{S}(\mathbf{b}_a,\mathbf{b}_b)\ u^a v^b \equiv S_{ab}u^a v^b \tag{5.94}$$

where the set of n^2 quantities $S_{ab}=\mathbf{S}(\mathbf{b}_a,\mathbf{b}_b)$ are the **components** of $\mathbf{S}$ *relative* to this basis. On the other hand, if the set of one-forms $\{\beta_k\}$ is a basis for $T_p^*(M)$, then by (5.93) and (5.84), we have

$$\mathbf{S}(\mathbf{u},\mathbf{v}) = S_{ab}u^a v^b = S_{ab}\omega(\mathbf{b}_a)\sigma(\mathbf{b}_b) = S_{ab}\left(\beta^a\otimes\beta^b\right)(\mathbf{u},\mathbf{v}), \tag{5.95}$$

and consequently,

$$\mathbf{S} = S_{ab}\ \beta^a\otimes\beta^b \tag{5.96}$$

So, any tensor of type $(0,2)$ can be written as a linear combination of the tensor products of basis one-forms, $\beta^a\otimes\beta^b$. Similarly, it's easy to show that such products are *linearly independent*, and hence, form a basis for the tensor product space $\otimes^2 T_p^*(M)$.

Proceeding as before, we can easily generalize the above results by defining a **tensor S of type** $(0,s)$ at $p\in M$ as a *multilinear mapping*

$$\mathbf{T} : T_p(M)\times T_p(M)\times\ \ldots\ \times T_p(M)\longrightarrow R$$

which maps each ordered s-tuple of vectors $(\mathbf{u}_1,\mathbf{u}_2,\ldots,\mathbf{u}_s)$ into a unique real number. Once again, when we say that $\mathbf{S}$ is *multilinear*, and we require that it is *linear in each one of its arguments*.

The tensor product space $\otimes^s T_p^*(M)$ is now the tensor product of $T_p^*(M)$ with itself, written s-times:

$$\otimes^s T_p^*(M) = T_p^*(M)\otimes T_p^*(M)\otimes\ \ldots\ \otimes T_p^*(M) \quad (s \text{ factors}) \tag{5.97}$$

Each element $\mathbf{T}\in\otimes^s T_p^*(M)$ is a tensor of type $(0,r)$, and $\otimes^s T_p^*(M)$ is a vector space with the operations of addition and scalar multiplication given by

$$\begin{aligned}(\mathbf{S}_1+\mathbf{S}_2)(\mathbf{u}_1,\mathbf{u}_2,\ldots,\mathbf{u}_s) &= \mathbf{S}_1(\mathbf{u}_1,\mathbf{u}_2,\ldots,\mathbf{u}_s)+\mathbf{S}_2(\mathbf{u}_1,\mathbf{u}_2,\ldots,\mathbf{u}_s) \\ (c\mathbf{S})(\mathbf{u}_1,\mathbf{u}_2,\ldots,\mathbf{u}_s) &= c\mathbf{S}(\mathbf{u}_1,\mathbf{u}_2,\ldots,\mathbf{u}_s) \quad (\text{ for all } c\in R\ )\end{aligned} \tag{5.98}$$

Following the same argument as above, it's easy to show that each $\mathbf{S} \in \otimes^s T_p^*(M)$ may be written in the form

$$\mathbf{S} = S_{k_1 k_2 \ldots k_s} \beta^{k_1} \otimes \beta^{k_2} \cdots \otimes \beta^{k_s} \tag{5.99}$$

where the set of tensor products of basis vectors $\left\{\beta^{k_1} \otimes \beta^{k_2} \cdots \otimes \beta^{k_s}\right\}$ forms a *basis* for the vector space $\otimes^s T_p^*(M)$. Furthermore, since $\dim(M) = n$, then we have $\dim \otimes^s T_p^*(M) = n^s$.

Assuming a change of *coordinate* basis, it's easy to show that under the corresponding coordinate transformation $\bar{x}^p = \bar{x}^p(x^k)$, the components of $\mathbf{S}$ must transform according to

$$\bar{S}_{k_1 k_2 \ldots k_r} = \frac{\partial x^{j_1}}{\partial \bar{x}^{k_1}} \frac{\partial x^{j_2}}{\partial \bar{x}^{k_2}} \cdots \frac{\partial x^{j_s}}{\partial \bar{x}^{k_s}} S_{j_1 j_2 \ldots j_s}, \tag{5.100}$$

and this law must apply to any tensor of type $(0, s)$. Any such tensor $\mathbf{S}$ is said to be $r-$times covariant.

Finally, the most general type of tensor product space can be formed by taking the product of the spaces in (5.87) and (5.97). We denote this new tensor product space by

$$\otimes^r T_p(M) \otimes^s T_p^*(M)$$

and each element $\mathbf{M} \in \otimes^r T_p(M) \otimes^s T_p^*(M)$ is a *multilinear* map

$$\mathbf{M} : \left(T_p^*(M)\right)^r \times \left(T_p(M)\right)^s \longrightarrow R \tag{5.101}$$

which maps each ordered $(r+s)$-tuple into a real number. More specifically,

$$\mathbf{M}(\omega_1, \omega_2, \ldots, \omega_r, \mathbf{u}_1, \mathbf{u}_2, \ldots, \mathbf{u}_s) \in R \tag{5.102}$$

where $\omega_1, \omega_2, \ldots, \omega_r$ are r one-forms, and $\mathbf{u}_1, \mathbf{u}_2, \ldots, \mathbf{u}_s$ are s tangent vectors, *taken precisely in that order*.

Similarly, it's easy to show that any such tensor $\mathbf{M}$ can be expanded in the form

$$\mathbf{M} = M^{\,j_1 j_2 \ldots j_r}_{\,k_1 k_2 \ldots k_s} \mathbf{b}_{j_1} \otimes \mathbf{b}_{j_2} \otimes \ldots \otimes \mathbf{b}_{j_r} \otimes \beta^{k_1} \otimes \beta^{k_2} \cdots \otimes \beta^{k_s} \tag{5.103}$$

and it is called a **mixed tensor** of type (r, s), and we call $(r+s)$ the **rank** of the tensor. When no confusion can result, rather than writing out $\mathbf{M}$ in full, as in (5.103), we shall often employ the shorthand notation

$$\mathbf{M} = \left(M^{\,j_1 j_2 \ldots . j_r}_{\,k_1 k_2 \ldots k_s}\right) \tag{5.103a}$$

Given a change of *coordinate* basis, it is easy to show that under the resulting coordinate transformation $\bar{x}^p = \bar{x}^p(x^k)$, the components of $\mathbf{M}$ must transform according to

$$\bar{M}^{\,j_1 j_2 \ldots j_r}_{\,k_1 k_2 \ldots k_s} = \left(\frac{\partial \bar{x}^{j_1}}{\partial x^{l_1}} \frac{\partial \bar{x}^{j_2}}{\partial x^{l_2}} \cdots \frac{\partial \bar{x}^{j_r}}{\partial x^{l_r}}\right) \cdot \left(\frac{\partial x^{m_1}}{\partial \bar{x}^{k_1}} \frac{\partial x^{m_2}}{\partial \bar{x}^{k_2}} \cdots \frac{\partial x^{m_s}}{\partial \bar{x}^{k_s}}\right) M^{\,l_1 l_2 \ldots l_r}_{\,m_1 m_2 \ldots m_s} \tag{5.104}$$

5.12. Operations with Tensors

In many physical applications, any operations defined on tensors which themselves produce tensors are of vital importance. We now define certain operations on tensors which comprise the basic algebra of tensors; in each instance, the result of each such operation is a tensor of the appropriate rank and type.

First, we define the **addition** of two tensors. If **S** and **T** are both tensors of type (r,s), we define the sum $\mathbf{R} = \mathbf{S} + \mathbf{T}$ as the tensor of type (r,s) which is given by the formula

$$\mathbf{R}(\omega_1,...,\omega_r,\mathbf{u}_1,...,\mathbf{u}_s) = \mathbf{S}(\omega_1,...,\omega_r,\mathbf{u}_1,...,\mathbf{u}_s) + \mathbf{T}(\omega_1,...,\omega_r,\mathbf{u}_1,...,\mathbf{u}_s) \tag{5.105}$$

where $\omega_1,\omega_2,...,\omega_r$ are r one-forms, and $\mathbf{u}_1,\mathbf{u}_2,...,\mathbf{u}_s$ are s tangent vectors. In terms of components, the last equation becomes

$$R^{\,i_1\,\dots\,i_r}_{\,j_1\,\dots\,j_s} = A^{\,i_1\,\dots\,i_r}_{\,j_1\,\dots\,j_s} + B^{\,i_1\,\dots\,i_r}_{\,j_1\,\dots\,j_s} \tag{5.105a}$$

Next, we define the operation of **scalar multiplication**. If **T** is a tensor of type (r,s), and λ is a real number (scalar), then $\lambda\mathbf{T}$ is given by

$$(\lambda\mathbf{T})\,(\omega_1,...,\omega_r,\mathbf{u}_1,...,\mathbf{u}_s) = \lambda\cdot\mathbf{T}(\omega_1,...,\omega_r,\mathbf{u}_1,...,\mathbf{u}_s) \tag{5.106}$$

where $\omega_1,\omega_2,...,\omega_r$ are r one-forms, and $\mathbf{u}_1,\mathbf{u}_2,...,\mathbf{u}_s$ are s tangent vectors. It is obvious that $\lambda\mathbf{T}$ is again a tensor of type (r,s). In terms of components, this is

$$(\lambda T)^{\,i_1\,\dots\,i_r}_{\,j_1\,\dots\,j_s} = \lambda\left(T^{\,i_1\,\dots\,i_r}_{\,j_1\,\dots\,j_s}\right) \tag{5.106a}$$

It should be equally apparent that at each point $p \in M$, the collection of all tensors of type (r,s), equipped with the operations of tensor addition and scalar multiplication, comprise a *vector space*.

If **S** is a tensor of type (p,q) and **T** is a tensor of type (r,s), we define the **tensor product** (or **outer product**) $\mathbf{S}\otimes\mathbf{T}$ to be the tensor of type $(p+r,q+s)$ defined by

$$(\mathbf{S}\otimes\mathbf{T})\,(\omega_1,...,\omega_{p+r},\mathbf{u}_1,...,\mathbf{u}_{q+s}) = \mathbf{S}(\omega_1,...,\omega_p,\mathbf{u}_1,...,\mathbf{u}_q)\;\mathbf{T}(\omega_1,...,\omega_r,\mathbf{u}_1,...,\mathbf{u}_s) \tag{5.107}$$

or in component form, the last equation becomes

$$(\mathbf{S}\otimes\mathbf{T})^{i_1\,\dots\,i_{p+r}}_{j_1\,\dots\,j_{q+s}} = S^{i_1\,\dots\,i_p}_{j_1\,\dots\,j_q}\cdot T^{i_1\,\dots\,i_r}_{j_1\,\dots\,j_s} \tag{5.107a}$$

We now turn our attention to a new operation, called **contraction**. For simplicity, assume that **T** is a tensor of type $(2,2)$. We can form a contraction of **T** by setting *any* contravariant (upper) index equal to *any* covariant (lower) index. Many such contractions are possible; for instance, a possible contraction of **T** is the object **C** with components

$$C^m_q = T^{mk}_{kq} \tag{5.108}$$

where we have set $n = p = k$, and where *summation* is implied on the repeated index k. Clearly, other contractions are possible. It turns out, however, that $\mathbf{C} = (C^m_q)$ is itself a *tensor* of type $(1,1)$.

In order to see this, we first observe that under a change of basis, the components of **T** must transform according to

$$\bar{T}^{\ mn}_{\ pq} = \frac{\partial \bar{x}^m}{\partial x^a}\frac{\partial \bar{x}^n}{\partial x^b}\frac{\partial x^c}{\partial \bar{x}^p}\frac{\partial x^d}{\partial \bar{x}^q}\, T^{\ ab}_{\ cd}$$

Setting $n = p = k$ in the last equation, and recalling (5.108), gives

$$\begin{aligned}
\bar{C}^m_q &= \bar{T}^{\ mk}_{\ kq} = \frac{\partial \bar{x}^m}{\partial x^a}\left(\frac{\partial \bar{x}^k}{\partial x^b}\frac{\partial x^c}{\partial \bar{x}^k}\right)\frac{\partial x^d}{\partial \bar{x}^q}\, T^{\ ab}_{\ cd} \\
&= \frac{\partial \bar{x}^m}{\partial x^a}\delta^c_b \frac{\partial x^d}{\partial \bar{x}^q}\, T^{\ ab}_{\ cd} \\
&= \frac{\partial \bar{x}^m}{\partial x^a}\frac{\partial x^d}{\partial \bar{x}^q}\, C^a_d
\end{aligned}$$

Thus, the components of $\bar{\mathbf{C}}$ are those of a tensor of type $(1,1)$, *We conclude that the operation of contraction is independent of the choice of basis*, and this argument is easily generalized to handle the case where **T** is a general tensor of type (p,q).

Now, assume that **T** is now a general tensor of type (p,q), so that

$$\mathbf{T} = \left(T^{\ j_1\, j_2\, \dots\, j_p}_{\ k_1 k_2 \dots k_q}\right)$$

Furthermore, let a and b be whole numbers such that $1 \le a \le p$, and $1 \le b \le q$. If one superscript of **T** is set equal to one subscript of **T**, then according to the Einstein summation convention, we must sum over the resulting repeated index, and the resulting object is a tensor **C** of type $(p-1, q-1)$, whose rank is two less than the original tensor.

In particular, if we set $i_a = j_b = m$, then we obtain the tensor

$$\mathbf{C} = (T^{\ i_1\, i_2\, i_3\, \dots\, m\, \dots\, i_p}_{\ j_1\, j_2\, j_3\, \dots\, m\, \dots\, j_q}) \tag{5.109}$$

which is called a *contraction* of **T**.

Finally, we can perform another operation, called the inner product of two tensors. Let

$$\mathbf{S} = (S^{\ i_1\, i_2\, i_3\, \dots\, i_p}_{\ j_1\, j_2\, j_3\, \dots\, j_q}) \text{ and } \mathbf{T} = (T^{\ k_1\, k_2\, k_3\, \dots\, k_r}_{\ l_1\, l_2\, l_3\, \dots\, l_s})$$

be tensors of type (p,q) and (r,s) respectively. Then an **inner product** of **S** and **T**, is a tensor **C** of type $(p+r-1, q+s-1)$ which is obtained by contracting the outer product of **S** and **T**.

If, for example, we set $i_a = l_b = t$ where $1 \le a \le p$ and $1 \le b \le s$, then such a contracted inner product is given by

$$\mathbf{C} = S^{\ i_1\, i_2\, i_3\, \dots\, t\, \dots\, i_p}_{\ j_1\, j_2\, j_3\, \dots\, j_q}\, T^{\ k_1\, k_2\, k_3\, \dots\, k_r}_{\ l_1\, l_2\, l_3\, \dots\, t \dots\, l_s}$$

Clearly, *several* inner products are possible, and when confusion may arise, we shall agree to display the corresponding components of a given inner product in explicit form. *It is clear that an inner product of two tensors is again a tensor* since such a product is obtained by first forming the outer product of the given tensors (which produces a tensor), and then contracting the previous result, which again produces a tensor of lower rank.

5.13. Vector and Tensor Fields

Given a differentiable manifold M, a **vector field V** on M is a rule which assigns a *unique* tangent vector $\mathbf{V}(p) \in T_p(M)$ to each point $p \in M$ in a *smooth* manner. That is, if we are given a coordinate chart (U, ϕ) containing the point p, where $x^k = \phi^k(p)$, then in terms of the coordinate basis at p, we can write

$$\mathbf{V}(p) = V^k(\mathbf{x}) \left.\frac{\partial}{\partial x^k}\right|_p \tag{5.110}$$

where we *require* that the functions $V^k(\mathbf{x})$ are *smooth functions* of the coordinates; that is $V^k(\mathbf{x}) \in C^\infty(M)$ for all k.

Similarly, we define a **tensor field T** of type (r,s) as a rule which *smoothly* assigns a *unique* tensor to each point $p \in M$. Following the general prescription given in (5.103), we know that can write **T** in terms of the coordinate basis vectors and basis one-forms at p as

$$\mathbf{T}(p) = \left(T^{j_1 j_2 \dots j_r}_{k_1 k_2 \dots k_s}(\mathbf{x})\right) \left.\frac{\partial}{\partial x^{j_1}}\right|_p \otimes \dots \otimes \left.\frac{\partial}{\partial x^{j_r}}\right|_p \otimes \left.dx^{k_1}\right|_p \otimes \cdots \otimes \left.dx^{k_s}\right|_p \tag{5.111}$$

If this result which holds good at any point $p \in M$, and if the expansion coefficients are *smooth functions* of the coordinates, then we say that **T** is a tensor field of type (r,s) on M.

The set of all tensor fields of type (r,s) on M shall be denoted by $\Im^r_s(M)$. Observe that at each fixed point $p \in M$, the collection of tensor fields $\Im^r_s(M)$ is itself a *vector space*, with the operations of tensor addition and scalar multiplication given by (5.105) and (5.106), respectively.

In particular, $\Im^1_0(M)$ is the set of all vector fields on M, $\Im^0_1(M)$ is the set of all **covector fields** (or one-form fields) on M, and $\Im^0_0(M) \equiv C^\infty(M)$ is the set of all **scalar fields** (or smooth functions) on M.

5.14. Invariance of Tensor Equations

Suppose that $A \in \Im^r_s(M)$ and $B \in \Im^r_s(M)$ are tensor fields of the *same type* on a differentiable manifold M. We shall say that these tensor fields are **equal**, and write $\mathbf{A} = \mathbf{B}$, if and only if

$$\mathbf{A}(\omega_1, \dots, \omega_r, \mathbf{u}_1, \dots, \mathbf{u}_s) = \mathbf{B}(\omega_1, \dots, \omega_r, \mathbf{u}_1, \dots, \mathbf{u}_s) \tag{5.112}$$

for all one-forms $\omega_1, \omega_2, \dots, \omega_r$ and all tangent vectors $\mathbf{u}_1, \mathbf{u}_2, \dots, \mathbf{u}_s$ at each point $p \in M$. When written out in terms of components, (5.112) becomes

$$A^{\,i_1\; i_2\; i_3\; \dots\; i_r}_{\,j_1\; j_2\; j_3\; \dots\; j_s} = B^{\,i_1\; i_2\; i_3\; \dots\; i_r}_{\,j_1\; j_2\; j_3\; \dots\; j_s}, \tag{5.112a}$$

at p, an expression which is called a **tensor equation**. It turns out that such tensor equations are *invariant* under coordinate transformations.

In order to see this, recall that if (U, ϕ) and $(V, \bar{\phi})$ are overlapping charts containing the fixed point $p \in U \bigcap V$, then the coordinates of p will be given by both $x^k = \phi^k(p)$ and $\bar{x}^k = \bar{\phi}^k(p)$, and there will exist a coordinate transformation $\bar{x}^m = \bar{x}^m(x^k)$, which relates the two sets of coordinates. Now, due to the *linearity* of transformation law for tensor components as expressed in (5.104), it is obvious that if the components of a given tensor are identically zero relative to the local coordinates (x^i), then they are also zero in the local coordinate system $(\bar{x}^i)$.

In particular, if the tensor equation (5.112a) holds true, when written in terms of the local coordinates (x^i) constructed at p, then we must also have

$$C^{i_1\,i_2\,i_3\,\ldots\,i_r}_{j_1\,j_2\,j_3\,\ldots\,j_s} = A^{i_1\,i_2\,i_3\,\ldots\,i_r}_{j_1\,j_2\,j_3\,\ldots\,j_s} - B^{i_1\,i_2\,i_3\,\ldots\,i_r}_{j_1\,j_2\,j_3\,\ldots\,j_s} = 0$$

relative to the same local coordinates; however, since the difference **C** of the two tensors is again a tensor of the same type, its components must also be identically zero in any other local coordinate system $(\bar{x}^i)$, so that

$$\overline{C}^{i_1\,i_2\,i_3\,\ldots\,i_r}_{j_1\,j_2\,j_3\,\ldots\,j_s} = \overline{A}^{i_1\,i_2\,i_3\,\ldots\,i_r}_{j_1\,j_2\,j_3\,\ldots\,j_s} - \overline{B}^{i_1\,i_2\,i_3\,\ldots\,i_r}_{j_1\,j_2\,j_3\,\ldots\,j_s} = 0$$

Consequently, the tensor equation

$$\overline{A}^{i_1\,i_2\,i_3\,\ldots\,i_r}_{j_1\,j_2\,j_3\,\ldots\,j_s} = \overline{B}^{i_1\,i_2\,i_3\,\ldots\,i_r}_{j_1\,j_2\,j_3\,\ldots\,j_s} \tag{5.112b}$$

must hold true in the local coordinate system $(\bar{x}^i)$. In other words, the same tensor equation is must be valid in the barred coordinate system. *So, in general, if a tensor equation holds in a given coordinate system, then it must hold in all coordinate systems.* As we shall see, in the General Theory of Relativity, Einstein postulated what is commonly called the **Principle of General Covariance**. This principle asserts that *the mathematical statement of a physical law must take the same mathematical form in all possible coordinate systems.*

Given the invariance of tensor equations under general coordinate transformations, it is apparent why Einstein chose to formulate his theory in the language of tensors.

5.15. Tests for Tensor Character

In many cases, it turns out that it is useful to have an *indirect* method of determining whether a given set of quantities has tensor character. This method is afforded to us by what is commonly referred to as a *Quotient Rule*. As we will see, there are actually many special cases of the Quotient Rule, some of which we will now consider.

First, given a differentiable manifold M, let us assume that V^i are a set of n-quantities defined at each point $p \in M$. Suppose further, that for every *arbitrary* covector $\omega = (\omega_k) \in \mathfrak{I}^0_1(M)$, the quantity

$$\lambda = V^k \omega_k$$

is a scalar field on M, then it turns out that the quantities V^i are the components of a vector field on M. In order to see this, let (U, ϕ) and $(V, \bar{\phi})$ be overlapping charts such that the point $p \in U \bigcap V$, so the coordinates of p are given by both $x^k = \phi^k(p)$ and $\bar{x}^k = \bar{\phi}^k(p)$. Let $\bar{x}^m = \bar{x}^m(x^k)$ be the coordinate transformation relating the two sets of coordinates. Since λ is a scalar field on M, then we must have

$$\bar{\lambda}(\bar{x}^k) = \lambda(x^k)$$

at p. Consequently,

$$\bar{V}^k\bar{\omega}_k = V^m\omega_m$$

Recalling that ω is a covector, the last equation may be rewritten as

$$\bar{V}^k\bar{\omega}_k = \bar{V}^k\left(\frac{\partial x^m}{\partial \bar{x}^k}\omega_m\right) = V^m\omega_m$$

so that

$$\left(\frac{\partial \bar{x}^m}{\partial x^k}\bar{V}^k - V^m\right)\omega_m = 0$$

But since ω is arbitrary, then we must have

$$V^m = \frac{\partial \bar{x}^m}{\partial x^k}\bar{V}^k$$

and if we interchange barred and unbarred quantities, we obtain

$$\bar{V}^m = \frac{\partial \bar{x}^m}{\partial x^k}V^k$$

so that $\mathbf{V} = (V^k)$ is a vector field on M.

Secondly, let's consider n^2-quantities T^{ij} defined at each point $p \in M$ such that for every *arbitrary* covector $\omega = (\omega_j) \in \Im_1^0(M)$, the quantities given by

$$T^{ij}\omega_j = V^i \tag{5.113}$$

are the components of a vector at $p \in M$, then the quantities T^{ij} are the components of a tensor field of type $(2,0)$ on M.

The proof is as follows. Using the fact that $\mathbf{V}$ is a vector, we can write

$$\overline{T}^{ij}\overline{\omega}_j = \overline{V}^i = \frac{\partial \overline{x}^i}{\partial x^p}V_p = \frac{\partial \overline{x}^i}{\partial x^p}T^{pm}\omega_m \tag{5.113a}$$

But ω is a covector, so we have (under the inverse transformation)

$$\omega_m = \frac{\partial \overline{x}^j}{\partial x^m}\bar{\omega}_j \tag{5.113b}$$

Substituting (5.113b) into (5.113a) gives

$$\overline{T}^{ij}\overline{\omega}_j = \left(\frac{\partial \overline{x}^i}{\partial x^p}\frac{\partial \overline{x}^j}{\partial x^m}T^{pm}\right)\bar{\omega}_j \implies \left(\overline{T}^{ij} - \frac{\partial \overline{x}^i}{\partial x^p}\frac{\partial \overline{x}^j}{\partial x^m}T^{pm}\right)\bar{\omega}_j = 0\,.$$

Recalling that ω is arbitrary, we conclude that

$$\overline{T}^{ij} = \frac{\partial \overline{x}^i}{\partial x^p}\frac{\partial \overline{x}^j}{\partial x^m}T^{pm}$$

Consequently, (T^{ij}) are the components of a tensor field $\mathbf{T} \in \Im_0^2(M)$.

The thread which is common in both parts our discussion is that if at each point $p \in M$ we are given an *inner product* between a set of symbols A and an *arbitrary* covector field ω, and if this inner product always results in a tensor $\mathbf{B}$; then we can conclude that the set of symbols represented by $\mathbf{A}$ are the components of a tensor field of the appropriate type. It's easy to show that this is true in general, and we shall take this to be the general form of the Quotient Rule:

Quotient Rule

Given a differentiable manifold M, let the set of n^{p+q} quantities

$$A^{\,i_1\, i_2\, i_3\, \dots\, i_p}_{\;j_1\, j_2\, j_3\, \dots\, j_q}$$

be defined at each point $p \in M$. If for every *arbitrary* covector $\omega = (\omega_j) \in \Im^0_1(M)$, the quantities given by

$$A^{\,i_1\, i_2\, i_3\, \dots\dots\dots\dots i_p}_{\;j_1\, j_2\, j_3\, \dots\, j_{q-1}\, k}\, \omega^k = B^{\,i_1\, i_2\, i_3\, \dots\, i_p}_{\;j_1\, j_2\, j_3\, \dots\, j_{q-1}} \tag{5.114}$$

are the components of a tensor field $\mathbf{B} \in \Im^p_{q-1}(M)$, then $\mathbf{A} \in \Im^p_q(M)$.

5.16. Symmetric and Anti-Symmetric Tensors

Suppose that $\mathbf{S} = (S_{k_1 k_2 \dots k_p})$ is a tensor of type $(0, p)$ on M. We say that $\mathbf{S}$ is **symmetric** with respect to a given pair of indices, say k_i and k_j, if

$$S_{k_1 k_2 \dots k_i \dots k_j \dots k_p} = S_{k_1 k_2 \dots k_j \dots k_i \dots k_p}\,, \tag{5.115}$$

so that the components of $\mathbf{S}$ do not change when these two indices are interchanged. We shall say that $\mathbf{S}$ is **completely symmetric**, if its components remain the same whenever *any* two indices are interchanged.

Similarly, if $\mathbf{A} = (A_{k_1 k_2 \dots k_p})$ is a tensor field of type $(0, p)$ on M, then we say that $\mathbf{A}$ is **antisymmetric (or skew-symmetric)** with respect to the indices k_i and k_j if

$$A_{k_1 k_2 \dots k_i \dots k_j \dots k_p} = -A_{k_1 k_2 \dots k_j \dots k_i \dots k_p} \tag{5.116}$$

Observe that the components of $\mathbf{A}$ experience a change of sign when these two indices are interchanged. In the event that components of $\mathbf{A}$ exhibit a change of sign whenever *any* two indices are interchanged, then we say that $\mathbf{A}$ is **completely antisymmetric**.

It turns out that the properties of symmetry and anti-symmetry are invariant under a change of basis. To see this, assume that $\mathbf{S} = (S_{ab})$ is symmetric with respect to the indices a and b, so that $S_{ab} = S_{ba}$. Given a change of *coordinate* basis, we know that under the resulting change of coordinates $\bar{x}^p = \bar{x}^p(x^k)$, the components of $\mathbf{S}$ transform according to

$$\bar{S}_{cd} = \frac{\partial x^a}{\partial \bar{x}^c}\frac{\partial x^b}{\partial \bar{x}^d} S_{ab},$$

Consequently,

$$\bar{S}_{dc} = \frac{\partial x^a}{\partial \bar{x}^d}\frac{\partial x^b}{\partial \bar{x}^c} S_{ab} = \frac{\partial x^b}{\partial \bar{x}^c}\frac{\partial x^a}{\partial \bar{x}^d} S_{ba} = \bar{S}_{cd}$$

Thus $\bar{S}_{dc} = \bar{S}_{cd}$, and we see that this symmetry property is invariant under a change of basis. The same is true of antisymmetry, and this argument is easily extended to handle more general types of tensors, although the notation is quite messy!

Now, let $\mathbf{T} = (T_{ab})$ be a tensor of type $(0, 2)$. We define the **symmetric part** of $\mathbf{T}$ to be the *symmetric* tensor with components $T_{(ab)}$ given by

$$T_{(ab)} = \frac{1}{2}(T_{ab} + T_{ba}) \tag{5.117}$$

and the **anti-symmetric part** of **T** to be the *anti-symmetric* tensor whose components $T_{[ab]}$ are given by

$$T_{[ab]} = \frac{1}{2}(T_{ab} - T_{ba}) \tag{5.118}$$

It is obvious that

$$T_{ab} = T_{(ab)} + T_{[ab]} \,, \tag{5.119}$$

so every tensor of type $(0,2)$ may be expressed as the sum of symmetric tensor and an antisymmetric tensor. This simple fact will turn out to be quite useful later on.

Given a tensor field $\mathbf{T} = (T_{k_1 k_2 \ldots k_p})$ of type $(0,p)$ we define a new tensor, called the **symmetrization** of **T**, to be the tensor $Sym(\mathbf{T})$, whose components $T_{(k_1 k_2 \ldots k_p)}$ are given by

$$T_{(k_1 k_2 \ldots k_p)} = \frac{1}{p!} \sum_{\pi \in S_p} T_{k_{\pi(1)} k_{\pi(2)} \ldots k_{\pi(p)}} \tag{5.120}$$

where the sum is taken over all permutations π of the indices, and S_p is the collection of all $p!$ permutations of the elements in the set $\{1,2,\ldots,p\}$. It should be clear that $T_{(k_1 k_2 \ldots k_p)}$ is completely symmetric. For example, for $p = 3$, we have

$$T_{(abc)} = \frac{1}{3!}(T_{abc} + T_{cab} + T_{bca} + T_{cba} + T_{acb} + T_{bac}) \tag{5.121}$$

We define the **antisymmetrization** of **T**, to be the tensor $Anti(\mathbf{T})$, whose components $T_{[k_1 k_2 \ldots k_p]}$ are given by

$$T_{[k_1 k_2 \ldots k_p]} = \frac{1}{p!} \sum_{\pi \in S_p} sgn(\pi) \cdot T_{k_{\pi(1)} k_{\pi(2)} \ldots k_{\pi(p)}} \tag{5.122}$$

where $sgn(\pi)$ denotes the sign of the permutation π; that is,

$$sgn(\pi) = \begin{cases} +1, & \text{if } \pi \text{ is an even permutation} \\ -1, & \text{if } \pi \text{ is an odd permutation} \end{cases} \tag{5.122a}$$

Observe that by definition, the antisymmetrization of **T** is a completely antisymmetric tensor. For example, if $p = 3$, then

$$T_{[abc]} = \frac{1}{3!}(T_{abc} + T_{cab} + T_{bca} - T_{cba} - T_{acb} - T_{bac})$$

It may be shown that if T is a completely symmetric tensor, then $\mathbf{T} = Sym(\mathbf{T})$, while if **T** is a completely antisymmetric tensor, then $\mathbf{T} = Anti(\mathbf{T})$. Needless to say, all of the above definitions may be extended to encompass general tensors of type (r,s), provided that the symmetry or antisymmetry properties of such tensors are defined *separately* on their r-upper and s-lower indices.

5.17. The Metric Tensor

In order to do Physics, we need to add some additional structure to a manifold; this is the concept of distance. Given a *rectangular* coordinate system (x^i) in the n-dimensional space $\mathbf{R}^n$, we know that the distance between two neighboring points $P(x^k)$ and $Q(x^k + dx^k)$ may be written as

$$ds^2 = (dx^1)^2 + (dx^2) + ... + (dx^n)^2 = \delta_{ab} dx^a dx^b \tag{5.123}$$

where δ_{ab} is the Kronecker delta, and we have employed the somewhat sloppy notation ds^2 to mean $(ds)^2$. We call the distance function (5.123) a **line element** or **metric** for $\mathbf{R}^n$. The manifold R^n equipped with the above metric $\delta = (\delta_{ab})$ is precisely what we mean by the **Euclidean space** E^n.

We now extend the notion of distance to a general manifold. Consider a tensor field $\mathbf{g} = (g_{ab})$ of type *(0,2)* which is defined on a differentiable manifold M. By its very definition, $\mathbf{g}$ maps each pair of tangent vectors $\mathbf{u}, \mathbf{v} \in T_p(M)$ into a real number $\mathbf{g}(\mathbf{u}, \mathbf{v})$, and it does so in a *bilinear* fashion. We say that $\mathbf{g}$ is a **metric tensor** or that $\mathbf{g}$ is a **metric** on M, if g is *symmetric*, and $\mathbf{g}$ is *non-degenerate*; that is,

$$\mathbf{g}(\mathbf{u}, \mathbf{v}) = 0 \text{ for all } \mathbf{v} \in T_p(M) \Longrightarrow \mathbf{u} = \mathbf{0} \tag{5.124}$$

The non-degeneracy of $\mathbf{g}$ implies that the matrix G, whose elements are g_{ab}, is non-singular. To see this, assume that

$$\mathbf{g}(\mathbf{u}, \mathbf{v}) = g_{ab} u^a v^b = 0$$

for all vectors $\mathbf{v}$. So, in particular, if we take $\mathbf{v} = (v^b) = \mathbf{0}$, then we obtain

$$g_{ab} u^a = g_{ba} u^a = 0,$$

But the non-degeneracy of $\mathbf{g}$ requires that the system of equations $G\mathbf{u} = \mathbf{0}$ must have the *unique* solution $\mathbf{u} = \mathbf{0}$. Consequently $\det(G) \neq 0$ and G^{-1} exists.

A manifold which is equipped with a metric tensor $\mathbf{g}$ has the degree of additional structure which enables us to define distances, lengths, and angles. First, we define the **inner product** or **scalar product** of any two vectors $\mathbf{u}, \mathbf{v} \in T_p(M)$ by

$$\langle \mathbf{u}, \mathbf{v} \rangle = g(\mathbf{u}, \mathbf{v}) = g_{ab} u^a v^b \,, \tag{5.125}$$

a result which is always a real number. In particular, the **length** or **norm** of a vector $\mathbf{v} \in T_p(M)$ may now be defined in terms of the inner product as

$$\|\mathbf{v}\|^2 = \langle \mathbf{v}, \mathbf{v} \rangle = \mathbf{g}(\mathbf{v}, \mathbf{v}) = g_{ab} v^a v^b \tag{5.126}$$

We say that the metric $\mathbf{g}$ is **positive definite** (or **negative definite**) if $\|\mathbf{v}\|^2 > 0$ (or $\|\mathbf{v}\|^2 < 0$) for all *non-zero* vectors $\mathbf{v}$; otherwise, we say that the metric is **indefinite**.

A manifold M which is equipped with a positive definite metric is called a **Riemannian space,** while a manifold furnished with an indefinite metric is called a **semi-Riemannian** (or **psuedo-Riemannian**) **space**. Strictly speaking a Riemannian space (or a semi-Riemannian space) is an *ordered pair* $(M, \mathbf{g})$, since the same manifold M equipped with a different metric $\mathbf{g}'$ would comprise a *different* space.

For example, recall that the **spacetime interval** (4.21) of Special Relativity is given by

$$ds^2 = c^2dt^2 - dx^2 - dy^2 - dz^2$$

It is apparent that this quantity may be positive, negative, or zero. If we set $(x^0, x^1, x^2, x^3) = (ct, x, y, z)$, then the spacetime interval may be rewritten as

$$ds^2 = \eta_{ab}dx^a dx^b \quad (a, b = 0, 1, 2, 3) \tag{5.127}$$

where

$$\eta = (\eta_{ab}) = diag(1, -1, -1, -1) = \begin{bmatrix} 1 & 0 & 0 & 0 \\ 0 & -1 & 0 & 0 \\ 0 & 0 & -1 & 0 \\ 0 & 0 & 0 & -1 \end{bmatrix} \tag{5.128}$$

is the *indefinite* metric which characterizes the space of Special Relativity, called **Minkowski space**. Thus, Minkowski space is an example of a semi-Riemannian manifold. *From now on, unless otherwise stated, we shall assume that we are working with semi-Riemannian manifolds.*

Now, any two vectors $\mathbf{u}, \mathbf{v} \in T_p(M)$ are said to be **orthogonal** if

$$\langle \mathbf{u}, \mathbf{v} \rangle = \mathbf{g}(\mathbf{u}, \mathbf{v}) = g_{ab}u^a v^b = 0 \tag{5.129}$$

In particular, a vector $\mathbf{u}$ is said to be a **null vector** if $\langle \mathbf{u}, \mathbf{u} \rangle = 0$. If the metric is *indefinite*, then it may be shown that at each point $p \in M$, there will always exist at least one *non-zero* null vector $\mathbf{u} \in T_p(M)$.

Given a semi-Riemannian manifold M, we define the angle $\angle(\mathbf{u}, \mathbf{v})$ between two non-zero vectors $\mathbf{u}, \mathbf{v} \in T_p(M)$ as

$$\cos\angle(\mathbf{u}, \mathbf{v}) = \frac{\mathbf{g}(\mathbf{u}, \mathbf{v})}{\sqrt{|\mathbf{g}(\mathbf{u}, \mathbf{u})|} \cdot \sqrt{|\mathbf{g}(\mathbf{v}, \mathbf{v})|}} = \frac{g_{mn}u^m v^n}{\sqrt{|g_{pq}u^p u^q|} \cdot \sqrt{|g_{rs}v^r v^s|}} \tag{5.130}$$

In order to define the notion of distance, let p and q be two neighboring points of a semi-Riemannian manifold M such that the single chart (U, ϕ) contains both of these points. Assume further that the points are *infinitesimally close* in the sense that if the coordinates of p are given by $x^k = \phi^k(p)$, then the coordinates of of q may be written as $x^k + dx^k = \phi^k(q)$. We define the infinitesimal **distance** ds between these points as the number

$$\begin{aligned} ds^2 &= g_{ab}dx^a dx^b \\ &= g_{11}(dx^1)^2 + g_{12}dx^1 dx^2 + ... + g_{nn}(dx^n)^2 \end{aligned} \tag{5.131}$$

Now, let p be a fixed point of a semi-Riemannian space M, and assume the set of vectors $\{\mathbf{e}_1, \mathbf{e}_2, ..., \mathbf{e}_n\}$ comprise a *general* basis for $T_p(M)$. We can effect a change of basis $e_k \longrightarrow \bar{e}_k$:

$$\bar{\mathbf{e}}_k = \Lambda_k^m \mathbf{e}_m \quad \text{(where } \det(\Lambda) \neq 0)$$

by using *any* non-singular matrix Λ. Since $\mathbf{g}$ is a tensor of type $(0,2)$, then under a change of basis, its components must transform according to

$$\bar{g}_{ab} = \mathbf{g}(\bar{e}_a, \bar{e}_b) = \mathbf{g}(\Lambda_a^m \mathbf{e}_m, \Lambda_b^n \mathbf{e}_n) = \Lambda_a^m \Lambda_b^n \, \mathbf{g}(\mathbf{e}_m, \mathbf{e}_n) = \Lambda_a^m \, \Lambda_b^n \, g_{mn} \tag{5.132}$$

The last equation can be recast into matrix form as

$$\bar{G} = \Lambda^T G \Lambda \tag{5.133}$$

where $G = (g_{ab})$, and Λ^T is the transpose of $\Lambda = (\Lambda_a^m)$. A transformation like (5.133) is called a **congruence transformation**, and as long as the matrix Λ is non-singular, we can use any matrix Λ we like, and we will be guaranteed that (5.133) will remain true.

We know from linear algebra, that since G is a symmetric matrix, its eigenvectors are *orthogonal*, and they form a basis for $T_p(M)$. Furthermore, if we set

$$\bar{\mathbf{e}}_k = \frac{\mathbf{e}_k}{\|\mathbf{e}_k\|} \qquad (k = 1, 2, 3, ..., n) \tag{5.134}$$

then the vectors $\{\bar{\mathbf{e}}_k\}$ will be *orthonormal* relative to the Euclidean metric, and they will also be eigenvectors of G. Since the choice of Λ is completely at our disposal, let Λ be the matrix whose columns are formed by the vectors $\{\bar{\mathbf{e}}_k\}$, so that the k-th column of Λ is the vector $\bar{\mathbf{e}}_k$:

$$\text{col}_k\,(\Lambda) = \bar{\mathbf{e}}_k \tag{5.135}$$

Thus, the matrix Λ will be an *orthogonal* matrix by its very construction, and once again, we have $\bar{G} = \Lambda^T G \Lambda$. But under this orthogonal transformation, the matrix $\bar{G}$ will be a *diagonal* matrix

$$\bar{G} = \Lambda^T G \Lambda = diag(\lambda_1, \lambda_2, \lambda_3, ..., \lambda_n) \tag{5.136}$$

whose entries along its main diagonal are simply the eigenvalues of G. In order to see this, let $\mathbf{v} = v^a \bar{\mathbf{e}}_a$ be an arbitrary tangent vector written in terms of the new basis. Then we have

$$ds^2 = \lambda_1 (d\bar{x}^1)^2 + \lambda_2 (d\bar{x}^2)^2 + ... + \lambda_n (d\bar{x}^n)^2, \tag{5.137}$$

an expression we call the **semi-canonical form** of ds^2. Furthermore, since G is non-singular, then none of these eigenvalues can be zero.

Now, we can even go further, and cast the line element ds^2 in (5.137) into an even simpler form. We do this by making a *change of scale*; that is, we define the *diagonal* matrix

$$S = diag\left(\frac{1}{\sqrt{|\lambda_1|}}, \frac{1}{\sqrt{|\lambda_2|}}, \frac{1}{\sqrt{|\lambda_3|}}, ..., \frac{1}{\sqrt{|\lambda_n|}}\right)$$

and perform a change of basis using the matrix $S\Lambda$ where Λ is still given by (5.135). Since S is a diagonal matrix, then $S^T = S$, and S commutes with any other matrix. So, equation (5.136) now becomes

$$\begin{aligned} \bar{G} &= (S\Lambda)^T G (S\Lambda) = S^T \left(\Lambda^T G \Lambda\right) S \\ &= S^2 \, diag(\lambda_1, \lambda_2, \lambda_3, ..., \lambda_n) \\ &= diag\left(\frac{\lambda_1}{|\lambda_1|}, \frac{\lambda_2}{|\lambda_2|}, \frac{\lambda_3}{|\lambda_3|}, ..., \frac{\lambda_n}{|\lambda_n|}\right) \end{aligned}$$

and the line element assumes the **canonical form**

$$ds^2 = \frac{\lambda_1}{|\lambda_1|}(d\bar{x}^1)^2 + \frac{\lambda_2}{|\lambda_2|}(d\bar{x}^2)^2 + ... + \frac{\lambda_n}{|\lambda_n|}(d\bar{x}^n)^2 \tag{5.138}$$

Recapitulating, we have shown that at any point p of a semi- Riemannian manifold, we can always effect a change of basis which casts the line element into the canonical form (5.138).

If we examine the coefficients of the coordinate differentials in (5.138), we see that they must assume the values ± 1. As a matter of convention, we can choose Λ to ensure that the positive eigen-values appear first, so $\bar{G} = diag\ (1, ..., 1, ..., -1, ..., -1)$. The sum of these diagonal elements is called the **signature** of the metric. Thus, for example, the Minkowski metric η has signature -2.

Since $\det(G) \neq 0$, then its inverse $G^{-1} = (g^{ab})$, exists and is given by

$$g^{ab} = \frac{C^{ab}}{\det(G)} \tag{5.139}$$

where C^{ab} is the matrix whose elements are the cofactors of each element g_{ab} of G. Moreover, G^{-1} is a symmetric matrix since the cofactor matrix C^{ab} is itself symmetric, as a result of the symmetry of G.

We now show that the quantities g^{ab} are the components of a tensor of type $(2,0)$, denoted by $\mathbf{g}^{-1}$, called the **reciprocal metric tensor**. First, observe that by matrix multiplication, we must have

$$g_{bc}g^{ca} = \delta^a_b \tag{5.140}$$

where δ^a_b is the Kronecker delta. Secondly, note that the quantities (δ^a_b) are the components of a tensor of type $(1,1)$, since they obey the appropriate transformation law for a tensor of this type:

$$\frac{\partial \bar{x}^a}{\partial x^p}\frac{\partial x^q}{\partial \bar{x}^b}\delta^p_q = \frac{\partial \bar{x}^a}{\partial x^p}\frac{\partial x^p}{\partial \bar{x}^b} = \frac{\partial \bar{x}^a}{\partial \bar{x}^b} = \bar{\delta}^a_b \tag{5.141}$$

Consequently, by (5.140) and the Quotient Rule, $\mathbf{g}^{-1} = (g^{ab})$ must be a tensor of type $(2,0)$ as claimed.

The metric can be used to lower tensor indices. For example, given an arbitrary tangent vector $\mathbf{v} = (v^b) \in T_p(M)$, we can associate with it the covector whose components are given by

$$v_a = g_{ab}v^b \tag{5.142a}$$

As a matter of convention, we use the same letter to denote the covariant partner of v^b. At the same time, we can use the reciprocal metric tensor to raise tensor indices; for example,

$$v^b = g^{ba}v_a \tag{5.142b}$$

Moreover, the operations of index raising and lowering can be applied to a general tensor $\mathbf{T}$ of type (r,s) in an analogous fashion, as follows

$$g_{ai_1}T^{i_1\ i_2\ i_3\ ...\ i_r}_{\ \ j_1\ j_2\ j_3\ ...\ j_s} = T_{a\ \ j_1\ j_2\ j_3\ ...\ j_s}^{\ \ i_2\ i_3\ ...\ i_r} \tag{5.143a}$$

$$g^{\alpha j_1}T^{i_1\ i_2\ i_3\ ...\ i_r}_{\ \ j_1\ j_2\ j_3\ ...\ j_s} = T^{a\ i_1\ i_2\ i_3\ ...\ i_r}_{\ \ \ \ j_2\ j_3\ ...\ j_s} \tag{5.143b}$$

Observe that when we raise or lower the indices of mixed tensors, we usually leave a *blank space* in the place of the index that was raised or lowered; we shall do this whenever confusion may result.

We can even apply these operations repeatedly. For example, we can write

$$T_{abc\ldots} = g_{ai}g_{bj}g_{ck}T^{ijk\ldots} \tag{5.144a}$$
$$T^{abc\ldots} = g^{ai}g^{bj}g^{ck}T_{ijk\ldots} \tag{5.144b}$$

and so on. In all cases, note that the operations of index raising and index lowering do not change the *rank* of the original tensor, but only its *type*.

5.18. Exercises

1. *(Gradient of a Scalar Field)* Let $\varphi(x^1,\ldots,x^n) \in C^\infty(M)$ be a scalar field on M. Recall that the gradient of φ is given by $\nabla\varphi = \left(\partial\varphi/\partial x^k\right)$. Show that $\nabla\varphi$ is a tensor field of type $(0,1)$.

(Solution)

Let p be an arbitrary point of M, and let (U,ϕ) and $(V,\bar{\phi})$ be overlapping charts on M such that the point $p \in U \bigcap V$. So the coordinates of p are given by both $x^k = \phi^k(p)$ and $\bar{x}^k = \bar{\phi}^k(p)$. Let $\bar{x}^m = \bar{x}^m(x^k)$ be the corresponding coordinate transformation. Since φ is a scalar field on M, then $\bar{\varphi}(\bar{x}^m) = \varphi(x^k)$. Also, by the chain rule, we obtain

$$\frac{\partial\bar{\varphi}}{\partial\bar{x}^k} = \frac{\partial\varphi}{\partial\bar{x}^k} = \frac{\partial x^m}{\partial\bar{x}^k}\frac{\partial\varphi}{\partial x^m},$$

which is the transformation law for a tensor of type $(0,1)$.

2. Given a semi-Riemannian space $(M,\mathbf{g})$, show that for any vectors $\mathbf{x},\mathbf{y} \in T_p(M)$, we have

$$\langle\mathbf{x},\mathbf{y}\rangle = x_b y^b = x^a y_a \tag{5.145a}$$

and

$$\langle\mathbf{x},\mathbf{y}\rangle = \langle\mathbf{x},G\mathbf{y}\rangle = \langle G\mathbf{x},\mathbf{y}\rangle \tag{5.145b}$$

where G is the matrix whose elements are $(g_{\alpha b})$.

(Solution)

By (5.125), we have both

$$\langle\mathbf{x},\mathbf{y}\rangle = g_{ab}x^a y^b = x_b y^b \;\text{ and }\; \langle\mathbf{x},\mathbf{y}\rangle = x^a g_{ab} y^b = x^a y_a$$

so (5.145a) holds true. On the other hand,

$$\langle\mathbf{x},\mathbf{y}\rangle = (g_{ab}x^a)\,y^b = (G\mathbf{x})_b\; y^b = \langle G\mathbf{x},\mathbf{y}\rangle$$
$$\langle\mathbf{x},\mathbf{y}\rangle = x^a\left(g_{ab}y^b\right) = x^a\,(G\mathbf{y})_a = \langle\mathbf{x},G\mathbf{y}\rangle$$

So, (5.145b) follows as well.

3. *(Invariance of the Inner Product)* Given a semi-Riemannian space $(M, \mathbf{g})$, show that under a coordinate transformation $\bar{x}^m = \bar{x}^m(x^k)$ at the point $p \in M$,

$$\langle \bar{\mathbf{u}}, \bar{\mathbf{v}} \rangle = \langle \mathbf{u}, \mathbf{v} \rangle \tag{5.146}$$

for any vectors $\mathbf{u}, \mathbf{v} \in T_p(M)$.

(Solution)

Recalling the transformation laws for tensors of type $(0,2)$ and $(1,0)$, we have

$$\begin{aligned}
\langle \bar{\mathbf{u}}, \bar{\mathbf{v}} \rangle &= \bar{g}_{ab}\bar{u}^a\bar{v}^b = \left(\frac{\partial x^n}{\partial \bar{x}^a}\frac{\partial x^p}{\partial \bar{x}^b}g_{np}\right)\left(\frac{\partial \bar{x}^a}{\partial x^q}u^q\right)\left(\frac{\partial \bar{x}^b}{\partial x^r}v^r\right) \\
&= \left(\frac{\partial x^n}{\partial \bar{x}^a}\frac{\partial \bar{x}^a}{\partial x^q}\right)\left(\frac{\partial x^p}{\partial \bar{x}^b}\frac{\partial \bar{x}^b}{\partial x^r}\right)g_{np}u^q v^r \\
&= \frac{\partial x^n}{\partial x^q}\frac{\partial x^p}{\partial x^r}g_{np}u^q v^r \quad \text{[Chain Rule]} \\
&= \delta^n_q\, \delta^p_r\, g_{np}u^q v^r = g_{qr}u^q v^r = \langle \mathbf{u}, \mathbf{v} \rangle
\end{aligned}$$

4. If $\mathbf{T} = (T_{ab})$ is an antisymmetric tensor, show that $T_{ab}dx^a dx^b = 0$.

(Solution)

Let $\varphi = T_{ab}dx^a dx^b$, and observe that

$$\begin{aligned}
\varphi &= T_{ab}dx^a dx^b = T_{ba}dx^b dx^a \quad \text{[Since } a, b \text{ are dummy indices.]} \\
&= -T_{ab}dx^a dx^b \quad \text{[Since } \mathbf{T} \text{ is antisymmetric.]} \\
&= -\varphi
\end{aligned}$$

Consequently, $\varphi = 0$ as claimed.

5. *(Curl of a Vector)* Let $\mathbf{v} = (v^b)$ be a smooth vector field on M. Show that the quantities

$$F_{ab} = \partial_a v_b - \partial_b v_a \tag{5.147}$$

where $\partial_a = \partial/\partial x^a$, are the components of an *antisymmetric* tensor of type $(0,2)$. Here, $\mathbf{F} = (F_{ab}) = (\partial_a v_b - \partial_b v_a)$ is called the **curl** of $\mathbf{v}$.

(Solution)

Now, $\mathbf{F}$ is antisymmetric, since

$$F_{ab} = \partial_a v_b - \partial_b v_a = -(\partial_b v_a - \partial_a v_b) = -F_{ba}$$

Moreover, we can write:

$$\begin{aligned}
\bar{F}_{ab} &= \frac{\partial}{\partial \bar{x}^a}\bar{v}_b - \frac{\partial}{\partial \bar{x}^b}\bar{v}_a = \frac{\partial}{\partial \bar{x}^a}\left(\frac{\partial x^k}{\partial \bar{x}^b}v_k\right) - \frac{\partial}{\partial \bar{x}^b}\left(\frac{\partial x^m}{\partial \bar{x}^a}v_m\right) \\
&= \frac{\partial x^k}{\partial \bar{x}^b}\frac{\partial v_k}{\partial \bar{x}^a} - \frac{\partial x^m}{\partial \bar{x}^a}\frac{\partial v_m}{\partial \bar{x}^b} + \left(\frac{\partial^2 x^k}{\partial \bar{x}^a \partial \bar{x}^b}v_k - \frac{\partial^2 x^m}{\partial \bar{x}^a \partial \bar{x}^b}v_m\right) \quad \text{[Equality of mixed partials.]} \\
&= \frac{\partial x^k}{\partial \bar{x}^b}\frac{\partial v_k}{\partial \bar{x}^a} - \frac{\partial x^m}{\partial \bar{x}^a}\frac{\partial v_m}{\partial \bar{x}^b} \\
&= \frac{\partial x^k}{\partial \bar{x}^b}\left(\frac{\partial v_k}{\partial x^p}\frac{\partial x^p}{\partial \bar{x}^a}\right) - \frac{\partial x^m}{\partial \bar{x}^a}\left(\frac{\partial v_m}{\partial x^k}\frac{\partial x^k}{\partial \bar{x}^b}\right) \quad \text{[Chain Rule.]} \\
&= \frac{\partial x^k}{\partial \bar{x}^b}\left(\frac{\partial v_k}{\partial x^p}\frac{\partial x^p}{\partial \bar{x}^a}\right) - \frac{\partial x^p}{\partial \bar{x}^a}\left(\frac{\partial v_p}{\partial x^k}\frac{\partial x^k}{\partial \bar{x}^b}\right) \quad \text{[Replace } m \text{ by } p \text{ in 2nd term.]} \\
&= \frac{\partial x^p}{\partial \bar{x}^a}\frac{\partial x^k}{\partial \bar{x}^b}\left(\frac{\partial v_k}{\partial x^p} - \frac{\partial v_p}{\partial x^k}\right)
\end{aligned}$$

The last equality implies that

$$\bar{F}_{ab} = \frac{\partial x^p}{\partial \bar{x}^a}\frac{\partial x^k}{\partial \bar{x}^b}F_{pk}$$

and consequently, $\mathbf{F} = (F_{ab})$ is a tensor of type $(0,2)$.

6. *(Diagonalization of the Metric)* The metric $\mathbf{g}$ on a three-dimensional manifold M has the matrix form

$$G = \begin{bmatrix} 1 & 1 & 3 \\ 1 & 5 & 1 \\ 3 & 1 & 1 \end{bmatrix}$$

at some fixed point $p \in M$. By explicitly tracing the steps from equation (5.134) to equation (5.137), write the line element in semi-canonical, and canonical form.

(Solution)

A simple calculation shows that the eigenvalues of G are $\lambda_1 = 6$, $\lambda_2 = 3$, $\lambda_3 = -2$, with the corresponding eigenvectors

$$\mathbf{a}_1 = \begin{bmatrix} 1 \\ 2 \\ 1 \end{bmatrix}, \quad \mathbf{a}_2 = \begin{bmatrix} 1 \\ -1 \\ 1 \end{bmatrix}, \quad \mathbf{a}_3 = \begin{bmatrix} -1 \\ 0 \\ 1 \end{bmatrix}$$

We normalize these eigenvectors (relative to the *Euclidean metric*), to obtain

$$\mathbf{e}_1 = \begin{bmatrix} 1/\sqrt{6} \\ 2/\sqrt{6} \\ 1/\sqrt{6} \end{bmatrix}, \quad \mathbf{e}_2 = \begin{bmatrix} 1/\sqrt{3} \\ -1/\sqrt{3} \\ 1/\sqrt{3} \end{bmatrix}, \quad \mathbf{e}_3 = \begin{bmatrix} -1/\sqrt{2} \\ 0 \\ 1/\sqrt{2} \end{bmatrix}$$

and form the matrix Λ as in (5.135):

$$\Lambda = \begin{bmatrix} \frac{1}{\sqrt{6}} & \frac{1}{\sqrt{3}} & \frac{-1}{\sqrt{2}} \\ \frac{2}{\sqrt{6}} & \frac{-1}{\sqrt{3}} & 0 \\ \frac{1}{\sqrt{6}} & \frac{1}{\sqrt{3}} & \frac{1}{\sqrt{2}} \end{bmatrix}$$

From (5.136), we find that

$$\bar{G} = \Lambda^T G \Lambda = \begin{bmatrix} 6 & 0 & 0 \\ 0 & 3 & 0 \\ 0 & 0 & -2 \end{bmatrix}$$

which is the diagonal matrix advertised earlier. In terms of the new basis, the *semi-canonical form* of the line element is

$$ds^2 = 6(d\bar{x}^1)^2 + 3(d\bar{x}^2)^2 - 2(d\bar{x}^3)^2$$

and the *canonical form* is

$$ds^2 = (d\bar{x}^1)^2 + (d\bar{x}^2)^2 - (d\bar{x}^3)^2$$

Observe that once we had obtained the eigenvalues of G, we could have written down both forms immediately.

7. *(Conformal Metrics)* A manifold M is endowed with two different metrics to form the semi-Riemannian spaces $(M, \mathbf{g})$ and $(M, \bar{\mathbf{g}})$. We say that the metrics $\mathbf{g}$ and $\bar{\mathbf{g}}$ are said to be **conformal** (or **conformally related**) to each other if

$$\bar{g}_{ab} = \Omega^2 g_{ab} \tag{5.148}$$

where $\Omega = \Omega(\mathbf{x})$ is a smooth non-vanishing function on M. Show that the angle between any two non-zero vectors is the same for conformal metrics.

(Solution)

Let $\mathbf{u}, \mathbf{v}$ be non-zero vectors, so from (130) we have

$$\begin{aligned} \overline{(\cos\angle(\mathbf{u},\mathbf{v}))} &= \frac{\bar{g}_{mn}u^m v^n}{\sqrt{|\bar{g}_{pq}u^p u^q|}\cdot\sqrt{|\bar{g}_{rs}v^r v^s|}} = \frac{\Omega^2 g_{mn}u^m v^n}{\sqrt{|\Omega^2 g_{pq}u^p u^q|}\cdot\sqrt{|\Omega^2 g_{rs}v^r v^s|}} \\ &= \frac{\Omega^2 g_{mn}u^m v^n}{|\Omega|^2\sqrt{|g_{pq}u^p u^q|}\cdot\sqrt{|g_{rs}v^r v^s|}} = \frac{g_{mn}u^m v^n}{\sqrt{|g_{pq}u^p u^q|}\cdot\sqrt{|g_{rs}v^r v^s|}} \\ &= (\cos\angle(\mathbf{u},\mathbf{v})) \end{aligned}$$

8. *(Null Vectors)* Given a semi-Riemannian manifold $(M, \mathbf{g})$, show that at each point $p \in M$, there always exists at least one *non-zero* null vector $\mathbf{u} \in T_p(M)$.

(Solution)

We can perform a change of basis at p, and write the line element in the semi-canonical form (5.137) as

$$ds^2 = \lambda_1(d\bar{x}^1)^2 + \lambda_2(d\bar{x}^2)^2 + \ldots + \lambda_n(d\bar{x}^n)^2$$

Now, since $\bar{g}$ is an indefinite metric, then at least one of the eigen- values must be negative, and at least one of them must be positive. Without loss of generality, let's assume that $\lambda_1 < 0$ and $\lambda_2 > 0$. Now, consider the vectors

$$\mathbf{u}(t) = \bar{\mathbf{e}}_1 + t\bar{\mathbf{e}}_2$$

where $\bar{\mathbf{e}}_1$ and $\bar{\mathbf{e}}_2$ are the eigenvectors corresponding to λ_1 and λ_2, respectively, and t is a real. Since $T_p(M)$ is a vector space, $\mathbf{u}(t)$ is also a vector. Observe that

$$\begin{aligned} \|\mathbf{u}(t)\|^2 &= \bar{\mathbf{g}}(\bar{\mathbf{e}}_1 + t\,\bar{\mathbf{e}}_2\,, \bar{\mathbf{e}}_1 + t\,\bar{\mathbf{e}}_2) = \bar{\mathbf{g}}(\bar{\mathbf{e}}_1, \bar{\mathbf{e}}_1) + 2t\,\bar{\mathbf{g}}(\bar{\mathbf{e}}_1, \bar{\mathbf{e}}_2) + t^2\,\bar{\mathbf{g}}(\bar{\mathbf{e}}_2, \bar{\mathbf{e}}_2) \\ &= \bar{g}_{11} + t^2\,\bar{g}_{22} \quad \text{[Since } \bar{g}_{12} = \bar{\mathbf{g}}(\bar{\mathbf{e}}_1, \bar{\mathbf{e}}_2) = 0\,.] \\ &= \lambda_1 + \lambda_2 t^2 \qquad \text{[Since } \bar{g}_{11} = \lambda_1 \text{ and } \bar{g}_{22} = \lambda_2.] \end{aligned}$$

Now, if we set $t = t_0 = \sqrt{-\lambda_2/\lambda_1}$ [Note that t is real since by hypothesis, $\lambda_1 < 0$ and $\lambda_2 > 0$], then we will have $\|\mathbf{u}(t_0)\| = 0$, and $\mathbf{u}(t_0)$ will be a *non-zero* null vector.

Chapter 6

Tensor Calculus

[The tensor calculus] is the debauch of indices.

Elie Cartan

6.1. Introduction

In the last chapter, we introduced the concept of a manifold, and we confined ourselves to the basics of tensor algebra. As the laws of physics are usually expressed as differential equations, however, we need to specify how we can differentiate tensors.

In the discussion that follows, we will show that the partial derivative of a tensor is not a well defined operation on a general manifold, and we must replace it with a new operation, called *covariant differentiation*, to enable us to develop the tensor calculus.

Later on in this chapter, we will see how the tensor calculus provides us with the machinery we need to introduce an important concept called the *curvature of a manifold*. This is a crucial concept in the Theory of Relativity as the effects of a gravitational field may be under- stood in terms of the *curvature of spacetime*, while the absence of such a field demands that the corresponding spacetime, called Minkowski space, is *flat*.

6.2. Covariant Differentiation

Assume that we are working in the Euclidean space $E^n = (R^n, \delta)$; recall that this is simply the manifold R^n endowed with the Euclidean metric $\delta = (\delta_{ab})$. Assume further, that we introduce a *rectangular coordinate* system (x^k) on our space, along with the standard basis vectors

$$\mathbf{e}_1 = (1,0,0,...,0),\ \mathbf{e}_2 = (0,1,0,...,0),...,\ \mathbf{e}_n = (0,0,0,...,0,1).$$

Now, if $\mathbf{x} = \mathbf{x}(t)$ is a smooth curve in E^n, and $\mathbf{V} = V^a(\mathbf{x})\mathbf{e}_a$ is a smooth vector field defined at each point $\mathbf{x} \in E^n$, then in particular, $\mathbf{V}$ will be defined at each point of the given curve. From ordinary vector calculus, the rate of change of $\mathbf{V}$ along the curve is given by the familiar expression

$$\frac{d\mathbf{V}}{dt} = \left(\frac{\partial V^a}{\partial x^b} \frac{dx^b}{dt} \right) \mathbf{e}_a \tag{6.1}$$

Now, this equation contains two implicit assumptions: (1) the basis vectors $\{\mathbf{e}_a\}$ remain *constant* as we move along the curve α, and (2) the partial derivatives $\partial V^a/\partial x^b$ can be calculated. For a general manifold, however, both of these assumptions present serious problems.

First, for a general manifold M, we know that the basis vectors $\{\mathbf{e}_a\}$ *change* as we move along the manifold. This is a direct result of the fact that given two distinct points $p, q \in M$, the tangent spaces $T_p(M)$ and $T_q(M)$ are *distinct* vector spaces, so any basis constructed at the point p will usually differ from one erected at the point q.

Consequently, the partial derivative formula doesn't make any sense either. In order to see this, observe that the partial derivative of the a-th field component is given by

$$\frac{\partial V^a}{\partial x^b} = \lim_{\Delta x^b \longrightarrow 0} \frac{V^a(x^1, \ldots, x^b + \Delta x^b, \ldots, x^n) - V^a(x^1, \ldots, x^b, \ldots, x^n)}{\Delta x^b} \tag{6.2}$$

where, before passing to the limit, we must compare the value of the component V^a at the point $q(x^1, \ldots, x^b + \Delta x^b, \ldots, x^n)$ with that at the point $p(x^1, \ldots, x^b, \ldots, x^n)$. Since these vectors live in different tangent spaces, we can't perform this subtraction, unless we make the *assumption* that we can somehow *transport* the vector $\mathbf{V}(q) \in T_q(M)$ to the point p, *without changing it in any way*, and then perform the operation of subtraction in the tangent space $T_p(M)$, before passing to the limit.

This operation, which involves moving a vector - without change - from one point of a manifold to another, is called **parallel transport**. Now, for the Euclidean space $E^n = (R^n, \delta)$, the operation of parallel transport is performed by simply requiring that the *rectangular* components V^a of the vector *remain constant* as we move from p to q. For a general manifold, however, the operation is not quite so simple, and we must *define* what we mean by parallel transport.

To accomplish this task, consider a smooth curve on a general mani- fold M, given by $\alpha = \alpha(t)$ where $t \in [a, b]$, such that $\alpha(a) = p$, and $\alpha(b) = q$. Without loss of generality, assume that the image $\alpha(t)$ is contained in a single coordinate chart (U, ϕ) so the coordinate representation of α is given by $x^k = x^k(t)$. In addition, we assume that these points are infinite-simally close in the sense that if $x^k = \phi^k(p)$ are the coordinates of p, then the coordinates of q are given by $x^k + \Delta x^k = \phi^k(q)$, where Δx^k are small quantities. Also, let the vectors

$$\mathbf{e}_b(\mathbf{x}) = \left. \frac{\partial}{\partial x^b} \right|_{\mathbf{x}} \quad \text{and } \mathbf{e}_b(\mathbf{x} + \Delta \mathbf{x}) = \left. \frac{\partial}{\partial x^b} \right|_{\mathbf{x} + \Delta \mathbf{x}} \qquad (b = 1, 2, 3, \ldots, n)$$

denote the natural basis vectors for $T_p(M)$ and $T_q(M)$ respectively.

We now suppose, as shown in Figure 6.1, that the basis vector $\mathbf{e}_b(\mathbf{x} + \Delta \mathbf{x})$ (with b *fixed*), which lives at the point q, is parallel transported - *without change* - along the curve α to the point p. Since the parallel-transported vector $\hat{\mathbf{e}}_b(\mathbf{x})$ and the basis vector $\mathbf{e}_b(\mathbf{x})$ both live in the same tangent space $T_p(M)$, their difference can be written as a linear combination of the basis vectors at p; in other words,

$$\hat{\mathbf{e}}_b(\mathbf{x}) - \mathbf{e}_b(\mathbf{x}) = (\Delta x^a)\ \Gamma^c_{ab} \mathbf{e}_c(\mathbf{x}) \tag{6.3}$$

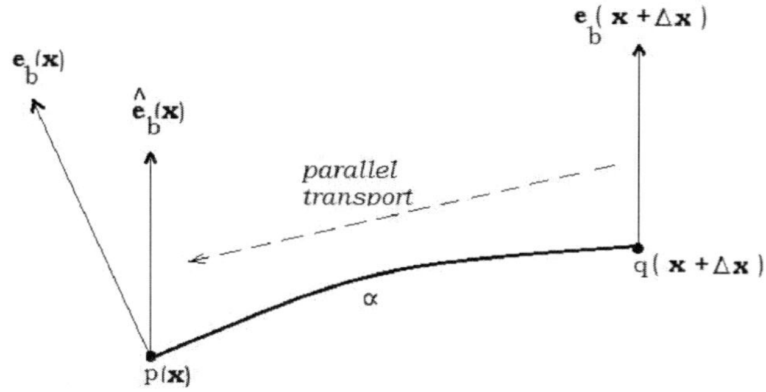

Figure 6.1. The basis vector $\mathbf{e}_b(\mathbf{x}+\Delta\mathbf{x})$ is parallel transported to p.

where the quantities Γ^c_{ab} serve as expansion coefficients, and we have assumed the presence of a linear factor Δx^a, since we would expect the difference between these two vectors to approach zero as $\Delta x \longrightarrow 0$.

Rearranging terms in (6.3), the **parallel-transported** basis vector is *defined* as[1]

$$\hat{\mathbf{e}}_b(\mathbf{x}) = \mathbf{e}_b(\mathbf{x}) + (\Delta x^a)\ \Gamma^c_{ab}\mathbf{e}_c(\mathbf{x}) \tag{6.4}$$

The expansion coefficients Γ^c_{ab} are called the **connection coefficients**. Although our notation might suggest that these quantities have tensor character, this is not the case, and later on, we will demonstrate that they do not represent the components of a tensor.

Given the rule for the parallel transport of basis vectors in (6.4), we can take any vector, say

$$\mathbf{V}(\mathbf{x}+\Delta\mathbf{x}) = V^b(\mathbf{x}+\Delta\mathbf{x})\,\mathbf{e}_b(\mathbf{x}+\Delta\mathbf{x}),$$

at the point q and parallel transport it along α to p to obtain the vector

$$\hat{\mathbf{V}}(\mathbf{x}) = V^b(\mathbf{x}+\Delta\mathbf{x})\,\hat{\mathbf{e}}_b(\mathbf{x}) = V^b(\mathbf{x}+\Delta\mathbf{x})\,\mathbf{e}_b(\mathbf{x}) + (\Delta x^a)\ \Gamma^c_{ab}V^b(\mathbf{x}+\Delta\mathbf{x})\,\mathbf{e}_c(\mathbf{x}) \tag{6.5}$$

We now parallel transport the vector $\mathbf{V}(\mathbf{x}+\Delta\mathbf{x})$ along the curve α from q to p, to obtain the vector $\hat{\mathbf{V}}(\mathbf{x})$ at p, and then we compute the difference quotient:

$$\begin{aligned}
\frac{\hat{\mathbf{V}}(\mathbf{x})-\mathbf{V}(\mathbf{x})}{\Delta t} &= \frac{1}{\Delta t}\left\{V^a(\mathbf{x}+\Delta\mathbf{x})\,\hat{\mathbf{e}}_a(\mathbf{x}) - V^a(\mathbf{x})\,\mathbf{e}_a(\mathbf{x})\right\}\\
&= \frac{1}{\Delta t}\left\{\left[V^a(\mathbf{x}+\Delta\mathbf{x}) - V^a(\mathbf{x})\right]\mathbf{e}_a(\mathbf{x}) + (\Delta x^a)\ \Gamma^c_{ab}V^b(\mathbf{x}+\Delta\mathbf{x})\,\mathbf{e}_c(\mathbf{x})\right\}\\
&= \left(\frac{\Delta x^a}{\Delta t}\right)\left\{\frac{V^c(\mathbf{x}+\Delta\mathbf{x}) - V^c(\mathbf{x})}{\Delta x^a} + \Gamma^c_{ab}V^b(\mathbf{x}+\Delta\mathbf{x})\right\}\mathbf{e}_c(\mathbf{x})
\end{aligned}$$

If we now let $\Delta t \longrightarrow 0$, then $\Delta x \longrightarrow 0$, and we obtain

$$\frac{D\mathbf{V}}{Dt} = \lim_{\Delta t\longrightarrow 0}\frac{\hat{\mathbf{V}}(\mathbf{x})-\mathbf{V}(\mathbf{x})}{\Delta t} = \frac{dx^a}{dt}\left\{\frac{\partial V^c}{\partial x^a} + \Gamma^c_{ab}V^b\right\}\mathbf{e}_c(\mathbf{x}) \tag{6.6}$$

which is called the **absolute derivative** of the vector $\mathbf{V} = (V^c)$ along the curve α. Observe that $D\mathbf{V}/Dt \in T_p(M)$, and consequently, it is a **vector**. Since $\mathbf{X} = (dx^a/dt)$ is just the tangent vector along α, *the absolute derivative tells us how fast V is changing in the direction of the tangent vector*. We shall also use the conventional notation

$$\frac{D\mathbf{V}}{Dt} \equiv \nabla_{\mathbf{X}}\mathbf{V} = X^a \left(\frac{\partial V^c}{\partial x^a} + \Gamma^c_{ab} V^b \right) \mathbf{e}_c\,(\mathbf{x}) \tag{6.7}$$

By the Quotient Rule, the quantities in the parentheses of (6.7) must form the components of a tensor of type $(1,1)$, called the **covariant derivative** $\nabla\mathbf{V}$ of the vector $\mathbf{V} = (V^c)$. This term derives from the fact that *the operation covariant differentiation increases the covariant (lower) index of the object differentiated by unity*. It is customary to write

$$\nabla\mathbf{V} = (\nabla_b V^a)\; \mathbf{e}_a \otimes \mathbf{d}x^b \tag{6.8}$$

where

$$\nabla_b V^a = \frac{\partial V^a}{\partial x^b} + \Gamma^a_{bc} V^c, \tag{6.9}$$

are the *components* of the covariant derivative $\nabla\mathbf{V}$.

We turn our attention to defining the absolute derivative of a scalar field. To this end, let $\varphi(\mathbf{x}) \in C^\infty(M)$ be a smooth scalar field on M, and recall that $\mathfrak{I}^0_0(M) \equiv C^\infty(M)$, so that φ may also be regarded as a tensor field of type $(0,0)$ on M. Obviously, in this case, we don't have to worry about parallel transport, and the rate of change of φ along the curve α, is simply given by its directional derivative:

$$\frac{D\varphi}{Dt} \equiv \nabla_{\mathbf{X}}(\varphi) = \frac{dx^k}{dt}\frac{\partial\varphi}{\partial x^k} = X^k \frac{\partial\varphi}{\partial x^k} \tag{6.10}$$

The last equation suggests that we should define the covariant derivative of a scalar field φ as the *one-form*

$$\nabla\varphi = (\nabla_k\varphi)\,\mathbf{d}x^k = \frac{\partial\varphi}{\partial x^k}\mathbf{d}x^k, \tag{6.11}$$

since then we have

$$\nabla_{\mathbf{X}}\,\varphi = \langle \nabla\varphi, \mathbf{X} \rangle = \frac{\partial\varphi}{\partial x^k}\left\langle \mathbf{d}x^k, \mathbf{X} \right\rangle = X^k \frac{\partial\varphi}{\partial x^k}, \tag{6.12}$$

which is an expression that clearly agrees with (6.10).

Now, we require that the operation of absolute differentiation satisfies the **Leibnitz rule**; that is,

$$\nabla_{\mathbf{X}}(\mathbf{S} \otimes \mathbf{T}) = \mathbf{S} \otimes (\nabla_{\mathbf{X}}\mathbf{T}) + (\nabla_{\mathbf{X}}\mathbf{S}) \otimes \mathbf{T} \tag{6.13}$$

for any smooth tensor fields $\mathbf{S}$ and $\mathbf{T}$. Moreover, we shall demand that this operation satisfies

$$\nabla_{\mathbf{X}}\langle \omega, \mathbf{V} \rangle = \langle \omega, \nabla_{\mathbf{X}}\mathbf{V} \rangle + \langle \nabla_{\mathbf{X}}\omega, \mathbf{V} \rangle \tag{6.14}$$

for any one-form field ω and vector field $\mathbf{V}$. Requirement (6.14) is equivalent to saying that the operator $\nabla_{\mathbf{X}}$ *commutes* with inner products.

An immediate consequence of the first requirement [See the Exercises] is that the absolute derivative of a one-form field $\omega(\mathbf{x}) = \omega_c \mathbf{d}x^c$ turns out to be the one-form given by

$$\frac{D\omega}{Dt} \equiv \nabla_{\mathbf{X}}\omega = X^a \left(\frac{\partial \omega_c}{\partial x^a} - \Gamma^b_{ac}\omega_b \right) \mathbf{d}x^c \tag{6.14a}$$

where the quantities

$$\nabla_a \omega_c = \frac{\partial \omega_c}{\partial x^a} - \Gamma^b_{ac}\omega_b \tag{6.15}$$

are the components of a tensor of type $(0,2)$, called the **covariant derivative** $\nabla\omega$ of the one-form $\omega = (\omega_c)$. That is, for a one-form ω we have

$$\nabla\omega = (\nabla_a \omega_c)\ \mathbf{d}x^a \otimes \mathbf{d}x^c \tag{6.16}$$

In a similar manner, it can be shown that the covariant derivative of a general tensor $\mathbf{T}$ of type (p,q) results in a tensor of type $(p,q+1)$. *So, taking the covariant derivative of a tensor increases its covariant character by unity.* Using the above definitions, we can compute the covariant derivative of any tensor by first writing the partial derivative of that tensor, adding a "plus term" for each contravariant (upper) index, and adding a "minus term" for each covariant (lower) index. In other words, if

$$\mathbf{T} = (T^{\alpha_1 \ldots a_p}_{b_1 \ldots b_q})\ \mathbf{e}_{a_1} \otimes \cdots \otimes \mathbf{e}_{a_p} \otimes \mathbf{d}x^{b_1} \cdots \otimes \mathbf{d}x^{b_q}$$

then we have

$$\nabla\mathbf{T} = \left(\nabla_k T^{a_1 \ldots a_p}_{b_1 \ldots b_q}\right) \mathbf{e}_{a_1} \otimes \cdots \otimes \mathbf{e}_{a_p} \otimes \mathbf{d}x^{b_1} \cdots \otimes \mathbf{d}x^{b_q} \otimes \mathbf{d}x^k$$

where the components of $\nabla\mathbf{T}$ are given by

$$\begin{aligned}\nabla_k \left(T^{a_1 \ldots a_p}_{b_1 \ldots b_q}\right) &= \frac{\partial}{\partial x^k}\left(\nabla_k T^{a_1 \ldots a_p}_{b_1 \ldots b_q}\right) + \Gamma^{a_1}_{kc}\left(T^{c \ldots a_p}_{b_1 \ldots b_q}\right) + \ \ldots\ + \Gamma^{a_p}_{kc}\left(T^{a_1 \ldots c}_{b_1 \ldots b_q}\right) \\ &\quad -\Gamma^c_{kb_1}\left(T^{\alpha_1 \ldots a_p}_{c \ \ldots b_q}\right) - \ \ldots\ - \Gamma^c_{kb_q}\left(T^{\alpha_1 \ldots a_p}_{b_1 \ldots c}\right)\end{aligned} \tag{6.17}$$

One final word about notation. *Physicists generally use a semi-colon to denote covariant differentiation, and a comma to denote partial differentian.* For example, under this convention, (9) and (15) become

$$\nabla_b V^a \equiv V^a_{;b} = V^a_{,b} + \Gamma^a_{bc} V^c \tag{6.18a}$$

$$\nabla_a \omega_c \equiv \omega_{c;a} = \omega_{c,a} - \Gamma^c_{ab}\omega_c \tag{6.18b}$$

We shall employ this notation whenever it is convenient to do so.

6.3. Exercises

1. Show that for any scalar field φ, and vector field $\mathbf{V}$

$$\nabla_{\mathbf{X}}(\varphi\mathbf{V}) = \varphi(\nabla_{\mathbf{X}}\mathbf{V}) + \mathbf{V}(\nabla_{\mathbf{X}}\varphi) \tag{6.19}$$

(Solution)

Let $\mathbf{W} = \varphi \mathbf{V}$ so that $W^c = \varphi V^c$ and (6.7) gives

$$\nabla_{\mathbf{X}}\mathbf{W} = X^a \left(\frac{\partial W^c}{\partial x^a} + \Gamma^c_{ab} W^b \right) \mathbf{e}_c = X^a \left\{ \left(\varphi \frac{\partial V^c}{\partial x^a} + V^c \frac{\partial \varphi}{\partial x^a} \right) + \Gamma^c_{ab} \varphi V^b \right\} \mathbf{e}_c$$

Consequently,

$$\begin{aligned} \nabla_{\mathbf{X}}(\varphi \mathbf{V}) &= \varphi \left\{ X^a \left(\frac{\partial V^c}{\partial x^a} + \Gamma^c_{ab} V^b \right) \mathbf{e}_c \right\} + (V^c \mathbf{e}_c) X^a \frac{\partial \varphi}{\partial x^a} \\ &= \varphi (\nabla_{\mathbf{X}} \mathbf{V}) + \mathbf{V} (\nabla_{\mathbf{X}} \varphi) \end{aligned}$$

2. *(Absolute Derivative of a One-form)* Given a one-form field $\omega(\mathbf{x})$, show that

$$\nabla_{\mathbf{X}} \omega = X^a \left(\frac{\partial \omega_p}{\partial x^a} - \Gamma^b_{ap} \omega_b \right) \mathbf{d}x^p$$

(Solution)

For an arbitrary vector field $\mathbf{V}(\mathbf{x}) = V^a \mathbf{e}_a(\mathbf{x})$, observe that $\langle \omega, \mathbf{V} \rangle = \omega_b V^b$ is a scalar field. Consequently, by (6.11),

$$\nabla_{\mathbf{X}} \langle \omega, \mathbf{V} \rangle = X^a \frac{\partial}{\partial x^a} \left(\omega_b V^b \right) = X^a \omega_b \frac{\partial V^b}{\partial x^a} + X^a V^b \frac{\partial \omega_b}{\partial x^a}$$

On the other hand, by (6.14),

$$\begin{aligned} \nabla_{\mathbf{X}} \langle \omega, \mathbf{V} \rangle &= \langle \omega, \nabla_{\mathbf{X}} \mathbf{V} \rangle + \langle \nabla_{\mathbf{X}} \omega, \mathbf{V} \rangle = \left\langle \omega_b \mathbf{d}x^b, \nabla_{\mathbf{X}} \mathbf{V} \right\rangle + \langle \nabla_{\mathbf{X}} \omega, \mathbf{V} \rangle \\ &= \left\langle \omega_b \mathbf{d}x^b, X^a \left(\frac{\partial V^c}{\partial x^a} + \Gamma^c_{ap} V^p \right) \mathbf{e}_c \right\rangle + \langle \nabla_{\mathbf{X}} \omega, \mathbf{V} \rangle \\ &= X^a \omega_c \left(\frac{\partial V^c}{\partial x^a} + \Gamma^c_{ap} V^p \right) + \langle \nabla_{\mathbf{X}} \omega, \mathbf{V} \rangle \end{aligned}$$

Combining the last two equations, gives

$$\begin{aligned} X^a \omega_b \frac{\partial V^b}{\partial x^a} + X^a V^b \frac{\partial \omega_b}{\partial x^a} &= X^a \omega_c \left(\frac{\partial V^c}{\partial x^a} + \Gamma^c_{ap} V^p \right) + \langle \nabla_{\mathbf{X}} \omega, \mathbf{V} \rangle \\ &\Longrightarrow X^a V^p \frac{\partial \omega_p}{\partial x^a} = X^a \omega_b \Gamma^b_{ap} V^p + V^p (\nabla_{\mathbf{X}} \omega)_p \\ &\Longrightarrow V^p \left[X^a \left(\frac{\partial \omega_p}{\partial x^a} - \omega_b \Gamma^b_{ap} \right) - (\nabla_{\mathbf{X}} \omega)_p \right] = 0 \end{aligned}$$

Since $\mathbf{V}$ is arbitrary, we find that

$$(\nabla_{\mathbf{X}} \omega)_p = X^a \left\{ \frac{\partial \omega_p}{\partial x^a} - \Gamma^b_{ap} \omega_b \right\}$$

and consequently,

$$\nabla_{\mathbf{X}}\omega = (\nabla_{\mathbf{X}}\omega)_p \, \mathbf{d}x^p = X^a \left\{ \frac{\partial \omega_p}{\partial x^a} - \omega_b \Gamma^b_{ap} \omega_b \right\} \mathbf{d}x^p$$

3. *(Absolute Derivative of Basis Vectors)* Show that for any vector field $\mathbf{X}$ and coordinate basis vectors $\mathbf{e}_a$ and $\mathbf{e}_b$:

$$\nabla_{\mathbf{X}}(\mathbf{e}_b) = X^a \, \Gamma^c_{ab} \mathbf{e}_c \tag{6.20a}$$

$$\nabla_a(\mathbf{e}_b) \equiv \nabla_{\mathbf{e}_a}(\mathbf{e}_b) = \Gamma^c_{ab} \mathbf{e}_c \tag{6.20b}$$

(Solution #1)

For any *fixed* index b, we can write $\mathbf{e}_b = (\delta^c_b)\,\mathbf{e}_c$, so that δ^c_b are the components of $\mathbf{e}_c$. We can write,

$$\begin{aligned} \nabla_{\mathbf{X}}(\mathbf{e}_b) &= X^a \left\{ \frac{\partial}{\partial x^a}(\delta^c_b) + \Gamma^c_{ad}\delta^d_b \right\} \mathbf{e}_c \\ &= X^a \{(0) + \Gamma^c_{ab}\}\, \mathbf{e}_c = X^a \, \Gamma^c_{ab} \mathbf{e}_c \end{aligned}$$

which establishes (6.20a). Now, for *fixed* k, we have $\mathbf{e}_k = (\delta^a_k)\,\mathbf{e}_a$. So, if we set $\mathbf{X} = \mathbf{e}_k$, then $X^a = \delta^a_k$, and the last equation becomes

$$\nabla_{\mathbf{e}_k}(\mathbf{e}_b) = \delta^a_k \, \Gamma^c_{ab} \mathbf{e}_c = \Gamma^c_{kb} \mathbf{e}_c$$

Finally, replacing free index k by a, the second result follows.

(Solution #2)

Recalling Figure 6.1, equation (6.3) implies that

$$\frac{\hat{\mathbf{e}}_b(\mathbf{x}) - \mathbf{e}_b(\mathbf{x})}{\Delta t} = \left(\frac{\Delta x^a}{\Delta t} \right) \Gamma^c_{ab} \mathbf{e}_c(\mathbf{x})$$

Now we let $\Delta t \longrightarrow 0$ to obtain

$$\frac{D}{Dt}(\mathbf{e}_b) = \nabla_{\mathbf{X}}(\mathbf{e}_b) = X^a \, \Gamma^c_{ab} \mathbf{e}_c$$

which is again, (6.20a). Once again, (6.20b) follows immediately by setting $\mathbf{X} = \mathbf{e}_a$ in (6.20a).

4. *(Absolute Derivative of Basis One-forms)* Show that for any coordinate basis one-forms $\mathbf{d}x^a$ and $\mathbf{d}x^b$, and for any vector field $\mathbf{X}$:

$$\nabla_{\mathbf{X}}(\mathbf{d}x^b) = -\,\Gamma^b_{ak} X^a \mathbf{d}x^k \tag{6.21a}$$

$$\nabla_a\left(\mathbf{d}x^b\right) \equiv \nabla_{\mathbf{e}_a}\left(\mathbf{d}x^b\right) = -\,\Gamma^b_{ak} \mathbf{d}x^k \tag{6.21b}$$

(Solution)

Recall that $\langle \mathbf{d}x^b, \mathbf{e}_a \rangle = \delta^b_a$; consequently,

$$\nabla_{\mathbf{X}} \langle \mathbf{d}x^a, \mathbf{e}_b \rangle = 0 = \langle \mathbf{d}x^a, \nabla_{\mathbf{X}} \mathbf{e}_b \rangle + \langle \nabla_{\mathbf{X}} \mathbf{d}x^a, \mathbf{e}_b \rangle,$$

so that

$$\begin{aligned} \langle \nabla_{\mathbf{X}} \mathbf{d}x^a, \mathbf{e}_b \rangle &= -\langle \mathbf{d}x^a, \nabla_{\mathbf{X}} \mathbf{e}_b \rangle = -\langle \mathbf{d}x^a, X^p \, \Gamma^c_{pb} \mathbf{e}_c \rangle \\ &\Longrightarrow \; (\nabla_{\mathbf{X}} \mathbf{d}x^a)_b = -X^p \, \Gamma^a_{pb} \end{aligned}$$

On the other hand, since $\nabla_{\mathbf{X}} \mathbf{d}x^a = (\nabla_{\mathbf{X}} \mathbf{d}x^a)_b \, \mathbf{d}x^b$ then

$$\nabla_{\mathbf{X}} \mathbf{d}x^a = (\nabla_{\mathbf{X}} \mathbf{d}x^a)_b \, \mathbf{d}x^b = -X^p \, \Gamma^a_{pb} \mathbf{d}x^b$$

Finally, (6.24b) follows immediately from setting $\mathbf{X} = \mathbf{e}_a$.

5. *(Absolute Derivative of a Mixed Tensor)* Let $\mathbf{T} = \left(T^a_b\right)$ be a tensor of type $(1,1)$. By using the Leibnitz rule, along with the relations (6.20a) and (6.21a), show that

$$\nabla_{\mathbf{X}} \mathbf{T} = X^s \left(\frac{\partial T^a_b}{\partial x^s} + \Gamma^a_{sq} T^q_b - \Gamma^p_{sb} T^a_p \right) \mathbf{e}_a \otimes \mathbf{d}x^b \tag{6.22}$$

(Solution)

We have

$$\begin{aligned} \nabla_{\mathbf{X}} \mathbf{T} &= \nabla_{\mathbf{X}} \left(T^a_b \mathbf{e}_a \otimes \mathbf{d}x^b \right) \\ &= (\nabla_{\mathbf{X}} T^a_b) \, \mathbf{e}_a \otimes \mathbf{d}x^b + T^q_b \, (\nabla_{\mathbf{X}} \, \mathbf{e}_q) \otimes \mathbf{d}x^b + T^a_p \, \mathbf{e}_a \otimes (\nabla_{\mathbf{X}} \, \mathbf{d}x^p) \\ &= X^s \frac{\partial T^a_b}{\partial x^s} \mathbf{e}_a \otimes \mathbf{d}x^b + T^q_b \, (X^s \, \Gamma^a_{sq} \mathbf{e}_a) \otimes \mathbf{d}x^b + T^a_p \, \mathbf{e}_a \otimes \left(-\Gamma^p_{sb} X^s \mathbf{d}x^b \right) \\ &= X^s \left(\frac{\partial T^a_b}{\partial x^s} + \Gamma^a_{sq} T^q_b - \Gamma^p_{sb} T^a_p \right) \mathbf{e}_a \otimes \mathbf{d}x^b \end{aligned}$$

Observe that the Leibnitz rule, along with the sets of relations (6.20) and (6.21), allow us to take the absolute derivative of any tensor.

6. *(Connection One-forms)* Assuming a *coordinate basis*, we define the **connection one-forms** by

$$\Gamma^a_c = \Gamma^a_{bc} \mathbf{d}x^b \tag{6.23}$$

Show that for any one-form field $\omega(\mathbf{x})$ and vector field $\mathbf{V}(\mathbf{x})$,

$$\nabla \omega = (\mathbf{d}\omega_a - \Gamma^c_a \omega_c) \mathbf{d}x^a \tag{6.24a}$$

$$\nabla \mathbf{V} = (\mathbf{d}V^a + \Gamma^a_c V^c) \, \mathbf{e}_a \tag{6.24b}$$

(Solution)

To establish (6.24a), we first observe that from (6.15) we have:

$$\frac{\partial \omega_a}{\partial x^b} = \nabla_b \omega_a + \Gamma^c_{ba}\omega_c$$

We now multiply both sides of this equation by $\mathbf{d}x^b$ to get

$$\begin{aligned}\frac{\partial \omega_a}{\partial x^b}\mathbf{d}x^b &= \nabla_b\omega_a\mathbf{d}x^b + \Gamma^c_{ba}\omega_c\mathbf{d}x^b \\ \mathbf{d}\omega_a &= \nabla_b\omega_a\mathbf{d}x^b + \Gamma^c_a\omega_c \\ &\Longrightarrow (\mathbf{d}\omega_a - \Gamma^c_a\omega_c) = \nabla_b\omega_a\mathbf{d}x^b\end{aligned}$$

Finally we multiply both sides of the last equation by $\mathbf{d}x^a$ to obtain

$$(\mathbf{d}\omega_a - \Gamma^c_a\omega_c)\,\mathbf{d}x^a = (\nabla_b\omega_a)\,\mathbf{d}x^b \otimes \mathbf{d}x^a = \nabla\omega$$

as claimed. The derivation of (6.24b) proceeds along similar lines and is left to the reader.

7. *(Transformation Rule for the Connection Components)* Given a one-form $\omega = (\omega_b)$, Use the fact that the quantities $\nabla_a\ \omega_b$ are the components of a tensor of type $(0,2)$ to show that under a change of coordinates, $x^i \longrightarrow \bar{x}^j = \bar{x}^j(x^i)$, the **connection components** transform according to the rule

$$\bar{\Gamma}^a_{bc} = \frac{\partial \bar{x}^a}{\partial x^m}\frac{\partial x^n}{\partial \bar{x}^b}\frac{\partial x^p}{\partial \bar{x}^c}\Gamma^m_{np} + \frac{\partial^2 x^q}{\partial \bar{x}^b \partial \bar{x}^c}\frac{\partial \bar{x}^a}{\partial x^q} \tag{6.25}$$

(Solution)

Since ω_b are the components of a tensor of type $(0,1)$, then under a change of coordinates, we have

$$\frac{\partial \bar{\omega}_b}{\partial \bar{x}^a} = \frac{\partial}{\partial \bar{x}^a}\left(\frac{\partial x^s}{\partial \bar{x}^b}\omega_s\right) = \frac{\partial^2 x^s}{\partial \bar{x}^a\partial \bar{x}^b}\omega_s + \frac{\partial \omega_s}{\partial \bar{x}^a}\frac{\partial x^s}{\partial \bar{x}^b}$$

Consequently,

$$\bar{\nabla}_a\bar{\omega}_b + \bar{\Gamma}^c_{ab}\bar{\omega}_c = \frac{\partial^2 x^s}{\partial \bar{x}^a\partial \bar{x}^b}\omega_s + \frac{\partial \omega_s}{\partial \bar{x}^a}\frac{\partial x^s}{\partial \bar{x}^b}$$

Since $\bar{\omega}_c = (\partial x^t/\partial \bar{x}^c)\,\omega_t$, we find

$$\begin{aligned}\bar{\nabla}_a\bar{\omega}_b + \bar{\Gamma}^c_{ab}\left(\frac{\partial x^t}{\partial \bar{x}^c}\omega_t\right) &= \frac{\partial^2 x^s}{\partial \bar{x}^a\partial \bar{x}^b}\omega_s + \frac{\partial \omega_s}{\partial \bar{x}^a}\frac{\partial x^s}{\partial \bar{x}^b} \\ &= \frac{\partial^2 x^s}{\partial \bar{x}^a\partial \bar{x}^b}\omega_s + \left(\frac{\partial x^k}{\partial \bar{x}^a}\frac{\partial \omega_s}{\partial x^k}\right)\frac{\partial x^s}{\partial \bar{x}^b} \\ &= \frac{\partial^2 x^s}{\partial \bar{x}^a\partial \bar{x}^b}\omega_s + \frac{\partial x^k}{\partial \bar{x}^a}\frac{\partial x^s}{\partial \bar{x}^b}\left(\nabla_k\omega_s + \Gamma^t_{ks}\omega_t\right)\end{aligned}$$

so that

$$\bar{\nabla}_a\bar{\omega}_b + \bar{\Gamma}^c_{ab}\left(\frac{\partial x^t}{\partial \bar{x}^c}\omega_t\right) = \frac{\partial^2 x^s}{\partial \bar{x}^a\partial \bar{x}^b}\omega_s + \left(\frac{\partial x^k}{\partial \bar{x}^a}\frac{\partial x^s}{\partial \bar{x}^b}\nabla_k\omega_s\right) + \frac{\partial x^k}{\partial \bar{x}^a}\frac{\partial x^s}{\partial \bar{x}^b}\Gamma^t_{ks}\omega_t$$

But since $\bar{\nabla}_a \bar{\omega}_b$ are the components of a tensor of type $(0,2)$, then they must transform according to the rule

$$\bar{\nabla}_a \bar{\omega}_b = \frac{\partial x^k}{\partial \bar{x}^a} \frac{\partial x^s}{\partial \bar{x}^b} \nabla_k \omega_s,$$

so, subtracting the last two equations, we get

$$\bar{\Gamma}^c_{ab} \left(\frac{\partial x^t}{\partial \bar{x}^c} \omega_t \right) = \frac{\partial x^k}{\partial \bar{x}^a} \frac{\partial x^s}{\partial \bar{x}^b} \Gamma^t_{ks} \omega_t + \frac{\partial^2 x^s}{\partial \bar{x}^a \partial \bar{x}^b} \omega_s$$

We replace the dummy index $s \longrightarrow t$ in the last term on the right to obtain

$$\bar{\Gamma}^c_{ab} \left(\frac{\partial x^t}{\partial \bar{x}^c} \omega_t \right) = \frac{\partial x^k}{\partial \bar{x}^a} \frac{\partial x^s}{\partial \bar{x}^b} \Gamma^t_{ks} \omega_t + \frac{\partial^2 x^t}{\partial \bar{x}^a \partial \bar{x}^b} \omega_t$$

Since $\omega = (\omega_t)$ is arbitrary, we must have

$$\bar{\Gamma}^c_{ab} \frac{\partial x^t}{\partial \bar{x}^c} = \frac{\partial x^k}{\partial \bar{x}^a} \frac{\partial x^s}{\partial \bar{x}^b} \Gamma^t_{ks} + \frac{\partial^2 x^t}{\partial \bar{x}^a \partial \bar{x}^b}$$

We now multiply both sides of this equation by $\partial \bar{x}^d / \partial x^t$ to get

$$\bar{\Gamma}^d_{ab} = \frac{\partial \bar{x}^d}{\partial x^t} \frac{\partial x^k}{\partial \bar{x}^a} \frac{\partial x^s}{\partial \bar{x}^b} \Gamma^t_{ks} + \frac{\partial^2 x^t}{\partial \bar{x}^a \partial \bar{x}^b} \frac{\partial \bar{x}^d}{\partial x^t}$$

Finally, by relabeling the indices in the last equation, we have

$$\bar{\Gamma}^a_{bc} = \frac{\partial \bar{x}^a}{\partial x^m} \frac{\partial x^n}{\partial \bar{x}^b} \frac{\partial x^p}{\partial \bar{x}^c} \Gamma^m_{np} + \frac{\partial^2 x^m}{\partial \bar{x}^b \partial \bar{x}^c} \frac{\partial \bar{x}^a}{\partial x^m}$$

as required. *Observe that due to the presence of the second term on the righthand side of this equation, the connection components Γ^m_{np} are not the components of a $(1,2)$ tensor.*

6.4. Formal Approach to Covariant Differentiation

In Section 6.2, we defined the absolute derivative $\nabla_{\mathbf{X}}\mathbf{V}$ of the vector field $\mathbf{V}(\mathbf{x})$ along a smooth curve α by the expression

$$\frac{D\mathbf{V}}{Dt} \equiv \nabla_{\mathbf{X}}\mathbf{V} = X^a \left(\frac{\partial V^c}{\partial x^a} + \Gamma^c_{ab} V^b \right) \mathbf{e}_c(\mathbf{x}) \tag{6.7}$$

Here, we assumed that in some chart (U, ϕ), the curve α had the coordinate representation $x^k = x^k(t)$ where $t \in [a,b]$. Consequently, the vector $\mathbf{X} = (X^a) = (dx^a/dt)$ is just the tangent vector to α and $\nabla_{\mathbf{X}}\mathbf{V}$ may be interpreted as the rate of change of $\mathbf{V}$ along the curve. In other words, the absolute derivative $\nabla_{\mathbf{X}}\mathbf{V}$ is nothing more than a *directional covariant derivative*.

This development was intuitive and along classical lines. We began our discussion by introducing the concept of parallel transport and showed that the connection components must satisfy the relation

$$\bar{\Gamma}^{a}_{bc} = \frac{\partial \bar{x}^{a}}{\partial x^{m}} \frac{\partial x^{n}}{\partial \bar{x}^{b}} \frac{\partial x^{p}}{\partial \bar{x}^{c}} \Gamma^{m}_{np} + \frac{\partial^{2} x^{m}}{\partial \bar{x}^{b} \partial \bar{x}^{c}} \frac{\partial \bar{x}^{a}}{\partial x^{m}} \tag{6.25}$$

under a change of coordinates, $x^a \longrightarrow \bar{x}^b = \bar{x}^b(x^a)$.

A more general approach is to define a **connection (or covariant derivative)** as a mapping $\nabla : \Im^0_1(M) \times \Im^0_1(M) \longrightarrow \Im^0_1(M)$ which maps each *ordered pair* of vector fields $(\mathbf{X}, \mathbf{V})$ into a *unique* vector field $\nabla_{\mathbf{X}}\mathbf{V}$ such that:

$$\begin{aligned} \nabla_{(\mathbf{X}+\mathbf{Y})}\mathbf{V} &= \nabla_{\mathbf{X}}\mathbf{V} + \nabla_{\mathbf{Y}}\mathbf{V} && (6.26a) \\ \nabla_{\mathbf{X}}(\mathbf{V}+\mathbf{W}) &= \nabla_{\mathbf{X}}\mathbf{V} + \nabla_{\mathbf{X}}\mathbf{W} && (6.26b) \\ \nabla_{(\varphi\mathbf{X})}\mathbf{V} &= \varphi\, \nabla_{\mathbf{X}}\mathbf{V} && (6.26c) \\ \nabla_{\mathbf{X}}(\varphi\mathbf{V}) &= \varphi\, \nabla_{\mathbf{X}}\mathbf{V} + (\nabla_{\mathbf{X}}\varphi)\,\mathbf{V}, && (6.26d) \end{aligned}$$

where $\mathbf{V}, \mathbf{W}, \mathbf{X}$, and $\mathbf{Y}$ are arbitrary vector fields and φ is any smooth function.

Observe that relations (6.26) are not only fully consistent with our previous results, but they generalize these results. That is, the absolute derivative $\nabla_{\mathbf{X}}\mathbf{V}$ now represents the derivative of $\mathbf{V}$ in the direction of *any* smooth vector field $\mathbf{X}$ which is defined on M.

In particular, if $\mathbf{X}$ and $\mathbf{V}$ are natural basis fields on M so that

$$\mathbf{X} = \mathbf{e}_a(\mathbf{x}) = \left.\frac{\partial}{\partial x^a}\right|_{\mathbf{x}} \quad \text{and } \mathbf{V} = \mathbf{e}_b(\mathbf{x}) = \left.\frac{\partial}{\partial x^b}\right|_{\mathbf{x}},$$

then there must exist n^3-smooth functions Γ^c_{ab} such that

$$\nabla_a(\mathbf{e}_b) \equiv \nabla_{\mathbf{e}_a}(\mathbf{e}_b) = \Gamma^c_{ab}\mathbf{e}_c \tag{6.27}$$

Observe that (6.27) agrees with equation (6.20b) which we derived earlier.

In addition, the transformation rule for the connection components can be derived directly from (6.27). To see this, assume a change of coordinates, $x^a \longrightarrow \bar{x}^b = \bar{x}^b(x^a)$ so that

$$\bar{\mathbf{e}}_a = \frac{\partial x^k}{\partial \bar{x}^a}\frac{\partial}{\partial x^k} = \left(\frac{\partial x^k}{\partial \bar{x}^a}\right)\mathbf{e}_k$$

Consequently, by relations (6.26), and (6.27), we can write

$$\begin{aligned} \bar{\Gamma}^c_{ab}\mathbf{e}_c &= \nabla_{\bar{\mathbf{e}}_a}(\bar{\mathbf{e}}_b) = \frac{\partial x^k}{\partial \bar{x}^a}\nabla_{\mathbf{e}_k}(\bar{\mathbf{e}}_b) = \frac{\partial x^k}{\partial \bar{x}^a}\nabla_{\mathbf{e}_k}\left(\frac{\partial x^p}{\partial \bar{x}^b}\mathbf{e}_p\right) \\ &= \frac{\partial x^k}{\partial \bar{x}^a}\frac{\partial x^p}{\partial \bar{x}^b}\nabla_{\mathbf{e}_k}(\mathbf{e}_p) + \frac{\partial x^k}{\partial \bar{x}^a}\left[\frac{\partial}{\partial x^k}\left(\frac{\partial x^p}{\partial \bar{x}^b}\right)\right]\mathbf{e}_p \\ &= \frac{\partial x^k}{\partial \bar{x}^a}\frac{\partial x^s}{\partial \bar{x}^b}\nabla_{\mathbf{e}_k}(\mathbf{e}_p) + \frac{\partial^2 x^p}{\partial \bar{x}^a \partial \bar{x}^b}\mathbf{e}_p \end{aligned}$$

Furthermore, if substitute (6.27) into the last equation, and change the dummy index $p \longrightarrow s$ in the last term, we get

$$\bar{\Gamma}^c_{ab}\left(\frac{\partial x^s}{\partial \bar{x}^c}\mathbf{e}_s\right) = \frac{\partial x^k}{\partial \bar{x}^a}\frac{\partial x^p}{\partial \bar{x}^b}\left(\Gamma^s_{kp}\mathbf{e}_s\right) + \frac{\partial^2 x^s}{\partial \bar{x}^a \partial \bar{x}^b}\mathbf{e}_s,$$

and consequently,

$$\bar{\Gamma}^c_{ab}\frac{\partial x^s}{\partial \bar{x}^c} = \frac{\partial x^k}{\partial \bar{x}^a}\frac{\partial x^p}{\partial \bar{x}^b}\Gamma^s_{kp} + \frac{\partial^2 x^s}{\partial \bar{x}^a \partial \bar{x}^b}$$

The last equation implies that

$$\bar{\Gamma}^c_{ab} = \frac{\partial \bar{x}^c}{\partial x^s}\frac{\partial x^k}{\partial \bar{x}^a}\frac{\partial x^p}{\partial \bar{x}^b}\Gamma^s_{kp} + \frac{\partial^2 x^s}{\partial \bar{x}^a \partial \bar{x}^b}\frac{\partial \bar{x}^c}{\partial x^s}$$

Finally, relabeling dummy indices ($s \longrightarrow m$, and $k \longrightarrow n$) in both terms on the righthand side of the last equation, we get

$$\bar{\Gamma}^c_{ab} = \frac{\partial \bar{x}^c}{\partial x^m}\frac{\partial x^n}{\partial \bar{x}^a}\frac{\partial x^p}{\partial \bar{x}^b}\Gamma^m_{np} + \frac{\partial^2 x^m}{\partial \bar{x}^a \partial \bar{x}^b}\frac{\partial \bar{x}^c}{\partial x^m},$$

which is exactly (6.25).

Once again, we can extend the definition of $\nabla_{\mathbf{X}}$ to handle any smooth function φ by defining

$$\nabla_{\mathbf{X}}(\varphi) = \mathbf{X}[\varphi] \tag{6.28}$$

and we require that the operator $\nabla_{\mathbf{X}}$ respects the Leibnitz rule:

$$\nabla_{\mathbf{X}}(\mathbf{S} \otimes \mathbf{T}) = \mathbf{S} \otimes (\nabla_{\mathbf{X}}\mathbf{T}) + (\nabla_{\mathbf{X}}\mathbf{S}) \otimes \mathbf{T} \tag{6.29}$$

for any smooth tensor fields **S** and **T**. As in the past, we also require that

$$\nabla_{\mathbf{X}}\langle\omega, \mathbf{V}\rangle = \langle\omega, \nabla_{\mathbf{X}}\mathbf{V}\rangle + \langle\nabla_{\mathbf{X}}\omega, \mathbf{V}\rangle \tag{6.30}$$

for any one-form field ω and vector field **V**, so that the operator $\nabla_{\mathbf{X}}$ commutes with contractions.

Now, as shown previously, equations (6.27) and (6.30) imply that

$$\nabla_a\left(\mathbf{d}x^b\right) \equiv \nabla_{\mathbf{e}_a}\left(\mathbf{d}x^b\right) = -\,\Gamma^b_{ak}\mathbf{d}x^k \tag{6.31}$$

This means that (6.27), (6.31), and the Leibnitz rule allow us to find the absolute derivative of any tensor; they allow us to extend the original definition of the operator $\nabla_{\mathbf{X}}$ to compute the covariant derivative of any *general* tensor field.

6.5. Lie Bracket and Torsion Tensor

Given smooth vector fields $\mathbf{U}(\mathbf{x})$ and $\mathbf{V}(\mathbf{x})$, we define their **Lie bracket** as the *vector field* $[\mathbf{U}, \mathbf{V}]$ such that

$$[\mathbf{U}, \mathbf{V}]\varphi = \mathbf{U}(\mathbf{V}\varphi) - \mathbf{V}(\mathbf{U}\varphi) \tag{6.32}$$

for any smooth scalar field φ. In other words, the Lie bracket is just the **commutator** of the vector fields **U** and **V**.

From (6.32), the Lie bracket is *antisymmetric*, since

$$[\mathbf{U}, \mathbf{V}] = -\,[\mathbf{V}, \mathbf{U}] \tag{6.33}$$

In addition, it is easy to show that the Lie bracket satisfies the relations

$$[c_1\mathbf{U}+c_2\mathbf{V},\mathbf{W}] = c_1[\mathbf{U},\mathbf{W}]+c_2[\mathbf{V},\mathbf{W}] \tag{6.34a}$$

$$[\mathbf{U},c_1\mathbf{V}+c_2\mathbf{W}] = c_1[\mathbf{U},\mathbf{V}]+c_2[\mathbf{U},\mathbf{W}] \quad \text{for all } c_1,c_2 \in R, \tag{6.34b}$$

as well as the **Jacobi identity**

$$[\mathbf{U},[\mathbf{V},\mathbf{W}]]+[\mathbf{W},[\mathbf{U},\mathbf{V}]]+[\mathbf{V},[\mathbf{W},\mathbf{U}]]=\mathbf{0} \tag{6.35}$$

for any smooth vector fields **U**, **V**, and **W**.

Now, any vector space which is equipped with a product $[,]$ which satisfies (6.33) - (6.35) is called a **Lie algebra.** Consequently, the collection of all vector fields on a manifold M, when equipped with the Lie bracket, forms a Lie algebra.

The *vector character* of $[\mathbf{U},\mathbf{V}]$ is most easily seen if we write **U** and **V** in terms of the coordinate basis. Given a coordinate chart (U,ϕ), we can always expand the vector fields **U** and **V** in terms of the coordinate basis vectors $\{\partial/\partial x^a\}$ as

$$\mathbf{U}=U^a\frac{\partial}{\partial x^a} \quad \text{and } \mathbf{V}=V^b\frac{\partial}{\partial x^b}$$

Consequently, for any smooth function φ, we have

$$\begin{aligned}
[\mathbf{U},\mathbf{V}]\varphi &= \left[U^a\frac{\partial}{\partial x^a},V^b\frac{\partial}{\partial x^b}\right]\varphi = U^a\frac{\partial}{\partial x^a}\left(V^b\frac{\partial\varphi}{\partial x^b}\right)-V^b\frac{\partial}{\partial x^b}\left(U^a\frac{\partial\varphi}{\partial x^a}\right)\\
&= U^a\frac{\partial V^b}{\partial x^a}\frac{\partial\varphi}{\partial x^b}-V^b\frac{\partial U^a}{\partial x^b}\frac{\partial\varphi}{\partial x^a}+U^aV^b\left(\frac{\partial^2\varphi}{\partial x^a\partial x^b}-\frac{\partial^2\varphi}{\partial x^b\partial x^a}\right)\\
&= U^a\frac{\partial V^b}{\partial x^a}\frac{\partial\varphi}{\partial x^b}-V^a\frac{\partial U^b}{\partial x^a}\frac{\partial\varphi}{\partial x^b}=\left(U^a\frac{\partial V^b}{\partial x^a}-V^a\frac{\partial U^b}{\partial x^a}\right)\frac{\partial}{\partial x^b}(\varphi)
\end{aligned}$$

where we have used the equality of mixed partial derivatives, and we relabeled indices in the second term. Since φ is arbitrary, we conclude that

$$[\mathbf{U},\mathbf{V}]=\left(U^a\frac{\partial V^b}{\partial x^a}-V^a\frac{\partial U^b}{\partial x^a}\right)\frac{\partial}{\partial x^b} \tag{6.36}$$

The vector character of $[\mathbf{U},\mathbf{V}]$ is apparent from this equation.

Now, for a **coordinate basis** $\{\partial/\partial x^a\}$ it is clear that

$$\left[\frac{\partial}{\partial x^a},\frac{\partial}{\partial x^b}\right]=\mathbf{0}, \tag{6.37}$$

owing to the equality of mixed partial derivatives. On the other hand, for a general basis or **non-coordinate basis** $\{\mathbf{b}_k\}$, it is not possible to write each basis vector as a partial derivative with respect to each coordinate, and consequently, we find that

$$[\mathbf{b}_k,\mathbf{b}_l]\neq\mathbf{0} \tag{6.38}$$

In fact, for a non-coordinate basis, if we recall that $[\mathbf{b}_k, \mathbf{b}_l]$ is a vector, there must exists constants C_{kl}^{m} (not all zero) such that

$$[\mathbf{b}_k, \mathbf{b}_l] = C_{kl}^{m}\mathbf{b}_m \tag{6.39}$$

We call the expansion coefficients C_{kl}^{m} the **structure constants** of the Lie algebra.
Now, we define the **torsion tensor T** by the relation

$$\mathbf{T}(\mathbf{U},\mathbf{V}) = \nabla_{\mathbf{U}}\mathbf{V} - \nabla_{\mathbf{V}}\mathbf{U} - [\mathbf{U},\mathbf{V}] \tag{6.40}$$

for arbitrary vector fields $\mathbf{U}(\mathbf{x})$ and $\mathbf{V}(\mathbf{x})$. Since the righthand of (6.40) is clearly a vector, then by the Quotient Rule, **T** must be a tensor of type $(1,2)$. Also, by its very definition, we see that **T** is *antisymmetric* in **U** and **V**.

We now show that **T** is linear in its first argument **U**. We first observe that:

$$\begin{aligned}\mathbf{T}(\varphi_1\mathbf{U}_1+\varphi_2\mathbf{U}_2,\mathbf{V}) &= \nabla_{(\varphi_1\mathbf{U}_1+\varphi_2\mathbf{U}_2)}\mathbf{V} - \nabla_{\mathbf{V}}(\varphi_1\mathbf{U}_1+\varphi_2\mathbf{U}_2) - [\varphi_1\mathbf{U}_1+\varphi_2\mathbf{U}_2,\mathbf{V}]\\ &= \varphi_1\nabla_{\mathbf{U}_1}\mathbf{V} + \varphi_2\nabla_{\mathbf{U}_2}\mathbf{V} - \varphi_1\nabla_{\mathbf{V}}\mathbf{U}_1 - \varphi_2\nabla_{\mathbf{V}}\mathbf{U}_2 - (\mathbf{V}[\varphi_1])\mathbf{U}_1\\ &\quad - (\mathbf{V}[\varphi_2])\mathbf{U}_2 - [\varphi_1\mathbf{U}_1+\varphi_2\mathbf{U}_2,\mathbf{V}]\ .\end{aligned} \tag{6.41}$$

But, when written in component form, the last term of this equation is simply

$$\begin{aligned}[\varphi_1\mathbf{U}_1+\varphi_2\mathbf{U}_2,\mathbf{V}]^a &= \left(\varphi_1 U_1^b + \varphi_2 U_2^b\right)\frac{\partial V^a}{\partial x^b} - V^b\frac{\partial}{\partial x^b}(\varphi_1 U_1^a + \varphi_2 U_2^a)\\ &= \varphi_1\left(U_1^b\frac{\partial V^a}{\partial x^b} - V^b\frac{\partial U_1^a}{\partial x^b}\right) + \varphi_2\left(U_2^b\frac{\partial V^a}{\partial x^b} - V^b\frac{\partial U_2^a}{\partial x^b}\right)\\ &\quad - \left(V^b\frac{\partial\varphi_1}{\partial x^b}\right)U_1^a - \left(V^b\frac{\partial\varphi_2}{\partial x^b}\right)U_2^a\ ,\end{aligned}$$

or, in other words,

$$[\varphi_1\mathbf{U}_1+\varphi_2\mathbf{U}_2,\mathbf{V}] = \varphi_1[\mathbf{U}_1,\mathbf{V}] + \varphi_2[\mathbf{U}_2,\mathbf{V}] - (\mathbf{V}[\varphi_1])\mathbf{U}_1 - (\mathbf{V}[\varphi_2])\mathbf{U}_2 \tag{6.42}$$

If we substitute (6.42) into (6.41), we find

$$\begin{aligned}\mathbf{T}(\varphi_1\mathbf{U}_1+\varphi_2\mathbf{U}_2,\mathbf{V}) &= \varphi_1(\nabla_{\mathbf{U}_1}\mathbf{V} - \nabla_{\mathbf{V}}\mathbf{U}_1 - [\mathbf{U}_1,\mathbf{V}])\\ &\quad +\varphi_2(\nabla_{\mathbf{U}_2}\mathbf{V} - \nabla_{\mathbf{V}}\mathbf{U}_2 - [\mathbf{U}_2,\mathbf{V}])\\ &= \varphi_1\mathbf{T}(\mathbf{U}_1,\mathbf{V}) + \varphi_2\mathbf{T}(\mathbf{U}_2,\mathbf{V})\end{aligned} \tag{6.43}$$

At the same time, **T** is linear in its second argument **V**, since if we exploit the antisymmetric character of **T**, we have immediately,

$$\begin{aligned}\mathbf{T}(\mathbf{U},\varphi_1\mathbf{V}_1+\varphi_2\mathbf{V}_2) &= -\mathbf{T}(\varphi_1\mathbf{V}_1+\varphi_2\mathbf{V}_2,\mathbf{U})\\ &= -\varphi_1\mathbf{T}(\mathbf{V}_1,\mathbf{U}) - \varphi_2\mathbf{T}(\mathbf{V}_2,\mathbf{U})\\ &= \varphi_1\mathbf{T}(\mathbf{U},\mathbf{V}_1) + \varphi_2\mathbf{T}(\mathbf{U},\mathbf{V}_2)\end{aligned} \tag{6.44}$$

Thus, **T** is a *multilinear* mapping of $\mathfrak{I}_0^1(M)\times\mathfrak{I}_0^1(M)$ into $\mathfrak{I}_0^1(M)$, which establishes the tensorial nature of **T** independently of the Quotient Rule.

We now determine the components of $\mathbf{T}$ relative to the natural basis $\{\mathbf{e}_k\}$ and the corresponding dual basis $\{\mathbf{d}x^k\}$. Since $\mathbf{T}$ is a tensor of type $(1,2)$, we can write $\mathbf{T} = T^a_{bc}\ \mathbf{e}_a \otimes \mathbf{d}x^b \otimes \mathbf{d}x^c$, so that

$$\mathbf{T}(\mathbf{e}_k, \mathbf{e}_l) = T^a_{bc}\ \mathbf{e}_a \otimes \left(\mathbf{d}x^b \otimes \mathbf{d}x^c\right)(\mathbf{e}_k, \mathbf{e}_l) = T^a_{bc}\ \mathbf{e}_a\ \delta^b_k \delta^c_l$$

In other words,

$$\mathbf{T}(\mathbf{e}_k, \mathbf{e}_l) = T^a_{kl}\ \mathbf{e}_a \tag{6.45}$$

and consequently, the components of $\mathbf{T}$ are given by

$$\begin{aligned} T^a_{bc} &= \langle \mathbf{T}(\mathbf{e}_b, \mathbf{e}_c), \mathbf{d}x^a \rangle \\ &= \langle \nabla_{\mathbf{e}_b} \mathbf{e}_c - \nabla_{\mathbf{e}_c} \mathbf{e}_b - [\mathbf{e}_b, \mathbf{e}_c]\ , \mathbf{d}x^a \rangle \\ &= \left\langle \left(\Gamma^p_{bc} - \Gamma^p_{cb}\right) \mathbf{e}_p, \mathbf{d}x^a \right\rangle \quad (\text{since } [\mathbf{e}_b, \mathbf{e}_c] = \mathbf{0}) \\ &= \Gamma^a_{bc} - \Gamma^a_{cb} \end{aligned}$$

In summary, we have shown that

$$T^a_{bc} = \Gamma^a_{bc} - \Gamma^a_{cb}, \tag{6.46}$$

which is an expression that agrees with the usual definition of the torsion tensor given in texts on classical differential geometry.

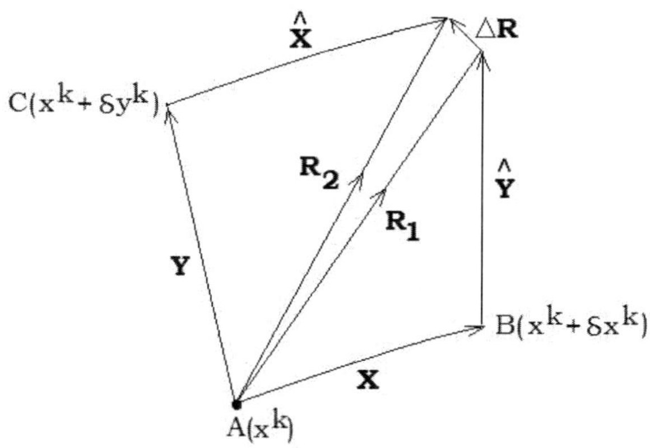

Figure 6.2.

The torsion tensor has an interesting geometric interpretation which we now discuss. Assume that the point $A \in M$ has coordinates (x^k). Also, let $\{\mathbf{e}_k\}$ denote the natural basis vectors for the tangent space $T_A(M)$, and assume, as shown in Figure–6.2, that the *infinitesimal* displacement vectors

$$\mathbf{X} = \left(\delta x^k\right) \mathbf{e}_k \ \text{ and } \ \mathbf{Y} = \left(\delta y^k\right) \mathbf{e}_k$$

both live in the tangent space $T_A(M)$. Now, we parallel transport the vector $\mathbf{X}$ along the path AC to obtain the vector $\hat{\mathbf{X}}$, and we parallel transport $\mathbf{Y}$ along the path AB to obtain the vector $\hat{\mathbf{Y}}$. According to the definition of parallel transport, these vectors are given by

$$\hat{\mathbf{X}} = \left(\delta x^k - \delta y^m \Gamma^k_{mq} \delta x^q\right) \mathbf{e}_k \text{ and } \hat{\mathbf{Y}} = \left(\delta y^k - \delta x^m \Gamma^k_{mq} \delta y^q\right) \mathbf{e}_k \,, \tag{6.47}$$

respectively. Next, we find that the *resultant* displacement vectors $\mathbf{R}_1$ and $\mathbf{R}_2$ are

$$\begin{aligned} \mathbf{R}_2 &= \hat{\mathbf{X}} + \mathbf{Y} = \left(\delta x^k + \delta y^k - \delta y^m \Gamma^k_{mq} \delta x^q\right) \mathbf{e}_k \\ \mathbf{R}_1 &= \mathbf{X} + \hat{\mathbf{Y}} = \left(\delta x^k + \delta y^k - \delta x^m \Gamma^k_{mq} \delta y^q\right) \mathbf{e}_k \end{aligned}$$

and consequently, their difference is

$$\begin{aligned} \Delta\mathbf{R} &= \mathbf{R}_2 - \mathbf{R}_1 = (\delta x^m \Gamma^k_{mq} \delta y^q - \delta y^m \Gamma^k_{mq} \delta x^q) \mathbf{e}_k \\ &= \left(\Gamma^k_{mq} - \Gamma^k_{qm}\right) \delta x^m \delta y^q \mathbf{e}_k \end{aligned}$$

In other words, from (6.46), we have

$$\Delta\mathbf{R} = T^k_{mq} \delta x^m \delta y^q \mathbf{e}_k \tag{6.48}$$

This equation tells us that the torsion is intimately related with the gap $\Delta\mathbf{R}$ in the infinitesimal paralleogram that we constructed under parallel translation.

Now, a connection ∇ is said to be a **symmetric connection** if the torsion tensor is identically zero. This term derives from the fact that the **torsion-free** condition $\mathbf{T} \equiv \mathbf{0}$ implies that the connection components must be *symmetric* (in their lower indices), so that

$$\Gamma^k_{mq} = \Gamma^k_{qm} \quad \text{(zero torsion condition)} \tag{6.49}$$

It is clear that for a symmetric connection, we will have $\Delta\mathbf{R} = \mathbf{0}$, *and any infinitesimal parallelogram constructed under the operation of parallel translation will close.*[2]

6.6. The Curvature Tensor

Given smooth vector fields $\mathbf{U}(\mathbf{x})$, $\mathbf{V}(\mathbf{x})$, and $\mathbf{W}(\mathbf{x})$ on a manifold M, we define the **Riemann curvature tensor R** by the expression

$$\begin{aligned} \mathbf{R}(\mathbf{U},\mathbf{V})\mathbf{W} &= \nabla_{\mathbf{U}} \nabla_{\mathbf{V}} \mathbf{W} - \nabla_{\mathbf{V}} \nabla_{\mathbf{U}} \mathbf{W} - \nabla_{[\mathbf{U},\mathbf{V}]} \mathbf{W} \\ &= [\nabla_{\mathbf{U}}, \nabla_{\mathbf{V}}] \mathbf{W} - \nabla_{[\mathbf{U},\mathbf{V}]} \mathbf{W} \end{aligned} \tag{6.50}$$

Clearly, **R** maps each ordered triple of vector fields into a vector field, and consequently, **R** must be a tensor of type $(1,3)$. Furthermore, it is obvious that **R** is *antisymmetric* with respect to its first two arguments; that is,

$$\mathbf{R}(\mathbf{V},\mathbf{U})\mathbf{W} = -\mathbf{R}(\mathbf{U},\mathbf{V})\mathbf{W} \tag{6.51}$$

What is not so obvious is that $\mathbf{R}$ is a *multilinear* object. That is, for smooth scalar fields $\varphi_1, \varphi_2 \in C^\infty(M)$, we have

$$\begin{aligned}
\mathbf{R}(\varphi_1\mathbf{U}_1+\varphi_2\mathbf{U}_2,\mathbf{V})\mathbf{W} &= \varphi_1\mathbf{R}(\mathbf{U}_1,\mathbf{V})\mathbf{W}+\varphi_2\mathbf{R}(\mathbf{U}_2,\mathbf{V})\mathbf{W} && (6.52a)\\
\mathbf{R}(\mathbf{U},\varphi_1\mathbf{V}_1+\varphi_2\mathbf{V}_2)\mathbf{W} &= \varphi_1\mathbf{R}(\mathbf{U},\mathbf{V}_1)\mathbf{W}+\varphi_2\mathbf{R}(\mathbf{U},\mathbf{V}_2)\mathbf{W} && (6.52b)\\
\mathbf{R}(\mathbf{U},\mathbf{V})(\varphi_1\mathbf{W}_1+\varphi_2\mathbf{W}_2) &= \varphi_1\mathbf{R}(\mathbf{U},\mathbf{V})\mathbf{W}_1+\varphi_2\mathbf{R}(\mathbf{U},\mathbf{V})\mathbf{W}_2 && (6.52c)
\end{aligned}$$

Before we establish (6.52a), it is worthwhile to note the identity

$$[\varphi\mathbf{U},\mathbf{V}] = \varphi\,[\mathbf{U},\mathbf{V}] - (\mathbf{V}\,[\varphi])\,\mathbf{U} \tag{6.53}$$

which holds for all $\mathbf{U},\mathbf{V} \in \mathfrak{I}^1_0(M)$, and for arbitrary $\varphi \in C^\infty(M)$. Now, the lefthand side of (6.52a) can be written as

$$\begin{aligned}
\mathbf{R}(\varphi_1\mathbf{U}_1+\varphi_2\mathbf{U}_2,\mathbf{V})\mathbf{W} &= \nabla_{(\varphi_1\mathbf{U}_1+\varphi_2\mathbf{U}_2)}\nabla_{\mathbf{V}}\mathbf{W}-\nabla_{\mathbf{V}}\nabla_{(\varphi_1\mathbf{U}_1+\varphi_2\mathbf{U}_2)}\mathbf{W}\\
&\quad -\nabla_{[\varphi_1\mathbf{U}_1+\varphi_2\mathbf{U}_2,\,\mathbf{V}]}\mathbf{W}\\
&= \nabla_{\varphi_1\mathbf{U}_1}\nabla_{\mathbf{V}}\mathbf{W}+\nabla_{\varphi_2\mathbf{U}_2}\nabla_{\mathbf{V}}\mathbf{W}-\nabla_{\mathbf{V}}\left(\nabla_{\varphi_1\mathbf{U}_1}\mathbf{W}+\nabla_{\varphi_2\mathbf{U}_2}\mathbf{W}\right)\\
&\quad -\nabla_{[\varphi_1\mathbf{U}_1+\varphi_2\mathbf{U}_2,\,\mathbf{V}]}\mathbf{W}\\
&= \varphi_1\nabla_{\mathbf{U}_1}\nabla_{\mathbf{V}}\mathbf{W}+\varphi_2\nabla_{\mathbf{U}_2}\nabla_{\mathbf{V}}\mathbf{W}-\nabla_{\mathbf{V}}\left(\varphi_1\nabla_{\mathbf{U}_1}\mathbf{W}+\varphi_2\nabla_{\mathbf{U}_2}\mathbf{W}\right)\\
&\quad -\nabla_{[\varphi_1\mathbf{U}_1+\varphi_2\mathbf{U}_2,\,\mathbf{V}]}\mathbf{W}
\end{aligned}$$

Now, we use (6.53) to simplify the last term of this equation to obtain

$$\begin{aligned}
\nabla_{[\varphi_1\mathbf{U}_1+\varphi_2\mathbf{U}_2,\,\mathbf{V}]}\mathbf{W} &= \nabla_{[\varphi_1\mathbf{U}_1,\,\mathbf{V}]}\mathbf{W}+\nabla_{[\varphi_2\mathbf{U}_2,\,\mathbf{V}]}\mathbf{W}\\
&= \nabla_{\varphi_1[\mathbf{U}_1,\mathbf{V}]-(\mathbf{V}[\varphi_1])\mathbf{U}}(\mathbf{W})+\nabla_{\varphi_2[\mathbf{U}_2,\mathbf{V}]-(\mathbf{V}[\varphi_2])\mathbf{U}}(\mathbf{W})\\
&= \varphi_1\nabla_{[\mathbf{U}_1,\,\mathbf{V}]}\mathbf{W}+\varphi_2\nabla_{[\mathbf{U}_2,\,\mathbf{V}]}\mathbf{W}-(\mathbf{V}\,[\varphi_1])\nabla_{\mathbf{U}_1}\mathbf{W}\\
&\quad -(\mathbf{V}\,[\varphi_2])\nabla_{\mathbf{U}_2}\mathbf{W},
\end{aligned}$$

from which, the original equation simplifies to

$$\begin{aligned}
&\mathbf{R}(\varphi_1\mathbf{U}_1+\varphi_2\mathbf{U}_2,\mathbf{V})\mathbf{W}\\
={}& \varphi_1\nabla_{\mathbf{U}_1}\nabla_{\mathbf{V}}\mathbf{W}+\varphi_2\nabla_{\mathbf{U}_2}\nabla_{\mathbf{V}}\mathbf{W}-\nabla_{\mathbf{V}}\left(\varphi_1\nabla_{\mathbf{U}_1}\mathbf{W}+\varphi_2\nabla_{\mathbf{U}_2}\mathbf{W}\right)\\
&-\left\{\varphi_1\nabla_{[\mathbf{U}_1,\,\mathbf{V}]}\mathbf{W}+\varphi_2\nabla_{[\mathbf{U}_2,\,\mathbf{V}]}\mathbf{W}-(\mathbf{V}\,[\varphi_1])\nabla_{\mathbf{U}_1}\mathbf{W}-(\mathbf{V}\,[\varphi_2])\nabla_{\mathbf{U}_2}\mathbf{W}\right\}\\
={}& \varphi_1\left\{\nabla_{\mathbf{U}_1}\nabla_{\mathbf{V}}\mathbf{W}-\nabla_{\mathbf{V}}\nabla_{\mathbf{U}_1}\mathbf{W}-\nabla_{[\mathbf{U}_1,\,\mathbf{V}]}\mathbf{W}\right\}\\
&+\varphi_2\left\{\nabla_{\mathbf{U}_2}\nabla_{\mathbf{V}}\mathbf{W}-\nabla_{\mathbf{V}}\nabla_{\mathbf{U}_2}\mathbf{W}-\nabla_{[\mathbf{U}_2,\,\mathbf{V}]}\mathbf{W}\right\}\\
={}& \varphi_1\mathbf{R}(\mathbf{U}_1,\mathbf{V})\mathbf{W}+\varphi_2\mathbf{R}(\mathbf{U}_2,\mathbf{V})\mathbf{W}
\end{aligned}$$

as claimed. In order to establish (6.53b), we use (6.53a), and exploit the antisymmetry of $\mathbf{R}$ (as expressed in (6.51)) to obtain

$$\begin{aligned}
\mathbf{R}(\mathbf{U},\varphi_1\mathbf{V}_1+\varphi_2\mathbf{V}_2)\mathbf{W} &= -\mathbf{R}(\varphi_1\mathbf{V}_1+\varphi_2\mathbf{V}_2,\mathbf{U})\mathbf{W}\\
&= -\varphi_1\mathbf{R}(\mathbf{V}_1,\mathbf{U})\mathbf{W}-\varphi_2\mathbf{R}(\mathbf{V}_2,\mathbf{U})\mathbf{W}\\
&= \varphi_1\mathbf{R}(\mathbf{U},\mathbf{V}_1)\mathbf{W}+\varphi_2\mathbf{R}(\mathbf{U},\mathbf{V}_2)\mathbf{W}
\end{aligned}$$

The proof (6.52c) is quite messy, and it appears in the Exercises that follow this section. The tensor character of **R** follows immediately from its multilinearity, since (6.52) implies that

$$\mathbf{R}(\mathbf{U},\mathbf{V})\mathbf{W} = U^b V^c W^d \, \mathbf{R}(\mathbf{e}_b,\mathbf{e}_c)\mathbf{e}_d \ , \tag{6.54}$$

which reaffirms that **R** is a tensor field of type $(1,3)$.

We now determine the components of **R** relative to the natural basis $\{\mathbf{e}_k\}$ and the corresponding dual basis $\{\mathbf{d}x^k\}$. We define[3]

$$R^a_{\ bcd}\mathbf{e}_a \equiv \mathbf{R}(\mathbf{e}_c,\mathbf{e}_d)\mathbf{e}_b$$

and consequently,

$$\begin{aligned} R^a_{\ bcd} &= \langle \mathbf{d}x^a, \mathbf{R}(\mathbf{e}_c,\mathbf{e}_d)\mathbf{e}_b\rangle \\ &= \langle \mathbf{d}x^a, \nabla_c\nabla_d\mathbf{e}_b - \nabla_d\nabla_c\mathbf{e}_b\rangle \quad [\text{Since } [\mathbf{e}_c,\mathbf{e}_d] = \mathbf{0}.] \\ &= \langle \mathbf{d}x^a, \nabla_c\left(\Gamma^r_{\ db}\mathbf{e}_r\right) - \nabla_d\left(\Gamma^s_{\ cb}\mathbf{e}_s\right)\rangle \\ &= \left\langle \mathbf{d}x^a, \left(\frac{\partial}{\partial x^c}\Gamma^r_{\ db}\right)\mathbf{e}_r + \Gamma^r_{\ db}\Gamma^s_{\ cr}\mathbf{e}_s - \left(\frac{\partial}{\partial x^d}\Gamma^s_{\ cb}\right)\mathbf{e}_s - \Gamma^s_{\ cb}\Gamma^r_{\ ds}\mathbf{e}_r\right\rangle \\ &= \frac{\partial}{\partial x^c}\Gamma^a_{\ db} - \frac{\partial}{\partial x^d}\Gamma^a_{\ cb} + \Gamma^r_{\ db}\Gamma^a_{\ cr} - \Gamma^s_{\ cb}\Gamma^a_{\ ds} \end{aligned}$$

Relabelling dummy indices, we now obtain

$$R^a_{\ bcd} = \frac{\partial}{\partial x^c}\Gamma^a_{\ db} - \frac{\partial}{\partial x^d}\Gamma^a_{\ cb} + \Gamma^a_{\ cp}\Gamma^p_{\ db} - \Gamma^a_{\ dp}\Gamma^p_{\ cb} \tag{6.56}$$

The curvature tensor has a remarkable geometric interpretation which we now discuss. As shown in Figure-6.3, consider an infinitesimal paralleogram *ABCD* whose vertices have the coordinates (x^k), $(x^k + \delta x^k)$, $(x^k + \delta x^k + \delta y^k)$, and $(x^k + \delta y^k)$, respectively.

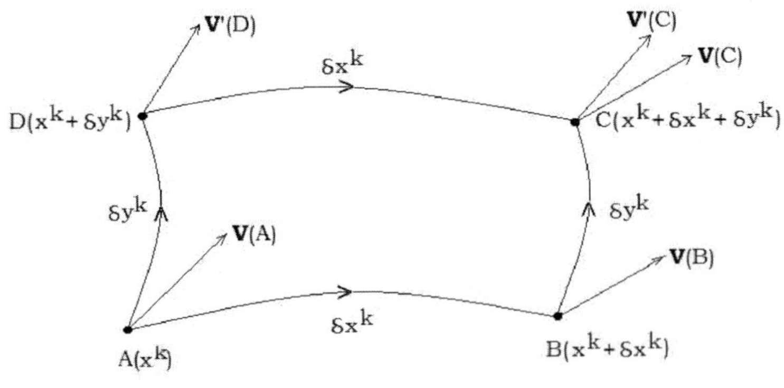

Figure 6.3.

The vector $\mathbf{V}(A) \in T_A(M)$ is first parallel transported along a counterclockwise path *ABC*, to obtain the vector $\mathbf{V}(C) \in T_C(M)$. The same vector $\mathbf{V}(A) \in T_A(M)$ is also parallel-transported along the clockwise path *ADC*, to obtain the vector $\mathbf{V}'(C) \in T_C(M)$. In general, it turns out that the parallel-transported vectors $\mathbf{V}(C)$ and $\mathbf{V}'(C)$ are not equal; in other words, the vector we get at the point C depends upon the path we take to get there. Whenever this happens, we say that the connection is **non-integrable**, and we immediately suspect that this non-integrability is related to the *curvature* of M.

This state of affairs is contrary to our experience in ordinary Euclidean space. For example, if the same parallelogram were to be constructed in the Euclidean *plane*, then it is apparent the resulting vectors $\mathbf{V}(C)$ and $\mathbf{V}'(C)$ would be the same.

On the other hand, for a general (non-Euclidean) manifold M, it turns out that the Riemann curvature tensor is directly related to this path dependence. To see this, let's actually compute the difference $\Delta\mathbf{V} = \mathbf{V}'(C) - \mathbf{V}(C)$.

First, assume that $\mathbf{V}(A)$ is parallel-transported along the counterclockwise path ABC. We have

$$\begin{aligned} V^k(B) &= V^k(A) - \delta x^r\,\Gamma^k_{rs}(A)V^s(A) \quad \text{(along the path } A \to B\text{)} \\ V^k(C) &= V^k(B) - \delta y^s\,\Gamma^k_{st}(B)V^t(B) \quad \text{(along the path } B \to C\text{)} \end{aligned}$$

where the connection coefficients are evaluated at the points A and B, respectively. By a Taylor expansion, however, we can express the connection coefficients at B in terms of those at A; in doing so, we have approximately,

$$\Gamma^k_{st}(B) = \Gamma^k_{st}(A) + \delta x^r \frac{\partial}{\partial x^r}\Gamma^k_{st}(A)$$

so that

$$\begin{aligned} V^k(C) &= \left\{V^k(A) - \delta x^r\,\Gamma^k_{rt}(A)V^t(A)\right\} \\ &\quad -\delta y^s\left\{\Gamma^k_{st}(A) + \delta x^r \frac{\partial}{\partial x^r}\Gamma^k_{st}(A)\right\}\left\{V^t(A) - \delta x^r\,\Gamma^t_{rs}(A)V^s(A)\right\} \end{aligned}$$

If we retain terms up to second order, we find that

$$V^k(C) = V^k - \delta x^r\,\Gamma^k_{rt}V^t - \delta y^s\,\Gamma^k_{st}V^t + \delta x^r\delta y^s\left(\Gamma^k_{st}\,\Gamma^t_{rp}\right)V^p - \delta x^r\delta y^s\left(\frac{\partial}{\partial x^r}\Gamma^k_{sp}\right)V^p$$

Similarly, when $\mathbf{V}(A)$ is parallel-transported along the clockwise path ADC, we find that

$$V'^k(C) = V^k - \delta x^r\,\Gamma^k_{rt}V^t - \delta y^s\,\Gamma^k_{st}V^t + \delta x^r\delta y^s\left(\Gamma^k_{rt}\Gamma^r_{sp}\right)V^p - \delta x^r\delta y^s\left(\frac{\partial}{\partial x^s}\Gamma^k_{rp}\right)V^p$$

Consequently,

$$\Delta V^k \equiv V'^k(C) - V^k(C) = \delta x^r\delta y^s\left\{\frac{\partial}{\partial x^r}\Gamma^k_{sp} - \frac{\partial}{\partial x^s}\Gamma^k_{rp} + \Gamma^k_{rt}\,\Gamma^t_{sp} - \Gamma^k_{st}\,\Gamma^t_{rp}\right\}V^p$$

Or, in other words, by (6.56), we have

$$\Delta V^k = R^k_{prs}\delta x^r\delta y^s V^p \tag{6.57}$$

This equation implies that *the connection is integrable whenever the curvature tensor vanishes identically*, $R^k_{prs} \equiv 0$.

6.7. Exercises

1. Show that the Lie bracket satisfies:

$$[\varphi_1 \mathbf{U}_1 + \varphi_2 \mathbf{U}_2, \mathbf{V}] = [\varphi_1 \mathbf{U}_1, \mathbf{V}] + [\varphi_2 \mathbf{U}_2, \mathbf{V}] \tag{6.58}$$

for all $\mathbf{U}_1, \mathbf{U}_2, \mathbf{V} \in \Im^1_0(M)$, and $\varphi_1, \varphi_2 \in C^\infty(M)$.

(Solution)

By the definition of the Lie bracket, we have

$$[\varphi_1 \mathbf{U}_1 + \varphi_2 \mathbf{U}_2, \mathbf{V}] = (\varphi_1 \mathbf{U}_1 + \varphi_2 \mathbf{U}_2)\mathbf{V} - \mathbf{V}(\varphi_1 \mathbf{U}_1 + \varphi_2 \mathbf{U}_2)$$

Expanding the righthand side of this equation gives

$$\begin{aligned}
[\varphi_1 \mathbf{U}_1 + \varphi_2 \mathbf{U}_2, \mathbf{V}] &= \varphi_1 \mathbf{U}_1 \mathbf{V} + \varphi_2 \mathbf{U}_2 \mathbf{V} - \varphi_1 \mathbf{V} \mathbf{U}_1 - (\mathbf{V}[\varphi_1])\mathbf{U}_1 \\
&\quad - \varphi_2 \mathbf{V} \mathbf{U}_2 - (\mathbf{V}[\varphi_2])\mathbf{U}_2 \\
&= (\varphi_1 \mathbf{U}_1 \mathbf{V} - \varphi_1 \mathbf{V} \mathbf{U}_1 - (\mathbf{V}[\varphi_1])\mathbf{U}_1) \\
&\quad + (\varphi_2 \mathbf{U}_2 \mathbf{V} - \varphi_2 \mathbf{V} \mathbf{U}_2 - (\mathbf{V}[\varphi_2])\mathbf{U}_2) \\
&= (\varphi_1 \mathbf{U}_1 \mathbf{V} - \mathbf{V}(\varphi_1 \mathbf{U}_1)) + (\varphi_2 \mathbf{U}_2 \mathbf{V} - \mathbf{V}(\varphi_2 \mathbf{U}_2)) \\
&= [\varphi_1 \mathbf{U}_1, \mathbf{V}] + [\varphi_2 \mathbf{U}_2, \mathbf{V}]
\end{aligned}$$

as claimed.

2. Establish the identity

$$[\varphi_1 \mathbf{U}, \varphi_2 \mathbf{V}] = \varphi_1 \varphi_2 [\mathbf{U}, \mathbf{V}] + \varphi_1 (\mathbf{U}[\varphi_2])\mathbf{V} - \varphi_2 (\mathbf{V}[\varphi_1])\mathbf{U} \tag{6.59}$$

which holds for all vector fields $\mathbf{U}, \mathbf{V} \in \Im^1_0(M)$, and for $\varphi_1, \varphi_2 \in C^\infty(M)$.

(Solution)

We have

$$\begin{aligned}
[\varphi_1 \mathbf{U}, \varphi_2 \mathbf{V}]^a &= \left(\varphi_1 U^b\right) \frac{\partial}{\partial x^b}(\varphi_2 V^a) - \left(\varphi_2 V^b\right) \frac{\partial}{\partial x^b}(\varphi_1 U^a) \\
&= \left(\varphi_1 U^b\right) \left\{\varphi_2 \frac{\partial V^a}{\partial x^b} + V^a \frac{\partial \varphi_2}{\partial x^b}\right\} - \left(\varphi_2 V^b\right) \left\{\varphi_1 \frac{\partial U^a}{\partial x^b} + U^a \frac{\partial \varphi_1}{\partial x^b}\right\}
\end{aligned}$$

Rearranging terms, we obtain

$$\begin{aligned}
[\varphi_1 \mathbf{U}, \varphi_2 \mathbf{V}]^a &= \varphi_1 \varphi_2 \left\{U^b \frac{\partial V^a}{\partial x^b} - V^b \frac{\partial U^a}{\partial x^b}\right\} + \varphi_1 V^a \left(U^b \frac{\partial \varphi_2}{\partial x^b}\right) - \varphi_2 \left(V^b \frac{\partial \varphi_1}{\partial x^b}\right) U^a \\
&= \varphi_1 \varphi_2 [U, V]^a + \varphi_1 (\mathbf{U}[\varphi_2]) V^a - \varphi_2 (\mathbf{V}[\varphi_1]) U^a
\end{aligned}$$

which is the component form of (6.59), as required. Observe that (6.53) is a special case of this equation, obtained by setting $\varphi_2 = 1$.

3. Show that **R** is *antisymmetric* in its first two arguments, i.e.,

$$\mathbf{R}(\mathbf{V},\mathbf{U})\mathbf{W} = -\mathbf{R}(\mathbf{U},\mathbf{V})\mathbf{W}$$

(Solution)

By the definition of **R**, we have

$$\begin{aligned}\mathbf{R}(\mathbf{V},\mathbf{U})\mathbf{W} &= \nabla_{\mathbf{V}}\nabla_{\mathbf{U}}\mathbf{W} - \nabla_{\mathbf{U}}\nabla_{\mathbf{V}}\mathbf{W} - \nabla_{[\mathbf{V},\mathbf{U}]}\mathbf{W} \\ &= \nabla_{\mathbf{V}}\nabla_{\mathbf{U}}\mathbf{W} - \nabla_{\mathbf{U}}\nabla_{\mathbf{V}}\mathbf{W} + \nabla_{[\mathbf{U},\mathbf{V}]}\mathbf{W} \\ &= -\left(\nabla_{\mathbf{U}}\nabla_{\mathbf{V}}\mathbf{W} - \nabla_{\mathbf{V}}\nabla_{\mathbf{U}}\mathbf{W} - \nabla_{[\mathbf{U},\mathbf{V}]}\mathbf{W}\right) = -\mathbf{R}(\mathbf{U},\mathbf{V})\mathbf{W}\end{aligned}$$

as required.

4. Establish equation (6.52c):

$$\mathbf{R}(\mathbf{U},\mathbf{V})(\varphi_1\mathbf{W}_1 + \varphi_2\mathbf{W}_2) = \varphi_1\mathbf{R}(\mathbf{U},\mathbf{V})\mathbf{W}_1 + \varphi_2\mathbf{R}(\mathbf{U},\mathbf{V})\mathbf{W}_2$$

for all smooth vector fields $\mathbf{U},\mathbf{V},\mathbf{W}_1,\mathbf{W}_2 \in \Im_0^1(M)$ and for all $\varphi_1,\varphi_2 \in C^\infty(M)$.

(Solution)

We have

$$\begin{aligned}\mathbf{R}(\mathbf{U},\mathbf{V})(\varphi_1\mathbf{W}_1 + \varphi_2\mathbf{W}_2) &= \nabla_{\mathbf{U}}\nabla_{\mathbf{V}}(\varphi_1\mathbf{W}_1 + \varphi_2\mathbf{W}_2) - \nabla_{\mathbf{V}}\nabla_{\mathbf{U}}(\varphi_1\mathbf{W}_1 + \varphi_2\mathbf{W}_2) \\ &\quad - \nabla_{[\mathbf{U},\mathbf{V}]}(\varphi_1\mathbf{W}_1 + \varphi_2\mathbf{W}_2)\end{aligned}$$

But, we find that

$$\begin{aligned}\nabla_{\mathbf{U}}\nabla_{\mathbf{V}}(\varphi_1\mathbf{W}_1 + \varphi_2\mathbf{W}_2) &= \varphi_1\nabla_{\mathbf{U}}\nabla_{\mathbf{V}}\mathbf{W}_1 + \mathbf{U}[\varphi_1]\nabla_{\mathbf{V}}\mathbf{W}_1 + \varphi_2\nabla_{\mathbf{U}}\nabla_{\mathbf{V}}\mathbf{W}_2 + \mathbf{U}[\varphi_2]\nabla_{\mathbf{V}}\mathbf{W}_2 \\ &\quad + \mathbf{V}[\varphi_1]\nabla_{\mathbf{U}}\mathbf{W}_1 + \mathbf{V}[\varphi_2]\mathbf{W}_2 + \mathbf{U}\mathbf{V}[\varphi_1]\mathbf{W}_1 + \mathbf{U}\mathbf{V}[\varphi_2]\mathbf{W}_2\end{aligned}$$

$$\begin{aligned}\nabla_{\mathbf{V}}\nabla_{\mathbf{U}}(\varphi_1\mathbf{W}_1 + \varphi_2\mathbf{W}_2) &= \varphi_1\nabla_{\mathbf{V}}\nabla_{\mathbf{U}}\mathbf{W}_1 + \mathbf{V}[\varphi_1]\nabla_{\mathbf{U}}\mathbf{W}_1 + \varphi_2\nabla_{\mathbf{V}}\nabla_{\mathbf{U}}\mathbf{W}_2 + \mathbf{V}[\varphi_2]\nabla_{\mathbf{U}}\mathbf{W}_2 \\ &\quad + \mathbf{U}[\varphi_1]\nabla_{\mathbf{V}}\mathbf{W}_1 + \mathbf{U}[\varphi_2]\mathbf{W}_2 + \mathbf{V}\mathbf{U}[\varphi_1]\mathbf{W}_1 + \mathbf{V}\mathbf{U}[\varphi_2]\mathbf{W}_2\end{aligned}$$

$$\begin{aligned}\nabla_{[\mathbf{U},\mathbf{V}]}(\varphi_1\mathbf{W}_1 + \varphi_2\mathbf{W}_2) &= \mathbf{U}\mathbf{V}[\varphi_1]\mathbf{W}_1 - \mathbf{V}\mathbf{U}[\varphi_1]\mathbf{W}_1 + \mathbf{U}\mathbf{V}[\varphi_2]\mathbf{W}_2 - \mathbf{V}\mathbf{U}[\varphi_2]\mathbf{W}_2 \\ &\quad + \varphi_1\nabla_{[\mathbf{U},\mathbf{V}]}\mathbf{W}_1 + \varphi_2\nabla_{[\mathbf{U},\mathbf{V}]}\mathbf{W}_2\end{aligned}$$

Substituting these three expressions into the first equation, and simplifying, gives

$$\begin{aligned}\mathbf{R}(\mathbf{U},\mathbf{V})(\varphi_1\mathbf{W}_1 + \varphi_2\mathbf{W}_2) &= \varphi_1\left\{\nabla_{\mathbf{U}}\nabla_{\mathbf{V}}\mathbf{W}_1 - \nabla_{\mathbf{V}}\nabla_{\mathbf{U}}\mathbf{W}_1 - \nabla_{[\mathbf{U},\mathbf{V}]}\mathbf{W}_1\right\} \\ &\quad + \varphi_2\left\{\nabla_{\mathbf{U}}\nabla_{\mathbf{V}}\mathbf{W}_2 - \nabla_{\mathbf{V}}\nabla_{\mathbf{U}}\mathbf{W}_2 - \nabla_{[\mathbf{U},\mathbf{V}]}\mathbf{W}_2\right\} \\ &= \varphi_1\mathbf{R}(\mathbf{U},\mathbf{V})\mathbf{W}_1 + \varphi_2\mathbf{R}(\mathbf{U},\mathbf{V})\mathbf{W}_2\end{aligned}$$

as required.

5. *(Ricci's Formula)* Show that in a coordinate basis, we have

$$[\nabla_a\,,\nabla_b]\,V^k = R^k_{\ mab}V^m \tag{6.70}$$

for any smooth vector field $\mathbf{V}$.

(Solution)

By definition of the curvature tensor, we have

$$\mathbf{R}(\mathbf{X},\mathbf{Y})\mathbf{V} = \nabla_{\mathbf{X}}\nabla_{\mathbf{Y}}\mathbf{V} - \nabla_{\mathbf{Y}}\nabla_{\mathbf{X}}\mathbf{V} - \nabla_{[\mathbf{X},\mathbf{Y}]}\mathbf{V}$$

Take a coordinate basis $\{\mathbf{e}_a\}$, and set $\mathbf{X} = \mathbf{e}_a$, $\mathbf{Y} = \mathbf{e}_b$ so that

$$\mathbf{R}(\mathbf{e}_a,\mathbf{e}_b)\mathbf{V} = \nabla_a\nabla_b\mathbf{V} - \nabla_b\nabla_a\mathbf{V} - \nabla_{[\mathbf{e}_a,\mathbf{e}_b]}\mathbf{V}$$

But $[\mathbf{e}_a,\mathbf{e}_b] = \mathbf{0}$ for a coordinate basis, so

$$\nabla_a\nabla_b\mathbf{V} - \nabla_b\nabla_a\mathbf{V} = \mathbf{R}(\mathbf{e}_a,\mathbf{e}_b)\mathbf{V}$$

In component form, the last equation is simply

$$[\nabla_a,\nabla_b]\,V^k = R^k_{\ mab}V^m$$

as required. *Observe that Ricci's formula shows that the curvature tensor is directly related to the failure of covariant derivatives to commute.*

6.8. The Levi-Civita Connection

Let $(M,\mathbf{g})$ be a semi-Riemannian space which is equipped with an affine connection ∇. The connection ∇ is said to be a **metric connection** or **metric compatible**, if

$$\nabla_{\mathbf{X}}\,\mathbf{g} = \mathbf{0} \quad \text{(metric compatibility condition)} \tag{6.71}$$

This means that for *any* smooth curve α with tangent vector $\mathbf{X}$, the rate of change of the metric along the curve is always zero. Here, the metric behaves like a *constant* with respect to absolute differentiation, and consequently, we say that for a metric connection, the metric is a **covariant constant.**

Let's see how the metric compatibility condition looks when it is written out in terms of components. We have

$$\begin{aligned}
\nabla_{\mathbf{X}}\,\mathbf{g} &= X^k\nabla_k\left(g_{ab}\,\mathbf{d}x^a\otimes\mathbf{d}x^b\right) \\
&= X^k\left\{\frac{\partial g_{ab}}{\partial x^k}\mathbf{d}x^a\otimes\mathbf{d}x^b + g_{pb}\left(\nabla_k\mathbf{d}x^p\right)\otimes\mathbf{d}x^b + g_{ap}\mathbf{d}x^a\otimes\left(\nabla_k\mathbf{d}x^p\right)\right\} \\
&= X^k\left\{\frac{\partial g_{ab}}{\partial x^k}\mathbf{d}x^a\otimes\mathbf{d}x^b - g_{pb}\left(\Gamma^p_{ka}\mathbf{d}x^a\right)\otimes\mathbf{d}x^b - g_{ap}\mathbf{d}x^a\otimes\left(\Gamma^p_{kb}\mathbf{d}x^b\right)\right\} \\
&= X^k\left(\frac{\partial g_{ab}}{\partial x^k} - \Gamma^p_{ka}g_{pb} - \Gamma^p_{kb}g_{ap}\right)\mathbf{d}x^a\otimes\mathbf{d}x^b = 0
\end{aligned}$$

But since the curve α is arbitrary, then the component form of the metric compatibility condition is simply

$$\partial_k g_{ab} - \Gamma^p_{ka} g_{pb} - \Gamma^p_{kb} g_{ap} = 0, \tag{6.72}$$

where for brevity, we set $\partial_k \equiv \partial/\partial x^k$.

Now, assuming the connection is a metric connection, let's see how the inner product of any two vectors is affected by parallel-transport. Let two non-zero vectors **U** and **V** be parallel-transported along a smooth curve $\alpha(t)$, from the point p to the point q, as shown in Figure 6.4. Also, let **X** be the tangent vector to the curve.

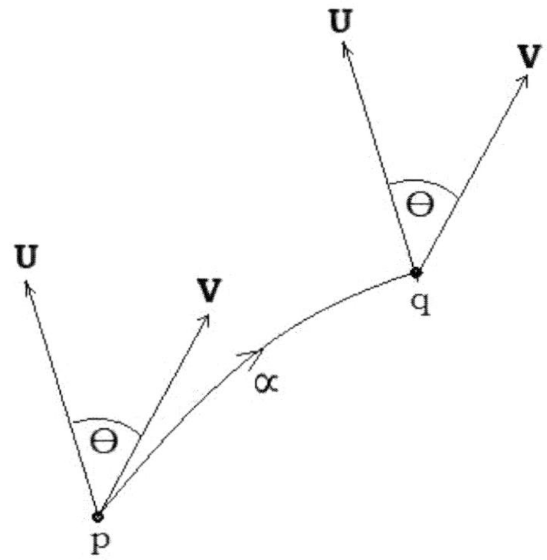

Figure 6.4. Two vectors are parallel-transported along α.

We observe that since the vectors **U** and **V** undergo parallel-transport, each vector must remain unchanged as it moves along the curve α, and we must have

$$\nabla_{\mathbf{X}}\mathbf{U} = \nabla_{\mathbf{X}}\mathbf{V} = \mathbf{0} \tag{6.73}$$

Consequently, the rate of change of their inner product along α is given by

$$\begin{aligned} \frac{D}{Dt}\langle \mathbf{U}, \mathbf{V}\rangle &= \nabla_{\mathbf{X}}\, \mathbf{g}(\mathbf{U}, \mathbf{V}) \\ &= (\nabla_{\mathbf{X}}\, \mathbf{g})\,(\mathbf{U}, \mathbf{V}) + \mathbf{g}(\nabla_{\mathbf{X}}\mathbf{U}, \mathbf{V}) + \mathbf{g}(\mathbf{U}, \nabla_{\mathbf{X}}\mathbf{V}) = 0\,, \end{aligned}$$

where we have used (6.71) and (6.72). We conclude that *if the metric is a covariant constant, then the inner product of any two vectors is preserved under the operation of parallel transport.*

Now, it turns out that there exists a *unique* connection, called the **Levi-Civita connection**, which is both *symmetric* (torsion-free) and which preserves the inner product of any two vectors under parallel transport. This remarkable statement is called the **Fundamental Theorem of (semi-) Riemannian Geometry**.

The proof of this statement is straightforward. We first suppose that ∇ is a *metric connection* which is also *torsion free*. We can perform two cyclic permutations of the indices in (6.72) to obtain the additional equations:

$$\partial_b g_{ka} - \Gamma^p_{bk} g_{pa} - \Gamma^p_{ba} g_{kp} = 0 \quad (6.74a)$$
$$\partial_a g_{bk} - \Gamma^p_{ab} g_{pk} - \Gamma^p_{ak} g_{bp} = 0 \quad (6.74b)$$

Taking $-(6.72)$ $+(6.74a)$ $+$ $(6.74b)$, and recalling that the metric is symmetric, i.e., $g_{ab} = g_{ba}$, we obtain

$$-\partial_k g_{ab} + \partial_b g_{ka} + \partial_a g_{bk} = \left(\Gamma^p_{ab} + \Gamma^p_{ba}\right) g_{kp} + \left(\Gamma^p_{ak} - \Gamma^p_{ka}\right) g_{bp} + \left(\Gamma^p_{bk} - \Gamma^p_{kb}\right) g_{ap}$$

Since, by hypothesis, the connection is torsion-free, the connection components are symmetric in their lower indices, and the last equation simplifies to

$$\Gamma^p_{ab} g_{kp} = \frac{1}{2}\left(-\partial_k g_{ab} + \partial_b g_{ka} + \partial_a g_{bk}\right)$$

We now solve for the connection components by simply multiplying both sides of this equation by the inverse metric g^{ck} to obtain

$$\Gamma^c_{ab} = \frac{1}{2} g^{ck}\left(-\partial_k g_{ab} + \partial_b g_{ka} + \partial_a g_{bk}\right) \quad (6.75)$$

We see that the connection components are *unique* by very nature of our construction, and the theorem is proved.

In all of the developments that follow, we shall assume that we will work with the Levi-Civita connection, unless stated otherwise. In passing, we note that it is customary to write

$$[ab,k] \equiv \Gamma_{abk} = \frac{1}{2}\left(-\partial_k g_{ab} + \partial_b g_{ka} + \partial_a g_{bk}\right) \quad (6.76)$$
$$\left\{ {c \atop ab} \right\} \equiv \Gamma^c_{ab} = g^{ck}\,[ab,k] \quad (6.77)$$

where $[ab,k]$ and $\left\{ {c \atop ab} \right\}$ are called the **Christoffel symbols** of the first and second kind, respectively.

6.9. Geodesic Coordinates

Assume that $(M,\mathbf{g})$ is a semi-Riemannian manifold which is equipped with the Levi-Civita connection. We will demonstrate that given overlapping coordinate charts (U,ϕ) and $(\bar{U},\bar{\phi})$ on M, for each point $p \in U \cap \bar{U}$, there exists a coordinate transformation $\bar{x}^i = \bar{x}^i(x^j)$ such that the Christoffel symbols vanish at p:

$$\bar{\Gamma}^i_{jk}\big|_p = 0 \quad (6.78)$$

Without loss of generality, we shall take the point p as the origin of the original coordinate system (x^i) so that

$$x^i\big|_p = \phi^i(p) = 0.$$

We now introduce a change of coordinates

$$\bar{x}^i = x^i + K^i_{jk} x^j x^k \tag{6.79}$$

where we assume that the symbols K^i_{jk} are *constants*, and they are *symmetric* with respect to their lower indices

$$K^i_{jk} = K^i_{kj} \,. \tag{6.80}$$

Under a coordinate transformation $\bar{x}^i = \bar{x}^i(x^j)$, we know that the Christoffel symbols obey the equation

$$\bar{\Gamma}^i_{jk} = \frac{\partial \bar{x}^i}{\partial x^t}\frac{\partial x^r}{\partial \bar{x}^j}\frac{\partial x^s}{\partial \bar{x}^k}\Gamma^t_{rs} + \frac{\partial^2 x^m}{\partial \bar{x}^j \partial \bar{x}^k}\frac{\partial \bar{x}^i}{\partial x^m} \,. \tag{6.25}$$

which was derived previously. Let's see how these symbols transform under the particular change of coordinates given by (6.79). We find that

$$\frac{\partial \bar{x}^i}{\partial x^t} = \delta^i_t + 2K^i_{tp} x^p \tag{6.81}$$

$$\frac{\partial^2 x^m}{\partial \bar{x}^j \partial \bar{x}^k} = 2K^m_{jk} \tag{6.82}$$

Now, at the point p (where $x^i = 0$), the last two equations become:

$$\left.\frac{\partial \bar{x}^i}{\partial x^t}\right|_p = \delta^i_t \text{ which implies } \left.\frac{\partial x^r}{\partial \bar{x}^j}\right|_P = \delta^r_j \text{ and } \left.\frac{\partial x^s}{\partial \bar{x}^k}\right|_P = \delta^s_k \tag{6.81a}$$

Also,

$$\left.\frac{\partial^2 x^m}{\partial \bar{x}^j \partial \bar{x}^k}\right|_P = 2\, K^m_{jk}\big|_P \tag{6.82a}$$

If we now substitute both (6.81a) and (6.82a) into (6.25), we obtain

$$\bar{\Gamma}^i_{jk}\Big|_p = \delta^i_t \delta^r_j \delta^s_k\, \Gamma^t_{rs}\big|_p + 2\, K^m_{kj}\big|_p\, \delta^i_m = \Gamma^i_{jk}\big|_p + 2\, K^i_{jk}\big|_p \tag{6.83}$$

Now, we observe that if we set

$$K^i_{jk}\big|_p = -\frac{1}{2}\,\Gamma^i_{jk}\big|_p$$

so that

$$\bar{x}^i = x^i - \frac{1}{2}\,\Gamma^i_{jk}\big|_p\, x^j x^k \tag{6.84}$$

then (6.83) reduces to

$$\bar{\Gamma}^i_{jk}\big|_p = 0 \tag{6.85}$$

as claimed. Any coordinate system $(\bar{x}^i)$ such that condition (6.85) is satisfied is called a **geodesic coordinate system** and the point p is called the **pole** of the coordinate system. Furthermore, we call (6.84) a **geodesic coordinate transformation**.

It is important to note that when the Christoffel symbols are expressed in terms of geodesic coordinates, they do not vanish identically, but only at the *single* point $\bar{x}^i = 0$.

Furthermore, since the Christoffel symbols vanish at $\bar{x}^i = 0$, then *covariant derivatives with respect to the barred coordinates reduce to partial derivatives at this point as well.*

Now we have to be careful, for although the Christoffel symbols vanish at the pole of a geodesic coordinate system one *cannot* infer that the partial derivatives of these symbols vanish there also; but the metric tensor is another story. If we express the metric compatibility condition (6.72) in terms of geodesic coordinates, we have

$$\frac{\partial \bar{g}_{ab}}{\partial \bar{x}^k} = \bar{\Gamma}^p_{ka}\bar{g}_{pb} + \bar{\Gamma}^p_{kb}\bar{g}_{ap}$$

From (6.78) we see that *at the pole of a geodesic coordinate system, the partial derivatives of the metric tensor vanish*

$$\left.\frac{\partial \bar{g}_{ab}}{\partial \bar{x}^k}\right|_p = 0 \tag{6.86}$$

As we will see, geodesic coordinates are extremely useful in establishing tensor identities that involve the Christoffel symbols. Also, in general relativity, we will find that the existence of geodesic coordinates is intimately related to the *Principle of Equivalence.*

6.10. Symmetries of the Curvature Tensor

Once again, we assume that a semi-Riemannian space $(M, \mathbf{g})$ is equipped with the *Levi-Civita connection*, so that the connection is *torsion-free*. It turns out that the Riemann curvature tensor possesses several useful symmetry properties.

Symmetries of the Curvature Tensor

Given arbitrary vector fields $\mathbf{U}, \mathbf{V}, \mathbf{W}, \mathbf{Y}$, the curvature tensor $\mathbf{R}$ satisfies:

1.

$$\begin{aligned} \mathbf{R}(\mathbf{U},\mathbf{V})\mathbf{W} &= -\mathbf{R}(\mathbf{V},\mathbf{U})\mathbf{W} \\ R^a_{\ bcd} &= -R^a_{\ bdc} \end{aligned} \tag{6.87}$$

2. *(First Bianchi Identity)*

$$\mathbf{R}(\mathbf{U},\mathbf{V})\mathbf{W} + \mathbf{R}(\mathbf{W},\mathbf{U})\mathbf{V} + \mathbf{R}(\mathbf{V},\mathbf{W})\mathbf{U} = \mathbf{0} \tag{6.88}$$

$$R^k_{\ abc} + R^k_{\ cab} + R^k_{\ bca} = 0$$

3. *(2nd Bianchi Identity)*

$$(\nabla_{\mathbf{X}}\mathbf{R})(\mathbf{U},\mathbf{V})\mathbf{W} + (\nabla_{\mathbf{U}}\mathbf{R})(\mathbf{V},\mathbf{X})\mathbf{W} + (\nabla_{\mathbf{V}}\mathbf{R})(\mathbf{X},\mathbf{U})\mathbf{W} = \mathbf{0} \tag{6.89}$$

$$\nabla_m R^k_{\ abc} + \nabla_b R^k_{\ acm} + \nabla_c R^k_{\ amb} = 0$$

Now, we have already proved (6.87); in order to establish (6.88), we first recall that the torsion tensor is given by (6.40) as

$$\mathbf{T}(\mathbf{U},\mathbf{V}) = \nabla_{\mathbf{U}}\mathbf{V} - \nabla_{\mathbf{V}}\mathbf{U} - [\mathbf{U},\mathbf{V}]$$

Since the Levi-Civita connection is torsion-free, then $\mathbf{T}(\mathbf{U},\mathbf{V}) = \mathbf{0}$, and consequently,

$$\nabla_{\mathbf{U}}\mathbf{V} - \nabla_{\mathbf{V}}\mathbf{U} = [\mathbf{U},\mathbf{V}]$$

The lefthand side of (6.88) becomes

$$\begin{aligned}
& \mathbf{R}(\mathbf{U},\mathbf{V})\mathbf{W} + \mathbf{R}(\mathbf{W},\mathbf{U})\mathbf{V} + \mathbf{R}(\mathbf{V},\mathbf{W})\mathbf{U} \\
= \; & \nabla_{\mathbf{U}}(\nabla_{\mathbf{V}}\mathbf{W} - \nabla_{\mathbf{W}}\mathbf{V}) + \nabla_{\mathbf{V}}(\nabla_{\mathbf{W}}\mathbf{U} - \nabla_{\mathbf{U}}\mathbf{W}) \\
& + \nabla_{\mathbf{W}}(\nabla_{\mathbf{U}}\mathbf{V} - \nabla_{\mathbf{V}}\mathbf{U}) - \nabla_{[\mathbf{U},\mathbf{V}]}\mathbf{W} - \nabla_{[\mathbf{W},\mathbf{U}]}\mathbf{V} \\
& - \nabla_{[\mathbf{V},\mathbf{W}]}\mathbf{U} \\
= \; & \left(\nabla_{\mathbf{U}}[\mathbf{V},\mathbf{W}] - \nabla_{[\mathbf{V},\mathbf{W}]}\mathbf{U}\right) + \left(\nabla_{\mathbf{V}}[\mathbf{W},\mathbf{U}] - \nabla_{[\mathbf{W},\mathbf{U}]}\mathbf{V}\right) \\
& + \left(\nabla_{\mathbf{W}}[\mathbf{U},\mathbf{V}] - \nabla_{[\mathbf{U},\mathbf{V}]}\mathbf{W}\right) \\
= \; & [\mathbf{U},[\mathbf{V},\mathbf{W}]] + [\mathbf{V},[\mathbf{W},\mathbf{U}]] + [\mathbf{W},[\mathbf{U},\mathbf{V}]] = \mathbf{0}
\end{aligned}$$

Observe that the Jacobi identity was used in the last step of the proof.

Before we establish (6.89), we introduce a useful formula for the absolute derivative. If $\mathbf{T}$ is a tensor of type (p,q) where either $p = 0$ or $p = 1$, then it may be shown that given a fixed vector field $\mathbf{X}$, and arbitrary vector fields $\mathbf{V}_1,\ldots,\mathbf{V}_q$ we can write

$$(\nabla_{\mathbf{X}}\mathbf{T})(\mathbf{V}_1,\ldots,\mathbf{V}_q) = \nabla_{\mathbf{X}}(\mathbf{T}(\mathbf{V}_1,\ldots,\mathbf{V}_q)) - \sum_{k=1}^{q}\mathbf{T}(\mathbf{V}_1,\ldots,\mathbf{V}_{k-1},\nabla_{\mathbf{X}}\mathbf{V}_k,\mathbf{V}_{k+1},\ldots,\mathbf{V}_q) \quad (6.90)$$

Assume that $\mathbf{U},\mathbf{V},\mathbf{W},\mathbf{X}$ are *coordinate basis* fields; since $\mathbf{R}$ is a multilinear object, there is no loss in generality, and any Lie brackets involving these fields will vanish. Now, employing (6.90), the each term on lefthand side of (6.89) may be rewritten:

$$\begin{aligned}
(\nabla_{\mathbf{X}}\mathbf{R})(\mathbf{U},\mathbf{V})\mathbf{W} \; = \; & \nabla_{\mathbf{X}}(\nabla_{\mathbf{U}}\nabla_{\mathbf{V}}\mathbf{W} - \nabla_{\mathbf{V}}\nabla_{\mathbf{U}}\mathbf{W}) - \mathbf{R}(\nabla_{\mathbf{X}}\mathbf{U},\mathbf{V})\mathbf{W} \\
& - \mathbf{R}(\mathbf{U},\nabla_{\mathbf{X}}\mathbf{V})\mathbf{W} - \nabla_{\mathbf{U}}\nabla_{\mathbf{V}}\nabla_{\mathbf{X}}\mathbf{W} + \nabla_{\mathbf{V}}\nabla_{\mathbf{U}}\nabla_{\mathbf{X}}\mathbf{W}
\end{aligned}$$

$$\begin{aligned}
(\nabla_{\mathbf{U}}\mathbf{R})(\mathbf{V},\mathbf{X})\mathbf{W} \; = \; & \nabla_{\mathbf{U}}(\nabla_{\mathbf{V}}\nabla_{\mathbf{X}}\mathbf{W} - \nabla_{\mathbf{X}}\nabla_{\mathbf{V}}\mathbf{W}) - \mathbf{R}(\nabla_{\mathbf{U}}\mathbf{V},\mathbf{X})\mathbf{W} \\
& - \mathbf{R}(\mathbf{V},\nabla_{\mathbf{U}}\mathbf{X})\mathbf{W} - \nabla_{\mathbf{V}}\nabla_{\mathbf{X}}\nabla_{\mathbf{U}}\mathbf{W} + \nabla_{\mathbf{X}}\nabla_{\mathbf{V}}\nabla_{\mathbf{U}}\mathbf{W}
\end{aligned}$$

$$\begin{aligned}
(\nabla_{\mathbf{V}}\mathbf{R})(\mathbf{X},\mathbf{U})\mathbf{W} \; = \; & \nabla_{\mathbf{V}}(\nabla_{\mathbf{X}}\nabla_{\mathbf{U}}\mathbf{W} - \nabla_{\mathbf{U}}\nabla_{\mathbf{X}}\mathbf{W}) - \mathbf{R}(\nabla_{\mathbf{V}}\mathbf{X},\mathbf{U})\mathbf{W} \\
& - \mathbf{R}(\mathbf{X},\nabla_{\mathbf{V}}\mathbf{U})\mathbf{W} - \nabla_{\mathbf{X}}\nabla_{\mathbf{U}}\nabla_{\mathbf{V}}\mathbf{W} + \nabla_{\mathbf{U}}\nabla_{\mathbf{X}}\nabla_{\mathbf{V}}\mathbf{W}
\end{aligned}$$

If we add these equations, then the terms involving successive covariant derivatives cancel in pairs, and we obtain

$$\begin{aligned}
& (\nabla_{\mathbf{X}}\mathbf{R})(\mathbf{U},\mathbf{V})\mathbf{W} + (\nabla_{\mathbf{U}}\mathbf{R})(\mathbf{V},\mathbf{X})\mathbf{W} + (\nabla_{\mathbf{V}}\mathbf{R})(\mathbf{X},\mathbf{U})\mathbf{W} \\
= \; & -\mathbf{R}(\nabla_{\mathbf{X}}\mathbf{U},\mathbf{V})\mathbf{W} - \mathbf{R}(\mathbf{U},\nabla_{\mathbf{X}}\mathbf{V})\mathbf{W} - \mathbf{R}(\nabla_{\mathbf{U}}\mathbf{V},\mathbf{X})\mathbf{W} - \mathbf{R}(\mathbf{V},\nabla_{\mathbf{U}}\mathbf{X})\mathbf{W} \\
& -\mathbf{R}(\mathbf{X},\nabla_{\mathbf{V}}\mathbf{U})\mathbf{W} - \mathbf{R}(\mathbf{X},\nabla_{\mathbf{V}}\mathbf{U})\mathbf{W}
\end{aligned}$$

Recalling that **R** is multilinear as well as antisymmetric in its first two arguments, we can rearrange the righthand side of the last equation to get

$$\begin{aligned}&(\nabla_{\mathbf{X}}\mathbf{R})(\mathbf{U},\mathbf{V})\mathbf{W}+(\nabla_{\mathbf{U}}\mathbf{R})(\mathbf{V},\mathbf{X})\mathbf{W}+(\nabla_{\mathbf{V}}\mathbf{R})(\mathbf{X},\mathbf{U})\mathbf{W}\\ =\;&\mathbf{R}(\mathbf{U},\nabla_{\mathbf{V}}\mathbf{X}-\nabla_{\mathbf{X}}\mathbf{V})\mathbf{W}+\mathbf{R}(\mathbf{V},\nabla_{\mathbf{X}}\mathbf{U}-\nabla_{\mathbf{U}}\mathbf{X})+\mathbf{R}(\mathbf{X},\nabla_{\mathbf{U}}\mathbf{V}-\nabla_{\mathbf{V}}\mathbf{U})\end{aligned}$$

Since the connection is torsion-free, we have

$$\nabla_{\mathbf{V}}\mathbf{X}-\nabla_{\mathbf{X}}\mathbf{V}=\nabla_{\mathbf{X}}\mathbf{U}-\nabla_{\mathbf{U}}\mathbf{X}=\nabla_{\mathbf{U}}\mathbf{V}-\nabla_{\mathbf{V}}\mathbf{U}=\mathbf{0}$$

and consequently,

$$(\nabla_{\mathbf{X}}\mathbf{R})(\mathbf{U},\mathbf{V})\mathbf{W}+(\nabla_{\mathbf{U}}\mathbf{R})(\mathbf{V},\mathbf{X})\mathbf{W}+(\nabla_{\mathbf{V}}\mathbf{R})(\mathbf{X},\mathbf{U})\mathbf{W}=\mathbf{0}$$

as claimed.

We now demonstrate how the introduction of geodesic coordinates often simplifies proofs of this nature. For example, to establish (6.89), we first note that the components of **R** may be written as

$$R^k_{\ abc}=\partial_b\Gamma^k_{\ ac}-\partial_c\Gamma^k_{\ ab}+\Gamma^k_{\ bp}\Gamma^p_{\ ac}-\Gamma^k_{\ cp}\Gamma^p_{\ ab} \tag{6.91}$$

It is clear that this equation holds throughout M, and in particular, it will hold at an arbitrary point $p\in M$. We now switch to geodesic coordinates; that is we perform the coordinate transformation $\bar{x}^i=\bar{x}^i(x^j)$ given by (6.84), where the point p is the pole of the geodesic coordinate system. In terms of geodesic coordinates, the components of **R** at p are simply

$$\bar{R}^k_{\ abc}=\bar{\partial}_b\bar{\Gamma}^k_{\ ac}-\bar{\partial}_c\bar{\Gamma}^k_{\ ab}$$

We now take the appropriate covariant derivatives of the this expression, while keeping in mind that at the point p, covariant derivatives reduce to partial derivatives, to obtain the following three equations:

$$\begin{aligned}\bar{\nabla}_m\bar{R}^k_{\ abc}&=\bar{\partial}_m\bar{\partial}_b\left(\bar{\Gamma}^k_{\ ac}\right)-\bar{\partial}_m\bar{\partial}_c\left(\bar{\Gamma}^k_{\ ab}\right)\\ \bar{\nabla}_b\bar{R}^k_{\ acm}&=\bar{\partial}_b\bar{\partial}_c\left(\bar{\Gamma}^k_{\ am}\right)-\bar{\partial}_b\bar{\partial}_m\left(\bar{\Gamma}^k_{\ ac}\right)\\ \bar{\nabla}_c\bar{R}^k_{\ amb}&=\bar{\partial}_c\bar{\partial}_m\left(\bar{\Gamma}^k_{\ ab}\right)-\bar{\partial}_c\bar{\partial}_b\left(\bar{\Gamma}^k_{\ am}\right)\end{aligned}$$

Adding these three equations, and utilizing the equality of mixed partial derivatives, we obtain:

$$\begin{aligned}\bar{\nabla}_m\bar{R}^k_{\ abc}+\bar{\nabla}_b\bar{R}^k_{\ acm}+\bar{\nabla}_c\bar{R}^k_{\ amb}&=\left[\bar{\partial}_m,\bar{\partial}_b\right]\bar{\Gamma}^k_{\ ac}+\left[\bar{\partial}_b,\bar{\partial}_c\right]\bar{\Gamma}^k_{\ am}+\left[\bar{\partial}_c,\bar{\partial}_m\right]\bar{\Gamma}^k_{\ ab}\\&=0\end{aligned}$$

Since this equation is a *tensor equation* which is valid at the pole of the geodesic coordinate system, then it is also valid at the point p of the original (unbarred) coordinate system. Furthermore, since p is an *arbitrary point* in M, we conclude that this equation must hold for *all* such points.

We define the **covariant curvature tensor** as the tensor of type $(0,4)$ given by

$$R_{abcd} = g_{ak}R^k_{bcd} = \langle \partial_a, \mathbf{R}(\partial_c, \partial_d)\partial_b \rangle \ , \tag{6.92}$$

which is obtained by lowering the upper index of the Riemann tensor. When written out in terms of components, the covariant curvature tensor may be expressed in the form

$$R_{abcd} = \partial_c \Gamma_{bda} - \partial_d \Gamma_{bca} + \Gamma_{adp}\Gamma^p_{bc} - \Gamma_{acp}\Gamma^p_{bd} \tag{6.93}$$

In order to see this, we first write

$$\begin{aligned} R_{abcd} &= g_{ap}R^p_{bcd} = g_{ap}\left[\partial_c \Gamma^p_{bd} - \partial_d \Gamma^p_{bc} + \Gamma^p_{cm}\Gamma^m_{bd} - \Gamma^p_{dm}\Gamma^m_{bc}\right] \\ &= g_{ap}\partial_c \Gamma^p_{bd} - g_{ap}\partial_d \Gamma^p_{bc} + \Gamma_{cma}\Gamma^m_{bd} - \Gamma_{dma}\Gamma^m_{bc} \end{aligned}$$

Now, we rewrite the partial derivative terms in the last equation, and replace the dummy index m by p in the last two terms, to obtain

$$\begin{aligned} R_{abcd} &= \partial_c(g_{ap}\Gamma^p_{bd}) - \partial_d(g_{ap}\Gamma^p_{bc}) - \Gamma^p_{bd}(\partial_c g_{ap}) + \Gamma^p_{bc}(\partial_d g_{ap}) + \Gamma_{cpa}\Gamma^p_{bd} - \Gamma_{dpa}\Gamma^p_{bc} \\ &= \partial_c(g_{ap}\Gamma^p_{bd}) - \partial_d(g_{ap}\Gamma^p_{bc}) - \Gamma^p_{bd}(\partial_c g_{ap} - \Gamma_{cpa}) + \Gamma^p_{bc}\left(\partial_d g_{ap} - \Gamma_{dpa}\right) \end{aligned} \tag{6.94}$$

By the metric compatibility condition (6.72), however, we have

$$\begin{aligned} \partial_c g_{ap} - \Gamma_{cpa} &= \Gamma_{acp} \\ \partial_d g_{ap} - \Gamma_{dpa} &= \Gamma_{adp} \end{aligned} \tag{6.95}$$

Substituting (6.95) into (6.94), we obtain

$$\begin{aligned} R_{abcd} &= \partial_c(g_{ap}\Gamma^p_{bd}) - \partial_d(g_{ap}\Gamma^p_{bc}) + \Gamma_{adp}\Gamma^p_{bc} - \Gamma_{acp}\Gamma^p_{bd} \\ &= \partial_c \Gamma_{bda} - \partial_d \Gamma_{bca} + \Gamma_{adp}\Gamma^p_{bc} - \Gamma_{acp}\Gamma^p_{bd} \end{aligned}$$

as claimed.

Perhaps, the most convenient form of the covariant curvature tensor is given by

$$R_{abcd} = \frac{1}{2}\left(\partial_b\partial_c g_{ad} + \partial_a\partial_d g_{bc} - \partial_a\partial_c g_{bd} - \partial_b\partial_d g_{ac}\right) + \Gamma_{adp}\Gamma^p_{bc} - \Gamma_{acp}\Gamma^p_{bd} \ , \tag{6.96}$$

which we now seek to establish. If we compare (6.96) with (6.93), we see that in order to establish (6.96), it is suffices to show that

$$\partial_c \Gamma_{bda} - \partial_d \Gamma_{bca} = \frac{1}{2}\left(\partial_b\partial_c g_{ad} + \partial_a\partial_d g_{bc} - \partial_a\partial_c g_{bd} - \partial_b\partial_d g_{ac}\right) \tag{6.97}$$

By the definition of the Christoffel symbols, we have

$$\begin{aligned} \Gamma_{bda} &= \tfrac{1}{2}\left(\partial_b g_{da} + \partial_d g_{ab} - \partial_a g_{bd}\right) \\ \Gamma_{bca} &= \tfrac{1}{2}\left(\partial_b g_{ca} + \partial_c g_{ab} - \partial_a g_{bc}\right) \end{aligned}$$

Taking the partial derivatives of these two expressions, we find

$$\begin{aligned} \partial_c \Gamma_{bda} - \partial_d \Gamma_{bca} &= \frac{1}{2}[\partial_c\partial_b g_{da} + \partial_c\partial_d g_{ab} - \partial_c\partial_a g_{bd} \\ &\quad -\partial_d\partial_b g_{ca} - \partial_d\partial_c g_{ab} + \partial_d\partial_a g_{bc}] \end{aligned}$$

Note that the second and the fifth terms on the righthand side of this equation cancel due to the equality of mixed partial derivatives, so that

$$\partial_c \Gamma_{bda} - \partial_d \Gamma_{bca} = \frac{1}{2}\left(\partial_c \partial_b g_{da} - \partial_c \partial_a g_{bd} - \partial_d \partial_b g_{ca} + \partial_d \partial_a g_{bc}\right)$$

Using the fact that the metric is symmetric, and the order of partial differentiation is immaterial, we obtain

$$\partial_c \Gamma_{bda} - \partial_d \Gamma_{bca} = \frac{1}{2}\left(\partial_b \partial_c g_{ad} + \partial_a \partial_d g_{bc} - \partial_a \partial_c g_{bd} - \partial_b \partial_d g_{ac}\right)$$

as claimed. If we now substitute (6.97) into (6.93), then (6.96) follows immediately.

Like the curvature tensor, the covariant curvature tensor obeys certain symmetry properties which will prove to be useful in our later work:

Symmetries of the Covariant Curvature Tensor

Given arbitrary vector fields $\mathbf{U}, \mathbf{V}, \mathbf{W}, \mathbf{X}, \mathbf{Y}$, the covariant curvature tensor has the following symmetries:

1. The covariant curvature tensor is *skew-symmetric* with respect to its first two arguments (indices):

$$\begin{aligned} \langle \mathbf{R}(\mathbf{U},\mathbf{V})\mathbf{W},\mathbf{X}\rangle &= -\langle \mathbf{R}(\mathbf{U},\mathbf{V})\mathbf{X},\mathbf{W}\rangle \\ R_{abcd} &= -R_{bacd} \end{aligned} \tag{6.98}$$

2. The covariant curvature tensor is *skew-symmetric* with respect to its last two arguments (indices):

$$\begin{aligned} \langle \mathbf{R}(\mathbf{U},\mathbf{V})\mathbf{W},\mathbf{X}\rangle &= -\langle \mathbf{R}(\mathbf{V},\mathbf{U})\mathbf{W},\mathbf{X}\rangle \\ R_{abcd} &= -R_{abdc} \end{aligned} \tag{6.99}$$

3. The covariant curvature tensor exhibits *block symmetry* with respect to its two pairs of arguments (indices):

$$\begin{aligned} \langle \mathbf{R}(\mathbf{U},\mathbf{V})\mathbf{W},\mathbf{X}\rangle &= \langle \mathbf{R}(\mathbf{W},\mathbf{X}),\mathbf{U},\mathbf{V}\rangle \\ R_{abcd} &= R_{cdab} \end{aligned} \tag{6.100}$$

4. *(First Bianchi Identity)*

$$\begin{aligned} \langle \mathbf{R}(\mathbf{U},\mathbf{V})\mathbf{W},\mathbf{X}\rangle + \langle \mathbf{R}(\mathbf{W},\mathbf{U})\mathbf{V},\mathbf{X}\rangle + \langle \mathbf{R}(\mathbf{V},\mathbf{W})\mathbf{U},\mathbf{X}\rangle &= 0 \\ R_{abcd} + R_{adbc} + R_{acdb} &= 0 \end{aligned} \tag{6.101}$$

5. *(Second Bianchi Identity)*

$$\begin{aligned} \nabla_{\mathbf{Y}}\langle \mathbf{R}(\mathbf{U},\mathbf{V})\mathbf{W},\mathbf{X}\rangle + \nabla_{\mathbf{U}}\langle \mathbf{R}(\mathbf{V},\mathbf{Y})\mathbf{W},\mathbf{X}\rangle + \nabla_{\mathbf{V}}\langle \mathbf{R}(\mathbf{Y},\mathbf{U})\mathbf{W},\mathbf{X}\rangle &= 0 \\ \nabla_p R_{abcd} + \nabla_c R_{abdp} + \nabla_d R_{abpc} &= 0 \end{aligned} \tag{6.102}$$

The proofs of (6.98) - (6.100) are straightforward, and are left as exercises. Similarly, the proofs of (6.101) and (6.102) follow immediately from (6.88) and (6.89), respectively. As we will see, the Bianchi identities are of crucial importance in the General Relativity.

We devote the remainer of this section to determining the number of *algebraically independent* components of the covariant curvature tensor. By the above symmetry properties, we suspect that the number of independent components of this tensor must be greatly reduced, since some of its components are related to others.

This is, in fact true, and it happens that in an n-dimensional semi- Riemannian space, the covariant curvature tensor has a total of

$$N = \frac{n^2(n^2-1)}{12} \tag{6.105}$$

algebraically independent components. Consequently, in the four-dimensional spacetime of Relativity, this tensor has 20 independent components. In order to see this, we divide the components of the covariant curvature tensor into three basic categories.

The first category we shall consider is that in which the first pair of indices have the same values as the last pair of indices:

Category-1: R_{abab} (no summation)

Now the two indices of each pair must be *different* since if they were not, then by (6.100) and (6.101), the given component would vanish. The total number n_1 of independent components falling into this category must be

$$n_1 = \binom{n}{2} = \frac{n(n-1)}{2} \tag{6.106}$$

The second category which we consider is the case where three of the indices are distinct. In this case, we can use (6.100) and (6.101) to write the given component in the form

Category-2: R_{abac} (no summation)

Now we can choose the first component in n-different ways; once this is done, however, their are $(n-1)$ ways of choosing b, and $(n-2)$ ways of choosing c. Due to the block symmetry property, however, the total number of ways of choosing b and c must be cut in half; therefore, the total number of independent components falling into this category is:

$$n_2 = \frac{n(n-1)(n-2)}{2} \tag{6.107}$$

Finally, we examine the category where all of the indices are distinct:

Category-3: R_{abcd}

Here, we see that we can choose the first index pair in a total of

$$\binom{n}{2} = \frac{n(n-1)}{2}$$

ways. Once this is done there remain $(n-2)$ symbols from which to choose the second index pair, or there a total of

$$\binom{n-2}{2} = \frac{(n-2)(n-3)}{2}$$

ways of choosing the second index pair. Thus, it would appear that the total number of ways of choosing four distinct indices should be

$$\binom{n}{2}\binom{n-2}{2}$$

But due to (6.101), any one component can be written in terms of the other two, so the last quantity must be multiplied by $2/3$. We conclude that the total number of algebraically independent components in this category must be

$$n_3 = \frac{2}{3}\binom{n}{2}\binom{n-2}{2} = \frac{n(n-1)(n-2)(n-3)}{12} \tag{6.108}$$

Finally, if we add the above results, we obtain

$$N = n_1 + n_2 + n_3 = \frac{n^2(n^2-1)}{12}$$

as claimed.

6.11. Exercises

1. If $\mathbf{T} = (T_{ab})$ is a tensor of type $(0,2)$, then show that

$$(\nabla_{\mathbf{X}}\mathbf{T})(\mathbf{U},\mathbf{V}) = \nabla_{\mathbf{X}}(\mathbf{T}(\mathbf{U},\mathbf{V})) - \mathbf{T}(\nabla_{\mathbf{X}}\mathbf{U},\mathbf{V}) - \mathbf{T}(\mathbf{U},\nabla_{\mathbf{X}}\mathbf{V}) \tag{6.109}$$

Observe that this is a special case of (6.90).

(Solution)

We have

$$\nabla_{\mathbf{X}}\mathbf{T} = X^s\left(\frac{\partial T_{ab}}{\partial x^s} - \Gamma^p_{sa}T_p - \Gamma^p_{sb}T_p\right)\mathbf{d}x^a \otimes \mathbf{d}x^b$$

Let the vectors $\mathbf{U} = U^k\mathbf{e}_k$ and $\mathbf{V} = V^m\mathbf{e}_m$ so that

$$(\nabla_{\mathbf{X}}\mathbf{T})(\mathbf{U},\mathbf{V}) = X^sU^kV^m\left(\partial_s T_{km} - \Gamma^p_{sk}T_{pm} - \Gamma^p_{sm}T_{kp}\right)$$

Now, we examine the righthand side of (6.109). We have,

$$\begin{aligned}\nabla_{\mathbf{X}}(\mathbf{T}(\mathbf{U},\mathbf{V})) &= X^s\partial_s\left(T_{ab}U^aV^b\right)\\ &= X^s(\partial_s T_{ab})U^aV^b + X^sT_{ab}(\partial_s U^a)V^b + X^sT_{ab}U^a\left(\partial_s V^b\right)\end{aligned}$$

while

$$\begin{aligned}
\mathbf{T}(\nabla_{\mathbf{X}}\mathbf{U},\mathbf{V}) &= \mathbf{T}\left(\left(X^s\partial_s U^a + X^s\Gamma^a_{sp}U^p\right)\mathbf{e}_a, V^b\mathbf{e}_b\right) \\
&= T_{ab}V^bX^s\partial_s U^a + T_{ab}X^sU^pV^b\Gamma^a_{sp} \\
\mathbf{T}(\mathbf{U},\nabla_{\mathbf{X}}\mathbf{V}) &= \mathbf{T}\left(U^a\mathbf{e}_a, \left(X^s\partial_s V^b + X^s\Gamma^b_{sp}V^p\right)\mathbf{e}_b\right) \\
&= T_{ab}U^aX^s\partial_s V^b + T_{ab}X^sU^aV^p\Gamma^b_{sp}
\end{aligned}$$

Consequently, the righthand side of (6.109) becomes

$$\begin{aligned}
\nabla_{\mathbf{X}}\left(\mathbf{T}(\mathbf{U},\mathbf{V})\right) - \mathbf{T}(\nabla_{\mathbf{X}}\mathbf{U},\mathbf{V}) - \mathbf{T}(\mathbf{U},\nabla_{\mathbf{X}}\mathbf{V}) &= X^s\left(\partial_s T_{ab}\right)U^aV^b \\
&\quad -T_{ab}X^sU^pV^b\Gamma^a_{sp} - T_{ab}X^sU^aV^p\Gamma^b_{sp} \\
&= X^s\left(\partial_s T_{km}\right)U^kV^m - T_{pm}X^sU^kV^m\Gamma^p_{sk} \\
&\quad -T_{km}X^sU^kV^p\Gamma^m_{sp} \\
&= X^sU^kV^m\left(\partial_s T_{km} - \Gamma^p_{sp}T_{pm} - U^kV^m\Gamma^p_{sm}T_{kp}\right)
\end{aligned}$$

where we relabeled dummy indices in the last line. We see that this expression is identical with the lefthand side of (6.109).

2. Assume that $(M,\mathbf{g})$ is a semi-Riemannian space where the metric $\mathbf{g} = (g_{ab})$ is a *diagonal* matrix. Show that for *fixed* indices a and b,

$$\Gamma^a_{ab} = \Gamma^a_{ba} = \frac{1}{2}\partial_b\left(\ln|g_{aa}|\right) \quad \text{(no summation)} \tag{6.110a}$$

$$\Gamma^a_{bb} = -\left(\frac{1}{2g_{aa}}\right)\partial_a g_{bb} \qquad (a \neq b, \text{ no summation}) \tag{6.110b}$$

while all other Christoffel symbols vanish.

(Solution)

We first establish (6.110a). We have

$$\begin{aligned}
\Gamma^a_{ba} &= \Gamma^a_{ab} = g^{ak}[ab,k] \\
&= g^{aa}[ab,a] \quad [\text{since } g^{ak} = 0 \text{ for } a \neq k] \\
&= \frac{1}{2}g^{aa}\left(\partial_b g_{aa}\right) \\
&= \frac{1}{2}\partial_b\left(\ln|g_{aa}|\right)
\end{aligned}$$

Also, (6.110b) follows by observing that

$$\begin{aligned}
\Gamma^a_{bb} &= g^{ak}[bb,k] = g^{aa}[bb,a] \\
&= -\frac{1}{2}g^{aa}\left(\partial_a g_{bb}\right) \\
&= -\left(\frac{1}{2g_{aa}}\right)\partial_a g_{bb}
\end{aligned}$$

Finally, if the indices are all distinct, then all of the off diagonal elements of the metric are zero, so

$$\begin{aligned}\Gamma^a_{bc} &= g^{ak}[bc,k] = g^{aa}[bc,a] = g^{aa}(-\partial_a g_{bc} + \partial_c g_{ab} + \partial_b g_{ac}) \\ &= g^{aa}(0) = 0\end{aligned}$$

as claimed.

3. Given the Euclidean space $(R^2, \mathbf{g})$, where $\mathbf{g}$ is expressed in terms of polar coordinates $\left(x^1, x^2\right) \equiv (r, \theta)$ as

$$\mathbf{g} = \mathbf{d}r \otimes \mathbf{d}r + r^2 \mathbf{d}\theta \otimes \mathbf{d}\theta \tag{6.111}$$

Show that the only non-zero Christoffel symbols are

$$\Gamma^1_{22} = -r,\ \Gamma^2_{12} = \Gamma^2_{21} = \frac{1}{r} \tag{6.111a}$$

(Solution)

By inspection, we first observe that $g_{11} = 1$, $g_{22} = r^2$. By (6.110),

$$\begin{aligned}\Gamma^1_{22} &= -\frac{1}{2g_{11}}\partial_1 g_{22} = -\frac{1}{2}\partial_r\left(r^2\right) = -r \\ \Gamma^2_{12} &= \Gamma^2_{21} = \frac{1}{2}\partial_1\left(\ln|g_{22}|\right) = \frac{1}{2}\partial_r\left(\ln r^2\right) = \frac{1}{r}\end{aligned}$$

The remaining symbols vanish by inspection.

4. Consider the Euclidean space $(R^3, \mathbf{g})$, where $\mathbf{g}$ is given in terms of spherical polar coordinates $\left(x^1, x^2, x^3\right) \equiv (r, \theta, \phi)$ as

$$\mathbf{g} = \mathbf{d}r \otimes \mathbf{d}r + r^2 \mathbf{d}\theta \otimes \mathbf{d}\theta + \left(r^2 \sin^2\theta\right) \mathbf{d}\phi \otimes \mathbf{d}\phi \tag{6.112}$$

Determine all of the non-zero Christoffel symbols.

(Solution)

By inspection, we find that $g_{11} = 1$, $g_{22} = r^2$, $g_{33} = r^2 \sin^2\theta$. By (6.110a),

$$\begin{aligned}\Gamma^2_{21} &= \Gamma^2_{12} = \frac{1}{2}\partial_1\left(\ln|g_{22}|\right) = \frac{1}{2}\frac{\partial}{\partial r}\left(\ln r^2\right) = \frac{1}{r} \\ \Gamma^3_{31} &= \Gamma^3_{13} = \frac{1}{2}\partial_1\left(\ln|g_{33}|\right) = \frac{1}{2}\frac{\partial}{\partial r}\ln\left(r^2 \sin^2\theta\right) = \frac{1}{r} \\ \Gamma^3_{32} &= \Gamma^3_{23} = \frac{1}{2}\partial_2\left(\ln|g_{33}|\right) = \frac{1}{2}\frac{\partial}{\partial\theta}\ln\left(r^2 \sin^2\theta\right) = \frac{\cos\theta}{\sin\theta} = \cot\theta\end{aligned}$$

Similarly, by (6.110b), we obtain

$$\begin{aligned}
\Gamma^a_{bb} &= -\left(\frac{1}{2g_{aa}}\right)\partial_a g_{bb} \\
\Gamma^1_{22} &= -\left(\frac{1}{2g_{11}}\right)\partial_1 g_{22} = -\left(\frac{1}{2}\right)\frac{\partial}{\partial r}\left(r^2\right) = -r \\
\Gamma^1_{33} &= -\left(\frac{1}{2g_{11}}\right)\partial_1 g_{33} = -\left(\frac{1}{2}\right)\frac{\partial}{\partial r}\left(r^2\sin^2\theta\right) = -r\sin^2\theta \\
\Gamma^2_{33} &= -\left(\frac{1}{2g_{22}}\right)\partial_2 g_{33} = -\frac{1}{2r^2}\frac{\partial}{\partial\theta}\left(r^2\sin^2\theta\right) = -\sin\theta\cos\theta
\end{aligned}$$

All of the remaining Christoffel symbols vanish.

5. *(Bianchi's 2nd Identity)* Use geodesic coordinates to establish (6.102):

$$\nabla_p R_{abcd} + \nabla_c R_{abdp} + \nabla_d R_{abpc} = 0$$

(Solution)

We perform the coordinate transformation $\bar{x}^i = \bar{x}^i(x^j)$, where the point p is the pole of the geodesic coordinate system. We first recall that

$$R_{abcd} = \partial_c\Gamma_{bda} - \partial_d\Gamma_{bca} + \Gamma_{adp}\Gamma^p_{bc} - \Gamma_{acp}\Gamma^p_{bd}$$

So, in terms of geodesic coordinates, the lefthand side of (6.102) becomes

$$\begin{aligned}
\bar{\nabla}_p\bar{R}_{abcd} + \bar{\nabla}_c\bar{R}_{abdp} + \bar{\nabla}_d\bar{R}_{abpc} &= \bar{\partial}_p\left(\bar{\partial}_c\bar{\Gamma}_{bda} - \bar{\partial}_d\bar{\Gamma}_{bca}\right) \\
&\quad + \bar{\partial}_c\left(\bar{\partial}_d\bar{\Gamma}_{bpa} - \bar{\partial}_p\bar{\Gamma}_{bda}\right) + \bar{\partial}_d\left(\bar{\partial}_p\bar{\Gamma}_{bca} - \bar{\partial}_c\bar{\Gamma}_{bpa}\right) \\
&= \left[\bar{\partial}_p, \bar{\partial}_c\right]\bar{\Gamma}_{bda} + \left[\bar{\partial}_c, \bar{\partial}_d\right]\bar{\Gamma}_{bpa} + \left[\bar{\partial}_d, \bar{\partial}_p\right]\bar{\Gamma}_{bca} = 0
\end{aligned}$$

as required.

6. Show that

$$\partial_a g_{bc} = \Gamma_{abc} + \Gamma_{acb} \tag{6.113}$$

(Solution)

By (6.76), and the symmetry of $\mathbf{g}$, we have

$$\begin{aligned}
\Gamma_{abc} + \Gamma_{acb} &= \frac{1}{2}\left(-\partial_c g_{ab} + \partial_b g_{ca} + \partial_a g_{bc}\right) + \frac{1}{2}\left(-\partial_b g_{ac} + \partial_c g_{ba} + \partial_a g_{cb}\right) \\
&= \partial_a g_{bc}
\end{aligned}$$

as required.

7. *(Derivative of a Determinant)* Let $A = (a_{ij}(\mathbf{x}))$ be a symmetric $n\times n$ non-singular matrix whose elements are smooth scalar fields; that is each element $a_{ij}(\mathbf{x}) \in C^\infty(M)$. Show that

$$\frac{\partial}{\partial x^a}\ln|(\det(A)| = Tr\left(A^{-1}\frac{\partial}{\partial x^a}A\right), \tag{6.114}$$

where Tr denotes the trace.

(Solution)

Since A is symmetric, we can diagonalize A by performing a similarlity transformation

$$\bar{A} = \Lambda^T A \Lambda = diag(\lambda_1, \lambda_2, \lambda_3, ..., \lambda_n)$$

where Λ is the orthogonal matrix whose columns are made up of the normalized eigenvectors of A. Now, observe that

$$\begin{aligned} \det(\bar{A}) &= \det\left(\Lambda^T A \Lambda\right) = \det\left(\Lambda^T\right)\det(A)\det(\Lambda) \\ &= 1 \cdot \det(A) \cdot 1 = \det(A), \end{aligned}$$

which implies

$$\det(A) = \det(\bar{A}) = \lambda_1\lambda_2\lambda_3 \cdots \lambda_n$$

Consequently, we can write

$$\frac{\partial}{\partial x^a}\det(A) = \sum_{k=1}^{n} \lambda_1\lambda_2\lambda_3 \cdots \frac{\partial \lambda_k}{\partial x^a} \cdots \lambda_n = \det(A)\sum_{k=1}^{n}\frac{1}{\lambda_k}\cdot\frac{\partial \lambda_k}{\partial x^a}$$

so that

$$\begin{aligned} \frac{\partial}{\partial x^a}\ln|(\det(A)| &= \frac{1}{\det(A)}\frac{\partial}{\partial x^a}\det(A) = \sum_{k=1}^{n}\frac{1}{\lambda_k}\cdot\frac{\partial \lambda_k}{\partial x^a} \\ &= Tr\left(A^{-1}\frac{\partial}{\partial x^a}A\right) \end{aligned}$$

as claimed.

8. *(A Useful Formula)* Given a semi-Riemannian manifold $(M, \mathbf{g})$, equipped with the Levi-Civita connection, show that

$$\frac{\partial}{\partial x^a}\left(\ln\sqrt{|g|}\right) = \left\{ {b \atop ab} \right\} = \Gamma^b_{ab} \qquad (6.115)$$

where $g = \det(\mathbf{g})$.

(Solution)

By (6.113), we have

$$\partial_a g_{bc} = \Gamma_{abc} + \Gamma_{acb}$$

Now, multiply both sides of this equation by g^{bc} to obtain

$$\begin{aligned} g^{bc}\partial_a g_{bc} &= g^{bc}\left(\Gamma_{abc} + \Gamma_{acb}\right) = \Gamma^b_{ab} + \Gamma^c_{ac} = 2\Gamma^b_{ab} \\ &\Longrightarrow \; Tr(\mathbf{g}^{-1}\partial_a \mathbf{g}) = 2\left\{ {b \atop ab} \right\} \end{aligned}$$

By (6.114), we can now write

$$\frac{\partial}{\partial x^a}\left(\ln\sqrt{|g|}\right) = \frac{1}{2}\frac{\partial}{\partial x^k}\left(\ln|g|\right) = \left\{ {b \atop ab} \right\}$$

as required.

6.12. The Ricci and Einstein Tensors

Let $(M, \mathbf{g})$ be a semi-Riemannian space which is equipped with the Levi-Civita connection. Given the curvature tensor $\mathbf{R}$, it is possible to obtain tensors of lower rank by the operation of contraction. First, we recall that

$$R^k_{abc} = \partial_b \Gamma^k_{ac} - \partial_c \Gamma^k_{ab} + \Gamma^k_{bp}\Gamma^p_{ac} - \Gamma^k_{cp}\Gamma^p_{ab}$$

By contracting the indices k and c, we obtain the **Ricci tensor,** R_{ab} whose components are given by

$$R_{ab} \equiv R^k_{abk} = \partial_b \Gamma^k_{ak} - \partial_k \Gamma^k_{ab} + \Gamma^k_{bp}\Gamma^p_{ak} - \Gamma^k_{kp}\Gamma^p_{ab} \tag{6.116}$$

Now, this equation can be put into a more convenient form. First, we recall that by (6.115),

$$\Gamma^k_{ak} = \partial_a \left(\ln\sqrt{|g|}\right) \text{ and } \Gamma^k_{kp} = \partial_p \left(\ln\sqrt{|g|}\right). \tag{6.117}$$

Next, we substitute (6.117) into the previous equation, to get

$$R_{ab} = \partial_b\partial_a\left(\ln\sqrt{|g|}\right) - \partial_k\Gamma^k_{ab} - \Gamma^p_{ab}\,\partial_p\left(\ln\sqrt{|g|}\right) + \Gamma^k_{bp}\Gamma^p_{ak}$$

Now, we switch the dummy index $(p \longrightarrow k)$ in the third term of this equation to obtain

$$R_{ab} = \partial_b\partial_a\left(\ln\sqrt{|g|}\right) - \partial_k\Gamma^k_{ab} - \Gamma^k_{ab}\partial_k\left(\ln\sqrt{|g|}\right) + \Gamma^k_{bp}\Gamma^p_{ak},$$

which may be rewritten as

$$\begin{aligned} R_{ab} &= \partial_a\partial_b\left(\ln\sqrt{|g|}\right) - \frac{1}{\sqrt{|g|}}\left\{\sqrt{|g|}\,\partial_k\Gamma^k_{ab} + \Gamma^k_{ab}\partial_k\sqrt{|g|}\right\} + \Gamma^k_{bp}\Gamma^p_{ak} \\ &= \partial_a\partial_b\left(\ln\sqrt{|g|}\right) - \frac{1}{\sqrt{|g|}}\partial_k\left(\sqrt{|g|}\,\Gamma^k_{ab}\right) + \Gamma^k_{bp}\Gamma^p_{ak} \end{aligned}$$

So, we finally obtain

$$R_{ab} = \frac{\partial^2}{\partial x^a \partial x^b}(\log\sqrt{|g|}) - \frac{1}{\sqrt{|g|}}\frac{\partial}{\partial x^k}\left(\Gamma^k_{ab}\sqrt{|g|}\right) + \Gamma^p_{ak}\Gamma^k_{bp}\,. \tag{6.118}$$

This equation is useful for computational purposes, and it clearly shows that the Ricci tensor is symmetric:

$$R_{ab} = R_{ba} \tag{6.119}$$

If we now raise the first index of the Ricci tensor, we obtain the **mixed Ricci tensor**:

$$R^a_b \equiv g^{am}R_{mb} = g^{am}\left\{\frac{\partial^2}{\partial x^m \partial x^b}(\log\sqrt{|g|}) - \frac{1}{\sqrt{|g|}}\frac{\partial}{\partial x^k}\left(\Gamma^k_{mb}\sqrt{|g|}\right) + \Gamma^k_{bp}\Gamma^p_{mk}\right\} \tag{6.120}$$

By contracting the mixed Ricci tensor, we obtain the **Ricci scalar**:

$$R \equiv R^a_a = g^{ab}\left\{\frac{\partial^2}{\partial x^a \partial x^b}(\log\sqrt{|g|}) - \frac{1}{\sqrt{|g|}}\frac{\partial}{\partial x^k}\left(\Gamma^k_{ab}\sqrt{|g|}\right) + \Gamma^k_{ap}\Gamma^p_{bk}\right\} \tag{6.121}$$

Finally, we define the **Einstein tensor** $\mathbf{G} = (G^a_b)$ in terms of the mixed Ricci tensor and the Ricci scalar as:

$$G^a_b = R^a_b - \left(\frac{1}{2}R\right)\delta^a_b \tag{6.122}$$

As we will see later on, the Einstein tensor plays a pivotal role in General Relativity, and it turns out that this tensor exhibits a property that is crucial to the development of this theory:

$$\nabla_a G^a_b = 0 \tag{6.123}$$

Equation (6.123) may be summarized by saying that the *divergence* of the Einstein tensor is zero, or equivalently, that the Einstein tensor is *divergence-free*.

In order to establish (6.123), we first recall Bianchi's second identity

$$\nabla_p R_{abcd} + \nabla_c R_{abdp} + \nabla_d R_{abpc} = 0,$$

and note that since the covariant curvature tensor is antisymmetric in each pair of its indices, the last equation may be rewritten as

$$\nabla_p R_{abcd} - \nabla_c R_{abpd} - \nabla_d R_{bapc} = 0$$

We now multiply both sides of the last equation in by g^{da} to obtain

$$g^{da}\nabla_p R_{abcd} - g^{da}\nabla_c R_{abpd} - g^{da}\nabla_d R_{bapc} = 0$$

Since $\mathbf{g} = (g_{ab})$ is a covariant constant, then $\mathbf{g}^{-1} = (g^{ab})$ must be a covariant constant as well, so the last equation may be rewritten as

$$\nabla_p\left(g^{da}R_{abcd}\right) - \nabla_c\left(g^{da}R_{abpd}\right) - \nabla_d\left(g^{da}R_{bapc}\right) = 0$$

which, in turn, implies

$$\nabla_p R_{bc} - \nabla_c R_{bp} - \nabla_d\left(g^{da}R_{bapc}\right) = 0$$

Next, we multiply both sides of the last equation by g^{cb} to obtain

$$\begin{aligned}
\nabla_p R - \nabla_c R^c_p - \nabla_d\left(g^{da}g^{cb}R_{bapc}\right) &= 0 \\
&\Longrightarrow \nabla_p R - \nabla_c R^c_p - \nabla_d\left(g^{da}R^c_{apc}\right) = 0 \\
&\Longrightarrow \nabla_p R - \nabla_c R^c_p - \nabla_d\left(g^{da}R_{ap}\right) = 0 \\
&\Longrightarrow \nabla_p R - \nabla_c R^c_p - \nabla_d R^d_p = 0 \\
&\Longrightarrow 2\left(\nabla_c R^c_p - \frac{1}{2}\delta^c_p R\right) = 0
\end{aligned}$$

Since the mixed Kronecker tensor $\delta = \left(\delta^c_p\right)$ is also a covariant constant, we have

$$\begin{aligned}
\left(\nabla_c R^c_p - \frac{1}{2}\delta^c_p \nabla_c R\right) &= \nabla_c\left(R^c_p - \frac{1}{2}\delta^c_p R\right) \\
&= \nabla_c G^c_p = 0
\end{aligned}$$

as claimed.

6.13. Calculus of Variations

Now, we shall discuss the basic elements of the Calculus of Variations in preparation for our future work. Readers already familiar with variational methods may skip this section and proceed to the next.

We will be working in the Euclidean plane E^2, and our goal is to find a curve, say $y = y(x)$, which passes through the fixed points $P_1(a, y(a))$ and $P_2(b, y(b))$, and which *extremizes* the integral:

$$S = \int_a^b F(x, y, y')dx \tag{6.124}$$

where $F(x, y, y')$ is some arbitrary function and $y' = dy/dx$. This integral may be viewed as mapping each function $y = y(x)$ into a *real number S*, and is called a **functional**.

It is important to note that we use the term "extremize" to mean that the solution $y = y(x)$ is a function which causes the above integral to assume either a maximum, minimum, or stationary value. In order to find the extremal $y = y(x)$, we first consider the functions given by:

$$Y(x, \varepsilon) = y(x) + \varepsilon g(x) \tag{6.125}$$

where $g(x)$ is any function which satisfies the end-point conditions

$$g(a) = g(b) = 0 \tag{6.126}$$

and ε is an arbitrary parameter.

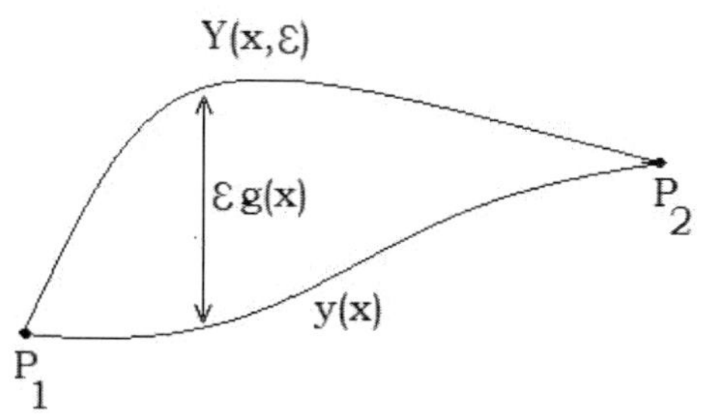

Figure 6.5.

If ε is sufficiently small, however, then $Y(x, \varepsilon)$ represents a neighboring curve which passes through the fixed points P_1 and P_2 (Figure 6.5). Along such a neighboring curve, the above integral becomes:

$$S(\varepsilon) = \int_a^b F(x, y + \varepsilon g, y' + \varepsilon g')dx \tag{6.127}$$

where the notation $S(\varepsilon)$ has been used to emphasize the fact that this new integral depends upon the parameter ε. By Leibnitz's rule, we have

$$\frac{d}{d\varepsilon}S(\varepsilon) = \int_a^b \left\{\frac{\partial}{\partial\varepsilon}F(x, y+\varepsilon g, y'+\varepsilon g')\right\}dx = \int_a^b \left\{\frac{\partial F}{\partial y}g + \frac{\partial F}{\partial y'}g'\right\}dx \tag{6.128}$$

or equivalently, we can write

$$\frac{d}{d\varepsilon}S(\varepsilon) = \int_a^b \left(\frac{\partial F}{\partial y}g\right)dx + \int_a^b \left(\frac{\partial F}{\partial y'}g'\right)dx \tag{6.129}$$

Next, we integrate the second integral by parts, to obtain

$$\int_a^b \left(\frac{\partial F}{\partial y'}g'\right)dx = \left(\frac{\partial F}{\partial y'}g\right)\Bigg|_{x=a}^{x=b} - \int_a^b g\frac{d}{dx}\left(\frac{\partial F}{\partial y'}\right)dx = -\int_a^b g\frac{d}{dx}\left(\frac{\partial F}{\partial y'}\right)dx \tag{6.130}$$

since the first term vanishes due to the end point conditions (6.126). If we now combine the last two equations, we obtain

$$\frac{d}{d\varepsilon}S(\varepsilon) = \int_a^b \left[\frac{\partial F}{\partial y} - \frac{d}{dx}\left(\frac{\partial F}{\partial y'}\right)\right]g\,dx \tag{6.131}$$

By assumption, the required solution curve is $Y(x,0) = y(x)$, so the function $S(\varepsilon)$ must achieve an extremum when $\varepsilon = 0$; consequently,

$$\frac{d}{d\varepsilon}S(\varepsilon) = \int_a^b \left[\frac{\partial F}{\partial y} - \frac{d}{dx}\left(\frac{\partial F}{\partial y'}\right)\right]g\,dx = 0 \tag{6.132}$$

Since $g(x)$ is *arbitrary*, (6.132) must hold for all such functions, so that

Euler-Lagrange Equation

$$\frac{\partial F}{\partial y} - \frac{d}{dx}\left(\frac{\partial F}{\partial y'}\right) = 0 \tag{6.133}$$

which is called **Euler-Lagrange equation**. We observe that this equation is a necessary but *not* sufficient condition for the original integral given by (6.124), to be extremized.

The above method may be easily generalized to obtain the solution curve $y^a = y^a(t)$ which extremizes the integral:

$$S = \int_a^b F(t, y^1, y^2, \ldots y^n, \ldots \dot{y}^1, \dot{y}^2, \ldots, \dot{y}^n)dt \equiv \int_a^b F(t, y^a, \dot{y}^a)dt \tag{6.134}$$

where $\dot{y}^a \equiv dy^a/dt$. In this case, it is easy to show that a necessary condition for this integral to be extremized is that the following n equations be satisfied:

Euler-Lagrange Equations

$$\frac{\partial F}{\partial y^a} - \frac{d}{dt}\left(\frac{\partial F}{\partial \dot{y}^a}\right) = 0 \text{ for } a = 1, 2, ..., n \tag{6.135}$$

We now turn our attention to the use of what is commonly called *variational notation*. We first consider the Taylor series expansion of the function $F(x, y+\varepsilon g, y'+\varepsilon g')$ which we previously encountered above; here, we can write

$$F(x, y+\varepsilon g, y+\varepsilon g) = F(x, y, y) + \varepsilon g \frac{\partial F}{\partial y} + \varepsilon g' \frac{\partial F}{\partial y'} + O(\varepsilon^2)$$

so that

$$\Delta F = F(x, y+\varepsilon g, y+\varepsilon g) - F(x, y, y) = \varepsilon g \frac{\partial F}{\partial y} + \varepsilon g' \frac{\partial F}{\partial y'} + O(\varepsilon^2)$$

We define the **variation** of F, denoted by δF, to be the *linear part* (in ε) of this expression; that is

$$\delta F \equiv \varepsilon g \frac{\partial F}{\partial y} + \varepsilon g' \frac{\partial F}{\partial y'} \tag{6.136}$$

In particular, if we take $F = y$ or $F = y'$, then the variations in y and y' are given respectively by

$$\delta y = \varepsilon g \text{ and } \delta y' = \varepsilon g' \tag{6.137}$$

Consequently, we can write:

$$\delta F \equiv \frac{\partial F}{\partial y}\delta y + \frac{\partial F}{\partial y'}\delta y' \tag{6.138}$$

It will be seen that the variational symbol δ enjoys many of the properties of a derivative operator, and the most useful of its properties have been summarized below:

Variational Notation

Let $F = F(x, y, y')$, $G = G(x, y, y')$ and let $a, b \in R$, then the following relationships hold:

1. The operators δ and d/dx commute:

$$\delta y' = \frac{d}{dx}(\delta y) \tag{6.139}$$

2. The operator δ is linear:

$$\delta(aF + bG) = a(\delta F) + b(\delta G) \tag{6.140}$$

3. The operator δ obeys the Liebnitz rule:

$$\delta(FG) = F(\delta G) + G(\delta F) \tag{6.141}$$

4. The variation of an integral is equal to the integral of its variation:

$$\delta \int_a^b F(x,y,y')dx = \int_a^b \delta F(x,y,y')dx \tag{6.143}$$

The proofs of the above properties are straightforward, and are left as exercise for the reader. We note, in passing, that in terms of variational notation, a necessary condition for (6.124) to be extremal is that

$$\delta \int_a^b F(x,y,y')dx = 0 \tag{6.144}$$

In order to see this, observe that we can write

$$\delta \int_a^b F(x,y,y')dx = \int_a^b \delta F(x,y,y')dx = \int_a^b \left[\frac{\partial F}{\partial y}\delta y + \frac{\partial F}{\partial y'}\delta y'\right] dx = 0$$

But $\delta y' = \frac{d}{dx}(\delta y)$, so the last equation becomes:

$$\int_a^b \left[\frac{\partial F}{\partial y}\delta y + \frac{\partial F}{\partial y'}\left(\frac{d}{dx}\delta y\right)\right] dx = 0$$

If we now integrate the second term by parts, while keeping in mind that $\delta y = 0$ at the end points, we obtain

$$\int_a^b \left[\frac{\partial F}{\partial y} - \frac{d}{dx}\left(\frac{\partial F}{\partial y'}\right)\right](\delta y)\, dx = 0$$

and since $\delta y = \varepsilon g(x)$ is arbitrary, we again arrive at Euler-Lagrange equation

$$\frac{\partial F}{\partial y} - \frac{d}{dx}\left(\frac{\partial F}{\partial y'}\right) = 0$$

6.14. Geodesics

In the Euclidean space (R^n, δ), we are all familiar with the fact that the shortest distance between two points is given by the length of the straight line joining these points; consequently, we say that straight lines are the *geodesics* for Euclidean space. In some sense, geodesics are also the *straightest possible curves* in this space, since the direction of the tangent vector to a straight line never changes as we move along the line.

In a semi-Riemannian space $(M, \mathbf{g})$, the situation is a bit messy, owing to indefiniteness of the metric $\mathbf{g}$, but geodesics still have both of the above properties. Suppose that in some coordinate patch on M, a smooth curve C has the coordinate representation

$$C : x^a = x^a(\lambda) \text{ where } \lambda_0 \leq \lambda \leq \lambda_1 \tag{6.145}$$

where λ is the curve parameter. If $\{\mathbf{e}_a\}$ are the coordinate basis vectors at p, then the tangent vector $\mathbf{X}$ to the curve at p is given by

$$\mathbf{X} = \left.\frac{dx^a}{d\lambda}\right|_{\lambda=\lambda_0} \mathbf{e}_a \tag{6.146}$$

and the **length** of the tangent vector is given by

$$\|\mathbf{X}\| = |\mathbf{g}(\mathbf{X},\mathbf{X})|^{1/2} = \left|g_{ab}X^aX^b\right|^{1/2} = \left|\frac{ds}{d\lambda}\right| \tag{6.147}$$

where as in (5.131), ds is the infinitesimal distance along the curve which corresponds to an infinitesimal change $d\lambda$ in the curve parameter.

We now characterize the tangent vector $\mathbf{X} \in \mathbf{T}_p(M)$ by the sign of the quantity $\mathbf{g}(\mathbf{X},\mathbf{X})$, which appears in (6.147). At the point p, we say that $\mathbf{X}$ is a **timelike vector** if $\mathbf{g}(\mathbf{X},\mathbf{X}) > 0$, while $\mathbf{X}$ is a **spacelike vector** if $\mathbf{g}(\mathbf{X},\mathbf{X}) < 0$. In the event that $\mathbf{g}(\mathbf{X},\mathbf{X}) = 0$, then of course, $\mathbf{X}$ is said to be a **null vector**, or a **lightlike vector**.

Now, whether the tangent vector $\mathbf{X}$ is timelike, spacelike, or null, we say that the curve C is a **geodesic** if its tangent vector $\mathbf{X}$ is parallel-transported along C; that is,

$$\nabla_{\mathbf{X}}\mathbf{X} = \mathbf{0}. \tag{6.148}$$

In terms of components, the geodesic equation is simply

Geodesic Equation

$$\frac{d^2x^i}{d\lambda^2} + \Gamma^i_{jk}\frac{dx^j}{d\lambda}\frac{dx^k}{d\lambda} = 0 \tag{6.149}$$

We call (6.149) the **standard form** of the geodesic equation.

Now, observe that (6.149) is a second order (quasi-linear) system of differential equations for the unknown functions $x^a = x^a(\lambda)$ and from the basic existence and uniqueness theorem which governs such systems, it is known that this system will have a *unique solution* which is valid in some open interval, provided that we specify the n-pairs of initial conditions

$$x_0^i = x^i\big|_{\lambda=\lambda_0} \quad \text{and} \quad \dot{x}_0^i = \left.\frac{dx^i}{d\lambda}\right|_{\lambda=\lambda_0} \qquad (i = 1,2,3,...,n) \tag{6.150}$$

at the initial point where $\lambda = \lambda_0$. This means that given any point $p \in M$ and an initial direction specified by any tangent vector $\mathbf{X}$, there always exists a *unique geodesic* which passes through p and whose tangent possesses the given direction.

Since neither the length nor the direction of any vector can change under parallel transport, we see that even in a semi-Riemannian space, geodesics may still be viewed as the *straightest possible curves* in such a general space. Also, since (6.148) indicates that for a geodesic, its tangent vector is parallel-transported parallel to itself, then geodesics are also called **autoparallel curves**.

Furthermore, since the character of the tangent vector **X** cannot change as it moves along a geodesic curve, we can classify such a curve as being a **timelike geodesic**, a **spacelike geodesic**, or a **null geodesic** according to whether its tangent vector is timelike, spacelike, or null, respectively.[4] In any event, we see that (6.150) must hold for both non-null geodesics and null geodesics.

We now consider non-null geodesics in a semi-Riemannian manifold $(M, \mathbf{g})$. We will show that a *non-null geodesic* is equivalent to any non-null curve $C : x^a = x^a(\lambda)$ with $\lambda_0 \leq \lambda \leq \lambda_1$, for which the integral

$$s = \int_{\lambda_0}^{\lambda_1} \sqrt{\left| g_{ij} \frac{dx^i}{d\lambda} \frac{dx^j}{d\lambda} \right|} d\lambda \tag{6.151}$$

is an *extremum* (i.e., either stationary, maximum or minimum). We call any curve which satisfies this property an **extremal**; thus, a we will show that any *non-null geodesic* in $(M, \mathbf{g})$ is an extremal.

We know that if the integral given by (6.151) is to achieve an extremum, then the Euler-Lagrange equations must be satisfied. Thus, if we set

$$\varphi = \varphi(\lambda, x^i, \dot{x}^i) = g_{ij} \frac{dx^i}{d\lambda} \frac{dx^j}{d\lambda} \tag{6.152}$$

(where we have used an overdot to denote differentiation with respect to λ), then the Euler-Lagrange equations become

$$\frac{d}{d\lambda} \left[\frac{\partial}{\partial \dot{x}^k} \left(\sqrt{|\varphi|} \right) \right] - \frac{\partial}{\partial x^k} \left(\sqrt{|\varphi|} \right) = 0 \qquad (k = 1, 2, 3, \ldots, n) \tag{6.153}$$

We find that

$$\frac{\partial}{\partial \dot{x}^k} \left(\sqrt{|\varphi|} \right) = \frac{1}{\sqrt{|\varphi|}} g_{kj} \dot{x}^j \text{ and } \frac{\partial}{\partial x^k} \left(\sqrt{|\varphi|} \right) = \frac{1}{\sqrt{|\varphi|}} (\partial_k g_{ij}) \dot{x}^i \dot{x}^j, \tag{6.154}$$

so, the Euler-Lagrange equations become

$$\frac{d}{d\lambda} \left(\frac{1}{\sqrt{|\varphi|}} g_{kj} \dot{x}^j \right) - \frac{1}{2\sqrt{|\varphi|}} (\partial_k g_{ij}) \dot{x}^i \dot{x}^j = 0 \tag{6.155}$$

Thus far, the curve parameter λ has been perfectly arbitrary; however, if we set $\lambda = s$ so that the curve parameter is now the arc length, then $|\varphi| = 1$ along the curve, and the last equation simplifies to

$$\frac{d}{ds} \left[g_{kj} \frac{dx^j}{ds} \right] - \frac{1}{2} (\partial_k g_{ij}) \frac{dx^i}{ds} \frac{dx^j}{ds} = 0$$

or equivalently,

$$g_{kj} \frac{d^2 x^j}{ds^2} + \frac{dx^j}{ds} \frac{d}{ds} (g_{kj}) - \frac{1}{2} (\partial_k g_{ij}) \frac{dx^i}{ds} \frac{dx^j}{ds} = 0 \tag{6.156}$$

By the chain rule and (6.113), the third factor in the second term of this equation may be rewritten as

$$\frac{d}{ds}\left(g_{kj}\right) = \frac{\partial g_{kj}}{\partial x^p}\frac{dx^p}{ds} = \left(\Gamma_{kpj} + \Gamma_{jpk}\right)\frac{dx^p}{ds} \tag{6.157}$$

while the second factor in the second term of (6.156) becomes

$$\partial_k g_{ij} = \Gamma_{ikj} + \Gamma_{jki} \tag{6.158}$$

Now, if we substitute (6.157) and (6.158) into (6.156), and simplify, we obtain

$$g_{kj}\frac{d^2x^j}{ds^2} + \Gamma_{jpk}\frac{dx^j}{ds}\frac{dx^p}{ds} = 0$$

Finally, we multiply both sides of this equation by g^{ik}, and replace the dummy index $(p \to k)$ to obtain

$$\frac{d^2x^i}{ds^2} + \Gamma^i_{jk}\frac{dx^j}{ds}\frac{dx^k}{ds} = 0,$$

which is identical to (6.149). *Thus, we have shown that non-null curves for which the integral in (6.151) is an extremum (usually a minimum) are, in fact, geodesics.*

Before we proceed to a discussion of null geodesics, it will be helpful to say something about of choice of curve parameter. Let us now suppose that we are given the geodesic equation (6.149) in standard form, and we introduce a different curve parameter, say $\alpha = \alpha(\lambda)$ to describe the original geodesic curve. Assume further, that the reparametrized curve in given by

$$x^i = x^i(\alpha)$$

where $d\alpha/d\lambda \neq 0$. Then it is easy to show (exercise) that (6.149) assumes the form

$$\frac{d^2x^i}{d\alpha^2} + \Gamma^i_{jk}\frac{dx^j}{d\alpha}\frac{dx^k}{d\alpha} = -\left(\frac{d^2\alpha/d\lambda^2}{d\alpha/d\lambda}\right)\frac{dx^i}{d\alpha}, \tag{6.159}$$

or, in other words,

$$\frac{D\mathbf{X}}{D\alpha} = K(\lambda)\mathbf{X} \tag{6.160}$$

where $\mathbf{X} = \left(dx^i/d\alpha\right)$ is the tangent vector to the curve, and the proportionality factor $K(\lambda)$ is given by:

$$K(\lambda) = -\left(\frac{d^2\alpha/d\lambda^2}{d\alpha/d\lambda}\right) \tag{6.160a}$$

We note that if we confine ourselves to changes of parameter of the the form

$$\alpha = A\lambda + B \quad (A \neq 0) \tag{6.151}$$

where A and B are constants, then (6.159) reduces again to the *standard form*

$$\frac{d^2x^i}{d\alpha^2} + \Gamma^i_{jk}\frac{dx^j}{d\alpha}\frac{dx^k}{d\alpha} = 0$$

We call any curve parameter α of the form (6.151) an **affine parameter**. For example, we have already seen that for non-null geodesics, the arc length s is an affine parameter. The main point is that the standard form of equation (6.149) is unaffected by a change of affine parameter. Stated differently, every affinely parametrized geodesic satisfies the standard form of the geodesic equation.

We now turn our attention to null curves, and investigate the stationary properties of null geodesics. We will show that like their counterparts, null geodesics may also be characterized as extremals.

We first note that the approach we used for non-null geodesics won't work since $\varphi = 0$ for null geodesics, and consequently, both (6.154) and (6.155) become meaningless. In order to avoid this problem, we will discard the square root symbol in (6.151); that is, we will show that a *null geodesic* is equivalent to any null curve $C : x^a = x^a(\lambda)$ with $\lambda_0 \leq \lambda \leq \lambda_1$, for which the quantity

$$T = \int_{\lambda_0}^{\lambda_1} \left(g_{ij} \frac{dx^i}{d\lambda} \frac{dx^j}{d\lambda} \right) d\lambda \tag{6.152}$$

is an extremal. We emphasize here that the curve parameter λ is *not* the arclength s. The Euler-Lagrange equations become

$$\frac{d}{d\lambda}\left[\frac{\partial}{\partial \dot{x}^{\,k}}\left(g_{ij}\dot{x}^i\dot{x}^j\right)\right] - \frac{\partial}{\partial x^k}\left(g_{ij}\dot{x}^i\dot{x}^j\right) = 0 \qquad (k = 1, 2, 3, ..., n) \tag{6.153}$$

where for brevity, we set $\dot{x}^i = dx^i/d\lambda$. We find that

$$\frac{\partial}{\partial \dot{x}^{\,k}}\left(g_{ij}\frac{dx^i}{d\lambda}\frac{dx^j}{d\lambda}\right) = 2g_{ik}\dot{x}^k \;\text{ and }\; \frac{\partial}{\partial x^k}\left(g_{ij}\dot{x}^i\dot{x}^j\right) = \left(\partial_k g_{ij}\right)\dot{x}^i\dot{x}^j \tag{6.154}$$

Consequently, proceeding as above, we obtain

$$\frac{d}{d\lambda}\left[\frac{\partial}{\partial \dot{x}^{\,k}}\left(g_{ij}\dot{x}^i\dot{x}^j\right)\right] = \frac{d}{d\lambda}\left(2g_{ik}\dot{x}^k\right) = 2g_{ik}\ddot{x}^k + 2\dot{x}^k\dot{x}^l\left(\Gamma_{ilk} + \Gamma_{kli}\right) \tag{6.155a}$$

$$\frac{\partial}{\partial x^k}\left(g_{ij}\dot{x}^i\dot{x}^j\right) = \dot{x}^i\dot{x}^j\left(\Gamma_{ikj} + \Gamma_{jki}\right) \tag{6.155b}$$

Next, we substitute equations (6.155) into (6.153) to get

$$2g_{ik}\ddot{x}^k + 2\dot{x}^k\dot{x}^l\left(\Gamma_{ilk} + \Gamma_{kli}\right) - \dot{x}^i\dot{x}^j\left(\Gamma_{ikj} + \Gamma_{jki}\right) = 0 \tag{6.156}$$

If we examine the terms involving the Christoffel symbols, we find

$$\begin{aligned} \dot{x}^k\dot{x}^l\Gamma_{ilk} - \dot{x}^i\dot{x}^j\Gamma_{ikj} &= \dot{x}^l\dot{x}^k\Gamma_{ilk} - \dot{x}^i\dot{x}^j\Gamma_{kij} = 0 \\ \dot{x}^k\dot{x}^l\Gamma_{ilk} - \dot{x}^k\dot{x}^l\Gamma_{kli} &= 0 \end{aligned}$$

Consequently, (6.156) simplifies to

$$g_{ik}\ddot{x}^k + \Gamma_{kli}\dot{x}^k\dot{x}^l = 0$$

Multiplying both sides of this equation by g^{ji} gives

$$\frac{d^2x^j}{d\lambda^2}+\Gamma^j_{kl}\frac{dx^k}{d\lambda}\frac{dx^l}{d\lambda}=0 \tag{6.157}$$

which is again the geodesic equation. We conclude that null geodesics are precisely those curves which extremize the integral given by (6.152). *In other words, even null geodesics may be viewed as extremals.*

We can summarize the above results by saying all geodesics, both null and non-null, may be viewed both as extremals and as the straightest possible curves in a semi-Riemannian manifold. Furthermore, all affinely parametrized geodesic curves $C : x^a = x^a(\lambda)$ must satisfy the equation (6.157) and exactly one of the conditions given by

$$g_{ij}\frac{dx^i}{d\lambda}\frac{dx^j}{d\lambda}=\begin{cases}-1, & \text{if } C \text{ is spacelike}\\ 0, & \text{if } C \text{ is lightlike}\\ +1, & \text{if } C \text{ is timelike}\end{cases} \tag{6.158}$$

The conditions in (6.158) follow directly from the definition of space-like, null, and timelike vectors, and they are *first integrals* that often help to simplify the solution of (6.157).

6.15. Exercises

1. *(Einstein Tensor)* Show that the covariant form of the Einstein tensor is given by:

$$G_{ab}=R_{ab}-\left(\frac{1}{2}R\right)g_{ab} \tag{6.159}$$

(Solution)

We have

$$G^c_b=R^c_b-\left(\frac{1}{2}R\right)\delta^c_b$$

Multiplying both sides of this equation by g_{ac} we obtain

$$\begin{aligned}G_{ab} &= g_{ac}G^c_b=g_{ac}R^c_b-\left(\frac{1}{2}R\right)g_{ac}\delta^c_b\\ &= R_{ab}-\left(\frac{1}{2}R\right)g_{ab}\end{aligned}$$

2. *(Geodesics on a Cylinder)* In terms of cylindrical coordinates, the infinitesimal distance ds^2 between two neighboring points on the surface of a cylinder in E^3 of radius a is

$$ds^2=a^2d\theta^2+dz^2 \tag{6.160}$$

Determine the geodesics for this two-dimensional manifold.

(Solution)

Here, $(\theta, z) = (x^1, x^2)$. Let's take the arclength s as the curve parameter, so we must extremize the integral

$$S = \int_{s_0}^{s_1} \left(a^2\dot{\theta}^2 + \dot{z}^2\right)^{1/2} ds$$

where $F = \left(a^2\dot{\theta}^2 + \dot{z}^2\right)^{1/2}$. We find that

$$\frac{\partial F}{\partial \dot{\theta}} = \frac{a^2\dot{\theta}}{\sqrt{a^2\dot{\theta}^2 + \dot{z}^2}}, \quad \frac{\partial F}{\partial \theta} = 0, \quad \frac{\partial F}{\partial \dot{z}} = \frac{\dot{z}}{\sqrt{a^2\dot{\theta}^2 + \dot{z}^2}}, \quad \frac{\partial F}{\partial z} = 0$$

So, the Euler-Lagrange equations are simply

$$\frac{d}{ds}\left(\frac{a^2\dot{\theta}}{\sqrt{a^2\dot{\theta}^2 + \dot{z}^2}}\right) = 0 \text{ and } \frac{d}{ds}\left(\frac{\dot{z}}{\sqrt{a^2\dot{\theta}^2 + \dot{z}^2}}\right) = 0$$

along with a first integral (6.158) of $a^2\dot{\theta}^2 + \dot{z}^2 = 1$. Consequently, we have

$$\frac{d}{ds}\left(a^2\dot{\theta}\right) = 0 \text{ and } \frac{d}{ds}(\dot{z}) = 0$$

with the solution $\theta = As + B$ and $z = Cs + D$, where A, B, C, D are arbitrary constants. The parametric equations of the geodesic then become

$$\begin{aligned} x &= a\cos(As + B) \\ y &= a\sin(As + B) \\ z &= Cs + D \end{aligned} \tag{6.161}$$

These are the parametric equations of a **right circular helix**.

3. *(Geodesic Equations on a Sphere)* In terms of spherical coordinates, the infinitesimal distance ds^2 between two neighboring points on the surface of a sphere of radius a is given by

$$ds^2 = a^2 d\theta^2 + \left(a^2 \sin^2\theta\right) d\phi^2 \tag{6.162}$$

Derive the geodesic equations and determine the Christoffel symbols for this manifold.

(Solution)

Here, $(\theta, \phi) = (x^1, x^2)$. Take s as the curve parameter, so we have to extremize the integral

$$S = \int_{s_0}^{s_1} \left\{a^2\dot{\theta}^2 + \left(a^2\sin^2\theta\right)\dot{\phi}^2\right\}^{1/2} ds$$

where $F=\left\{a^2\dot{\theta}^2+\left(a^2\sin^2\theta\right)\dot{\phi}^2\right\}^{1/2}$. We find that

$$\begin{aligned}\frac{\partial F}{\partial\dot{\theta}} &= \frac{a^2\dot{\theta}}{\sqrt{a^2\dot{\theta}^2+\left(a^2\sin^2\theta\right)\dot{\phi}^2}},\quad \frac{\partial F}{\partial\theta}=\frac{\left(a^2\sin\theta\cos\theta\right)\dot{\phi}^2}{\sqrt{a^2\dot{\theta}^2+\left(a^2\sin^2\theta\right)\dot{\phi}^2}}\\ \frac{\partial F}{\partial\dot{\phi}} &= \frac{\left(a^2\sin^2\theta\right)\dot{\phi}}{\sqrt{a^2\dot{\theta}^2+\left(a^2\sin^2\theta\right)\dot{\phi}^2}},\quad \frac{\partial F}{\partial\phi}=0\end{aligned}$$

From (6.162), a first integral is $a^2\dot{\theta}^2+\left(a^2\sin^2\theta\right)\dot{\phi}^2=1$, so the geodesic equations become

$$\frac{d^2\theta}{ds^2}-\sin\theta\cos\theta\left(\frac{d\phi}{ds}\right)^2 = 0 \qquad (6.162a)$$

$$\frac{d^2\phi}{ds^2}+2\cot\theta\frac{d\theta}{ds}\frac{d\phi}{ds} = 0 \qquad (6.162b)$$

as required. We can read off the Christoffel symbols as

$$\Gamma^1_{22}=-\sin\theta\cos\theta,\ \Gamma^2_{12}=\Gamma^2_{21}=\cot\theta \qquad (6.163)$$

4. *(Geodesics on a Sphere)* Find a solution to (6.162), and interpret the result.

(Solution)

First, we set $p=\dot{\phi}=d\phi/ds$ in (6.162b), so this equation becomes

$$\frac{1}{p}dp+2\cot\theta d\theta=0\Longrightarrow p=\frac{d\phi}{ds}=\dot{\phi}=\frac{A}{\sin^2\theta} \qquad (6.164)$$

Recalling the first integral of (6.158) is $a^2\dot{\theta}^2+\left(a^2\sin^2\theta\right)\dot{\phi}^2=1$, we can express $\dot{\theta}$ in terms of ϕ to get

$$\dot{\theta}=\frac{d\theta}{ds}=\frac{\sqrt{\sin^2\theta-a^2A^2}}{a\sin\theta} \qquad (6.165)$$

Dividing (6.164) by (6.165), we obtain

$$\frac{d\phi}{d\theta}=\frac{A}{\sin^2\theta}\frac{a\sin\theta}{\sqrt{\sin^2\theta-a^2A^2}}=\frac{aA\csc^2\theta}{\sqrt{1-a^2A^2\csc^2\theta}}$$

which integrates to

$$\phi=\cos^{-1}\left[\frac{aA\cot\theta}{\sqrt{1-a^2A^2}}\right]+B$$

In other words,

$$\cos B\sin\theta\cos\phi+\sin B\sin\theta\sin\phi=\frac{aA\cos\theta}{\sqrt{1-a^2A^2}},$$

or,

$$z=\alpha x+\beta y\ \text{ where } \alpha=\frac{\sqrt{1-a^2A^2}}{aA}\cos B,\ \beta=\frac{\sqrt{1-a^2A^2}}{aA}\sin B$$

This is the equation of a plane which passes through the origin, so the geodesic is a **great circle** (with center at the origin) which lies in the given plane.

5. Find the components of the Ricci tensor R_{ab}, and the Ricci curvature scalar R for the two-dimensional manifold which consists of a sphere of radius a.

(Solution)

By (6.162), we have $ds^2 = a^2 d\theta^2 + \left(a^2 \sin^2\theta\right) d\phi^2$, so $\sqrt{|g|} = a^2 \sin\theta$, so by (6.118) we find

$$R_{cd} = \frac{\partial^2}{\partial x^c \partial x^d} \log(a^2 \sin\theta) - \frac{1}{\sin\theta} \frac{\partial}{\partial x^k} \left(\Gamma^k_{cd} \sin\theta\right) + \Gamma^p_{ck} \Gamma^k_{dp}$$

where $\Gamma^1_{22} = -\sin\theta\cos\theta$, $\Gamma^2_{12} = \Gamma^2_{21} = \cot\theta$, and all other Christoffel symbols are zero. We find

$$R_{11} = \frac{\partial^2}{\partial\theta^2} \log(a^2 \sin\theta) - \frac{1}{\sin\theta}(0) + \Gamma^1_{11}\Gamma^1_{11} + \Gamma^1_{12}\Gamma^2_{11} + \Gamma^2_{11}\Gamma^1_{12} + \Gamma^2_{12}\Gamma^2_{12}\,,$$

so that

$$R_{11} = -\csc^2\theta - (0) + (0\cdot 0) + (0\cdot 0) + (0\cdot 0) + \left(\cot^2\theta\right) = -1$$

$$\begin{aligned}
R_{12} &= R_{21} = \frac{\partial^2}{\partial\theta\partial\phi} \log(a^2 \sin\theta) - \frac{1}{\sin\theta} \frac{\partial}{\partial\theta} \left(\Gamma^1_{12} \sin\theta\right) - \frac{1}{\sin\theta} \frac{\partial}{\partial\phi} \left(\Gamma^2_{12} \sin\theta\right) \\
&\quad + \Gamma^1_{11}\Gamma^1_{21} + \Gamma^1_{12}\Gamma^2_{22} + \Gamma^2_{11}\Gamma^1_{21} + \Gamma^2_{12}\Gamma^2_{22} \\
&= (0) - (0) - (0) - (0\cdot 0) + (0\cdot 0) + (0\cdot 0) + (\cot\theta \cdot 0) = 0
\end{aligned}$$

$$\begin{aligned}
R_{22} &= \frac{\partial^2}{\partial\phi^2} \log(a^2 \sin\theta) - \frac{1}{\sin\theta} \frac{\partial}{\partial\theta} \left(-\sin^2\theta\cos\theta\right) \\
&\quad + \Gamma^1_{21}\Gamma^1_{21} + \Gamma^1_{22}\Gamma^2_{21} + \Gamma^2_{21}\Gamma^1_{22} + \Gamma^2_{22}\Gamma^2_{22} \\
&= (0) + \left(-3\sin^2\theta + 2\right) + (0) + \left(-\cos^2\theta\right) + \left(-\cos^2\theta\right) + (0) = -\sin^2\theta
\end{aligned}$$

We can write the components of $\mathbf{R} = (R_{cd})$ in matrix form as

$$(R_{cd}) = \begin{pmatrix} -1 & 0 \\ 0 & -\sin^2\theta \end{pmatrix} \tag{6.166}$$

To find the Ricci curvature scalar R, we must find the trace of the mixed Ricci tensor R^c_d; we have

$$R^c_d \equiv g^{cm} R_{md} = g^{c1} R_{1d} + g^{c2} R_{2d}$$

so

$$\begin{aligned}
R^1_1 &= g^{11} R_{11} + g^{12} R_{21} = -\frac{1}{a^2} \\
R^2_2 &= g^{21} R_{12} + g^{22} R_{22} = -\frac{1}{a^2}
\end{aligned}$$

Finally, the Ricci curvature scalar is

$$R = R_1^1 + R_2^2 = -\frac{2}{a^2} \tag{6.167}$$

6. Determine the components of the Ricci tensor R_{ab}, and the Ricci curvature scalar R for the two-dimensional manifold which has a metric of the form

$$ds^2 = du^2 + [F(u,v)]^2\, dv^2 \tag{6.168}$$

where $F(u,v)$ is a function of u and v.

(Solution)

We must first calculate the Christoffel symbols. Since the metric is diagonal, we can use (6.110), and we find that the only non-zero Christoffel symbols are

$$\Gamma^2_{21} = \Gamma^2_{12} = \frac{1}{F}\frac{\partial F}{\partial u}, \quad \Gamma^1_{22} = -F\frac{\partial F}{\partial u}, \quad \Gamma^2_{22} = \frac{1}{F}\frac{\partial F}{\partial v}$$

We have

$$R_{ab} = \frac{\partial^2}{\partial x^a \partial x^b}(\log F) - \frac{1}{F}\frac{\partial}{\partial x^k}\left(\Gamma^k_{ab}F\right) + \Gamma^p_{ak}\Gamma^k_{bp}$$

Consequently,

$$\begin{aligned}
R_{11} &= \frac{\partial^2}{\partial u^2}(\log F) - \frac{1}{F}\frac{\partial}{\partial x^k}\left(\Gamma^k_{11}F\right) + \Gamma^p_{1k}\Gamma^k_{1p} \\
&= \frac{\partial}{\partial u}\left(\frac{1}{F}\frac{\partial F}{\partial u}\right) - \frac{1}{F}(0) + \Gamma^1_{1k}\Gamma^k_{11} + \Gamma^2_{1k}\Gamma^k_{12} \\
&= \frac{1}{F}\frac{\partial^2 F}{\partial u^2} - \frac{1}{F^2}\left(\frac{\partial F}{\partial u}\right)^2 + \Gamma^1_{11}\Gamma^1_{11} + \Gamma^1_{12}\Gamma^2_{11} + \Gamma^2_{11}\Gamma^1_{12} + \Gamma^2_{12}\Gamma^2_{12} \\
&= \frac{1}{F}\frac{\partial^2 F}{\partial u^2} - \frac{1}{F^2}\left(\frac{\partial F}{\partial u}\right)^2 + \frac{1}{F^2}\left(\frac{\partial F}{\partial u}\right)^2 = \frac{1}{F}\frac{\partial^2 F}{\partial u^2}
\end{aligned}$$

Similarly,

$$\begin{aligned}
R_{22} &= \frac{\partial^2}{\partial v^2}(\log F) - \frac{1}{F}\frac{\partial}{\partial x^k}\left(\Gamma^k_{22}F\right) + \Gamma^p_{2k}\Gamma^k_{2p} \\
&= \frac{1}{F}\frac{\partial^2 F}{\partial v^2} - \frac{1}{F^2}\left(\frac{\partial F}{\partial v}\right)^2 + F\frac{\partial^2 F}{\partial u^2} + 2\left(\frac{\partial F}{\partial u}\right)^2 - \frac{1}{F}\frac{\partial^2 F}{\partial v^2} \\
&\quad + \Gamma^1_{21}\Gamma^1_{21} + \Gamma^1_{22}\Gamma^2_{21} + \Gamma^2_{21}\Gamma^1_{22} + \Gamma^2_{22}\Gamma^2_{22} \\
&= -\frac{1}{F^2}\left(\frac{\partial F}{\partial v}\right)^2 + F\frac{\partial^2 F}{\partial u^2} + 2\left(\frac{\partial F}{\partial u}\right)^2 - 2\left(\frac{\partial F}{\partial u}\right)^2 + \frac{1}{F^2}\left(\frac{\partial F}{\partial v}\right)^2 \\
&= F\frac{\partial^2 F}{\partial u^2}
\end{aligned}$$

Also, we find that $R_{12} = R_{21} = 0$, so

$$(R_{ab}) = \begin{pmatrix} \frac{1}{F}\frac{\partial^2 F}{\partial u^2} & 0 \\ 0 & F\frac{\partial^2 F}{\partial u^2} \end{pmatrix} \tag{6.169}$$

In order to find the Ricci curvature scalar R, we write

$$R^a_b \equiv g^{am} R_{mb} = g^{a1} R_{1b} + g^{a2} R_{2b}$$

So,

$$\begin{aligned} R^1_1 &= g^{11} R_{11} + g^{12} R_{21} = \frac{1}{F} \frac{\partial^2 F}{\partial u^2} \\ R^2_2 &= g^{21} R_{12} + g^{22} R_{22} = \frac{1}{F} \frac{\partial^2 F}{\partial u^2} \end{aligned}$$

We conclude that

$$R = R^1_1 + R^2_2 = \frac{2}{F} \frac{\partial^2 F}{\partial u^2} \tag{6.170}$$

7. *(Divergence of a Vector Field)* We define the divergence of a vector field $\mathbf{V} = (V^a)$ by the contraction

$$\operatorname{div} \mathbf{V} = \nabla_a V^a \tag{6.171}$$

Show that

$$\operatorname{div} \mathbf{V} = \frac{1}{\sqrt{|g|}} \frac{\partial}{\partial x^p} (\sqrt{|g|} V^p) \tag{6.172}$$

(Solution)

We have

$$\begin{aligned} \operatorname{div} \mathbf{V} &= \delta^a_b \nabla_a V^b = \delta^a_b \left\{ \partial_a V^b + \Gamma^b_{ap} V^p \right\} \\ &= \partial_a V^a + \Gamma^a_{ap} V^p \\ &= \partial_a V^a + V^p \partial_p \left(\ln \sqrt{|g|} \right) \quad \text{[By (6.115)]} \\ &= \partial_p V^p + \frac{1}{\sqrt{|g|}} V^p \partial_p \left(\sqrt{|g|} \right) \\ &= \frac{1}{\sqrt{|g|}} \left\{ \sqrt{|g|} (\partial_p V^p) + V^p \partial_p \left(\sqrt{|g|} \right) \right\} \\ &= \frac{1}{\sqrt{|g|}} \frac{\partial}{\partial x^p} (\sqrt{|g|} V^p) \end{aligned}$$

as claimed.

6.16. Tensor Densities

Consider a semi-Riemannian space $(M, \mathbf{g})$, equipped with the Levi-Civita connection. It turns out that under a general coordinate transformation $\overline{x}^i = \overline{x}^i(x^j)$, there exist quantities whose transformation laws deviate from those for ordinary (or general) tensors.

For example, under such a coordinate transformation, the components of the metric tensor $\mathbf{g} = (g_{ab})$ transform according to

$$\overline{g}_{ab} = \frac{\partial x^p}{\partial \overline{x}^a} \frac{\partial x^q}{\partial \overline{x}^b} g_{pq} \, ,$$

as **g** is a tensor of type $(0,2)$. We may view this equation as one involving the multiplication of matrices; for if we set

$$G = (g_{pq}),\ D = (d_a^p) \equiv \frac{\partial x^p}{\partial \overline{x}^a}$$

then the last equation may be written in matrix form as

$$\overline{G} = D^T G D.$$

Now, if we take the determinant of both sides, letting $g = \det(G)$, and recalling that $\det(D^T) = \det(D)$, we get

$$\overline{g} = J^2 g \tag{6.173}$$

where J is the Jacobian of the *inverse* transformation: $x^i = x^i(\overline{x}^j)$. We see that due to the presence of the factor J^2, the determinant of the metric tensor does not transform as an ordinary scalar field; we call g a *scalar density of weight 2*.

In general, we define a **scalar density Ψ of weight** W as an object which transforms according to the rule

$$\overline{\Psi} = J^W \Psi \tag{6.174}$$

In particular, an ordinary scalar field is a scalar density of weight zero.

We can extend this thinking to handle tensors of arbitrary type. Under a coordinate transformation $\overline{x}^a = \overline{x}^a(x^b)$, let the Jacobian of the *inverse* transformation be given by

$$J = \det\left(\frac{\partial x^b}{\partial \overline{x}^a}\right), \tag{6.175}$$

where it is assumed that the determinant $J > 0$. We define a **tensor density** (or **relative tensor**) of type (p,q) *and weight* W as any object **T** whose components transform according to the rule

$$\overline{T}^{a_1 \ldots a_p}_{b_1 \ldots b_q} = J^W \left(\frac{\partial \overline{x}^{a_1}}{\partial x^{m_1}} \cdots \frac{\partial \overline{x}^{a_p}}{\partial x^{m_p}}\right)\left(\frac{\partial x^{n_1}}{\partial \overline{x}^{b_1}} \cdots \frac{\partial x^{n_q}}{\partial \overline{x}^{b_q}}\right) T^{m_1 \ldots m_p}_{n_1 \ldots n_q} \tag{6.176}$$

We see that a tensor density of type (p,q) and weight W transforms in the same manner as an ordinary tensor with the exception of the factor J^W. Furthermore, in order to distinguish ordinary tensors from relative tensors, ordinary tensors are often called **absolute tensors**.

We now turn our attention to the differentiation of tensor densities. Although there is a certain amount of arbitrariness connected with the definition of the covariant derivatives of such objects, considerations similar to those involved with the covariant differentiation of ordinary tensors lead us to define the **covariant derivative** of a tensor density $\mathbf{T} = (T^{a\ldots}_{b\ldots})$ of type (p,q) and of weight W, as

$$\nabla_k T^{a\ldots}_{b\ldots} = \frac{\partial}{\partial x^k}(T^{a\ldots}_{b\ldots}) + \Gamma^a_{kp} T^{p\ldots}_{b\ldots} - \Gamma^p_{kb} T^{a\ldots}_{p\ldots} - W \cdot \Gamma^p_{pk} T^{a\ldots}_{b\ldots} \tag{6.177}$$

From (6.177), we see that *the covariant derivative of a tensor density contains the usual terms one would expect for an ordinary tensor, plus an additional term which has W as a factor.*

Example 1. If $T = (T^a)$ is a tensor density of weight $+1$; that is, a **vector density**, then

$$\nabla_k T^a = \partial_k T^a + \Gamma^a_{kp} T^p - \Gamma^p_{pk} T^a$$

Example 2. Given that Ψ is a scalar density of weight $+2$, we can write

$$\nabla_k \Psi = \partial_k \Psi - 2\Psi \Gamma^p_{pk}$$

From the above discussion, one can show that the covariant derivative of the determinant g of the metric tensor is zero; in other words, g is a *covariant constant.*

$$\nabla_k g = 0 \tag{6.178}$$

To see this, we first note that the elements of the inverse matrix G^{-1} are given by

$$g^{ab} = \frac{G^{ab}}{g}$$

where G^{ij} is the cofactor corresponding to the element g_{ij}. Since the cofactor of doesn't contain the element g_{ab} explicitly, we have

$$\frac{\partial g}{\partial g_{ab}} = G^{ab}$$

On the other hand, the elements g_{ab} are functions of the coordinates, so by the chain rule, we can write

$$\frac{\partial g}{\partial x^k} = \frac{\partial g}{\partial g_{ab}} \frac{\partial g_{ab}}{\partial x^k},$$

and consequently,

$$\frac{\partial g}{\partial x^k} = \frac{\partial g}{\partial g_{ab}} \frac{\partial g_{ab}}{\partial x^k} = G^{ab} \frac{\partial g_{ab}}{\partial x^k} = g g^{ab} (\Gamma_{akb} + \Gamma_{bka})$$

where we used (6.113) in the last step. Rearranging terms, we obtain

$$\begin{aligned} \frac{\partial g}{\partial x^k} &= g g^{ab} (\Gamma_{akb} + \Gamma_{bka}) = g \left(g^{ab} \Gamma_{akb} + g^{ba} \Gamma_{bka} \right) \\ &= 2g \Gamma^a_{ak} \end{aligned}$$

Consequently, we find that

$$\nabla_k g = \frac{\partial g}{\partial x^k} - 2g \Gamma^a_{ak} = 0$$

as claimed.

The algebraic rules for ordinary tensors can be easily extended to tensor densities in an obvious manner:

Basic Algebra of Tensor Densities

1. The *outer product* of a tensor density of type (p_1, q_1) and weight W_1 with a tensor density of type (p_2, q_2) and weight W_2 is a tensor density of type $(p_1 + p_2, q_1 + q_2)$ and weight $W_1 + W_2$.

2. If we *contract* a tensor density of type (p, q) and weight W, we get a tensor density of type $(p-1, q-1)$ and of the same weight (provided that both $p \geq 1$ and $q \geq 1$).

3. Two tensor densities of the same type and weight may be added (or subtracted) to give a tensor density of the same type and weight.

Similar extensions of the algebra of ordinary tensors to tensor densities appear in the Exercises.

6.17. Numerical Tensor Densities

In this section, we introduce objects called *numerical tensor densities* that are useful in many types of applications. As we will see, they enable us to deal with antisymmetric sets of quantities in a computationally efficient manner.

We begin by defining the m-dimensional **(generalized) Kronecker delta** as the determinant of order m which is given by

$$\delta^{a_1 \ldots a_m}_{b_1 \ldots b_m} = \begin{vmatrix} \delta^{a_1}_{b_1} & \delta^{a_1}_{b_2} & \ldots & \delta^{a_1}_{b_m} \\ \delta^{a_2}_{b_1} & \delta^{a_2}_{b_2} & \ldots & \delta^{a_2}_{b_m} \\ \ldots & \ldots & \ldots & \ldots \\ \delta^{a_m}_{b_1} & \delta^{a_m}_{b_2} & \ldots & \delta^{a_m}_{b_m} \end{vmatrix} \tag{6.179}$$

In (6.179), we note that the upper indices of the generalized Kronecker delta correspond to the successive rows of the determinant, while its lower indices correspond to successive columns.

The ordinary Kronecker delta, δ^a_b, is just the one-dimensional version of the generalized Kronecker delta. Similarly, for the special case $m = 2$, we have

$$\delta^{ab}_{cd} = \begin{vmatrix} \delta^a_c & \delta^a_d \\ \delta^b_c & \delta^b_d \end{vmatrix} = \delta^a_c \delta^b_d - \delta^b_c \delta^a_d$$

Since the one-dimensional Kronecker delta δ^a_b forms the components of a tensor of type $(1,1)$, the m-dimensional Kronecker delta must form the components of a tensor of type (m,m). This follows immediately from the fact that the expansion of (6.179) involves a total of $m!$ terms, where each term contains the product of m one-dimensional Kronecker deltas.

The m-dimensional Kronecker delta is *totally* antisymmetric in both its upper and lower indices. That is, if we interchange any two upper or any two lower indices, its sign will change; that is,

$$\begin{aligned} \delta^{a_1 \ldots a_p \ldots a_q \ldots a_m}_{b_1 b_2 b_3 \ldots b_m} &= -\delta^{a_1 \ldots a_q \ldots a_p \ldots a_m}_{b_1 b_2 b_3 \ldots b_m} \\ \delta^{a_1 a_2 a_3 \ldots a_m}_{b_1 \ldots b_p \ldots b_q \ldots b_m} &= -\delta^{a_1 a_2 a_3 \ldots a_m}_{b_1 \ldots b_q \ldots b_p \ldots b_m} \end{aligned} \tag{6.180}$$

for all indices where $1 \leq p < q \leq m$. The antisymmetry exhibited in (6.180) follows immediately from the fact that the sign of a determinant changes whenever we interchange any two of its rows, or any two of its columns.

Now, suppose in (6.180), that the two upper indices a_p and a_q, are equal, then the m-dimensional Kronecker delta must vanish since

$$\delta^{a_1 \ldots a_p \ldots a_q \ldots a_m}_{b_1 b_2 b_3 \;\ldots\; b_m} = -\,\delta^{a_1 \ldots a_p \ldots a_q \ldots a_m}_{b_1 b_2 b_3 \;\ldots\; b_m} \Longrightarrow \delta^{a_1 \ldots a_p \ldots a_q \ldots a_m}_{b_1 b_2 b_3 \;\ldots\; b_m} = 0$$

Similarly, the generalized Kronecker delta will vanish if any two of its subscripts are equal. This simple observation implies that if the underlying manifold M is of dimension n, then

$$\delta^{a_1 \ldots a_m}_{b_1 \ldots b_m} \equiv 0 \;\text{ for } m > n = \dim(M)\,; \tag{6.181}$$

since for $m > n$, at least two of the indices must be repeat.

From the above discussion, it is apparent that for $m \leq n = \dim(M)$, we have

$$\delta^{a_1 \ldots a_m}_{b_1 \ldots b_m} = \begin{cases} -1, \text{ if } b_1 \ldots b_m \text{ is an odd permutation of } a_1 \ldots a_m \\ +1, \text{ if } b_1 \ldots b_m \text{ is an even permutation of } a_1 \ldots a_m \\ 0, \text{ otherwise} \end{cases} \tag{6.182}$$

This expression serves as an alternative definition of the generalized Kronecker delta which is easily seen to be equivalent to (6.179).

Contractions of the m-dimensional Kronecker delta have several important properties that we now discuss. First, we claim that

$$\delta^{a_1 \ldots a_k a_{k+1} \ldots a_m}_{b_1 \ldots b_k a_{k+1} \ldots a_m} = \frac{(n-k)!}{(n-m)!}\delta^{a_1 \ldots a_k}_{b_1 \ldots b_k} \quad (\text{for } m \leq n) \tag{6.186}$$

We shall use a combinatorial argument to establish this claim. At the outset, let

$$P(n,r) = \frac{n!}{(n-r)!} \tag{6.187}$$

denote the number of different permutations of n distinct objects, taken r at a time. Now, assume that the numerical values of the k-symbols $a_1 \ldots a_k$ have been chosen, and are *fixed*. Thus, there remain exactly $(n-k)$ numbers that can be assigned to the the set $a_{k+1} \ldots a_m$ of $(m-k)$ symbols. The numbers in the latter set of symbols can be arranged in a total of $P(n-k, m-k)$ different ways, where

$$P(n-k, m-k) = \frac{(n-k)!}{(n-m)!}.$$

Thus, the sum in (6.186) must consist of exactly $(n-k)!/(n-m)!$ terms.

Now, the numbers assigned to $b_1 \ldots b_k$ are either a permutation of the numbers assigned to $a_1 \ldots a_k$ or they are not. In the latter case, (6.186) is trivially satisfied, since both sides vanish. If the numbers assigned to $b_1 \ldots b_k$ are an *even* permutation of the numbers $a_1 \ldots a_k$ then each term in the sum is $+1$, so that

$$\delta^{a_1 \ldots a_k a_{k+1} \ldots a_m}_{b_1 \ldots b_k a_{k+1} \ldots a_m} = \frac{(n-k)!}{(n-m)!}(1) = \frac{(n-k)!}{(n-m)!}\delta^{a_1 \ldots a_k}_{b_1 \ldots b_k}$$

On the other hand, if the numbers $b_1...b_k$ are an *odd* permutation of the numbers $a_1...a_k$ then each term in the sum is -1, so that

$$\delta^{a_1...a_k a_{k+1}...a_m}_{b_1...b_k a_{k+1}...a_m} = \frac{(n-k)!}{(n-m)!}(-1) = \frac{(n-k)!}{(n-m)!}\delta^{a_1...a_k}_{b_1...b_k}$$

In all cases, we see that (6.186) is satisfied. By a similar argument, it's easy to show that

$$\delta^{a_1...a_m}_{a_1...a_m} = \frac{n!}{(n-m)!} \quad (\text{for } m \leq n) \tag{6.188}$$

Now, let $\mathbf{S} = (S_{a_1...a_m})$ be a completely antisymmetric tensor. We claim that

$$\delta^{a_1...a_m}_{b_1...b_m} S_{a_1...a_m} = m! \cdot S_{b_1...b_m} \tag{6.189}$$

Once again, we will prove this claim by a simple argument. We define the **multi-indices** $A = (a_1...a_m)$ and $B = (b_1...b_m)$, and assume that numerical values of the m-symbols $b_1...b_m$ have been chosen, and are *fixed*. This means that there are exactly $m!$ different permutations of numbers that can be assigned to the symbols $a_1...a_m$. [Note that if the numbers assigned to the symbols $a_1...a_m$ are not a permutation of the numbers $b_1...b_m$, then this contributes nothing to the sum on the lefthand side of (6.189).]

Let $\pi(B)$ and $\pi'(B)$ denote even and odd permutations of $B = (b_1...b_m)$, respectively. Now, half of these permutations will be even, and the other half will be odd. Consequently, we can separate the sum in (6.189) into two distinct terms, and we find that

$$\begin{aligned}
\delta^{a_1...a_m}_{b_1...b_m} S_{a_1...a_m} &= \delta^{\pi(B)}_{B} S_{\pi(B)} + \delta^{\pi'(B)}_{B} S_{\pi'(B)} \\
&= \frac{m!}{2}(+1)S_{\pi(B)} + \frac{m!}{2}(-1)S_{\pi'(B)} \\
&= \frac{m!}{2}S_B - \frac{m!}{2}(-S_B) \quad (\text{since } S \text{ is antisymmetric}) \\
&= m!S_B = m! \cdot S_{b_1...b_m}
\end{aligned}$$

as claimed.

We define the n-dimensional **Levi-Civita (or permutation) symbols** by

$$\begin{aligned}
\varepsilon^{a_1...a_n} &= \delta^{a_1...a_n}_{123...\,n} \\
\varepsilon_{b_1...b_n} &= \delta^{123...\,n}_{b_1...b_n}
\end{aligned} \tag{6.190}$$

It should be noted that these symbols are only defined when $n = \dim(M)$. By (6.180), we see that each of these objects is totally anti-symmetric; that is,

$$\begin{aligned}
\varepsilon^{a_1...a_p...a_q...a_n} &= -\,\varepsilon^{a_1...a_q...a_p...a_n} \\
\varepsilon_{b_1...b_p...b_q...b_n} &= -\,\varepsilon_{b_1...b_q...b_p...b_n}
\end{aligned} \tag{6.191}$$

for all indices where $1 \leq p < q \leq n$. Consequently, they both vanish if any two indices are equal. Also, by (6.179) and (6.190), we have

$$\varepsilon^{123...n} = \varepsilon_{123...n} = \delta^{123...n}_{123...n} = \det(I) = 1 \tag{6.192}$$

since the determinant of the $n \times n$ identity matrix I is unity. From the above discussion, we conclude that:

$$\varepsilon_{a_1...a_n} = \varepsilon^{a_1...a_n} = \begin{cases} -1, \text{ if } a_1...a_n \text{ is an odd permutation of } 123...n \\ +1, \text{ if } a_1...a_n \text{ is an even permutation of } 123...n \\ \ \ 0, \text{ otherwise} \end{cases} \tag{6.193}$$

Note that (6.193) is the usual definition of the permutation symbols which is found in most texts.

Sometimes, we can write completely antisymmetric quantities in terms of the permutation symbols. For example, observe that

$$\begin{aligned} A^{a_1...a_n}_{123...n} &= \varepsilon^{a_1...a_n} A^{123...n}_{123...n} \\ A^{123...n}_{b_1...b_n} &= \varepsilon_{b_1...b_n} A^{123...n}_{123...n} \end{aligned} \tag{6.194}$$

provided $A^{a_1...a_n}_{b_1...b_n}$ is completely antisymmetric in both sets of indices.

From this simple observation, we can deduce that

$$\delta^{a_1...a_n}_{b_1...b_n} = \varepsilon^{a_1...a_n} \cdot \varepsilon_{b_1...b_n} \tag{6.195}$$

To see this, consider the quantity

$$A^{a_1...a_n}_{b_1...b_n} = \delta^{a_1...a_n}_{b_1...b_n} - \varepsilon^{a_1...a_n} \cdot \varepsilon_{b_1...b_n}$$

This quantity is completely antisymmetric in both its upper and lower indices. Consequently, we can write

$$\varepsilon^{a_1...a_n} \cdot A^{123...n}_{123...n} = A^{a_1...a_n}_{123...n} = \delta^{a_1...a_n}_{123...n} - \varepsilon^{a_1...a_n} \cdot \varepsilon_{123...n} = \varepsilon^{a_1...a_n} - \varepsilon^{a_1...a_n} = 0$$

We conclude that

$$A^{123...n}_{123...n} = 0\,,$$

from which it follows that $A^{a_1...a_n}_{b_1...b_n} \equiv 0$, and so, (6.195) follows as well.

We now show that the determinant of an $n \times n$ matrix such as $A = (A^a_b)$, may be expressed in terms of the n-dimensional Kronecker delta.

First, we recall that $\det A$ may be written in the following two ways:

$$\begin{aligned} \det A &= \varepsilon_{b_1...b_n} A^{b_1}_1 A^{b_2}_2 \cdots A^{b_n}_n \\ &= \varepsilon^{a_1...a_n} A^1_{a_1} A^2_{a_2} \cdots A^n_{a_n} \end{aligned} \tag{6.196}$$

From (6.193), it's apparent that the first expression may be rewritten in the form

$$\varepsilon_{a_1...a_n} (\det A) = \varepsilon_{b_1...b_n} A^{b_1}_{a_1} A^{b_2}_{a_2} \cdots A^{b_n}_{a_n} \tag{6.197}$$

Now, if we multiply both sides of (6.197) by $\varepsilon^{a_1...a_n}$, we obtain

$$\begin{aligned} \varepsilon^{a_1...a_n} \cdot \varepsilon_{a_1...a_n} (\det A) &= \varepsilon^{a_1...a_n} \cdot \varepsilon_{b_1...b_n} A^{b_1}_{a_1} A^{b_2}_{a_2} \cdots A^{b_n}_{a_n} \\ &= \delta^{a_1...a_n}_{b_1...b_n} A^{b_1}_{a_1} A^{b_2}_{a_2} \cdots A^{b_n}_{a_n}\,, \end{aligned}$$

where we used (6.195). But clearly,

$$\varepsilon^{a_1\ldots a_n}\cdot\varepsilon_{a_1\ldots a_n}=n!, \tag{6.198}$$

so the previous equation may be rewritten as

$$\det A=\frac{1}{n!}\,\delta^{a_1\ldots a_n}_{b_1\ldots b_n}\,A^{b_1}_{a_1}A^{b_2}_{a_2}\cdots A^{b_n}_{a_n}\,. \tag{6.199}$$

Now, let's investigate the tensor character of the permutation symbols. Under a change of coordinates $\bar{x}^a=\bar{x}^a(x^b)$, we have

$$\bar{\delta}^{a_1\ldots a_n}_{b_1\ldots b_n}=\left(\frac{\partial\bar{x}^{a_1}}{\partial x^{m_1}}\cdots\frac{\partial\bar{x}^{a_n}}{\partial x^{m_n}}\right)\left(\frac{\partial x^{l_1}}{\partial\bar{x}^{b_1}}\cdots\frac{\partial x^{l_n}}{\partial\bar{x}^{b_n}}\right)\delta^{m_1\ldots m_n}_{l_1\ldots l_n} \tag{6.200}$$

since δ is a mixed tensor of type (n,n). We now set

$$b_1=1,b_2=2,\ldots,b_n=n$$

so that the last expression becomes

$$\begin{aligned}\bar{\varepsilon}^{a_1\ldots a_n}&=\bar{\delta}^{a_1\ldots a_n}_{123\ldots n}=\left(\frac{\partial\bar{x}^{a_1}}{\partial x^{m_1}}\cdots\frac{\partial\bar{x}^{a_n}}{\partial x^{m_n}}\right)\left(\frac{\partial x^{l_1}}{\partial\bar{x}^{1}}\cdots\frac{\partial x^{l_n}}{\partial\bar{x}^{n}}\right)\delta^{m_1\ldots m_n}_{l_1\ldots l_n}\\&=\left(\frac{\partial\bar{x}^{a_1}}{\partial x^{m_1}}\cdots\frac{\partial\bar{x}^{a_n}}{\partial x^{m_n}}\right)\varepsilon^{m_1\ldots m_n}\left(\frac{\partial x^{l_1}}{\partial\bar{x}^{1}}\cdots\frac{\partial x^{l_n}}{\partial\bar{x}^{n}}\right)\varepsilon_{l_1\ldots l_n}\end{aligned}$$

where we used (6.195) in the second line. But, by (6.196) we have

$$J=\left(\frac{\partial x^{l_1}}{\partial\bar{x}^{1}}\cdots\frac{\partial x^{l_n}}{\partial\bar{x}^{n}}\right)\varepsilon_{l_1\ldots l_n}$$

where J is the Jacobian of the inverse transformation. We conclude that

$$\bar{\varepsilon}^{a_1\ldots a_n}=J\cdot\left(\frac{\partial\bar{x}^{a_1}}{\partial x^{m_1}}\cdots\frac{\partial\bar{x}^{a_n}}{\partial x^{m_n}}\right)\varepsilon^{m_1\ldots m_n} \tag{6.201}$$

so the quantities $\varepsilon^{m_1\ldots m_n}$ are the components of a tensor density of type $(n,0)$ and weight $+1$. Likewise, we can show that the quantities $\varepsilon_{l_1\ldots ll_n}$ form the components of a tensor density of type $(0,n)$ and weight -1.

6.18. Introduction to Flows

Recall that a curve α on a manifold M is a *smooth* map $\alpha:(-\varepsilon,\varepsilon)\longrightarrow M$. If we are given a smooth vector field $\mathbf{V}(\mathbf{x})$ on M, we say that α is an **integral curve** of $\mathbf{V}(\mathbf{x})$ if the tangent vector to α at each point $\mathbf{x}$, coincides with the vector $\mathbf{V}(\mathbf{x})$.

Now, let's find a coordinate representation for such an integral curve. To this end, let (U,ϕ) be a coordinate chart on M, and assume that $\alpha(t)\in U$ for all $t\in(-\varepsilon,\varepsilon)$. Then the curve α has the coordinate representation $x^k=\phi^k(\alpha(t))$, where ϕ^k denotes the k-th coordinate of the point $\phi(\alpha(t))\in R^n$. So, α is an integral curve of $\mathbf{V}(\mathbf{x})$ if

$$\frac{d}{dt}x^k(t)=V^k(\alpha(t)) \tag{6.202a}$$

If we supplement (6.202a) with the initial condition

$$x^k(0) \equiv \phi^k(\alpha(0)) = x_0^k \tag{6.202b}$$

then from the theory of ordinary differential equations, we are guaranteed that there exists a *unique* solution to the initial value problem (6.202) which is valid in the interval $(-\varepsilon, \varepsilon)$, provided ε is sufficiently small.

Now, let $\sigma(t, x_0)$ denote the *unique* integral curve of $\mathbf{V}$ which passes through the point x_0 when $t = 0$. For some chart (U, ϕ) containing σ, we must have

$$\begin{aligned} \frac{d}{dt}\sigma^k(t, x_0) &= V^k(\sigma(t, x_0)) \\ \sigma^k(0, x_0) &= x_0^k \end{aligned} \tag{6.203}$$

where of course, $\sigma^k(t, x_0) = \phi^k(\sigma(t, x_0))$.

Once again, we emphasize that the solution to (6.203) is unique, and is valid over some interval $(-\varepsilon, \varepsilon)$. Thus, σ may be viewed as a map

$$\sigma : (-\varepsilon, \varepsilon) \times M \longrightarrow M \tag{6.204}$$

which for each fixed $t \in (-\varepsilon, \varepsilon)$, is a diffeomorphism of M into itself. We call σ the **flow** *generated* by the vector field $\mathbf{V}$.

Now, a flow σ is completely analogous to the steady flow of a real fluid. Here $\sigma(t, x_0)$ may be viewed as the point occupied by a fluid particle which was initially at the point $\sigma(0, x_0) = x_0$ at time $t = 0$. We can exploit this analogy to discover an important property of flows.

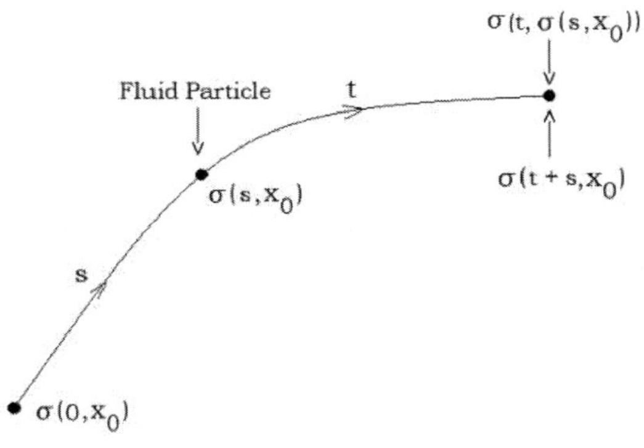

Figure 6.6. A fluid particle moving along a flow $\sigma(t, x_0)$.

Consider a fluid particle initially at the point $\sigma(0, x_0) = x_0$, as shown in Figure 6.6. After an elapsed time of $(t+s)$-seconds, this particle is at the point $\sigma(t+s, x_0) = x_{s+t}$. During its journey, however, the *same* particle must have passed through the point $x_s = \sigma(s, x_0)$ along the way. So, we must have

$$\sigma(t+s, x_0) = \sigma(t, x_s) = \sigma(t, \sigma(s, x_0)), \tag{6.205}$$

for s,t, and $t+s$ sufficiently small.

Since we can't always trust physical intuition, let's prove (6.205) mathematically. Let the parameter s be *fixed*, then $\sigma^k(t,\sigma(s,x_0))$ must satisfy the initial value problem

$$\begin{aligned} \frac{d}{dt}\sigma^k(t,\sigma(s,x_0)) &= V^k(\sigma(t,\sigma(s,x_0)) \\ \sigma^k(0,\sigma(s,x_0)) &= \sigma^k(s,x_0) \end{aligned} \tag{6.206a}$$

On the other hand, however, $\sigma^k(s+t,x_0)$ must satisfy the initial value problem

$$\begin{aligned} \frac{d}{dt}\sigma^k(s+t,x_0) &= V^k(\sigma(s+t,x_0)) \\ \sigma^k(0+s,x_0) &= \sigma^k(s,x_0) \end{aligned} \tag{6.206b}$$

where in (6.206b), we used the fact that $d/dt = d/d(s+t)$ since s is fixed. We see that both $\sigma^k(s+t,x_0)$ and $\sigma^k(t,\sigma(s,x_0))$ satisfy the same initial value problem, and since the solution to such a problem is *unique*, then (6.205) follows immediately.

Since the point $x_0 \in M$ is arbitrary, we can denote the initial point by x, and for any point $x \in M$, we have shown that there exists a flow $\sigma(t,x)$ which satisfies the properties:

1. $\sigma(t,x)$ is the *unique* solution of the initial value problem

$$\begin{aligned} \frac{d}{dt}\sigma^k(t,x) &= V^k(\sigma(t,x)) \\ \sigma^k(0,x) &= x_0^k \end{aligned} \tag{6.207}$$

which is valid over some interval $(-\varepsilon,\varepsilon)$.

2.

$$\sigma(0,x) = x \tag{6.208}$$

3. For $t,s,t+s \in (-\varepsilon,\varepsilon)$, we have

$$\sigma(t+s,x) = \sigma(t,\sigma(s,x)) \tag{6.209}$$

We now demonstrate these properties with a simple example.

Example 1. Let $M = R^2$ and consider the smooth vector field

$$\mathbf{V}((x,y)) = x\frac{\partial}{\partial x} + y\frac{\partial}{\partial y}$$

Let's find the flow σ generated by $\mathbf{V}$. We introduce rectangular coordinates, and let

$$\begin{aligned} \sigma^1(t,x_0) &= x,\ \sigma^2(t,x_0) = y \\ \sigma^1(0,x_0) &= X_0,\ \sigma^2(0,x_0) = Y_0 \end{aligned}$$

So, we obtain the system of differential equations

$$\begin{aligned} \frac{dx}{dt} &= x, \quad x(t=0) = X_0 \\ \frac{dy}{dt} &= y, \quad y(t=0) = Y_0 \end{aligned}$$

whose unique solution is

$$x = X_0 e^t, \quad y = Y_0 e^t.$$

Observe that this solution is valid for all real t. With some small abuse of notation, the corresponding flow in vector form is

$$\sigma(t, \mathbf{x}_0) = \sigma(t, (X^0, Y^0)) = (X_0, Y_0)e^t$$

We note that $\sigma(0, \mathbf{x}_0) = \sigma(0, (X^0, Y^0)) = (X_0, Y_0) = \mathbf{x}_0$ as it should. Also, the flow lines or integral curves of $\mathbf{V}$ are rays which pass through the origin of the xy-plane:

$$y = \frac{Y_0}{X_0} x, \qquad (X_0 \neq 0)$$

We now verify that the flow obeys (6.209). We have

$$\begin{aligned} \sigma(t, \sigma(s, \mathbf{x}_0)) &= \sigma(t, (X_0 e^s, Y_0 e^s)) = (X_0 e^{t+s}, Y_0 e^{t+s}) \\ &= \sigma(t+s, (X_0, Y_0)) \\ &= \sigma(t+s, \mathbf{x}_0) \end{aligned}$$

as required.

If for each fixed $t \in (-\varepsilon, \varepsilon)$, we set

$$\sigma_t(x) = \sigma(t, x) \tag{6.210}$$

then $\sigma_t : M \longrightarrow M$ is a *diffeomorphism* of M into itself. This situation is shown in Figure 6.7, where each point $x \in M$ is mapped to the point $\sigma_t(x)$ which is a parameter distance t away from x.

If $t \in (-\varepsilon, \varepsilon)$ is sufficiently small, then the flow σ_t maps any given point $x = (x^k) \in M$ into the point $\sigma_t(x)$ with coordinates

$$\sigma_t^k(x) = \sigma_0^k(x) + t \left. \frac{d}{dt} \sigma_t^k(x) \right|_{t=0} + O(t^2). \tag{6.211}$$

Furthermore, if we set

$$V^k(x) = \left. \frac{d}{dt} \sigma_t^k(x) \right|_{t=0}$$

then the last equation becomes

$$\sigma_t^k(x) = x^k + tV^k(x), \tag{6.212}$$

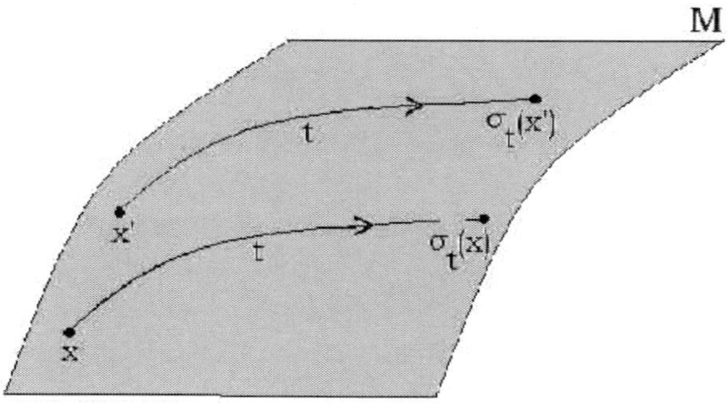

Figure 6.7. The diffeomorphism σ_t maps each $x \in M$ to $\sigma_t(x) \in M$.

up to first order terms. We call the vector field **V** the **infinitesimal generator** of the map $\sigma_t(x)$. Here, **V** is just the tangent vector to the curve $\sigma_t(x) = \sigma(t,x)$ at the initial point x, whose coordinates are given by $x^k = \sigma^k(0,x) = \sigma_0^k(x)$.

If we continue to view $\sigma_t : M \longrightarrow M$ as a *diffeomorphism* of M into itself, then the set of flows $\{\sigma_t\}$ forms a *group* under the operation of composition of mappings, provided we specify that:

1. For each $x \in M$,

$$(\sigma_t \circ \sigma_s)(x) = \sigma_t(\sigma_s(x)) = \sigma_{t+s}(x) \Longrightarrow \sigma_t \cdot \sigma_s = \sigma_{t+s} \tag{6.213}$$

provided $t, s, t+s \in (-\varepsilon, \varepsilon)$.

2. For each $x \in M$,

$$\sigma_0(x) = x \tag{6.214}$$

so, σ_0 is the *identity element* of the group.

3. For each $t \in (-\varepsilon, \varepsilon)$, the *inverse* of the element σ_t is given by

$$(\sigma_t)^{-1} = \sigma_{-t} \tag{6.215}$$

We denote this group by $\langle \{\sigma_t\}, \circ \rangle$, and from (6.215a), we see that it is a *commutative group*. We call $\langle \{\sigma_t\}, \circ \rangle$ the **local one-parameter group of transformations** on M. We emphasize the local nature of this group, since the group operation is only defined when the group parameter t is sufficiently small.

6.19. The Lie Derivative

Assume that we are given a smooth vector field **Y** on M, and a flow $\sigma_t(x) = \sigma(t,x)$ which is generated by the smooth vector field **X** on M. We can use the flow to define a new type of derivative, called the **Lie derivative** of **Y** in the direction **X** along the flow, by

$$L_{\mathbf{X}}\mathbf{Y} = \lim_{t \to 0} \frac{1}{t} \left\{ \mathbf{Y}\big|_{\sigma_t(x)} - (\sigma_t)_* \, \mathbf{Y}\big|_x \right\} \tag{6.216}$$

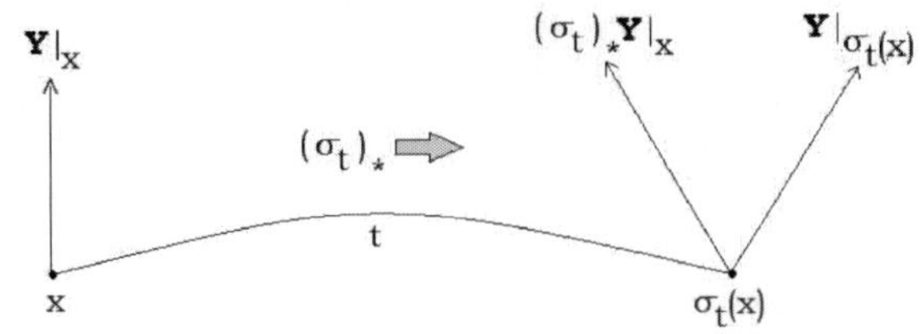

Figure 6.8. The geometry of the Lie derivative.

This definition makes sense, as may be seen by examining Figure 6.8.

The basic idea is that we compare the vector $\mathbf{Y}|_{\sigma_t(x)}$ with the vector $(\sigma_t)_* \mathbf{Y}|_x$ which is the *push-forward* of the vector $\mathbf{Y}|_x$ by the differential map $(\sigma_t)_*$. Since the vectors $\mathbf{Y}|_{\sigma_t(x)}$ and $(\sigma_t)_* \mathbf{Y}|_x$ live in the *same* tangent space, we can legitimately subtract them, and then calculate the indicated limit.

Now, let's determine the coordinate representation of (6.216). Assume that the set $\{\mathbf{e}_k\}$ forms a basis for the tangent space $T_{\sigma_t(x)}$. We have by (6.212),

$$\mathbf{Y}|_{\sigma_t(x)} = Y^k(x^m + tX^m)\mathbf{e}_k = \left\{Y^k(x) + tX^m\partial_m Y^k\right\}\mathbf{e}_k + O(t^2) \tag{6.217}$$

On the other hand, by the definition of the push-forward in (5.74),

$$\begin{aligned}(\sigma_t)_* \mathbf{Y}|_x &= \left\{\frac{\partial}{\partial x^m}\left(x^k + tX^k\right)\right\}Y^m\,\mathbf{e}_k = \left(\delta^k_m + t\partial_m X^k\right)Y^m\mathbf{e}_k \\ &= \left(Y^k + tY^m\partial_m X^k\right)\mathbf{e}_k\end{aligned} \tag{6.218}$$

From (6.217) and (6.218), we find

$$L_{\mathbf{X}}\mathbf{Y} = \lim_{t\to 0}\frac{1}{t}\left\{t\left(X^m\partial_m Y^k - Y^m\partial_m X^k\right)\mathbf{e}_k + O(t^2)\right\},$$

and consequently,

$$L_{\mathbf{X}}\mathbf{Y} = (X^m\partial_m Y^k - Y^m\partial_m X^k)\mathbf{e}_k, \tag{6.219}$$

an expression which gives us the component form of the Lie derivative of the vector field $\mathbf{Y}$ along the direction $\mathbf{X}$. If we recall the expression for the Lie bracket given in (6.36), we see that

$$L_{\mathbf{X}}\mathbf{Y} = [\mathbf{X}, \mathbf{Y}] \tag{6.220}$$

Clearly, the Lie derivative of a vector field is again a vector field; so we say that the Lie derivative is **type-preserving**.

We now turn our attention to task of defining the Lie derivative of a one-form field $\omega = \omega(\mathbf{x})$. Recall that we usually push forward tangent vectors, while we pull back one-forms. With this in mind, we define the Lie derivative of a one-form field ω as

$$L_{\mathbf{X}}\omega = \lim_{t\to 0}\frac{1}{t}\left\{(\sigma_t)^*\,\omega|_{\sigma_t(x)} - \omega|_x\right\}. \tag{6.221}$$

That is, using the dual tangent map $(\sigma_t)^*$, we pull back the one-form ω which is situated at the point $\sigma_t(x)$, to the point x, so it can be compared with the one-form $\omega|_x$ which is already this point. We then perform the the indicated operations, and take the limit.

We now determine the coordinate expression of (6.221). Let the sets $\{\mathbf{d}x^m\}$, and $\{\mathbf{d}\bar{x}^m\}$ comprise bases for the cotangent spaces $T_x^*(M)$ and $T_{\sigma_t(x)}^*(M)$, respectively. We have

$$\begin{aligned}\omega|_{\sigma_t(x)} &= \omega_m(x^k+tX^k)\mathbf{d}\bar{x}^m \\ &= \left\{\omega_m(x)+tX^k\partial_k\omega_m(x)\right\}\mathbf{d}\bar{x}^m \\ &= \omega_m(x)\mathbf{d}\bar{x}^m+tX^k\partial_k\omega_m(x)\mathbf{d}\bar{x}^m \end{aligned} \tag{6.222}$$

Consequently, by (5.77), the pull-back of $\omega|_{\sigma_t(x)}$ to $T_x^*(M)$ is given by

$$\begin{aligned}(\sigma_t)^*\,\omega|_{\sigma_t(x)} &= \left\{\frac{\partial\bar{x}^m}{\partial x^k}\left(\omega|_{\sigma_t(x)}\right)_m\right\}\mathbf{d}x^k=\left\{\frac{\partial}{\partial x^k}\left(x^m+tX^m\right)\left(\omega|_{\sigma_t(x)}\right)_m\right\}\mathbf{d}x^k \\ &= \left(\delta_k^m+t\partial_kX^m\right)\left\{\omega_m(x)+tX^p\partial_p\omega_m(x)\right\}\mathbf{d}x^k \\ &= \omega_m(x)\mathbf{d}x^m+t\left(X^p\partial_p\omega_m(x)+\omega_p(x)\partial_mX^p\right)\mathbf{d}x^m+O(t^2) \end{aligned} \tag{6.223}$$

Now, by (6.222) and (6.223), we obtain

$$L_{\mathbf{X}}\omega=\lim_{t\to 0}\frac{1}{t}\left\{t\left(X^p\partial_p\omega_m(x)+\omega_p(x)\partial_mX^p\right)\mathbf{d}x^m+O(t^2)\right\},$$

so that

$$L_{\mathbf{X}}\omega=\left(X^p\partial_p\omega_m+\omega_p\partial_mX^p\right)\mathbf{d}x^m\,. \tag{6.224}$$

Observe that $L_{\mathbf{X}}\omega$ is a one-form, in keeping with the type preserving property of the Lie derivative.

Given a function $f\in C^\infty(M)$, we define its Lie derivative by the rule

$$L_{\mathbf{X}}\,f=\lim_{t\to 0}\frac{1}{t}\left\{f\left(\sigma_t(x)\right)-f(x)\right\} \tag{6.225}$$

Consequently,

$$\begin{aligned}L_{\mathbf{X}}\,f &= \lim_{t\to 0}\frac{1}{t}\left\{f\left(x^k+tX^k\right)-f(x^k)\right\} \\ &= \lim_{t\to 0}\frac{1}{t}\left\{\left[f(x^k)+tX^k\partial_kf+O(t^2)\right]-f(x^k)\right\}=X^k\partial_k\,f\end{aligned}$$

In other words, the Lie derivative of a smooth function is given by its *directional derivative*:

$$L_{\mathbf{X}}\,f=\mathbf{X}[f] \tag{6.226}$$

In order to extend the definition of Lie derivative to general tensor fields, we require that it obeys the Leibnitz rule; that is for general tensor fields $\mathbf{S}$ and $\mathbf{T}$ of any type, we require that

$$L_{\mathbf{X}}\left(\mathbf{S}\otimes\mathbf{T}\right)=\left(L_{\mathbf{X}}\mathbf{S}\right)\otimes\mathbf{T}\,+\,\mathbf{S}\otimes\left(L_{\mathbf{X}}\mathbf{T}\right) \tag{6.227}$$

A simple example shows how this can be accomplished.

Example 1. Given a tensor field $\mathbf{T} = (T^{ij})$ of type $(2,0)$, let's compute its Lie derivative. By (6.227), we have

$$\begin{aligned} L_{\mathbf{X}}\mathbf{T} &= L_{\mathbf{X}}\left(T^{ij}\mathbf{e}_i \otimes \mathbf{e}_j\right) \\ &= \left(L_{\mathbf{X}}T^{ij}\right)\mathbf{e}_i \otimes \mathbf{e}_j + T^{ij}\left(L_{\mathbf{X}}\mathbf{e}_i\right)\otimes \mathbf{e}_j + T^{ij}\mathbf{e}_i \otimes \left(L_{\mathbf{X}}\mathbf{e}_j\right) \end{aligned} \tag{6.228}$$

Now, observe that

$$\begin{aligned} L_{\mathbf{X}}T^{ij} &= X^a \partial_a T^{ij} \\ L_{\mathbf{X}}\mathbf{e}_i &= L_{\mathbf{X}}\left(\delta_i^k \mathbf{e}_k\right) = -\,\delta_i^p \partial_p X^k \mathbf{e}_k \\ L_{\mathbf{X}}\mathbf{e}_j &= L_{\mathbf{X}}\left(\delta_j^k \mathbf{e}_k\right) = -\,\delta_j^p \partial_p X^k \mathbf{e}_k \end{aligned} \tag{6.229}$$

Substituting (6.229) into (6.228), we find

$$\begin{aligned} L_{\mathbf{X}}\mathbf{T} &= X^a \partial_a T^{ij}\mathbf{e}_i \otimes \mathbf{e}_j + T^{ij}\left(-\,\delta_i^p \partial_p X^k \mathbf{e}_k\right)\otimes \mathbf{e}_j + T^{ij}\mathbf{e}_i \otimes \left(-\,\delta_j^p \partial_p X^k \mathbf{e}_k\right) \\ &= \left(X^a \partial_a T^{ij}\right)\mathbf{e}_i \otimes \mathbf{e}_j - T^{ij}\,\partial_i X^k \mathbf{e}_k \otimes \mathbf{e}_j - T^{ij}\mathbf{e}_i \otimes \partial_j X^k \mathbf{e}_k \\ &= \left(X^a \partial_a T^{ij} - T^{aj}\,\partial_a X^i - T^{ib}\partial_b X^j\right)\mathbf{e}_i \otimes \mathbf{e}_j \end{aligned} \tag{6.230}$$

In a similar manner, we can show that given a general tensor field $\mathbf{T} = (T^{a_1 \ldots a_p}_{b_1 \ldots b_q})$ of type (p,q), that its Lie derivative is given by the rule

$$\begin{aligned} \left(L_{\mathbf{X}}\mathbf{T}\right)^{a_1 \ldots a_p}_{b_1 \ldots b_q} &= X^c \partial_c T^{a_1 \ldots a_p}_{b_1 \ldots b_q} - \left(T^{c \ldots a_p}_{b_1 \ldots b_q}\partial_c X^{a_1} + T^{a_1 c \ldots a_p}_{b_1 \ldots b_q}\partial_c X^{a_2} + \ldots + T^{a_1 \ldots c}_{b_1 \ldots b_q}\partial_c X^{a_p}\right) \\ &+ \left(T^{a_1 \ldots a_p}_{c \ldots b_q}\partial_{b_1} X^c + T^{a_1 \ldots a_p}_{b_1 c \ldots b_q}\partial_{b_2} X^c + \ldots + T^{a_1 \ldots a_p}_{b_1 b_2 \ldots c}\partial_{b_q} X^c\right) \end{aligned} \tag{6.231}$$

Observe that $L_{\mathbf{X}}\mathbf{T}$ is again a tensor of type (p,q).

6.20. Exercises

1. Show that:

$$\nabla_k \sqrt{-g} = 0 \tag{6.232}$$

where $g = \det(g_{ab}) < 0$.

(Solution)

Since g is a scalar density of weight $+2$, then $\sqrt{-g}$ must be a scalar density of weight $+1$. Consequently,

$$\nabla_k \sqrt{-g} = \partial_k\left(\sqrt{-g}\right) - \Gamma^p_{pk}\sqrt{-g} = \frac{-1}{2\sqrt{-g}}\left\{\partial_k(g) - \Gamma^p_{pk}g\right\} = 0\,,$$

since by (6.178), it follows that $\nabla_k g = 0$.

2. Given that $\mathbf{T} = (T^a)$ is a vector density of weight $+1$, show that its divergence is

$$\operatorname{div}\mathbf{T} \equiv \nabla_k T^k = \frac{\partial T^k}{\partial x^k} \tag{6.233}$$

(Solution)

We have

$$\begin{aligned}\operatorname{div}\mathbf{T} &\equiv \nabla_k T^k = \delta^a_k (\nabla_k T^a) \\ &= \delta^a_k \left(\partial_k T^a + \Gamma^a_{kp} T^p - \Gamma^p_{pk} T^a\right) \\ &= \partial_k T^k + \Gamma^k_{kp} T^p - \Gamma^p_{pk} T^k \\ &= \partial_k T^k\end{aligned}$$

3. Show that if $A = (A^a)$ is a vector density of weight W, then the quantities given by

$$B^a = |g|^{W/2} A^a \tag{6.234}$$

are the components of an absolute tensor of type $(1,0)$.

(Solution)

Recalling that $\bar{g} = J^2 g$, and that $J > 0$, under a coordinate transformation $\bar{x}^i = \bar{x}^i(x^j)$, we must have

$$\begin{aligned}\bar{B}^a &= |\bar{g}|^{W/2} \bar{A}^a = \left(J^2 |g|\right)^{W/2} \left(J^W \frac{\partial \bar{x}^a}{\partial x^b} A^b\right) = \frac{\partial \bar{x}^a}{\partial x^b} \left(|g|^{W/2} A^b\right) \\ &= \frac{\partial \bar{x}^a}{\partial x^b} B^b\end{aligned}$$

Consequently, $\mathbf{B} = (B^a)$ is an ordinary vector, or tensor of type $(1,0)$.

4. Show that if $\mathbf{A} = (A^a)$ is an absolute tensor of type $(1,0)$, then then the quantities

$$B^a = \sqrt{|g|} A^a \tag{6.235}$$

are the components of a vector density $\mathbf{B}$ of weight $+1$.

(Solution)

Recalling that $J > 0$, under a coordinate transformation $\bar{x}^a = \bar{x}^a(x^b)$, we have

$$\begin{aligned}\bar{B}^a &= \sqrt{|\bar{g}|} \bar{A}^a = \sqrt{J^2 |g|} \left(\frac{\partial \bar{x}^a}{\partial x^b} A^b\right) = J \cdot \frac{\partial \bar{x}^a}{\partial x^b} \left(\sqrt{|g|} A^b\right) \\ &= J \cdot \frac{\partial \bar{x}^a}{\partial x^b} B^b\end{aligned}$$

We conclude that **B** is a vector density of weight $+1$.

5. Let $M = R^2$; find the flow generated by the smooth vector field

$$\mathbf{V}(x,y) = -y\frac{\partial}{\partial x} + x\frac{\partial}{\partial y}$$

(Solution)

Using rectangular coordinates, we set

$$\begin{aligned} \sigma^1(t,x_0) &= x,\ \sigma^2(t,x_0) = y \\ \sigma^1(0,x_0) &= X_0,\ \sigma^2(0,x_0) = Y_0 \end{aligned}$$

So, we must solve the system of differential equations

$$\begin{aligned} \frac{dx}{dt} &= -y, \quad x(t=0) = X_0 \\ \frac{dy}{dt} &= x, \quad y(t=0) = Y_0 \end{aligned}$$

Dividing the second equation by the first, we obtain $dy/dx = -x/y$, which integrates to $x^2 + y^2 = C^2$. From the initial conditions, we find that $C^2 = X_0^2 + Y_0^2$, so we can write the solution in the parametric form:

$$\sigma(t,(X_0,Y_0)) = \sqrt{X_0^2 + Y_0^2}\,(\cos t, \sin t).$$

We see that the flow through the point (X_0, Y_0) is a circle centered at the origin with a radius of $\sqrt{X_0^2 + Y_0^2}$. Observe that the above solution is valid for all real t.

6. If **X** and **Y** are smooth vector fields on M, and $f \in C^\infty(M)$, then show that:

$$L_{f\mathbf{X}}\mathbf{Y} = f[\mathbf{X},\mathbf{Y}] - \mathbf{Y}[f]\mathbf{X} \qquad (6.236a)$$

$$L_{\mathbf{X}}(f\mathbf{Y}) = f[\mathbf{X},\mathbf{Y}] + \mathbf{X}[f]\mathbf{Y} \qquad (6.236b)$$

(Solution)

Let $g \in C^\infty(M)$ be an arbitrary function. To establish (6.236a), observe that

$$\begin{aligned} (L_{f\mathbf{X}}\mathbf{Y})g &= [f\mathbf{X},\mathbf{Y}]g = f\mathbf{X}(\mathbf{Y}[g]) - \mathbf{Y}(f\mathbf{X}[g]) \\ &= f\mathbf{X}(\mathbf{Y}[g]) - \mathbf{Y}[f]\mathbf{X}[g] - f\mathbf{Y}(\mathbf{X}[g]) \\ &= f[\mathbf{X},\mathbf{Y}] - \mathbf{Y}[f]\mathbf{X}g \end{aligned}$$

Since g is arbitrary, the first relation follows. To establish the second relation, let $g \in C^\infty(M)$ be arbitrary. We have

$$\begin{aligned} (L_{\mathbf{X}}(f\mathbf{Y}))g &= [\mathbf{X},f\mathbf{Y}]g = \mathbf{X}(f\mathbf{Y}[g]) - f\mathbf{Y}(\mathbf{X}[g]) \\ &= \mathbf{X}[f]\mathbf{Y}[g] + f\mathbf{X}(\mathbf{Y}[g]) - f\mathbf{Y}(\mathbf{X}[g]) \\ &= f[\mathbf{X},\mathbf{Y}] + \mathbf{X}[f]\mathbf{Y}g \end{aligned}$$

Since g is arbitrary, the second relation follows as well.

7. Given that $\mathbf{X}$ and $\mathbf{Y}$ are smooth vector fields on M, and $f : M \longrightarrow N$, show that

$$f_*[\mathbf{X},\mathbf{Y}] = [f_*\mathbf{X}, f_*\mathbf{Y}] \tag{6.237}$$

(Solution)

Let $\{\bar{\mathbf{e}}_k\}$ comprise a coordinate basis field on N. Then

$$f_*[\mathbf{X},\mathbf{Y}] = \frac{\partial \bar{x}^a}{\partial x^b}[\mathbf{X},\mathbf{Y}]^b \bar{\mathbf{e}}_a = \frac{\partial \bar{x}^a}{\partial x^b}\left\{X^c \frac{\partial Y^b}{\partial x^c} - Y^c \frac{\partial X^b}{\partial x^c}\right\}\bar{\mathbf{e}}_a$$

On the other hand,

$$\begin{aligned}
[f_*\mathbf{X}, f_*\mathbf{Y}] &= \left[\frac{\partial \bar{x}^a}{\partial x^b}X^b, \frac{\partial \bar{x}^a}{\partial x^b}Y^b\right]\bar{\mathbf{e}}_a \\
&= \left[\left(\frac{\partial \bar{x}^a}{\partial x^c}X^c\right)\frac{\partial}{\partial \bar{x}^a}\left(\frac{\partial \bar{x}^a}{\partial x^b}Y^b\right) - \left(\frac{\partial \bar{x}^a}{\partial x^c}Y^c\right)\frac{\partial}{\partial \bar{x}^a}\left(\frac{\partial \bar{x}^a}{\partial x^b}X^b\right)\right]\bar{\mathbf{e}}_a \\
&= \left[X^c \frac{\partial}{\partial x^c}\left(\frac{\partial \bar{x}^a}{\partial x^b}Y^b\right) - Y^c \frac{\partial}{\partial x^c}\left(\frac{\partial \bar{x}^a}{\partial x^b}X^b\right)\right]\bar{\mathbf{e}}_a \\
&= \frac{\partial \bar{x}^a}{\partial x^b}\left\{X^c \frac{\partial Y^b}{\partial x^c} - Y^c \frac{\partial X^b}{\partial x^c}\right\}\bar{\mathbf{e}}_a
\end{aligned}$$

Comparing the last two equations, we see that the result follows.

Chapter 7

Introduction to Minkowski Space

Henceforth space by itself, and time by itself, are doomed to fade away into mere shadows, and only a kind of union of the two will preserve an independent reality.

Hermann Minkowski

7.1. Introduction

We formally define **Minkowski space** M^4, as the four-dimensional, semi-Riemannian space (R^4, η) which is equipped with the indefinite metric η of signature -2. In other words, Minkowski space is the realm of Special Relativity, and it is just the space R^4 endowed with a special type of metric, called the *Minkowski metric*, η.

It is apparent that M^4 can be covered by a *single chart* of the form (R^4, ϕ) where ϕ is the identity map; so there exists a *single* Cartesian coordinate system $(x^\mu) = (x^0, x^1, x^2, x^3)$ covering all of M^4, where

$$x^0 = ct,\ x^1 = x,\ x^2 = y,\ x^3 = z. \tag{7.1}$$

We call the special coordinates (x^μ) **Minkowski coordinates**, and the coordinate chart (R^4, I), the **Minkowski coordinate system**.

As a matter of convention, when we are working in a spacetime, rather than in a general manifold, we shall assume that Greek indices run from 0 to 3, while lowercase Latin letters run from 1 to 3.

Each point of Minkowski space is called an **event**. Given any event $p \in M^4$, it is apparent that the tangent space $T_p(M^4)$ is just a copy of R^4 whose origin coincides with p. Any tangent vector $\mathbf{X} \in T_p(M^4)$ is called a **four-vector**. We denote a typical four-vector $\mathbf{X}$ by

$$\mathbf{X} = (X^\mu) \equiv (X^0, \mathbf{x}) = \left[X^0, X^1, X^2, X^3\right]^T = \begin{bmatrix} X^0 \\ X^1 \\ X^2 \\ X^3 \end{bmatrix} \tag{7.2}$$

In the above *(1+3) notation*, we must be careful to note that $\mathbf{x}$ is an ordinary *three-vector*; that is, $\mathbf{x} = \left[X^1, X^2, X^3\right]^T \in R^3$.

In terms of Minkowski coordinates, the **spacetime interval** ds^2 assumes the simple form

$$ds^2 = \eta_{\mu\nu} dx^\mu dx^\nu = c^2 dt^2 - dx^2 - dy^2 - dz^2 \tag{7.3}$$

where the **Minkowski metric** $\eta = (\eta_{\mu\nu})$ is given by

$$\eta = (\eta_{\mu\nu}) = diag(1,-1,-1,-1) = \begin{bmatrix} 1 & 0 & 0 & 0 \\ 0 & -1 & 0 & 0 \\ 0 & 0 & -1 & 0 \\ 0 & 0 & 0 & -1 \end{bmatrix} \tag{7.4}$$

Also, as in the past, we can use the metric to raise and lower indices. For example, if $\mathbf{X} = (X^\mu) = (X^0, \mathbf{x})$ then

$$(X_\mu) = (\eta_{\mu\nu} X^\nu) = \left[X^0, -X^1, -X^2, -X^3\right]^T \equiv (X^0, -\mathbf{x}) \tag{7.5}$$

It is worthwhile to observe that (X_μ) has the same zero component as its contravariant counterpart, but the sign of its *space component* has been reversed.

In terms of Minkowski coordinates, it is clear that the components of the metric η are constant; consequently, all of the Christoffel symbols vanish, and from (6.91), we see that

$$R^\mu_{\nu\rho\sigma} \equiv 0 \tag{7.6}$$

Since the curvature tensor vanishes identically, we say that Minkowski space is **flat**, or **psuedo-Euclidean**. Furthermore, since (7.6) is a *tensor equation*, it must hold in *all* coordinate systems.

In order to fully appreciate the significance of (7.6), assume that instead of Minkowksi coordinates, we use spherical polar coordinates (x^μ) where

$$(x^\mu) = (ct, r\sin\theta\cos\phi, r\sin\theta\sin\phi, r\cos\theta) \tag{7.7}$$

then the line element assumes the form

$$ds^2 = c^2 dt^2 - dr^2 - r^2 d\theta^2 - r^2 \sin^2\theta d\phi^2 \tag{7.8}$$

In this case, however, the Christoffel symbols are not constants, and the non-zero Christoffel symbols turn out to be

$$\begin{aligned} \Gamma^1_{22} &= -r,\ \Gamma^1_{33} = r\sin^2\theta, \\ \Gamma^2_{12} &= \frac{1}{r},\ \Gamma^2_{33} = -\sin\theta\cos\theta, \\ \Gamma^3_{13} &= \frac{1}{r},\ \Gamma^3_{23} = \cot\theta. \end{aligned} \tag{7.9}$$

If we now compute the curvature tensor in this new coordinate system, we will still obtain $R^\mu_{\nu\rho\sigma} \equiv 0$. This is a direct consequence of the *form-invariant* character of tensor equations under coordinate transformations, as discussed in Section 5.14.

7.2. The Light Cone

When written in terms of Minkowski coordinates, the **inner product** of the four-vectors $\mathbf{X} = (X^0, \mathbf{x})$, and $\mathbf{Y} = (Y^0, \mathbf{y})$ is given by

$$\langle \mathbf{X}, \mathbf{Y} \rangle = \eta_{\mu\nu} X^\mu Y^\nu = X^0Y^0 - X^1Y^1 - X^2Y^2 - X^3Y^3 = X^0Y^0 - \langle \mathbf{x}, \mathbf{y} \rangle \tag{7.10}$$

where $\langle \mathbf{x}, \mathbf{y} \rangle$ denotes the Euclidean inner product of the three-vectors $\mathbf{x}$ and $\mathbf{y}$. Observe that (7.10) may also be conveniently written in matrix form as

$$\langle \mathbf{X}, \mathbf{Y} \rangle = \mathbf{X}^T \eta \mathbf{X} \tag{7.10a}$$

Naturally, we say that the four-vectors $\mathbf{X}$ and $\mathbf{Y}$ are **orthogonal** if their inner product is zero; that is,

$$\langle \mathbf{X}, \mathbf{Y} \rangle = \eta_{\mu\nu} X^\mu Y^\nu = 0 \tag{7.11}$$

We can use the inner product to define the square of the **norm** of a four-vector $\mathbf{X}$ as

$$\|\mathbf{X}\|^2 = \langle \mathbf{X}, \mathbf{X} \rangle = \eta_{\mu\nu} X^\mu X^\nu = \left(X^0\right)^2 - \left(X^1\right)^2 - \left(X^2\right)^2 - \left(X^3\right)^2 \tag{7.12}$$

and we say that the vector $\mathbf{X}$ is

$$\left.\begin{array}{cl} \textbf{spacelike}, & \text{if } \|\mathbf{X}\|^2 < 0 \\ \textbf{lightlike} \text{ (or null)}, & \text{if } \|\mathbf{X}\|^2 = 0 \\ \textbf{timelike}, & \text{if } \|\mathbf{X}\|^2 > 0 \end{array}\right\} \tag{7.13}$$

Due to the invariance of the spacetime interval, this classification of four-vectors has an *absolute character* which is *independent* of any choice of coordinates in M^4.

Now, given any event $p \in M^4$, the set of all null vectors $\mathbf{X} \in T_p(M^4)$ must satisfy the relation

$$\eta_{\mu\nu} X^\mu X^\nu = \left(X^0\right)^2 - \left(X^1\right)^2 - \left(X^2\right)^2 - \left(X^3\right)^2 = 0\,. \tag{7.14}$$

This is the equation of a three-dimensional *hypersurface* in M^4 which is called a **light-cone** or **null cone**. As shown in Figure 7.1, this is just a double cone and lightlike vectors lie along its surface.

From the diagram, we see that timelike vectors lie inside the lightcone, lightlike vectors lie on its surface, while spacelike vectors lie outside it. Furthermore, if we define the timelike four-vector

$$\mathbf{T} = (T^\mu) = [1, 0, 0, 0]^T$$

then we say that any lightlike or timelike four-vector $\mathbf{V} \in T_p(M^4)$ is

$$\left.\begin{array}{cl} \textbf{future-pointing}, & \text{if } \langle \mathbf{V}, \mathbf{T} \rangle > 0 \\ \textbf{past-pointing}, & \text{if } \langle \mathbf{V}, \mathbf{T} \rangle < 0 \end{array}\right\} \tag{7.15}$$

Now, let $\mathbf{U}, \mathbf{V} \in T_p(M^4)$ where $\mathbf{U}$ is timelike, and where $\mathbf{U}$ and $\mathbf{V}$ are orthogonal. It's easy to show that $\mathbf{V}$ must be spacelike. To see this, we let $\mathbf{U} = (U^0, \mathbf{u})$, $\mathbf{V} = (V^0, \mathbf{v})$, and observe that owing to the orthogonality of the two vectors, we must have

$$U^0V^0 = U^1V^1 + U^2V^2 + U^3V^3 = \langle \mathbf{u}, \mathbf{v} \rangle$$

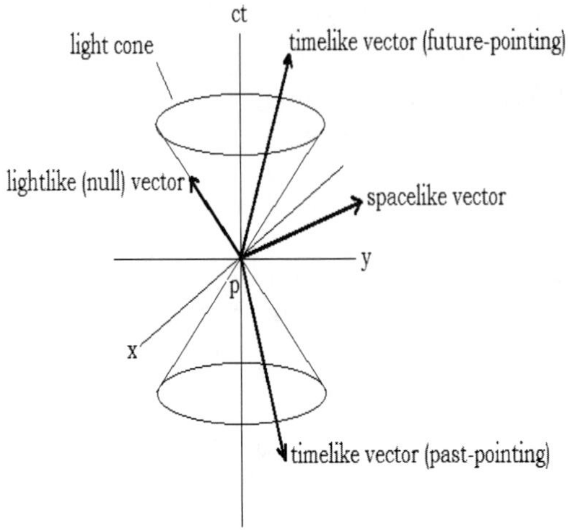

Figure 7.1. The lightcone with vertex at $p \in M^4$ (z-axis not shown).

Also, since **U** is timelike, then we must have $(U^0)^2 > \langle \mathbf{u}, \mathbf{u} \rangle$, and consequently,

$$\frac{(V^0)^2}{\langle \mathbf{v}, \mathbf{v} \rangle} = \frac{(U^0V^0)^2}{\langle \mathbf{v}, \mathbf{v} \rangle (U^0)^2} < \frac{(U^0V^0)^2}{\langle \mathbf{v}, \mathbf{v} \rangle \langle \mathbf{u}, \mathbf{u} \rangle} = \frac{\langle \mathbf{u}, \mathbf{v} \rangle^2}{\langle \mathbf{v}, \mathbf{v} \rangle \langle \mathbf{u}, \mathbf{u} \rangle} \leq 1$$

where we used the Cauchy-Schwartz inequality for ordinary three-vectors in the last step. We conclude that

$$\|\mathbf{V}\|^2 = \langle \mathbf{V}, \mathbf{V} \rangle = (V^0)^2 - \langle \mathbf{v}, \mathbf{v} \rangle < 0$$

so **V** is spacelike, as claimed. So, we have shown that *any four-vector which is orthogonal to a timelike vector must be spacelike.* Similarly, it's easy to show that *any four-vector which is orthogonal to a lightlike vector must be a scalar multiple of a timelike vector.*

Now, we can use the classification scheme in (7.14) to help us describe the spacetime relationship between any pair of events in M^4. If we are given any two events p and q, with Minkowski coordinates (X^0, X^1, X^2, X^3) and (Y^0, Y^1, Y^2, Y^3) respectively, we define the **separation** between p and q as the four-vector $\overrightarrow{pq} = \mathbf{Y} - \mathbf{X}$, and the **interval** between them as

$$\|\mathbf{Y} - \mathbf{X}\|^2 = c^2 \left(Y^0 - X^0\right)^2 - (Y^1 - X^1)^2 - (Y^2 - X^2)^2 - (Y^3 - X^3)^2 \ . \qquad (7.16)$$

We say that the events p and q are

$$\left.\begin{array}{ll} \textbf{spacelike separated,} & \text{if } \|\mathbf{Y} - \mathbf{X}\|^2 < 0 \\ \textbf{lightlike separated,} & \text{if } \|\mathbf{Y} - \mathbf{X}\|^2 = 0 \\ \textbf{timelike separated,} & \text{if } \|\mathbf{Y} - \mathbf{X}\|^2 > 0 \end{array}\right\} \qquad (7.17)$$

7.3. The Lorentz and Poincare' Groups

We define a **general Lorentz transformation** as any linear homogeneous transformation of the form

$$\bar{x}^\mu = \Lambda^\mu_\nu x^\nu \qquad (7.18)$$

which leaves the inner product of any two four-vectors *invariant*. In order to explore the consequences of this definition, consider two arbitrary four vectors $\mathbf{X}$ and $\mathbf{Y}$. If we require that their inner product (7.10a) is invariant under a general Lorentz transformation, then we must have

$$\langle \bar{\mathbf{X}}, \bar{\mathbf{Y}} \rangle = \bar{\mathbf{X}}^T \eta \bar{\mathbf{Y}} = (\Lambda \mathbf{X})^T \eta (\Lambda \mathbf{Y}) = \mathbf{X}^T \left(\Lambda^T \eta \Lambda\right) \mathbf{Y} = \langle \mathbf{X}, \mathbf{Y} \rangle$$

Since the four-vectors $\mathbf{X}$ and $\mathbf{Y}$ are arbitrary, then this equation implies that the matrix Λ of a general Lorentz transformation must satisfy the relation

$$\Lambda^T \eta \Lambda = \eta \tag{7.19}$$

Now, if we take the determinant of both sides of (7.19), and recall that $\det(\eta) = -1$, we find that

$$\det(\Lambda) = \pm 1 \tag{7.20}$$

General Lorentz transformations with $\det(\Lambda) = +1$ comprise the set of *proper homogeneous Lorentz transformations*, denoted by L_+, while those satisfying the relation $\det(\Lambda) = -1$ make up the set of *improper Lorentz transformations*, L_-. Observe that the 4×4 identity matrix $I \in L_+$ but $I \notin L_-$.

The set of general Lorentz transformations forms a group under the operation of matrix multiplication, called the **homogeneous Lorentz group**, L. In order to see this, assume that $\Lambda, \Lambda' \in L$, and observe that

$$(\Lambda\Lambda')^T \eta (\Lambda\Lambda') = \Lambda'^T \left(\Lambda^T \eta \Lambda\right) \Lambda' = \Lambda'^T \eta \Lambda' = \eta.$$

So, $\Lambda\Lambda' \in L$, and the set is *closed* under matrix multiplication. Also, the 4×4 identity matrix $I \in L$ since

$$I^T \eta I = \eta$$

and each element Λ has an inverse Λ^{-1} since $\det(\Lambda) \neq 0$. Clearly, $\Lambda^{-1} \in L$ owing to the fact that

$$\eta = \Lambda^T \eta \Lambda \Longrightarrow \left(\Lambda^T\right)^{-1} \eta \left(\Lambda^{-1}\right) = \eta \Longrightarrow \left(\Lambda^{-1}\right)^T \eta \left(\Lambda^{-1}\right) = \eta$$

The associativity of group multiplication follows from the associativity of matrix multiplication.

It's easy to see that the set of proper Lorentz transformations also forms a group L_+, which is a *subgroup* of L. Furthermore, we see immediately that ordinary *spatial* rotations Ω of the form

$$\Omega = \begin{bmatrix} 1 & 0 & 0 & 0 \\ 0 & R_{11} & R_{12} & R_{13} \\ 0 & R_{21} & R_{22} & R_{23} \\ 0 & R_{31} & R_{32} & R_{33} \end{bmatrix} \tag{7.21}$$

where $R = (R_{mn})$ is a 3×3 orthogonal matrix, satisfy (7.19), and they in turn form a subgroup of the proper Lorentz group L_+.

On the other hand, *the set of improper Lorentz transformations does not form a group* since $I \notin L_-$. Examples of improper Lorentz transformations include the *discrete* operations of **spatial inversion** Π and **temporal reflection** Θ where

$$\Pi = \begin{bmatrix} 1 & 0 & 0 & 0 \\ 0 & -1 & 0 & 0 \\ 0 & 0 & -1 & 0 \\ 0 & 0 & 0 & -1 \end{bmatrix} \tag{7.22}$$

and

$$\Theta = \begin{bmatrix} -1 & 0 & 0 & 0 \\ 0 & 1 & 0 & 0 \\ 0 & 0 & 1 & 0 \\ 0 & 0 & 0 & 1 \end{bmatrix} \tag{7.23}$$

Now, we can further analyze the homogeneous Lorentz group by first writing (7.19) in component form as

$$\eta_{\mu\nu}\Lambda^{\mu}_{\rho}\Lambda^{\nu}_{\sigma} = \eta_{\rho\sigma} \tag{7.24}$$

and then setting $\rho = \sigma = 0$ to obtain

$$\left(\Lambda^0_0\right)^2 - \left\{\left(\Lambda^1_0\right)^2 + \left(\Lambda^2_0\right)^2 + \left(\Lambda^3_0\right)^2\right\} = 1. \tag{7.24a}$$

The last equation implies that $\left(\Lambda^0_0\right)^2 \geq 1$ so we must have either $\Lambda^0_0 \geq 1$ or $\Lambda^0_0 \leq 1$. This permits us to formally divide general Lorentz transformations into four classes, according as to whether $\det(\Lambda) = \pm 1$ and $\Lambda^0_0 \geq 1$ or $\Lambda^0_0 \leq 1$. The explicit decomposition of L into four sets is

Decomposition of the Lorentz Group

$$\begin{aligned} L^{\uparrow}_{+} &= \left\{\Lambda \in L : \det(\Lambda) = 1 \ \text{ and } \Lambda^0_0 \geq 1\right\} \\ L^{\uparrow}_{-} &= \left\{\Lambda \in L : \det(\Lambda) = -1 \ \text{ and } \Lambda^0_0 \geq 1\right\} \\ L^{\downarrow}_{+} &= \left\{\Lambda \in L : \det(\Lambda) = 1 \ \text{ and } \Lambda^0_0 \leq 1\right\} \\ L^{\downarrow}_{-} &= \left\{\Lambda \in L : \det(\Lambda) = -1 \ \text{ and } \Lambda^0_0 \leq 1\right\} \end{aligned} \tag{7.25}$$

Proper Lorentz transformations which satisfy the condition $\Lambda^0_0 \geq 1$ are said to be **orthochronous**, or **time-orientation preserving**. The group of proper orthochronous Lorentz transformations is denoted by $L^{\uparrow}_{+}$ and is a subgroup of the group of general Lorentz transformations, L. Observe that the remaining three sets in (7.25) division do not contain the identity element I, and hence when taken individually, they cannot form subgroups of L.

Poincare' transformations are *inhomogeneous* Lorentz transformations of the form

$$\bar{x}^{\mu} = \Lambda^{\mu}_{\nu}x^{\nu} + A^{\mu} \tag{7.26}$$

where $\mathbf{A} = (A^\mu)$ is an arbitrary translation, and $\Lambda = \left(\Lambda^\mu_\nu\right) \in L$. Thus, any Poincare' transformation can be written as the product of a translation and a homogeneous Lorentz transformation. It's easy to show that the set of Poincare' transformations forms a group, called the **Poincare' group**, which we shall denote by P.

An arbitrary Poincare' transformation, say $\mathcal{L} \in P$ can be written in the form

$$\mathcal{L} = (\mathbf{A}, \Lambda) \tag{7.27}$$

where $\Lambda \in L$ and $\mathbf{A}$ is a four-vector. In fact, if we consider two successive Poincare' transformations, say $\mathcal{L}_1 = (\Lambda_1, \mathbf{A}_1)$ and $\mathcal{L}_2 = (\Lambda_2, \mathbf{A}_2)$, such that

$$\mathbf{X}' = \mathcal{L}_1\mathbf{X} \text{ and } \mathbf{X}'' = \mathcal{L}_2\mathbf{X}'$$

then

$$\mathbf{X}'' = \Lambda_2\mathbf{X}' + \mathbf{A}_2 = \Lambda_2\left(\Lambda_1\mathbf{X} + \mathbf{A}_1\right) + \mathbf{A}_2 = \left(\Lambda_2\Lambda_1\right)\mathbf{X} + \left(\Lambda_2\mathbf{A}_1 + \mathbf{A}_2\right)$$

Since $\Lambda_2\Lambda_1 \in L$ and $\mathbf{B} = \Lambda_2\mathbf{B}_1 + \mathbf{B}_2$ is again a four-vector, we see that P is *closed* under composition of mappings. Evidently, the rule for the product of two Poincare transformations is given by

$$\left(\Lambda_2, \mathbf{A}_2\right)\left(\Lambda_1, \mathbf{A}_1\right) = \left(\Lambda_2\Lambda_1,\ \Lambda_2\mathbf{A}_1 + \mathbf{A}_2\right) \tag{7.28}$$

Also, it is clear that the 4×4 identity matrix $I \in P$ since we can write $I = (I, \mathbf{0})$. On the other hand, the inverse of $\mathcal{L} = (\Lambda, \mathbf{A}) \in P$ is given by

$$\mathcal{L}^{-1} = (\Lambda, \mathbf{A})^{-1} = \left(\Lambda^{-1}, -\Lambda^{-1}\mathbf{A}\right) \tag{7.29}$$

since by (7.28) and (7.29), we have

$$\begin{aligned}
\mathcal{L}\mathcal{L}^{-1} &= (\Lambda, \mathbf{A})\left(\Lambda^{-1}, -\Lambda^{-1}\mathbf{A}\right) = \left(\Lambda\Lambda^{-1}, \Lambda\left(-\Lambda^{-1}\mathbf{A}\right) + \mathbf{A}\right) = (I, 0) = I \\
\mathcal{L}^{-1}\mathcal{L} &= \left(\Lambda^{-1}, -\Lambda^{-1}\mathbf{A}\right)(\Lambda, \mathbf{A}) = \left(\Lambda^{-1}\Lambda, \Lambda^{-1}\mathbf{A} - \Lambda^{-1}\mathbf{A}\right) = (I, 0) = I
\end{aligned}$$

It's worthwhile to note that the Poincare' group is a *ten-parameter* group, since we must specify six Lorentz parameters and the four components of a translation four-vector. It is a *continuous group* which enables us to relate the spacetime coordinates of any two inertial reference frames, regardless of their relative orientation.

7.4. Four-Velocity

Assume that we introduce an inertial reference frame $S(x^0, x^1, x^2, x^3)$ with Minkowski coordinates (x^a) into M^4. Assume further, that a massive particle travels along a curve C in spacetime, which passes through the point (event) $p \in M^4$. Such a curve is called a **worldline** in spacetime, and we can use any convenient curve parameter to des- cribe it.

For a material particle, however, it's most convenient to take the *proper time* τ as the curve parameter, so relative to the inertial frame S, the equation of this curve assumes the form $x^\mu = x^\mu(\tau)$. We define the **four-velocity** $\mathbf{U} = (U^\mu)$ of the particle to be the tangent vector to the worldline C at any point. In particular, at the point p we have

$$U^\mu = \frac{dx^\mu}{d\tau} \tag{7.30}$$

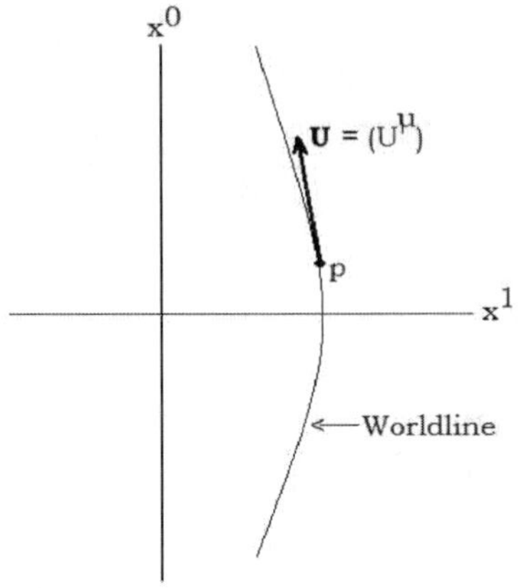

Figure 7.2. The four-velocity **U** of a particle at the point p.

where U^μ represent the components of the four-velocity relative to the inertial frame S.

Let's see how the components of the four-velocity transform under a coordinate transformation. To this end, we introduce another inertial frame $\bar{S}(\bar{x}^0,\bar{x}^1,\bar{x}^2,\bar{x}^3)$ at p relative to which the particle is *instantaneously* at rest. Since the inertial frames S and $\bar{S}$ must be related by a Poincare' transformation, we have by (4.26)

$$\bar{x}^\mu = \Lambda^\mu_\nu x^\nu + A^\mu \,,$$

where $\mathbf{A} = (A^\mu)$ is an arbitrary translation, and $\Lambda = \left(\Lambda^\mu_\nu\right) \in L$.

Taking the differential of both sides of this equation gives

$$d\bar{x}^\mu = \Lambda^\mu_\nu dx^\nu \tag{7.31}$$

Recalling that the proper time interval $d\tau$ is *invariant* under a Lorentz transformation, we have

$$\bar{U}^\mu = \frac{d\bar{x}^\mu}{d\bar{\tau}} = \Lambda^\mu_\nu \frac{dx^\nu}{d\tau} = \Lambda^\mu_\nu U^\nu$$

or,

$$\bar{U}^\mu = \Lambda^\mu_\nu U^\nu \tag{7.32}$$

and we see that *the components of the four-velocity in the two inertial frames must be related by a Lorentz transformation*. Incidently, as we can repeat the above argument for *any* tangent vector in $T_p(M^4)$, we conclude that a **four-vector** $\mathbf{X} = (X^\mu)$ may also be characterized as *any* geometrical object whose components transform according to the rule

$$\bar{X}^\mu = \Lambda^\mu_\nu X^\nu \,. \tag{7.34}$$

Now, recalling the fact that $d\tau = dt/\gamma(u)$, and $x^0 = ct$, we have

$$U^\mu = \frac{dx^\mu}{d\tau} = \frac{dt}{d\tau}\frac{dx^\mu}{dt} = \gamma(u)\frac{dx^\mu}{dt}$$

so that the four-velocity **U** may be written as

$$\mathbf{U} = \gamma(u)(c, \mathbf{u}) = \gamma(u)\left[c, u^1, u^2, u^3\right]^T \tag{7.35}$$

where $\mathbf{u} = (u^1, u^2, u^3)$ is the ordinary 3-velocity of the particle.

In addtion, from (7.35), we see that the Minkowski norm of **U** is given by

$$\|\mathbf{U}\|^2 = (c\gamma(u))^2 - \left(u^1\right)^2 - \left(u^2\right)^2 - \left(u^3\right)^2 = \gamma^2(u)\left(c^2 - u^2\right) = c^2 \tag{7.36}$$

so *the four-velocity of a material particle is a time-like vector with a constant Minkowski norm at each point of its worldline.* It is important to note that once the 3-velocity $\mathbf{u} = (u^1, u^2, u^3)$ of a massive particle is known at any point p, then the temporal component U^0 of its four-velocity is completely determined by (7.36).

7.5. The Energy-Momentum Four-Vector

We now examine how the ordinary Newtonian concept of momentum can be generalized in a manner which is consistent with our four-vector formalism. We must be guided by the philosophy that any dynamical quantity which we define must be covariant with respect to Lorentz transformations.

We first make the simple observation that if we write any physical quantity in terms of known four-vectors and invariants, then this quantity will automatically be covariant with respect to Lorentz transformations. Any such quantity is said to be **manifestly covariant**. That is, as a direct result of its definition, a manifestly covariant object must maintain the same mathematical form under a Lorentz transformation.

To this end, suppose that we introduce an inertial reference frame $S(x^0, x^1, x^2, x^3)$ with Minkowski coordinates (x^μ) into M^4, and consider a material particle whose **rest mass** (or proper mass) is m_0 relative to S. Now, the rest mass is a *scalar invariant* which is an intrinsic property of the given particle.

If this particle now has a four-velocity $\mathbf{U} = (U^\mu)$ relative to the reference frame S, then it is apparent that the set of quantities

given by

$$P^\mu = m_0 U^\mu \tag{7.37}$$

will automatically form the components of a four-vector, called the **energy-momentum four vector** $\mathbf{P} = (P^\mu)$.

Recall from (7.35) that the four-velocity $\mathbf{U} = \gamma(u)(c, \mathbf{u})$ so we can write

$$P^\mu = m_0\gamma(u)(c, \mathbf{u}) = (m_0\gamma(u)c, m_0\gamma(u)\mathbf{u}) \tag{7.38}$$

Since $\gamma(u) \longrightarrow 1$ as $u/c \to 0$ we see that the spatial component of (7.38) approaches the Newtonian momentum $m_0\mathbf{u}$ in the limit. This leads us to define the **relativistic mass** m of the particle by

$$m = m_0\gamma(u) = \frac{m_0}{\sqrt{1 - u^2/c^2}} \tag{7.39}$$

It is important to note that the relativistic mass of a particle is *velocity-dependent*, while its rest mass is not. Note further, that for a massive particle, its relativistic mass $m \longrightarrow \infty$ as $v \longrightarrow c$; thus, *no material particle can achieve the speed of light*, as it would take an infinite force to accelerate it to this speed.

Now, if we combine (7.38) and (7.39), we obtain

$$P^{\mu} = (mc, m\mathbf{u}) = (mc, \mathbf{p}) \tag{7.40}$$

where $\mathbf{p} = m\mathbf{u}$ is called the **relativistic 3-momentum** of the particle. We now turn our attention to the temporal component of (7.40). To interpret its meaning, we first write

$$P^0 c = mc^2 = m_0 c^2 (1 - \frac{u^2}{c^2})^{-1/2}.$$

For $u << c$, the binomial theorem gives

$$P^0 c = m_0 c^2 \left[1 + \frac{u^2}{2c^2} + \ldots \right] \approx m_0 c^2 + \frac{1}{2} m_0 u^2 \tag{7.41}$$

Since the second term in (7.41) is the non-relativistic kinetic energy of the particle, we can interpret $P^0 c$ as the sum of the particle's **rest energy** $m_0 c^2$ and its kinetic energy. Thus, in the relativistic case, our analysis suggests that

$$P^0 c = E = m_0 \gamma(u) c^2 = mc^2 \tag{7.42}$$

and consequently,

$$E = mc^2. \tag{7.43}$$

This is, of course, Einstein's famous equation, and we call E the total **relativistic energy** of the particle.

Using (7.43), the four-momentum may be rewritten as

$$P^a = (E/c, \mathbf{p}) \tag{7.44}$$

and its Minkowski norm becomes

$$\|\mathbf{P}\|^2 = \|(E/c, \mathbf{p})\|^2 = \frac{E^2}{c^2} - \mathbf{p} \cdot \mathbf{p} = \frac{E^2}{c^2} - p^2 \tag{7.45}$$

On the other hand, recalling (7.36), we can write

$$\|\mathbf{P}\|^2 = \|m_0 \mathbf{U}\|^2 = m_0^2 \|\mathbf{U}\|^2 = m_0^2 c^2 \tag{7.46}$$

Since $m_0 > 0$ for a massive particle, we see that its four-momentum $\mathbf{P}$ is a time-like vector. If we now equate the righthand sides of the last two equations, we deduce the important relation

$$E^2 = (m_0 c^2)^2 + (pc)^2 \tag{7.47}$$

Observe that even if the particle is at rest relative to our inertial frame, so that $p = 0$, the particle still possesses a rest energy of $m_0 c^2$. Thus, a massive particle's rest energy is an *intrinsic property* of the particle which is ever present.

7.6. The Minkowski Force

We generalize the notion of Newtonian force by defining the **Minkowski force** $\mathbf{F} = (F^\mu)$ as

$$F^\mu = \frac{d}{d\tau}(P^\mu) = m_0 \frac{dU^\mu}{d\tau} \tag{7.48}$$

Observe that we have defined the Minkowski force in a *manifestly covariant* manner, since F^μ may be viewed as the ratio of the four-vector dP^μ and the scalar invariant $d\tau$. Consequently, $\mathbf{F}$ is a true four-vector whose components in any two inertial reference frames must be related by a Lorentz transformation.

The Minkowski force is just the rate of change of the momentum four-vector per unit of proper time. Recall that $dt = \gamma(u)d\tau$, so the last equation can be rewritten as

$$\mathbf{F} = \frac{dt}{d\tau}\frac{d\mathbf{P}}{dt} = \gamma(u)\frac{d}{dt}\left(\frac{E}{c}, \mathbf{p}\right) = \gamma(u)\left(\frac{1}{c}\frac{dE}{dt}, \frac{d\mathbf{p}}{dt}\right),$$

or in other words,

$$\mathbf{F} = \gamma(u)\left(\frac{1}{c}\frac{dE}{dt}, \mathbf{f}\right) \tag{7.49}$$

where we define the **relativistic 3-force f** by

$$\mathbf{f} = \frac{d\mathbf{p}}{dt} = \frac{d}{dt}(m\mathbf{u}) \tag{7.50}$$

The Minkowski force $\mathbf{F} = (F^\mu)$ has several interesting properties which we now examine. First, it turns out that $\mathbf{F}$ is *orthogonal* to the four-velocity $\mathbf{U} = (U^\mu)$. That is,

$$\langle \mathbf{F}, \mathbf{U} \rangle = F^\mu U_\mu = 0 \tag{7.51}$$

To see this, we first recall from (7.36), that the four-velocity has a *constant* Minkowki norm $\|\mathbf{U}\|^2 = U^\mu U_\mu = c^2$. Consequently,

$$\begin{aligned} \frac{d}{d\tau}(U^\mu U_\mu) &= 2U_\mu \frac{d}{d\tau}(U^\mu) = 0 \\ &\Longrightarrow m_0 U_\mu \frac{d}{d\tau}(U^\mu) = 0 \\ &\Longrightarrow U_\mu F^\mu = 0 \end{aligned}$$

Secondly, the temporal factor dE/dt which appears in (7.49) is directly related to the *power* which is supplied to a material particle in producing its motion. In fact,

$$\frac{dE}{dt} = \mathbf{f} \cdot \mathbf{u} \tag{7.52}$$

This result follows directly from (7.51), since

$$\langle \mathbf{F}, \mathbf{U} \rangle = \left\langle \gamma(u)\left(\frac{1}{c}\frac{dE}{dt}, \mathbf{f}\right), \gamma(u)(c, \mathbf{u}) \right\rangle = \gamma(u)\left\{\frac{dE}{dt} - \mathbf{f}\cdot\mathbf{u}\right\} = 0,$$

from which (7.52) follows immediately. Substituting (7.52) into (7.49) gives

$$\mathbf{F} = \gamma(u)\left(\frac{\mathbf{f}\cdot\mathbf{u}}{c}, \mathbf{f}\right) \tag{7.53}$$

which is a useful, alternative expression for the Minkowski force.

We now explore the consequences of (7.52). Assume that a force acts upon a massive particle and accelerates it from rest relative to some inertial reference frame. The energy dE supplied to the particle during the time interval dt is given by

$$dE = (\mathbf{f}\cdot\mathbf{u})dt$$

Integrating both sides of this equation gives

$$\Delta E = \int_{t_1}^{t_2} (\mathbf{f}\cdot\mathbf{u})dt = \int_{t_1}^{t_2} \frac{d}{dt}(m\mathbf{u})\cdot\mathbf{u}dt = \int_{t_1}^{t_2} d(m\mathbf{u})\cdot\mathbf{u} = \int_{t_1}^{t_2} \left(m\mathbf{u}\cdot d\mathbf{u}+u^2dm\right) \tag{7.54}$$

In order to evaluate this integral, we first take the differential of both sides of (7.39), and simplify to obtain

$$m(\mathbf{u}\cdot d\mathbf{u}) = (c^2 - u^2)dm \tag{7.55}$$

Next, we substitute (7.55) into (7.54) to obtain

$$\Delta E = \int_{m_0}^{m} c^2 dm = mc^2 - m_0c^2$$

or the additional energy acquired by the particle is simply

$$\Delta E = mc^2 - m_0c^2 = m_0c^2\left[\frac{1}{\sqrt{1-u^2/c^2}} - 1\right] \tag{7.56}$$

But what is the nature of this energy gain ΔE ? To answer this question, we simply observe that as the particle's velocity $u \longrightarrow 0$, we have $\Delta E \longrightarrow 0$ as well. *Thus, we must identify ΔE with the kinetic energy T of the particle*; that is,

$$T = E - m_0c^2 = m_0c^2\left[\frac{1}{\sqrt{1-u^2/c^2}} - 1\right] \tag{7.57}$$

so, the kinetic energy T acquired by the particle is just the difference between its total relativistic energy E and its rest energy m_0c^2. Note that this derivation rigorously supports the intuitive argument given earlier, following the interpretation of (7.41).

7.7. Relativistic Collisions

We know that for an individual particle, the relativistic version of Newton's second law is given by (7.48) as

$$F^{\mu} = \frac{d}{d\tau}(P^{\mu})$$

In the absence of external forces, this equation reduces to

$$\frac{d}{d\tau}(P^{\mu}) = 0 \Longrightarrow P^{\mu} = C^{\mu} \tag{7.58}$$

where C^{μ} is a constant 4-vector. This equation expresses the **conservation of 4-momentum**. As $\mathbf{P} = (P^{\mu})$ is a 4-vector, we are guaranteed that both the particle's total energy and momentum are separately conserved.

Since no assumptions were made about the *internal composition* of such a particle, we would expect that a similar conservation law must hold for an isolated *system* of particles. The truth of this simple observation has been established by numerous laboratory experiments.

Thus, given a system of n particles, we can define the total momentum by

$$\mathbf{P}_{total} = \sum_{k=1}^{n} \mathbf{P}_k = (\frac{E_{total}}{c}, \mathbf{p}_{total}) \tag{7.59}$$

where E_{total} is the total energy of the system, and $\mathbf{p}_{total}$ is the total 3-momentum of the system. In the absence of external forces, we have $\Delta\mathbf{P}_{total} = \mathbf{0}$, and consequently,

$$\Delta E_{total} = 0 \text{ and } \Delta\mathbf{p}_{total} = \mathbf{0}, \tag{7.60}$$

so the total energy and total momentum of the system are each separately conserved. It is precisely equation (7.60) that allows us to analyze relativistic collisions.

As an example of such a collision, we shall examine Compton scattering. Here, a high-energy photon with frequency f and 4-momentum $\mathbf{P}$ collides with an electron of 4-momentum $\mathbf{P}_e$.

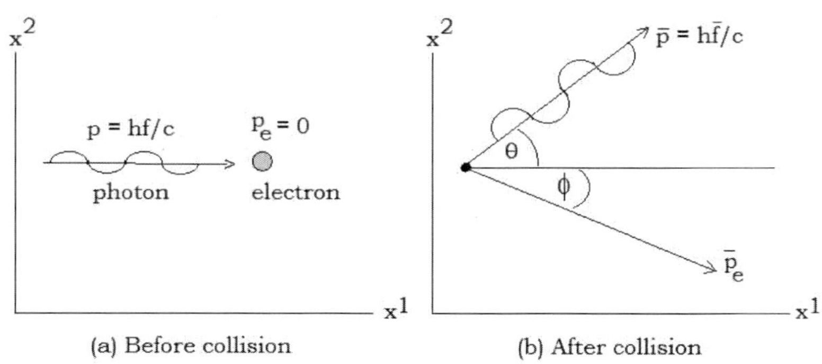

Figure 7.3. The Compton Effect.

In order to simplify the analysis, we choose an inertial frame in which the electron is initially at rest, and the incident photon moves parallel to the x^1-axis. This situation is shown in part (a) of Figure 7.3. Prior to the collision, the 4-momenta of the photon and the electron are

$$\mathbf{P} = \left(\frac{hf}{c}, \frac{hf}{c}, 0, 0\right) \text{ and } \mathbf{P}_e = (m_0 c, 0, 0, 0),$$

where f is the initial frequency of the photon, and m_0 is the mass of the electron relative to its rest frame.

After the collision, let $\bar{\mathbf{P}}$ and $\bar{\mathbf{P}}_e$ denote the 4-momenta of the photon and electron, and we assume that their 3-momenta make angles θ and ϕ with respect to the x^1-axis. We also assume that their motion continues to be confined to the $x^3 = 0$ plane. Consequently, after the collision, we have

$$\begin{aligned} \bar{\mathbf{P}} &= \left(\frac{h\bar{f}}{c}, \frac{h\bar{f}}{c}\cos\theta, \frac{h\bar{f}}{c}\sin\theta, 0\right) \\ \bar{\mathbf{P}}_e &= \left(\frac{\bar{E}_e}{c}, \bar{\mathbf{p}}_e\right) = \left(\frac{\bar{E}_e}{c}, \bar{p}_e\cos\phi, \bar{p}_e\sin\phi, 0\right) \end{aligned}$$

By conservation of 3-momentum, we can write

$$\begin{aligned} \frac{hf}{c} + 0 &= \frac{h\bar{f}}{c}\cos\theta + \bar{p}_e\cos\phi \\ 0 &= \frac{h\bar{f}}{c}\sin\theta - \bar{p}_e\sin\phi \end{aligned}$$

or equivalently,

$$\begin{aligned} \bar{p}_e c\cos\phi &= hf - h\bar{f}\cos\theta \\ \bar{p}_e c\sin\phi &= h\bar{f}\sin\theta \end{aligned}$$

If we now square and add the last two equations, we obtain

$$(\bar{p}_e c)^2 = (hf)^2 + (h\bar{f})^2 - 2(hf)(h\bar{f})\cos\theta \tag{7.61}$$

We now turn our attention to energy conservation. After the collision, the total energy of the electron is given by

$$E_e^2 = (T_e + m_0c^2)^2 = T_e^{\,2} + 2m_0c^2T_e + m_0^2c^4 \tag{7.62}$$

where T_e is the kinetic energy of the electron. On the other hand, by (7.47), we can write

$$E_e^2 = (p_e c)^2 + m_0^2c^4 \tag{7.63}$$

Equating (7.62) and (7.63), we obtain

$$(p_e c)^2 = T_e^{\,2} + 2m_0c^2T_e \tag{7.64}$$

If we now enforce energy conservation, then the kinetic energy of the recoiling electron is

$$T_e = hf - h\bar{f} \tag{7.65}$$

Substituting (7.65) into (7.64), we find

$$(p_e c)^2 = h^2\left(f - \bar{f}\right)^2 + 2m_0c^2h\left(f - \bar{f}\right) \tag{7.66}$$

Finally, equating (7.66) and (7.61), and simplifying, we obtain

$$hc^2\left(\frac{1}{\bar{f}}-\frac{1}{f}\right)=\frac{h^2}{m_0}(1-\cos\theta)$$

or, equivalently,

$$\bar{\lambda}-\lambda=\frac{h}{m_0c}(1-\cos\theta) \tag{7.67}$$

which is the equation governing Compton scattering. The quantity

$$\lambda_c=\frac{h}{m_0c} \tag{7.68}$$

is called the Compton wavelength, and for an electron, we find that

$$\lambda_c\approx 2.43\times 10^{-12}\text{ meter}$$

A cursory examination of (7.67) reveals that the recoil kinetic energy of the electron achieves a maximum value $T_e=2\lambda_c$ when $\theta=\pi$; this occurs when the incident photon is back-scattered.

7.8. Exercises

1. Show that the relativistic 3-force **f** may be written in the form

$$\mathbf{f}=m_0\gamma\frac{d\mathbf{u}}{dt}+\frac{m_0\gamma^3}{c^2}\left(\mathbf{u}\cdot\frac{d\mathbf{u}}{dt}\right)\mathbf{u} \tag{7.69}$$

where $\gamma=\gamma(u)$, and **u** is the 3-velocity.

(Solution)

We have

$$\begin{aligned}\mathbf{f} &= \frac{d}{dt}(m\mathbf{u})=m_0\left(\gamma\frac{d\mathbf{u}}{dt}+\mathbf{u}\frac{d\gamma}{dt}\right)=m_0\gamma\frac{d\mathbf{u}}{dt}+\mathbf{u}\frac{d}{dt}\left(1-\frac{u^2}{c^2}\right)^{-1/2}\\ &= m_0\gamma\frac{d\mathbf{u}}{dt}+\frac{m_0\gamma^3}{c^2}\left(\mathbf{u}\cdot\frac{d\mathbf{u}}{dt}\right)\mathbf{u}\end{aligned}$$

as claimed.

2. Show that if the 3-acceleration **a** of a body is parallel to its 3-velocity, then

$$\mathbf{f}=m_0\gamma^3\mathbf{a} \tag{7.70}$$

(Solution)

We can resolve the 3-acceleration into components parallel to, and perpendicular to the 3-velocity, as

$$\mathbf{a}=\mathbf{a}_{\parallel}+\mathbf{a}_{\perp}$$

Substituting this expression into (7.69), we obtain

$$\mathbf{f} = m_0\gamma\left(\mathbf{a}_{\parallel} + \mathbf{a}_{\perp}\right) + \frac{m_0\gamma^3}{c^2}\left[\mathbf{u}\cdot\left(\mathbf{a}_{\parallel} + \mathbf{a}_{\perp}\right)\right]\mathbf{u}$$

By hypothesis, $\mathbf{a}_{\perp} = \mathbf{0}$, so the last equation simplifies to

$$\mathbf{f} = m_0\gamma\mathbf{a}_{\parallel} + \frac{m_0\gamma^3}{c^2}u^2\mathbf{a}_{\parallel} = m_0\gamma^3\left(\frac{1}{\gamma^2} + \frac{u^2}{c^2}\right)\mathbf{a}_{\parallel} = m_0\gamma^3\mathbf{a}_{\parallel}$$

as required.

3. In the Compton effect, show that the maximum kinetic energy of the recoiling electron is given by

$$T_{\text{max}} = \frac{\varepsilon}{1+\varepsilon} \quad \text{where } \varepsilon = \frac{2hf}{m_0c^2} \tag{7.71}$$

(Solution)

From (7.67), we have

$$\bar{\lambda} - \lambda = \frac{h}{m_0c}(1 - \cos\theta)\,,$$

from which it follows that

$$\Delta f = f - \bar{f} = \frac{h}{m_0c^2}f\,\bar{f}(1 - \cos\theta)$$

On the other hand, we see that the maximum value of Δf occurs when $\theta = \pi$, so that

$$T_{\text{max}} = h\Delta f_{\text{max}} = \frac{2hf}{m_0c^2}\left(h\bar{f}\right)$$

By conservation of energy, however, $h\bar{f} = hf - T_{\text{max}}$, so the last equation becomes

$$T_{\text{max}} = \frac{2hf}{m_0c^2}\left(hf - T_{\text{max}}\right),$$

which is easily solved for T_{max} to give

$$T_{\text{max}} = \frac{\frac{2hf}{m_0c^2}}{1 + \frac{2hf}{m_0c^2}} = \frac{\varepsilon}{1+\varepsilon} \quad \text{where } \varepsilon = \frac{2hf}{m_0c^2}$$

4. Show that the kinetic energy of a massive particle may be written as

$$T = \frac{p^2}{m_0(\gamma+1)} \tag{7.72}$$

where $\mathbf{p}$ is its 3-momentum.

(Solution)

We have

$$E^2 = (pc)^2 + (m_0c^2)^2 = (pc)^2 + E_0^2$$

where E_0 is the particle's rest energy. Rearranging terms, we get

$$E^2 - E_0^2 = (E + E_0)(E - E_0) = p^2c^2$$

But, recalling that $T = E - E_0$, the last equation gives

$$T = \frac{p^2c^2}{E + E_0} = \frac{p^2c^2}{(m_0\gamma + m_0)\,c^2} = \frac{p^2}{m_0(\gamma + 1)}$$

as required. Observe that since $\gamma \longrightarrow 1$ as $u \longrightarrow 0$, we have

$$\lim_{u \longrightarrow 0} T = \frac{p^2}{2m_0}$$

which is the familiar Newtonian expression for T.

5. *(Zero Momentum Frame)* Given a system of massive particles, show that we can always find a reference frame $\bar{S}$, called the **zero-momentum reference frame**, relative to which the total 3-momentum of the system is zero.

(Solution)

Given a system of n particles, the total 3-momentum is given by

$$\mathbf{p}_{total} = \sum_{k=1}^{n} \mathbf{p}_k$$

where $\mathbf{p}_k$ is the 3-momentum of the kth particle, relative to the laboratory frame S. We choose a reference frame $\bar{S}$ such that the $\bar{x}$-axis of $\bar{S}$ has the same direction as $\mathbf{p}_{total}$. We suppose, further, that the frame $\bar{S}$ moves with a velocity $\mathbf{V}$ relative to our laboratory frame S, such that relative to $\bar{S}$, the total 3-momentum is zero. We now seek to determine $\mathbf{V}$.

In the zero-momentum frame $\bar{S}$, the total 4-momentum must be

$$\bar{\mathbf{P}}_{total} = (\frac{\bar{E}_{total}}{c}, 0, 0, 0) \tag{7.73a}$$

Consequently, relative to the zero-momentum frame, the total 3-momentum is simply

$$\bar{\mathbf{p}}_{total} = \sum_{k=1}^{n} \bar{\mathbf{p}}_k = \mathbf{0}$$

But, from the Lorentz transformation (with the respective axes in the standard position) we see that the 3-momenta in each frame are related by

$$\bar{\mathbf{p}}_k = -\gamma \frac{\mathbf{V}}{c^2} E_k + \gamma \mathbf{p}_k$$

From which it follows that

$$\sum_{k=1}^{n} \bar{\mathbf{p}}_k = -\gamma\frac{\mathbf{V}}{c^2}\sum_{k=1}^{n} E_k + \gamma\sum_{k=1}^{n}\mathbf{p}_k = -\gamma\frac{\mathbf{V}}{c^2}E_{total} + \gamma\mathbf{p}_{total} = \mathbf{0}$$

Solving the last equation for $\mathbf{V}$, we obtain

$$\mathbf{V} = \left(\frac{c^2}{E_{total}}\right)\mathbf{p}_{total} \tag{7.73}$$

Thus, if the frame $\bar{S}$ moves with velocity $\mathbf{V}$ relative to the laboratory frame S, then we are guaranteed that $\bar{\mathbf{p}}_{total} = 0$.

6. For a system of massive particles, show that

$$E_{total} = \gamma(V)\bar{E}_{total} \tag{7.74}$$

$$E^2_{total} = (p_{total}c)^2 + \bar{E}^2_{total} \tag{7.75}$$

where $\bar{E}_{total}$ and E_{total} are the total energies of the system, relative to the zero-momentum frame $\bar{S}$, and the laboratory frame S, respectively, and V is given by (7.73).

(Solution)

From (7.73a), we find

$$\left\|\bar{\mathbf{P}}_{total}\right\|^2 = \frac{1}{c^2}\bar{E}^2_{total}$$

On the other hand, in the laboratory frame, we have

$$\left\|\mathbf{P}_{total}\right\|^2 = \frac{1}{c^2}E^2_{total} - p^2_{total}$$

Due to the invariance of the Minkowski norm, we have

$$\frac{1}{c^2}\bar{E}^2_{total} = \frac{1}{c^2}E^2_{total} - p^2_{total}$$

from which (7.75) follows immediately. Equation (7.74) follows trivially from the inverse Lorentz transformation.

7. *(Elastic Collision)* Particle A collides with an *identical* particle B, which is *initially at rest*, relative to the laboratory frame. As shown in Figure 7.4, after impact, particle #1 moves off at an angle θ_A while the target particle recoils at an angle θ_B. Show that

$$\tan\theta_A \tan\theta_B = 1 - \frac{V^2}{c^2} \tag{7.76}$$

where $\mathbf{V}$ is the velocity of the zero-momentum frame relative to the laboratory frame.

(Solution)

After impact, relative to the zero-momentum frame, the total momentum is zero, and consequently, the 3-momenta of each particle must be equal and opposite. Since the particles are identical, then they must have the same rest mass m_0, and their energies after impact must also be equal.

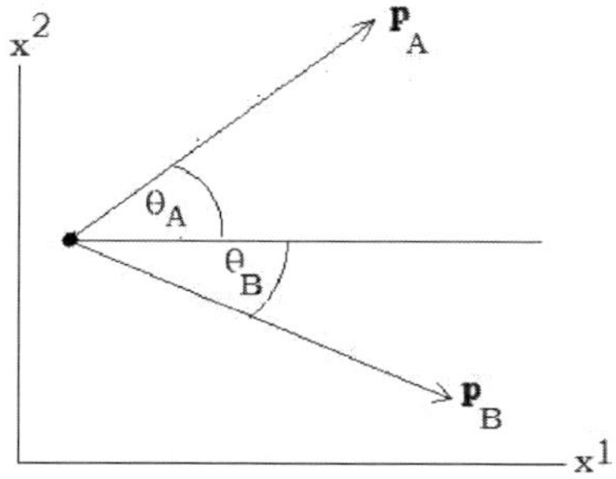

Figure 7.4. Collision as seen in the lab frame S.

Relative to the zero-momentum frame $\bar{S}$, let the energy of each particle (after impact) be $\bar{E}$ and let

$$\begin{aligned} \bar{\mathbf{P}}_A &= (\bar{E}/c, p_A^1, p_A^2, 0) \\ \bar{\mathbf{P}}_B &= (\bar{E}/c, p_B^1, p_B^2, 0) = (\bar{E}/c, -p_A^1, -p_A^2, 0) \end{aligned}$$

Using the Lorentz transformation, we can transform the 4-vector $\bar{\mathbf{P}}_A$ back to the lab frame S as:

$$\begin{bmatrix} E_A/c \\ p_A^1 \\ p_A^2 \\ 0 \end{bmatrix} = \begin{bmatrix} \gamma & \gamma V/c & 0 & 0 \\ \gamma V/c & \gamma & 0 & 0 \\ 0 & 0 & 1 & 0 \\ 0 & 0 & 0 & 1 \end{bmatrix} \begin{bmatrix} \bar{E}/c \\ \bar{p}_A^1 \\ \bar{p}_A^2 \\ 0 \end{bmatrix}$$

Consequently,

$$p_A^2 = \gamma\beta\frac{\bar{E}}{c} + \gamma\bar{p}_A^1 = \gamma\left(\beta\frac{\bar{E}}{c} + \bar{p}_A^1\right) \quad \text{and} \quad p_A^2 = \bar{p}_A^2$$

so that

$$\tan\theta_A = \frac{p_A^2}{p_A^1} = \frac{p_A^2}{\gamma\left(\beta\frac{\bar{E}}{c} + \bar{p}_A^1\right)}$$

Similarly, by transforming the 4-vector $\bar{\mathbf{P}}_B$ back to the lab frame S, we find

$$\tan\theta_B = \frac{p_B^2}{p_B^1} = \frac{\bar{p}_A^2}{\gamma\left(\beta\frac{\bar{E}}{c} - \bar{p}_A^1\right)}$$

From the last two equations, we obtain

$$\tan\theta_A \tan\theta_B = \frac{\left(\bar{p}_A^2\right)^2}{\gamma^2\left[\beta^2\frac{\bar{E}^2}{c^2} - \left(\bar{p}_A^1\right)^2\right]} \tag{7.77}$$

But observe that $\bar{p}_A c = (\gamma m_0)\beta c^2 = \beta \bar{E}$, so we can write

$$\frac{\beta^2 \bar{E}^2}{c^2} = \|\bar{\mathbf{p}}_A\|^2 = (p_A^1)^2 + (p_A^2)^2 \tag{7.78}$$

Substituting (7.78) into (7.77), we obtain

$$\tan\theta_A \tan\theta_B = \frac{1}{\gamma^2} = 1 - \frac{V^2}{c^2}$$

as required.

8. *(Photon Rocket)* A photon rocket emits high energy photons to generate thrust. Relative to the laboratory frame, if the rocket initially has rest mass M_0, and rest mass M after achieving the velocity V, then show that

$$\frac{M_0}{M} = \sqrt{\frac{1+\beta}{1-\beta}} \text{ where } \beta = \frac{V}{c} \tag{7.79}$$

(Solution)

Let E be the total energy radiated, so by energy conservation, we must have

$$M_0 c^2 = M\gamma c^2 + E$$

Also, by conservation of 3-momentum, we can write

$$M\gamma V = \frac{E}{c}$$

Substituting the second equation into the first, and simplifying, we get

$$M_0 c^2 = M\gamma c^2 + M\gamma V c$$

from which, we obtain

$$\frac{M_0}{M} = \gamma\left(1 + \frac{V}{c}\right) = \frac{1+\beta}{\sqrt{1-\beta^2}} = \sqrt{\frac{1+\beta}{1-\beta}}$$

as claimed. Note that if we invert this equation, then we get

$$\frac{M}{M_0} = \sqrt{\frac{1-\beta}{1+\beta}}$$

So if we want to achieve a modest final velocity of say, $V = .8c$, then

$$\frac{M}{M_0} = \sqrt{\frac{1-(.8)}{1+(.8)}} = \frac{1}{3}$$

We see that the photon rocket must lose 67% of its rest mass. This is not very practical, since we must slow it down prior to arriv- ing at our intended destination!

9. *(Relativistic Rocket)* Relative to the lab frame, a rocket which is initially at rest, expels high speed exhaust gases at a speed w. Show that

$$M = M_0 \left(\frac{c+u}{c-u} \right)^{-c/2\bar{w}} \tag{7.80}$$

where M and M_0 are the rest mass and initial rest mass of the rocket, respectively, relative to the lab frame, u is the speed of the rocket relative to the lab frame, and $\bar{w}$ is the speed of its exhaust gases, relative to the rocket frame.

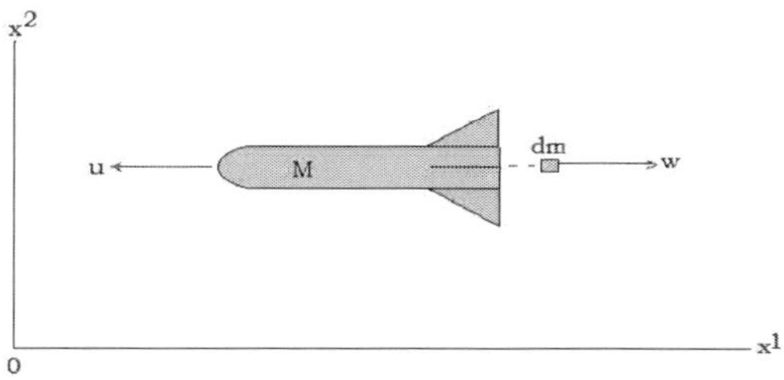

Figure 7.5. Motion of rocket relative to the lab frame.

(Solution)

Assume that the rocket is initially at rest relative to the lab frame, with rest mass M. A short time dt later, the rocket expels a small amount of mass dm at a speed w relative to the lab frame, and the rocket accelerates to the speed u. By conservation of 3-momentum,

$$\Delta \mathbf{p}_{total} = \Delta \mathbf{p}_{rocket} + \Delta \mathbf{p}_{exhaust} = \mathbf{0}$$

so that

$$-d\left[M\gamma(u)u\right] + (dm)\gamma(w)w = 0 \tag{7.81a}$$

Similarly, by conservation of energy, we have

$$\Delta E_{total} = d\left[M\gamma(u)c^2\right] + (dm)\gamma(w)c^2 = 0 \tag{7.81b}$$

Now, let $\bar{w}$ be the exhaust velocity of dm relative to the rocket frame, so by the Lorentz velocity transformation, we have

$$w = \frac{\bar{w} - u}{1 - \bar{w}u/c^2} \tag{7.82}$$

Consequently, by (7.81a) and (7.82),

$$\begin{aligned} d\left[M\gamma(u)u\right] &= \frac{\bar{w}-u}{1-\bar{w}u/c^2}(dm)\gamma(w) \\ \implies d\left[M\gamma(u)u\right] &= \frac{u-\bar{w}}{1-\bar{w}u/c^2} d\left[M\gamma(u)\right] \quad [\text{ by (7.81b) }] \end{aligned} \tag{7.83}$$

On the other hand, it is useful to note that

$$d\left[\gamma(u)\right] = \frac{\gamma(u)}{c^2 - u^2} u du \tag{7.83a}$$

Expanding both sides of (7.83) gives

$$\begin{aligned} & M\gamma(u)du + Mud\left[\gamma(u)\right] + \gamma(u)udM \\ = & \left(\frac{u - \bar{w}}{1 - \bar{w}u/c^2}\right) Md\left[\gamma(u)\right] + \left(\frac{u - \bar{w}}{1 - \bar{w}u/c^2}\right)\gamma(u)dM \end{aligned}$$

Using (7.83a), the last equation becomes

$$\begin{aligned} & M\gamma(u)du + Mu^2\frac{\gamma(u)du}{c^2 - u^2} + \gamma(u)udM \\ = & M\left(\frac{u - \bar{w}}{1 - \bar{w}u/c^2}\right)\frac{\gamma(u)udu}{c^2 - u^2} + \left(\frac{u - \bar{w}}{1 - \bar{w}u/c^2}\right)\gamma(u)dM \end{aligned}$$

We now eliminate $\gamma(u)$ from the last equation, and rearrange terms to obtain

$$\begin{aligned} & du + \frac{u^2 du}{c^2 - u^2} + u\frac{dM}{M} \\ = & \left(\frac{u - \bar{w}}{1 - \bar{w}u/c^2}\right)\frac{udu}{c^2 - u^2} + \left(\frac{u - \bar{w}}{1 - \bar{w}u/c^2}\right)\frac{dM}{M} \end{aligned}$$

or,

$$\left\{1 + \frac{u}{c^2 - u^2}\left(u - \frac{u - \bar{w}}{1 - \bar{w}u/c^2}\right)\right\} du = -\left(u - \frac{u - \bar{w}}{1 - \bar{w}u/c^2}\right)\frac{dM}{M}$$

After some algebra, the last equation simplifies to

$$\frac{du}{\bar{w}\left(1 - \frac{u^2}{c^2}\right)} = -\frac{dM}{M}$$

Assuming that the rest mass at $t = 0$ is M_0, integration gives

$$\ln\frac{M}{M_0} = -\frac{1}{\bar{w}}\int_0^u \frac{du}{\left(1 - \frac{u^2}{c^2}\right)} = -\frac{c}{2\bar{w}}\ln\left(\frac{c + u}{c - u}\right)$$

which, in turn, implies that

$$M = M_0\left(\frac{c + u}{c - u}\right)^{-c/2\bar{w}}$$

Observe that for $\bar{w} = c$, (7.80) reduces to

$$\frac{M_0}{M} = \sqrt{\frac{c + u}{c - u}} = \sqrt{\frac{1 + \beta}{1 - \beta}} \quad \text{where } \beta = u/c$$

This is just the equation we previously obtained for the photon rocket.

Chapter 8

Maxwell's Equations in Minkowski Space

A great deal of my work is just playing with equations, and seeing what they give.

Paul A.M. Dirac

8.1. Potential Formulation of Maxwell's Equations

In this chapter, we again turn our attention to the basic elements of electrodynamics and see how Maxwell's equations are woven into the into the spacetime fabric of Minkowski space. Our task is considerably simplified by the fact that Maxwell's Equations are covariant with respect to Lorentz transformations. This means that our development will be largely formal nature, expressing old concepts in terms of our new four-vector notation. Before doing so, however, it will be conven- ient to reformulate the basic elements of this theory in terms of scalar and vector potentials.

Recall that Maxwell's equations are

$$\nabla \cdot \mathbf{E} = 4\pi\rho \tag{8.1}$$

$$\nabla \cdot \mathbf{B} = 0 \tag{8.2}$$

$$\nabla \times \mathbf{E} = -\frac{1}{c}\frac{\partial \mathbf{B}}{\partial t} \tag{8.3}$$

$$\nabla \times \mathbf{B} = \frac{4\pi}{c}\mathbf{j} + \frac{1}{c}\frac{\partial \mathbf{E}}{\partial t} \tag{8.4}$$

where, **E** is the electric field, **B** is the magnetic field, ρ is the charge density, and **j** is the current density.

Now, equation (8.2) implies that there exists a vector field $\mathbf{A}(\mathbf{x},t)$, called the **vector potential**, such that

$$\mathbf{B} = \nabla \times \mathbf{A} \tag{8.5}$$

Recall that this equation is also a direct consequence of Biot-Savart law, as previously shown in Section 3.6. If we now substitute (8.5) into (8.3), we obtain

$$\nabla \times \mathbf{E} + \frac{1}{c}\frac{\partial}{\partial t}(\nabla \times \mathbf{A}) = \mathbf{0}$$

Exploiting the fact that the operators ∇ and $\partial/\partial t$ commute, the last equation can be rewritten as

$$\nabla \times \left(\mathbf{E} + \frac{1}{c}\frac{\partial \mathbf{A}}{\partial t}\right) = \mathbf{0} \tag{8.6}$$

Now, since the expression in the parentheses is curl-free, there must exist a scalar field $\Phi(\mathbf{x},t)$, called the **scalar potential**, such that

$$\mathbf{E} + \frac{1}{c}\frac{\partial \mathbf{A}}{\partial t} = -\nabla\Phi \tag{8.7}$$

where the minus sign in front of $\nabla\Phi$ is taken so that (8.7) reduces to (3.13b) in the special case of electrostatics. Rearranging terms in the last equation gives

$$\mathbf{E} = -\nabla\Phi - \frac{1}{c}\frac{\partial \mathbf{A}}{\partial t} \tag{8.8}$$

We see that the introduction of the vector and potentials guaran- tees that two of Maxwell's equations, namely, (8.2) and (8.3) are *identically satisfied.* Clearly, this is a major advantage of this new formulation.

The vector and scalar potentials are arbitrary to a certain degree. To see this, suppose that at the same time, the vector and scalar potentials undergo a transformation of the form

$$\mathbf{A} \longrightarrow \mathbf{A}' = \mathbf{A} + \nabla\Lambda \tag{8.9a}$$

$$\Phi \longrightarrow \Phi' = \Phi - \frac{1}{c}\frac{\partial \Lambda}{\partial t} \tag{8.9b}$$

where Λ is a smooth but otherwise *arbitrary* scalar field. A transformation of this type is called a **gauge transformation**. Now, if we insert (8.9a) into (8.5), and (8.9b) into (8.8), respectively, we find that

$$\mathbf{B}' = \nabla \times \mathbf{A}' = \nabla \times (\mathbf{A} + \nabla\Lambda) = \nabla \times \mathbf{A} + \mathbf{0} = \mathbf{B} \ ,$$

and

$$\begin{aligned} \mathbf{E}' &= -\nabla\Phi' - \frac{1}{c}\frac{\partial \mathbf{A}'}{\partial t} = -\nabla\left(\Phi - \frac{1}{c}\frac{\partial \Lambda}{\partial t}\right) - \frac{1}{c}\frac{\partial}{\partial t}(\mathbf{A} + \nabla\Lambda) \\ &= -\nabla\Phi - \frac{1}{c}\frac{\partial \mathbf{A}}{\partial t} = \mathbf{E}\,, \end{aligned}$$

so the gauge transformation has no effect on the pertinent fields.

It has no effect on the physical fields; the electric and magnetic fields are *invariant* under a gauge transformation. This state of affairs is called **gauge symmetry**. Since Λ is an arbitrary scalar field, we see that many other choices of gauge are possible. We can exploit this freedom of choice or *gauge freedom* to simplify the remaining two Maxwell's equations.

For example, since (8.5) only specifies the curl of the vector field $\mathbf{A}(\mathbf{x})$, we are still free to specify its divergence. So, we can set

$$\nabla \cdot \mathbf{A} = -\frac{1}{c}\frac{\partial \Phi}{\partial t} \tag{8.10}$$

This choice is called the **Lorentz gauge**, and (8.10) is often called the **Lorentz condition**. It's easy to show that under the Lorentz gauge, field equations (8.1) and (8.4) become:

$$\Box^2 \Phi = \frac{1}{c^2}\frac{\partial^2 \Phi}{\partial t^2} - \nabla^2 \Phi = 4\pi\rho \tag{8.11a}$$

$$\Box^2 \mathbf{A} = \frac{1}{c^2}\frac{\partial^2 \mathbf{A}}{\partial t^2} - \nabla^2 \mathbf{A} = \frac{4\pi}{c}\mathbf{j} \tag{8.11b}$$

where we have introduced the **D'Alembertian operator** $\Box^2$, defined by

$$\Box^2 \equiv \frac{1}{c^2}\frac{\partial^2}{\partial t^2} - \nabla^2 \tag{8.12}$$

As soon as (8.11) is solved for the potentials $\mathbf{A}$ and Φ, we see that all of Maxwell's equations are satisfied, and the physical fields $\mathbf{B}(\mathbf{x},t)$ and $\mathbf{E}(\mathbf{x},t)$ are then uniquely determined. We note in passing that in the case of steady fields, equations (8.11) reduce to the corresponding equations for electrostatics, and magnetostatics, respectively.

It may be shown that equations (8.11) have the classical solutions

$$\Phi(\mathbf{x},t) = \int_V \frac{\rho(\mathbf{x}',t-R/c)}{R} d^3\mathbf{x}' \tag{8.13a}$$

$$\mathbf{A}(\mathbf{x},t) = \frac{1}{c}\int_V \frac{\mathbf{j}(\mathbf{x}',t-R/c)}{R} d^3\mathbf{x}' \tag{8.13b}$$

where

$$R = \sqrt{(x-x')^2+(y-y')^2+(z-z')^2}$$

Observe that these equations are almost identical to the corresponding equations for steady fields, apart from the appearance of the argument $t - R/c$ which appears in both integrands. This expression is known as the **retarded time**

$$t_r = t - \frac{R}{c} \ . \tag{8.14}$$

For the case of unsteady fields, however, the current density and the charge density must be evaluated at the time $t - R/c$ since the effects of the potentials at the point (x',y',z') of the *source* don't arrive at the *field point* (x,y,z) until after the finite time R/c. For this reason, the potentials in (8.13) are often called **retarded potentials**.

8.2. Covariant Form of Maxwell's Equations

We now discuss the Lorentz covariant form of Maxwell's equations. First, we define the **four-potential** A^μ by the expression

$$(A^\mu) = (\Phi, \mathbf{A}) \tag{8.15}$$

where Φ and A are the scalar and the vector potentials, respectively. Recall that the magnetic and electric fields may be expressed in terms of these potentials as

$$\mathbf{B} = \nabla \times \mathbf{A} \text{ and } \mathbf{E} = -\nabla\Phi - \frac{1}{c}\frac{\partial \mathbf{A}}{\partial t}$$

Next, we define the **gradient operator** ∂_ρ by

$$(\partial_\rho) = \left(\frac{\partial}{\partial x^\rho}\right) = \left(\frac{\partial}{\partial x^0}, \frac{\partial}{\partial x^1}, \frac{\partial}{\partial x^2}, \frac{\partial}{\partial x^3}\right) \tag{8.16}$$

and its contravariant counterpart ∂^μ by

$$\partial^\mu = \eta^{\mu\rho}\partial_\rho = \left(\frac{\partial}{\partial x^0}, -\frac{\partial}{\partial x^1}, -\frac{\partial}{\partial x^2}, -\frac{\partial}{\partial x^3}\right) \tag{8.17}$$

Now, we define the **electromagnetic field tensor** $F^{\mu\nu}$ by

$$F^{\mu\nu} = \partial^\mu A^\nu - \partial^\nu A^\mu \tag{8.18}$$

It's easy to see that this second rank tensor is *antisymmetric*, since

$$F^{\nu\mu} = \partial^\nu A^\mu - \partial^\mu A^\nu = -F^{\mu\nu} \tag{8.19}$$

and consequently,

$$F^{00} = F^{11} = F^{22} = F^{33} = 0$$

Now, let's compute the remaining components of $F^{\mu\nu}$. We have

$$\begin{aligned}
F^{01} &= \partial^0 A^1 - \partial^1 A^0 = \frac{1}{c}\frac{\partial A^1}{\partial t} + \frac{\partial \Phi}{\partial x^1} = -E_x \\
F^{02} &= \partial^0 A^2 - \partial^2 A^0 = \frac{1}{c}\frac{\partial A^2}{\partial t} + \frac{\partial \Phi}{\partial x^2} = -E_y \\
F^{03} &= \partial^0 A^3 - \partial^3 A^0 = \frac{1}{c}\frac{\partial A^3}{\partial t} + \frac{\partial \Phi}{\partial x^3} = -E_z
\end{aligned}$$

Also,

$$\begin{aligned}
F^{12} &= \partial^1 A^2 - \partial^2 A^1 = -\frac{\partial A^2}{\partial x^1} + \frac{\partial A^1}{\partial x^2} = -B_z \\
F^{13} &= \partial^1 A^3 - \partial^3 A^1 = -\frac{\partial A^3}{\partial x^1} + \frac{\partial A^1}{\partial x^3} = B_y \\
F^{23} &= \partial^2 A^3 - \partial^3 A^2 = -\frac{\partial A^3}{\partial x^2} + \frac{\partial A^2}{\partial x^3} = -B_x
\end{aligned}$$

From our above calculations and the antisymmetry of $F^{\mu\nu}$, we find

$$F^{\mu\nu} = \begin{bmatrix} 0 & -E_x & -E_y & -E_z \\ E_x & 0 & -B_z & B_y \\ E_y & B_z & 0 & -B_x \\ E_z & -B_y & B_x & 0 \end{bmatrix} \tag{8.20}$$

We define the **current density four-vector** $\mathbf{J} = (J^\mu)$ by

$$J^\mu = \rho_0 U^\mu \tag{8.21}$$

where $\mathbf{U} = (U^\mu) = \gamma(c, \mathbf{u})$ is the four-velocity and ρ_0 is the **proper charge density**, so that

$$J^0 = \gamma\rho_0 c = \rho c \;\text{ and }\; J^a = \gamma\rho_0 u^a = \rho u^a \tag{8.21a}$$

Observe that

$$\partial_0 J^0 = \frac{1}{c}\frac{\partial}{\partial t}(\rho c) = \frac{\partial\rho}{\partial t}$$

while

$$\begin{aligned} \partial_a J^a &= \frac{\partial}{\partial x^1}J^1 + \frac{\partial}{\partial x^2}J^2 + \frac{\partial}{\partial x^3}J^3 \\ &= \frac{\partial}{\partial x^1}\rho u^1 + \frac{\partial}{\partial x^2}\rho u^2 + \frac{\partial}{\partial x^3}\rho u^3 \\ &= \nabla\cdot(\rho\mathbf{u}) \end{aligned}$$

Consequently,

$$\partial_\mu J^\mu = \partial_0 J^0 + \partial_a J^a = \frac{\partial\rho}{\partial t} + \nabla\cdot(\rho\mathbf{u}) = 0$$

so that

$$\partial_\mu J^\mu = 0 \tag{8.22}$$

is the covariant form of the charge-conservation equation.

We now turn our attention to the field equations themselves, which we list below for easy reference.

$$\nabla\cdot\mathbf{E} = 4\pi\rho \tag{8.1}$$

$$\nabla\cdot\mathbf{B} = 0 \tag{8.2}$$

$$\nabla\times\mathbf{E} = -\frac{\mathbf{1}}{\mathbf{c}}\frac{\partial\mathbf{B}}{\partial\mathbf{t}} \tag{8.3}$$

$$\nabla\times\mathbf{B} = \frac{4\pi}{c}\mathbf{j} + \frac{1}{c}\frac{\partial\mathbf{E}}{\partial t} \tag{8.4}$$

As shown in the Exercises, equations (8.2) and (8.3) can be written in terms of the field tensor as

$$\partial^\rho F^{\mu\nu} + \partial^\nu F^{\rho\mu} + \partial^\mu F^{\nu\rho} = 0 \;, \tag{8.23}$$

while (8.1) and (8.3) become

$$\partial_\mu F^{\mu\nu} = \frac{4\pi}{c}J^\nu \tag{8.24}$$

These equations are the manifestly covariant form of Maxwells equations. The reader will note that (8.24) also encodes the charge conservation equation (8.22).

8.3. Covariant Form of the Lorentz Force

We know that a particle with charge q experiences, in the presence of an electromagnetic field, a force which is given by

$$\mathbf{f} = q\mathbf{E} + \frac{q}{c}(\mathbf{u} \times \mathbf{B}) \tag{8.25}$$

where $\mathbf{u}$ is the 3-velocity of the particle. This is, of course, the familiar Lorentz force law. A question which naturally arises is how do we express this equation in a manifestly covariant manner?

In order to answer this question, assume that a particle with charge q moves with 3-velocity $\mathbf{u}$ under the influence of the field $F^{\mu\nu}$. Since we require that the resulting four-force F^{μ} is a genuine 4-vector, we assume

$$F^{\mu} = \alpha F^{\mu\nu} U_{\nu} \tag{8.26}$$

where α is a scalar invariant (constant) to be determined. This assumption makes sense since $F^{\mu\nu}$ already contains both the electric and magnetic field components, and contracting this field tensor with the four velocity U_{ν} guarantees that F^{μ} is a bonefide 4-vector.

Let $K(ct,x,y,z)$ be the *instantaneous rest frame* of the particle. Since the particle is momentarily at rest relative to K, it can only experience an electrostatic force $q\mathbf{E}$, so that

$$(F^{\mu}) = (0,\mathbf{f}) = (0,q\mathbf{E}) \tag{8.27}$$

However, by (8.20) and (8.26), we can write

$$(F^{\mu}) = \alpha \begin{bmatrix} 0 & -E_x & -E_y & -E_z \\ E_x & 0 & 0 & 0 \\ E_y & 0 & 0 & 0 \\ E_z & 0 & 0 & 0 \end{bmatrix} \begin{bmatrix} c \\ 0 \\ 0 \\ 0 \end{bmatrix} = \alpha(0,cE_x,cE_y,cE_z) = (0,\alpha c\mathbf{E}) \tag{8.28}$$

Comparing the righthand sides of the last two equations gives $\alpha = q/c$; thus, relative to the frame K, we have

$$F^{\mu} = \frac{q}{c} F^{\mu\nu} U_{\nu} \tag{8.29}$$

Since this is a tensor equation, it must hold in all such reference frames, and therefore (8.29) is the required generalization of the Lorentz force law.

We now show how (8.29) reduces to the three dimensional version of the Lorentz force law given by (8.25). When written in explicit form, (8.29) gives

$$\begin{bmatrix} F^0 \\ F^1 \\ F^2 \\ F^3 \end{bmatrix} = \frac{q}{c} \begin{bmatrix} 0 & -E_x & -E_y & -E_z \\ E_x & 0 & -B_z & B_y \\ E_y & B_z & 0 & -B_x \\ E_z & -B_y & B_x & 0 \end{bmatrix} \begin{bmatrix} \gamma c \\ -\gamma u_x \\ -\gamma u_y \\ -\gamma u_z \end{bmatrix}$$

If we carry out the matrix multiplication, then we find

$$\begin{aligned} F^0 &= \frac{q}{c}\gamma(\mathbf{E}\cdot\mathbf{u}) \\ F^1 &= \gamma\left[qE_x + \frac{q}{c}(\mathbf{u}\times\mathbf{B})_x\right] \\ F^2 &= \gamma\left[qE_y + \frac{q}{c}(\mathbf{u}\times\mathbf{B})_y\right] \\ F^3 &= \gamma\left[qE_z + \frac{q}{c}(\mathbf{u}\times\mathbf{B})_z\right] \end{aligned}$$

Consequently, the Lorentz 4-force is given by

$$\mathbf{F} = \gamma\left(\frac{q}{c}(\mathbf{E}\cdot\mathbf{u}),\ q\left[\mathbf{E} + \frac{1}{c}(\mathbf{u}\times\mathbf{B})\right]\right) \tag{8.30}$$

On the other hand, by the general definition of the Minkowski force as expressed in (7.53), and by (7.52), we must also have

$$\mathbf{F} = \gamma\left(\frac{1}{c}\frac{dE}{dt}, \mathbf{f}\right) \tag{8.31}$$

Equating the spatial components in (8.30) and (8.31), gives

$$\mathbf{f} = q\left[\mathbf{E} + \frac{1}{c}(\mathbf{u}\times\mathbf{B})\right]$$

which is the familiar three dimensional version of the Lorentz force.

If we now equate the temporal components of the (8.30) and (8.31), we obtain

$$\frac{dE}{dt} = q(\mathbf{E}\cdot\mathbf{u}) \tag{8.32}$$

This equation expresses the *rate of work done*, or power expended by the field in changing the energy of the particle. We note, in passing, the absence of any terms involving the magnetic field **B** in (8.32). This makes sense since the magnetic force experienced by a particle is always perpendicular to the direction of the particle's motion, and so, the work done by magnetic forces must be zero.

Finally, we can write down the covariant form of the equations of motion for a charged particle with rest mass m_0 as

$$m_0\frac{dU^\mu}{d\tau} = \frac{dP^\mu}{d\tau} = \frac{q}{c}F^{\mu\nu}U_\nu \tag{8.33}$$

It's easy to show that spatial components of equations (8.33) reduce to the familiar equation of motion

$$m_0\frac{d}{dt}(\gamma\mathbf{u}) = \frac{d\mathbf{p}}{dt} = q\left[\mathbf{E} + \frac{1}{c}(\mathbf{u}\times\mathbf{B})\right] \tag{8.34}$$

while the temporal component of (8.33) reduces to (8.32), which governs the particle's energy.

8.4. The Energy-Momentum Tensor

For the case of a *continuous* charge distribution, it turns out that (8.29) must be slightly modified to read

$$G^\mu = \frac{\rho_0}{c} F^{\mu\nu} U_\nu \tag{8.35}$$

where G^μ is now the **4-force density** and the discrete charge q has been replaced by the *proper* charge density ρ_0.

We will now show that the 4-force density can be written in terms of a new tensor, called the **electromagnetic energy momentum tensor**. We first note that by (8.21), the last equation can be rewritten as

$$G^\mu = \frac{1}{c}\eta_{\nu\rho} F^{\mu\nu} J^\rho,$$

which, in turn, by (8.24) becomes

$$G^\mu = \frac{1}{4\pi} F^\mu_\alpha \frac{\partial}{\partial x^\beta} F^{\beta\alpha} = \frac{1}{4\pi}\left[\frac{\partial}{\partial x^\beta}\left(F^\mu_\alpha F^{\beta\alpha}\right) - F^{\beta\alpha}\frac{\partial}{\partial x^\beta}\left(F^\mu_\alpha\right)\right] \tag{8.36}$$

Next, we take a closer look at the second term in (8.36). Observe that

$$\begin{aligned}
-F^{\beta\alpha}\frac{\partial}{\partial x^\beta}\left(F^\mu_\alpha\right) &= F^{\alpha\beta}\frac{\partial}{\partial x^\beta}\left(F^\mu_\alpha\right) = \eta^{\mu\nu} F^{\alpha\beta}\frac{\partial}{\partial x^\beta}\left(F_{\nu\alpha}\right) \\
&= \eta^{\mu\nu}\left[\frac{1}{2}F^{\alpha\beta}\frac{\partial}{\partial x^\beta}\left(F_{\nu\alpha}\right) + \frac{1}{2}F^{\alpha\beta}\frac{\partial}{\partial x^\beta}\left(F_{\nu\alpha}\right)\right] \\
&= \eta^{\mu\nu}\left[\frac{1}{2}F^{\alpha\beta}\frac{\partial}{\partial x^\beta}\left(F_{\nu\alpha}\right) + \frac{1}{2}F^{\beta\alpha}\frac{\partial}{\partial x^\beta}\left(F_{\alpha\nu}\right)\right]
\end{aligned}$$

Here we have exploited the anti-symmetry of the field tensor in the second term of the last expression. If we now interchange *dummy* variables α and β in the second term, we obtain

$$\begin{aligned}
-F^{\beta\alpha}\frac{\partial}{\partial x^\beta}\left(F^\mu_\alpha\right) &= \eta^{\mu\nu}\left[\frac{1}{2}F^{\alpha\beta}\frac{\partial}{\partial x^\beta}\left(F_{\nu\alpha}\right) + \frac{1}{2}F^{\alpha\beta}\frac{\partial}{\partial x^\alpha}\left(F_{\beta\nu}\right)\right] \\
&= \frac{1}{2}\eta^{\mu\nu}F^{\alpha\beta}\left[\frac{\partial}{\partial x^\beta}\left(F_{\nu\alpha}\right) + \frac{\partial}{\partial x^\alpha}\left(F_{\beta\nu}\right)\right] \\
&= -\frac{1}{2}\eta^{\mu\nu}F^{\alpha\beta}\frac{\partial}{\partial x^\nu}\left(F_{\alpha\beta}\right) \quad \text{[By (8.23)]} \\
&= -\frac{1}{4}\eta^{\mu\nu}\left[2F^{\alpha\beta}\frac{\partial}{\partial x^\nu}\left(F_{\alpha\beta}\right)\right] \\
&= -\frac{\partial}{\partial x^\nu}\left[\frac{1}{4}\eta^{\mu\nu}F^{\alpha\beta}F_{\alpha\beta}\right]
\end{aligned}$$

We conclude that

$$-F^{\beta\alpha}\frac{\partial}{\partial x^\beta}\left(F^\mu_\alpha\right) = -\frac{\partial}{\partial x^\nu}\left[\frac{1}{4}\eta^{\mu\nu}F^{\alpha\beta}F_{\alpha\beta}\right] \tag{8.37}$$

Next, we substitute (8.37) into (8.36) to obtain

$$G^{\mu} = \frac{\partial}{\partial x^{\nu}} \left\{ \frac{1}{4\pi} \left(F^{\mu}_{\alpha} F^{\nu\alpha} - \frac{1}{4} \eta^{\mu\nu} F^{\alpha\beta} F_{\alpha\beta} \right) \right\} \tag{8.38}$$

As a matter of convention, we take the *negative* of the quantity in the braces of (8.38) to be the **electromagnetic energy-momentum tensor**:

$$T^{\mu\nu} = -\frac{1}{4\pi} \left(F^{\mu}_{\alpha} F^{\nu\alpha} - \frac{1}{4} \eta^{\mu\nu} F^{\alpha\beta} F_{\alpha\beta} \right) \tag{8.39}$$

so that (8.38) simply becomes

$$G^{\mu} = -\, \partial_{\nu} T^{\mu\nu} \tag{8.40}$$

It is helpful to express the components of the energy-momentum tensor directly in terms of the fields **E** and **B**. First, it may be shown that its purely spatial components are given by

$$T^{ab} = -\frac{1}{4\pi} \left[E^a E^b + B^a B^b - \frac{1}{2} \delta^{ab} (E^2 + B^2) \right] \tag{8.41}$$

for $a, b = 1, 2, 3$. The set of quantities T^{ab} form the components of what is commonly called the **Maxwell stress tensor**.

From (8.41), it's easy to see that the Maxwell stress tensor is *symmetric*. Furthermore, as shown in the Exercises, we find that

$$w \equiv T^{00} = \frac{1}{8\pi} (\|\mathbf{E}\|^2 + \|\mathbf{B}\|^2) \tag{8.42}$$

and

$$T^{01} = T^{10} = \frac{1}{4\pi} (\mathbf{E} \times \mathbf{B})_x, \; T^{02} = T^{20} = \frac{1}{4\pi} (\mathbf{E} \times \mathbf{B})_y, \; T^{03} = T^{30} = \frac{1}{4\pi} (\mathbf{E} \times \mathbf{B})_z \tag{8.43}$$

In the discussion that follows, we will show that T^{00} is the **energy density** w (or energy per unit volume) of the electromagnetic field. On the other hand, from (8.43), it is evident that the quantities T^{0k} form the components of a 3-vector. So, if we define the 3-vector $\mathbf{S} = (S^k)$ by

$$\mathbf{S} = \frac{c}{4\pi} (\mathbf{E} \times \mathbf{B}) \tag{8.44}$$

then (8.43) may be neatly written as

$$T^{0k} = T^{k0} = \frac{1}{c} S^k \tag{8.45}$$

The object **S** is called the **Poynting vector**, and as we shall see, it is directly related to the flux of field energy.

Based upon the above discussion, the energy-momentum tensor can be written as

$$T^{\mu\nu} = \begin{bmatrix} w & \frac{1}{c} S_x & \frac{1}{c} S_y & \frac{1}{c} S_z \\ \frac{1}{c} S_x & & & \\ \frac{1}{c} S_y & & T^{ab} & \\ \frac{1}{c} S_z & & & \end{bmatrix} \tag{8.46}$$

From (8.46), and (8.41), one can show that the energy-momentum tensor is both *symmetric* and *trace-free*:

$$T^{\mu\nu} = T^{\nu\mu} \tag{8.47}$$

$$Tr(\mathbf{T}) \equiv T^{\mu}_{\mu} = 0 \tag{8.48}$$

We now determine the physical significance of T^{00}, and the Poynting vector $\mathbf{S}$. From (8.3) and (8.4) we can write

$$\frac{1}{c}\left[\mathbf{E}\cdot\frac{\partial\mathbf{E}}{\partial t}+\mathbf{B}\cdot\frac{\partial\mathbf{B}}{\partial t}\right] = \mathbf{E}\cdot(\nabla\times\mathbf{B}) - \mathbf{B}\cdot(\nabla\times\mathbf{E}) - \frac{4\pi}{c}(\mathbf{j}\cdot\mathbf{E})$$

Recalling the vector identity

$$\mathbf{a}\cdot(\nabla\times\mathbf{b}) = \mathbf{b}\cdot(\nabla\times\mathbf{a}) - \mathbf{a}\cdot(\nabla\times\mathbf{b}), \tag{8.49}$$

the last equation may be rewritten as

$$\frac{1}{2c}\frac{\partial}{\partial t}\left(\|\mathbf{E}\|^2+\|\mathbf{B}\|^2\right) = -\frac{4\pi}{c}(\mathbf{j}\cdot\mathbf{E}) - \nabla\cdot(\mathbf{E}\times\mathbf{B})$$

which, in turn, becomes

$$\frac{\partial}{\partial t}\left(\frac{\|\mathbf{E}\|^2+\|\mathbf{B}\|^2}{8\pi}\right) = -(\mathbf{j}\cdot\mathbf{E}) - \nabla\cdot\left[\frac{c}{4\pi}(\mathbf{E}\times\mathbf{B})\right] \tag{8.50}$$

Recalling (8.42), and (8.44), the last equation simplifies to

$$\frac{\partial w}{\partial t} = -(\mathbf{j}\cdot\mathbf{E}) - \nabla\cdot\mathbf{S} \tag{8.51}$$

Now consider a volume V bounded by a closed surface ∂V which contains a system of discrete charges $q_1, q_2, \ldots, q_n$ that are situated at the locations $\mathbf{x}_1, \mathbf{x}_2, \ldots, \mathbf{x}_n$, respectively. We can write the current density as

$$\mathbf{j} = \sum_{k=1}^{n} q_k\mathbf{u}_k\delta^3(\mathbf{x}-\mathbf{x}_k)$$

where $\mathbf{u}_k$ is the 3-velocity of the k-th charge, and $\delta^3(\mathbf{x}-\mathbf{x}_k)$ is the 3-dimensional Dirac delta function. Consequently, we can write

$$\mathbf{j}\cdot\mathbf{E} = \sum_{k=1}^{n} q_k(\mathbf{u}_k\cdot\mathbf{E})\,\delta^3(\mathbf{x}-\mathbf{x}_k) \tag{8.52}$$

From (8.32), however, the time rate of change of energy of the k-th charge is can be written as

$$\frac{dE_k}{dt} = q_k(\mathbf{u}_k\cdot\mathbf{E})$$

The previous equation now becomes

$$\mathbf{j}\cdot\mathbf{E}=\sum_{k=1}^{n}\delta^3(\mathbf{x}-\mathbf{x}_k)\frac{dE_k}{dt} \tag{8.53}$$

We now substitute (8.53) into (8.51), and integrate over V to obtain

$$\begin{aligned}\frac{d}{dt}\left(\int_V w\,d^3\mathbf{x}\right) &= -\frac{d}{dt}\sum_{k=1}^{n}\int_V E_k\,\delta^3(\mathbf{x}-\mathbf{x}_k)d^3\mathbf{x}-\int_V(\nabla\cdot\mathbf{S})\,d^3\mathbf{x}\\ &= -\frac{d}{dt}\sum_{k=1}^{n}E_k-\int_V(\nabla\cdot\mathbf{S})\,d^3\mathbf{x}\end{aligned}$$

If we convert the last volume integral on the righthand side into a surface integral, and simplify, we obtain

$$\frac{d}{dt}\left(\int_V w\,d^3\mathbf{x}+E_{total}\right)=-\oint_{\partial V}\mathbf{S}\cdot d\mathbf{\Sigma} \tag{8.54}$$

where $d\mathbf{\Sigma}$is the surface area element, and

$$E_{total}=\sum_{k=1}^{n}E_k \tag{8.55}$$

is the total energy (kinetic energy plus rest energy) of the system of charges. If we let the bounding surface ∂V go to infinity, then $\mathbf{E},\mathbf{B}\longrightarrow\mathbf{0}$, and (8.54) becomes

$$\frac{d}{dt}\left(\int_V w\,d^3\mathbf{x}+E_{total}\right)=0$$

and we see that the quantity in the parentheses is *conserved* for all time. Since E_{total} is the total energy of the system of charges, then the remainder of the energy must reside in the field itself; that is,

$$\text{Total Field Energy}=\int_V w\,d^3\mathbf{x} \tag{8.56}$$

and consequently,

$$w\equiv T^{00}=\frac{1}{8\pi}(\|\mathbf{E}\|^2+\|\mathbf{B}\|^2)=\text{ Energy Density of the Field} \tag{8.57}$$

On the other hand, if we keep the volume V finite, then lefthand side of (8.54) will, in general, not be constant in time. For example, if $\mathbf{S}$ points outward so that $\mathbf{S}\cdot d\mathbf{\Sigma}>\mathbf{0}$, then the sum of the total energy of the charges and the total field energy will decrease in time. In order to account for this missing energy, we must conclude that

$$\oint_{\partial V}\mathbf{S}\cdot d\mathbf{\Sigma}=\text{Energy flux thru the surface }\partial V\text{ per unit time} \tag{8.58}$$

and consequently, the k-th component of the Poynting vector

$$S^k = \left\{ \begin{array}{l} \text{Energy flux in the } k\text{-direction,} \\ \text{per unit area, per unit time.} \end{array} \right. \tag{8.59}$$

Finally, we examine the physical meaning of the Maxwell stress tensor. By (8.40),

$$G^a = -\partial_\nu T^{a\nu} = -\frac{1}{c}\frac{\partial}{\partial t}\left(T^{a0}\right) - \frac{\partial}{\partial x^b}\left(T^{ab}\right)$$

If we integrate over a closed volume V, then we get

$$\int_V G^a dV = -\frac{1}{c}\frac{d}{dt}\int_V T^{a0} dV - \int_V \frac{\partial}{\partial x^b}\left(T^{ab}\right) dV \tag{8.60}$$

Since (G^a) is the force *density*, the volume integral on the lefthand side must be equal to the component F^a of the total force. At the same time, however, the component F^a of the total force must be equal to the time rate of change of the component of total momentum P^a, so

$$F^a = \frac{dP^a}{dt} = -\frac{1}{c^2}\int_V S^a dV - \int_V \frac{\partial}{\partial x^b}\left(T^{ab}\right) dV$$

where we have used (8.45) to rewrite the first integral on the right. We now convert the second volume integral into a surface integral and rearrange terms, to obtain

$$\frac{d}{dt}\left[P^a + \frac{1}{c^2}\int_V S^a dV\right] = -\oint_{\partial V} T^{ab} d\Sigma_b \tag{8.61}$$

Once again, if we let the bounding surface ∂V go to infinity, then the surface integral vanishes, and (8.61) becomes

$$\frac{d}{dt}\left[P^a + \frac{1}{c^2}\int_V S^a dV\right] = 0 \tag{8.62}$$

Since the quantity in the brackets is then conserved, we conclude that

$$\frac{1}{c^2}\int_V S^a dV = a\text{-th component of field momentum} \tag{8.63}$$

and consequently,

$$\frac{1}{c^2} S^a = a\text{-th component of field-momentum density} \tag{8.64}$$

Now, if we allow the volume V to remain *finite*, then (8.61) will again apply to the situation, and by a similar accounting argument as above, we must conclude that

$$T^{ab} = \left\{ \begin{array}{l} \text{Flux of the } a\text{-th component of the momentum thru the} \\ \text{surface } \partial V \text{ in the direction } b\text{, per unit area, per unit time.} \end{array} \right. \tag{8.65}$$

Finally, we can summarize the results of the above analysis, as

$$T^{\mu\nu} = \left[\begin{array}{cc} \textit{(energy density)} \Leftarrow & \frac{1}{c}\textit{(energy flux)} \qquad \Longrightarrow \\ \Uparrow & \left[\begin{array}{c} ------------ \\ \textit{(momentum flux)} \\ ------------ \end{array} \right] \\ \frac{1}{c}\textit{(energy flux)} & \\ \Downarrow & \end{array} \right] \tag{8.66}$$

8.5. Transformation of Fields

How do the fields **E** and **B** transform under a Lorentz transformation? In order to answer this question, we first note that due to its tensor character, the field $F^{\mu\nu}$ must transform according to the rule

$$\bar{F}^{\mu\nu} = \Lambda^{\mu}_{\sigma}\Lambda^{\nu}_{\tau}F^{\sigma\tau}\,, \tag{8.67}$$

under a Lorentz transformation. As a matter of convenience, we can rewrite (8.67) in matrix form as

$$\bar{\mathbf{F}} = \Lambda\mathbf{F}\Lambda^T \tag{8.68}$$

Next, let's consider a simple Lorentz boost; that is, suppose we have two inertial frames, $K(x,y,z,t)$ and $\bar{K}(\bar{x},\bar{y},\bar{z},\bar{t})$, whose origins coincide at the time $t = \bar{t} = 0$, where the latter frame moves in the positive x-direction at a constant speed v relative to the former frame. Recall that the Lorentz transformation matrix for this state of affairs is

$$\Lambda = \begin{bmatrix} \gamma & -\frac{v}{c}\gamma & 0 & 0 \\ -\frac{v}{c}\gamma & \gamma & 0 & 0 \\ 0 & 0 & 1 & 0 \\ 0 & 0 & 0 & 1 \end{bmatrix} \tag{8.68a}$$

In explicit form, (8.68) now becomes

$$\bar{F}^{\mu\nu} = \begin{bmatrix} \gamma & -\frac{v}{c}\gamma & 0 & 0 \\ -\frac{v}{c}\gamma & \gamma & 0 & 0 \\ 0 & 0 & 1 & 0 \\ 0 & 0 & 0 & 1 \end{bmatrix} \begin{bmatrix} 0 & -E_x & -E_y & -E_z \\ E_x & 0 & -B_z & B_y \\ E_y & B_z & 0 & -B_x \\ E_z & -B_y & B_x & 0 \end{bmatrix} \begin{bmatrix} \gamma & -\frac{v}{c}\gamma & 0 & 0 \\ -\frac{v}{c}\gamma & \gamma & 0 & 0 \\ 0 & 0 & 1 & 0 \\ 0 & 0 & 0 & 1 \end{bmatrix}$$

Carrying out the matrix multiplication, and simplifying, we obtain

$$\bar{F}^{\mu\nu} = \begin{bmatrix} 0 & -E_x & -\gamma(E_y - \frac{v}{c}B_z) & -\gamma(E_z + \frac{v}{c}B_y) \\ E_x & 0 & -\gamma(B_z - \frac{v}{c}E_y) & \gamma(B_y + \frac{v}{c}E_z) \\ \gamma(E_y - \frac{v}{c}B_z) & \gamma(B_z - \frac{v}{c}E_y) & 0 & -B_x \\ \gamma(E_z + \frac{v}{c}B_y) & \gamma(B_y + \frac{v}{c}E_z) & B_x & 0 \end{bmatrix} \tag{8.69}$$

On the other hand, by (8.20), we have

$$\bar{F}^{\mu\nu} = \begin{bmatrix} 0 & -\bar{E}_x & -\bar{E}_y & -\bar{E}_z \\ \bar{E}_x & 0 & -\bar{B}_z & \bar{B}_y \\ \bar{E}_y & \bar{B}_z & 0 & -\bar{B}_x \\ \bar{E}_z & -\bar{B}_y & \bar{B}_x & 0 \end{bmatrix} \tag{8.70}$$

Equating the elements of (8.69) and (8.70) gives

$$\bar{E}_x = E_x, \;\; \bar{E}_y = \gamma(E_y - \frac{v}{c}B_z), \;\; \bar{E}_z = \gamma(E_z + \frac{v}{c}B_y) \tag{8.71}$$

and

$$\bar{B}_x = B_x, \;\; \bar{B}_y = \gamma(B_y + \frac{v}{c}E_z), \;\; \bar{B}_z = \gamma(B_z - \frac{v}{c}E_y) \tag{8.72}$$

We can recast (8.71) and (8.72) into a nicer form. If $\hat{\mathbf{x}}, \hat{\mathbf{y}}$, and $\hat{\mathbf{z}}$ are the standard unit vectors along the coordinate axes, then (8.71) gives

$$\begin{aligned}
\bar{\mathbf{E}} &= \bar{E}_x\hat{\mathbf{x}}+\bar{E}_y\hat{\mathbf{y}}+\bar{E}_z\hat{\mathbf{z}} = E_x\hat{\mathbf{x}}+\gamma(E_y-\frac{v}{c}B_z)\hat{\mathbf{y}}+\gamma(E_z+\frac{v}{c}B_y)\hat{\mathbf{z}} \\
&= (E_x\hat{\mathbf{x}})+\gamma\left\{(E_y\hat{\mathbf{y}}+E_z\hat{\mathbf{z}})+\frac{1}{c}\left|\begin{array}{ccc} \hat{\mathbf{x}} & \hat{\mathbf{y}} & \hat{\mathbf{z}} \\ v & 0 & 0 \\ B_x & B_y & B_z \end{array}\right|\right\}
\end{aligned}$$

or, in other words,

$$\bar{\mathbf{E}} = \mathbf{E}_{\|}+\gamma\left[\mathbf{E}_{\perp}+\frac{1}{c}(\mathbf{v}\times\mathbf{B})\right]$$

where

$$\mathbf{E}_{\|} = E_x\hat{\mathbf{x}} \text{ and } \mathbf{E}_{\perp} = E_y\hat{\mathbf{y}}+E_z\hat{\mathbf{z}} \tag{8.73}$$

are the components of the electric field parallel and perpendicular to the direction of $\mathbf{v}$, respectively. Clearly, we can perform the same sort of algebraic manipulation on (8.72), so the transformation of fields under a simple Lorentz boost $\mathbf{v} = v\hat{\mathbf{x}}$ can be neatly written as

$$\bar{\mathbf{E}}_{\|} = \mathbf{E}_{\|},\ \bar{\mathbf{E}}_{\perp} = \gamma\left[\mathbf{E}_{\perp}+\frac{1}{c}(\mathbf{v}\times\mathbf{B})\right] \tag{8.74a}$$

$$\bar{\mathbf{B}}_{\|} = \mathbf{B}_{\|},\ \bar{\mathbf{B}}_{\perp} = \gamma\left[\mathbf{B}_{\perp}-\frac{1}{c}(\mathbf{v}\times\mathbf{E})\right] \tag{8.74b}$$

Equations (8.74) are particularly useful in solving practical problems.

8.6. Exercises

1. *(Lorentz Gauge)* Show that in the Lorentz gauge, field equations (8.1) and (8.4) imply that:

$$\frac{1}{c^2}\frac{\partial^2\Phi}{\partial t^2}-\nabla^2\Phi = 4\pi\rho \tag{8.11a}$$

$$\frac{1}{c^2}\frac{\partial^2\mathbf{A}}{\partial t^2}-\nabla^2\mathbf{A} = \frac{4\pi}{c}\mathbf{j} \tag{8.11b}$$

(Solution)

From (8.1) and (8.8), we can write

$$\nabla\cdot\mathbf{E} = \nabla\cdot\left(-\nabla\Phi-\frac{1}{c}\frac{\partial\mathbf{A}}{\partial t}\right) = 4\pi\rho \Longrightarrow \nabla^2\Phi+\frac{1}{c}\frac{\partial}{\partial t}(\nabla\cdot\mathbf{A}) = -4\pi\rho$$

By the Lorentz condition, the last equation may be rewritten as

$$\nabla^2\Phi+\frac{1}{c}\frac{\partial}{\partial t}\left(-\frac{1}{c}\frac{\partial\Phi}{\partial t}\right) = -4\pi\rho \Longrightarrow \frac{1}{c^2}\frac{\partial^2\Phi}{\partial t^2}-\nabla^2\Phi = 4\pi\rho\ ,$$

as claimed. On the other hand, by (8.4) and (8.5), we have

$$\nabla\times\mathbf{B}=\nabla\times(\nabla\times\mathbf{A})=\frac{4\pi}{c}\mathbf{j}+\frac{1}{c}\frac{\partial\mathbf{E}}{\partial t}$$

and consequently,

$$\begin{aligned}\nabla\times(\nabla\times\mathbf{A}) &= -\nabla^2\mathbf{A}+\nabla(\nabla\cdot\mathbf{A})=\frac{4\pi}{c}\mathbf{j}+\frac{1}{c}\frac{\partial}{\partial t}\left(-\nabla\Phi-\frac{1}{c}\frac{\partial\mathbf{A}}{\partial t}\right)\\ \Longrightarrow &\quad -\nabla^2\mathbf{A}+\nabla\left[(\nabla\cdot\mathbf{A})+\frac{1}{c}\frac{\partial\Phi}{\partial t}\right]=\frac{4\pi}{c}\mathbf{j}-\frac{1}{c^2}\frac{\partial^2\mathbf{A}}{\partial t^2}\\ \Longrightarrow &\quad \left(\frac{1}{c^2}\frac{\partial^2\mathbf{A}}{\partial t^2}-\nabla^2\mathbf{A}\right)+\nabla\left[(\nabla\cdot\mathbf{A})+\frac{1}{c}\frac{\partial\Phi}{\partial t}\right]=\frac{4\pi}{c}\mathbf{j}\end{aligned}$$

Enforcing the Lorentz condition, the last equation becomes

$$\frac{1}{c^2}\frac{\partial^2\mathbf{A}}{\partial t^2}-\nabla^2\mathbf{A}=\frac{4\pi}{c}\mathbf{j}$$

as required.

2. *(Retarded Potential)* Show that

$$\Phi(\mathbf{x},t)=\int_V\frac{\rho(\mathbf{x}',t-R/c)}{R}d^3\mathbf{x}' \tag{8.13a}$$

is a solution of

$$\frac{1}{c^2}\frac{\partial^2\Phi}{\partial t^2}-\nabla^2\Phi=4\pi\rho \tag{8.11a}$$

(Solution)

We have

$$\begin{aligned}\nabla\Phi &= \int_V\nabla\left[\frac{\rho(\mathbf{x}',t_r)}{R}\right]d^3\mathbf{x}'\\ &= \int_V\left[\frac{1}{R}\nabla\rho+\rho\nabla\left(\frac{1}{R}\right)\right]d^3\mathbf{x}'\\ &= \int_V\left[-\frac{1}{c}\frac{\partial\rho}{\partial t_r}\frac{\hat{\mathbf{R}}}{R}-\frac{\rho}{R^2}\hat{\mathbf{R}}\right]d^3\mathbf{x}'\end{aligned}$$

where $\hat{\mathbf{R}}$ is a unit vector in the direction of $\mathbf{R}$, and we have used the fact that

$$\nabla\rho=-\frac{1}{c}\left(\frac{\partial\rho}{\partial t_r}\right)\hat{\mathbf{R}}$$

since $\rho(\mathbf{x}',t_r)$ depends on $\mathbf{x}$ through t_r. Taking the divergence of the integral gives

$$\nabla^2\Phi=\int_V\nabla\cdot\left[-\frac{1}{c}\frac{\partial\rho}{\partial t_r}\frac{\hat{\mathbf{R}}}{R}-\frac{\rho}{R^2}\hat{\mathbf{R}}\right]d^3\mathbf{x}'$$

Observe that

$$\begin{aligned} \nabla\cdot\left(-\frac{\rho}{R^2}\hat{\mathbf{R}}\right) &= -\rho\nabla\cdot\left(\frac{1}{R^2}\hat{\mathbf{R}}\right)-\frac{\hat{\mathbf{R}}}{R^2}\nabla\rho \\ &= -4\pi\rho\delta^3(\mathbf{R})+\frac{1}{cR^2}\frac{\partial\rho}{\partial t_r} \end{aligned}$$

where we have used (3.52). Also,

$$\begin{aligned} \nabla\cdot\left(-\frac{1}{c}\frac{\partial\rho}{\partial t_r}\frac{\hat{\mathbf{R}}}{R}\right) &= \frac{\hat{\mathbf{R}}}{R}\cdot\frac{-1}{c^2}\left(\frac{\partial^2\rho}{\partial t_r^2}\right)\hat{\mathbf{R}}-\frac{1}{c}\frac{\partial\rho}{\partial t_r}\left(\nabla\cdot\frac{\hat{\mathbf{R}}}{R}\right) \\ &= -\frac{1}{Rc^2}\left(\frac{\partial^2\rho}{\partial t_r^2}\right)+\frac{1}{cR^2}\frac{\partial^2\rho}{\partial t_r^2} \end{aligned}$$

Substituting the last two expressions into the previous integral yields

$$\begin{aligned} \nabla^2\Phi &= -\int_V 4\pi\rho\delta^3(\mathbf{R})d^3\mathbf{x}'+\frac{1}{c^2}\int_V\frac{1}{R}\frac{\partial^2\rho}{\partial t_r^2}d^3\mathbf{x}' \\ &= -4\pi\rho+\frac{1}{c^2}\frac{\partial^2}{\partial t^2}\left(\int_V\frac{\rho(\mathbf{x}',t_r)}{R}d^3\mathbf{x}'\right) \quad [\text{Since } \partial/\partial t\equiv\partial/\partial t_r] \\ \Longrightarrow\quad &\nabla^2\Phi=-4\pi\rho+\frac{1}{c^2}\frac{\partial^2\Phi}{\partial t^2} \end{aligned}$$

which is equivalent to (8.13a). Similarly, we can show that the vector potential given by (8.13b) is a solution of (8.11b).

3. Show that the scalar and vector potentials satisfy the Lorentz gauge condition.

(Solution)

Since $\mathbf{j}(\mathbf{x}',t_r)$ depends upon $\mathbf{x}'$ both directly, and through the retarded time t_r, we find that

$$\nabla'\cdot\mathbf{j}=\left(\nabla'\cdot\mathbf{j}\right)\big|_{t_r}-\frac{1}{c}\frac{\partial\mathbf{j}}{\partial t_r}\cdot\nabla R$$

where the expression $\left(\nabla'\cdot\mathbf{j}\right)\big|_{t_r}$ indicates that t_r is held constant.

Similarly, it's easy to show that

$$\nabla\cdot\mathbf{j}=\frac{1}{c}\frac{\partial\mathbf{j}}{\partial t_r}\cdot\nabla'\mathbf{R}$$

From the last two equations, we obtain the relation

$$\nabla\cdot\mathbf{j}=\left(\nabla'\cdot\mathbf{j}\right)\big|_{t_r}-\nabla'\cdot\mathbf{j}$$

We now take the divergence of (8.13b) to obtain

$$\begin{aligned} \nabla\cdot\mathbf{j} &= \frac{1}{c}\int_V\left(\nabla\cdot\frac{\mathbf{j}}{R}\right)d^3\mathbf{x}' \\ &= \frac{1}{c}\int_V\frac{1}{R}(\nabla\cdot\mathbf{j})\,d^3\mathbf{x}'-\frac{1}{c}\int_V\left(\mathbf{j}\cdot\frac{-1}{R^2}\nabla R\right)d^3\mathbf{x}' \\ &= \frac{1}{c}\int_V\frac{1}{R}\left(\left(\nabla'\cdot\mathbf{j}\right)\big|_{t_r}-\nabla'\cdot\mathbf{j}\right)d^3\mathbf{x}'-\frac{1}{c}\int_V\left(\mathbf{j}\cdot\frac{1}{R^2}\nabla'R\right)d^3\mathbf{x}' \end{aligned}$$

where in the last step, we used the fact that $\nabla' R = -\nabla R$. Now, the last integral may be rewritten more neatly as

$$\nabla \cdot \mathbf{j} = \frac{1}{c}\int_V \frac{1}{R}\left((\nabla' \cdot \mathbf{j})\big|_{t_r} \right) d^3\mathbf{x}' - \frac{1}{c}\int_V \left(\nabla' \cdot \frac{\mathbf{j}}{R}\right) d^3\mathbf{x}'$$

By the Divergence Theorem, the second integral becomes

$$\int_V \left(\nabla' \cdot \frac{\mathbf{j}}{R}\right) d^3\mathbf{x}' = \oint_{\partial V} \left(\frac{\mathbf{j}}{R} \cdot \mathbf{n}\right) dS \longrightarrow 0$$

since we can take V as a sphere of arbitrarily large radius containing all sources, so that on its boundary ∂V, we have $\|\mathbf{j}\| = 0$. Consequently, the last equation becomes

$$\nabla \cdot \mathbf{j} = \frac{1}{c}\int_V \frac{1}{R}\left((\nabla' \cdot \mathbf{j})\big|_{t_r} \right) d^3\mathbf{x}' \ ,$$

and we can write

$$\begin{aligned} \nabla \cdot \mathbf{j} + \frac{1}{c}\frac{\partial \Phi}{\partial t} &= \frac{1}{c}\int_V \frac{1}{R}\left((\nabla' \cdot \mathbf{j})\big|_{t_r} \right) d^3\mathbf{x}' + \frac{1}{c}\int_V \frac{1}{R}\left(\frac{\partial \rho_r}{\partial t_r}\right) d^3\mathbf{x}' \\ &= \frac{1}{c}\int_V \frac{1}{R}\left(\frac{\partial \rho_r}{\partial t_r} + (\nabla' \cdot \mathbf{j})\big|_{t_r}\right) d^3\mathbf{x}' = 0 \end{aligned}$$

since the integrand vanishes identically, due to charge conservation.

4. Show that the electromagnetic field tensor given (8.18) may also be written in the covariant form

$$F_{\mu\nu} = \partial_\mu A_\nu - \partial_\nu A_\mu \tag{8.75}$$

We call $F_{\mu\nu}$ the **covariant field tensor**.

(Solution)

We have

$$\begin{aligned} F_{\mu\nu} &= \eta_{\mu\rho}\eta_{\nu\sigma}F^{\rho\sigma} = \eta_{\mu\rho}\eta_{\nu\sigma}\left(\partial^\rho A^\sigma - \partial^\sigma A^\rho\right) \\ &= \partial_\mu A_\nu - \partial_\nu A_\mu \end{aligned}$$

as required.

5. Show that (8.2) and (8.3) can be written in terms of the field tensor as

$$\partial^\rho F^{\mu\nu} + \partial^\nu F^{\rho\mu} + \partial^\mu F^{\nu\rho} = 0 \tag{8.23}$$

(Solution)

For definiteness, let use rectangular coordinates. First, let's take $\rho = 1, \mu = 2$, and $\nu = 3$, so (8.23) reduces to

$$\begin{aligned} \frac{\partial}{\partial x}F^{23} + \frac{\partial}{\partial z}F^{12} + \frac{\partial}{\partial z}F^{31} &= -\frac{\partial B_x}{\partial x} - \frac{\partial B_z}{\partial z} - \frac{\partial B_y}{\partial y} = 0 \\ &\Longrightarrow \frac{\partial B_x}{\partial x} + \frac{\partial B_y}{\partial y} + \frac{\partial B_z}{\partial z} = 0 \end{aligned}$$

or, $\nabla\cdot\mathbf{B}=0$, which is simply (8.2). Now, let's take $\rho=0$, $\mu=1$, and $\nu=2$, so (8.23) reduces to

$$\frac{1}{c}\frac{\partial}{\partial t}(-B_z)+\frac{\partial}{\partial y}(-E_x)+\frac{\partial}{\partial z}(E_y)=0\Longrightarrow\frac{-1}{c}\left[\frac{\partial\mathbf{B}}{\partial t}\right]_z=[\nabla\times\mathbf{E}]_z$$

We see that this is just the z-component of equation (8.3); the derivation of the remaining components is similar.

6. Show that equations (8.1) and (8.4) may be written as

$$\partial_\mu F^{\mu\nu}=\frac{4\pi}{c}J^\nu \tag{8.24}$$

(Solution)

For simplicity, let's use rectangular coordinates. First, let's take $\nu=0$ in (8.24), to obtain

$$\begin{aligned}\partial_0F^{00}+\partial_1F^{10}+\partial_2F^{20}+\partial_3F^{30} &= \frac{4\pi}{c}J^0=4\pi\rho\\ &\Longrightarrow \frac{1}{c}\frac{\partial}{\partial t}(0)+\frac{\partial E_x}{\partial x}+\frac{\partial E_y}{\partial y}+\frac{\partial E_z}{\partial z}=4\pi\rho\\ &\Longrightarrow \nabla\cdot\mathbf{E}=4\pi\rho\end{aligned}$$

which is just (8.1). Now, let's set $\nu=1$ in (8.24) to obtain

$$\begin{aligned}\frac{1}{c}\frac{\partial}{\partial t}(-E_x)+\frac{\partial}{\partial x}(0)+\frac{\partial}{\partial y}(B_z)+\frac{\partial}{\partial z}(-B_y) &= \frac{4\pi}{c}J_x\\ &\Longrightarrow \frac{\partial B_z}{\partial y}-\frac{\partial B_y}{\partial z}=\frac{1}{c}\frac{\partial E_x}{\partial t}+\frac{4\pi}{c}J_x\\ &\Longrightarrow [\nabla\times\mathbf{B}]_x=\frac{1}{c}\left[\frac{\partial\mathbf{E}}{\partial t}\right]_x+\frac{4\pi}{c}J_x\end{aligned}$$

The last equation is just the x-component of (8.24). The verification of the remaining components is similar.

8. Show that the covariant field tensor $F_{\mu\nu}$ is antisymmetric and find its components.

(Solution)

We have $F_{\mu\nu}=\eta_{\mu\rho}\eta_{\nu\sigma}F^{\rho\sigma}$. We first show that $F_{\mu\nu}$ is antisymmetric. Observe that

$$F_{\nu\mu}=\eta_{\nu\rho}\eta_{\mu\sigma}F^{\rho\sigma}=-\eta_{\mu\sigma}\eta_{\nu\rho}F^{\sigma\rho}=-F_{\mu\nu}$$

Now, let's calculate its non-zero, independent components. We find

$$\begin{aligned}F_{01} &= \eta_{0\rho}\eta_{1\sigma}F^{\rho\sigma}=-F^{01}=E_x,\ F_{02}=\eta_{0\rho}\eta_{2\sigma}F^{\rho\sigma}=-F^{02}=E_y\\ F_{03} &= \eta_{0\rho}\eta_{3\sigma}F^{\rho\sigma}=-F^{03}=E_z,\ F_{12}=\eta_{1\rho}\eta_{2\sigma}F^{\rho\sigma}=F^{12}=-B_z,\\ F_{13} &= \eta_{1\rho}\eta_{3\sigma}F^{\rho\sigma}=F^{13}=B_y,\ \ F_{23}=\eta_{2\rho}\eta_{3\sigma}F^{\rho\sigma}=F^{23}=B_y\end{aligned}$$

We summarize our results in matrix form as

$$F_{\mu\nu} = \begin{bmatrix} 0 & E_x & E_y & E_z \\ -E_x & 0 & -B_z & B_y \\ -E_y & B_z & 0 & -B_x \\ -E_z & -B_y & B_x & 0 \end{bmatrix} \tag{8.76}$$

9. Determine the components of the mixed field tensor F_ν^μ.

(Solution)

We have

$$F_\nu^\mu = F^{\mu\sigma}\eta_{\sigma\nu} \,,$$

so let's work out these components in detail:

$$F_0^0 = F^{0\sigma}\eta_{\sigma 0} = 0, \; F_1^0 = F^{0\sigma}\eta_{\sigma 1} = E_x, \; F_2^0 = F^{0\sigma}\eta_{\sigma 2} = E_{y,} \; F_3^0 = F^{0\sigma}\eta_{\sigma 3} = E_z$$

$$F_0^1 = F^{1\sigma}\eta_{\sigma 0} = E_x, \; F_1^1 = F^{1\sigma}\eta_{\sigma 1} = 0, \; F_2^1 = F^{1\sigma}\eta_{\sigma 2} = B_z, \; F_3^1 = F^{1\sigma}\eta_{\sigma 3} = -B_y$$

$$F_0^2 = F^{2\sigma}\eta_{\sigma 0} = E_y, \; F_1^2 = F^{2\sigma}\eta_{\sigma 1} = -B_z, \; F_2^2 = F^{2\sigma}\eta_{\sigma 2} = 0, \; F_3^2 = F^{2\sigma}\eta_{\sigma 3} = B_x$$

$$F_0^3 = F^{3\sigma}\eta_{\sigma 0} = E_z, \; F_1^3 = F^{3\sigma}\eta_{\sigma 1} = B_y, \; F_2^3 = F^{3\sigma}\eta_{\sigma 2} = -B_x, \; F_3^3 = F^{3\sigma}\eta_{\sigma 3} = 0$$

We can summarize our results in matrix form as

$$F_\nu^\mu = \begin{bmatrix} 0 & E_x & E_y & E_z \\ E_x & 0 & B_z & -B_y \\ E_y & -B_z & 0 & B_x \\ E_z & B_y & -B_x & 0 \end{bmatrix} \tag{8.77}$$

The reader should be careful to note that the mixed field tensor is nether symmetric nor antisymmetric.

10. Show that

$$F_{\mu\nu}F^{\mu\nu} = 2\left(\|\mathbf{B}\|^2 - \|\mathbf{E}\|^2\right) \tag{8.78}$$

(Solution)

We have

$$\begin{aligned} F_{\mu\nu}F^{\mu\nu} &= 2(F_{01}F^{01} + F_{02}F^{02} + F_{03}F^{03}) + 2(F_{12}F^{12} + F_{13}F^{13} + F_{23}F^{23}) \\ &= 2(-E_x^2 - E_y^2 - E_z^2) + 2(B_z^2 + B_y^2 + B_x^2) \\ &= 2\left(\|\mathbf{B}\|^2 - \|\mathbf{E}\|^2\right) \end{aligned}$$

11. Show that

$$T^{00} = \frac{1}{8\pi}(\|\mathbf{E}\|^2 + \|\mathbf{B}\|^2)$$

(Solution)

By (8.39), we have

$$T^{00} = -\frac{1}{4\pi}\left(F^{0}_{\alpha}F^{0\alpha} - \frac{1}{4}\eta^{00}F^{\alpha\beta}F_{\alpha\beta}\right) = -\frac{1}{4\pi}\left(F^{0}_{\alpha}F^{0\alpha} - \frac{1}{4}F^{\alpha\beta}F_{\alpha\beta}\right)$$

But

$$\begin{aligned} F^{0}_{\alpha}F^{0\alpha} &= F^{0}_{1}F^{01} + F^{0}_{2}F^{02} + F^{0}_{3}F^{03} = -E_x^2 - E_y^2 - E_z^2 = -\|\mathbf{E}\|^2 \\ F^{\alpha\beta}F_{\alpha\beta} &= 2\left(\|\mathbf{B}\|^2 - \|\mathbf{E}\|^2\right) \quad \text{[By Exercise 10.]} \end{aligned}$$

Consequently,

$$T^{00} = -\frac{1}{4\pi}\left[-\|\mathbf{E}\|^2 - \frac{1}{2}\left(\|\mathbf{B}\|^2 - \|\mathbf{E}\|^2\right)\right] = \frac{1}{8\pi}(\|\mathbf{E}\|^2 + \|\mathbf{B}\|^2)$$

12. Show that the energy-momentum tensor is trace-free.

(Solution)

We can write

$$Tr(\mathbf{T}) \equiv T^{\mu}_{\mu} = T^{0}_{0} + T^{k}_{k}$$

Now,

$$T^{\mu}_{\nu} = \eta_{\nu\sigma}T^{\mu\sigma} \Longrightarrow T^{0}_{0} = T^{00} \text{ but } T^{a}_{b} = -T^{ab}$$

Using the result of Exercise-11, and (8.41), we obtain

$$\begin{aligned} Tr(\mathbf{T}) &\equiv T^{\mu}_{\mu} = T^{0}_{0} - T^{k}_{k} \\ &= \frac{1}{8\pi}(\|\mathbf{E}\|^2 + \|\mathbf{B}\|^2) + \frac{1}{4\pi}\left[\|\mathbf{E}\|^2 + \|\mathbf{B}\|^2 - \frac{3}{2}(\|\mathbf{E}\|^2 + \|\mathbf{B}\|^2)\right] = 0 \end{aligned}$$

13. Show that

$$T^{01} = \frac{1}{4\pi}(\mathbf{E}\times\mathbf{B})_x, \;\; T^{02} = \frac{1}{4\pi}(\mathbf{E}\times\mathbf{B})_y, \;\; T^{03} = \frac{1}{4\pi}(\mathbf{E}\times\mathbf{B})_z$$

(Solution)

From (8.39), we have

$$\begin{aligned} T^{01} &= -\frac{1}{4\pi}\left(F^{0}_{\alpha}F^{1\alpha} - \frac{1}{4}\eta^{01}F^{\alpha\beta}F_{\alpha\beta}\right) = -\frac{1}{4\pi}\left(F^{0}_{\alpha}F^{1\alpha}\right) \\ &= \frac{-1}{4\pi}\left(F^{0}_{0}F^{10} + F^{0}_{1}F^{11} + F^{0}_{2}F^{12} + F^{0}_{3}F^{13}\right) \\ &= \frac{-1}{4\pi}[(0) + (0) + (-E_yB_z) + (E_zB_y)] \\ &= \frac{1}{4\pi}(E_yB_z - E_zB_y) = \frac{1}{4\pi}(\mathbf{E}\times\mathbf{B})_x \end{aligned}$$

Also,

$$\begin{aligned}
T^{02} &= -\frac{1}{4\pi}\left(F^0_\alpha F^{2\alpha} - \frac{1}{4}\eta^{02}F^{\alpha\beta}F_{\alpha\beta}\right) = -\frac{1}{4\pi}\left(F^0_\alpha F^{2\alpha}\right) \\
&= \frac{-1}{4\pi}\left(F^0_0 F^{20} + F^0_1 F^{21} + F^0_2 F^{22} + F^0_3 F^{23}\right) \\
&= \frac{-1}{4\pi}\left[(0) + (E_x B_z) + (0) + (-E_z B_x)\right] \\
&= \frac{1}{4\pi}\left(E_z B_x - E_x B_z\right) = \frac{1}{4\pi}(\mathbf{E}\times\mathbf{B})_y
\end{aligned}$$

The derivation of the remaining relation is similar.

14. *(4-Force Density)* Show that the 4-force density **G** may be written as

$$\mathbf{G} = \gamma\left(\frac{\rho_0}{c}(\mathbf{E}\cdot\mathbf{u}),\ \rho_0\left[\mathbf{E} + \frac{1}{c}(\mathbf{u}\times\mathbf{B})\right]\right) \tag{8.79}$$

(Solution)

By (8.35), we have

$$G^\mu = \frac{\rho_0}{c}F^{\mu\nu}U_\nu$$

so that

$$\begin{bmatrix} G^0 \\ G^1 \\ G^2 \\ G^3 \end{bmatrix} = \frac{\rho_0}{c}\begin{bmatrix} 0 & -E_x & -E_y & -E_z \\ E_x & 0 & -B_z & B_y \\ E_y & B_z & 0 & -B_x \\ E_z & -B_y & B_x & 0 \end{bmatrix}\begin{bmatrix} \gamma c \\ -\gamma u_x \\ -\gamma u_y \\ -\gamma u_z \end{bmatrix}$$

Carrying out the matrix multiplication, and simplifying, we find

$$\begin{aligned}
G^0 &= \frac{\gamma\rho_0}{c}(\mathbf{E}\cdot\mathbf{u}),\ G^1 = \gamma\rho_0\left[E_x + \frac{1}{c}(\mathbf{u}\times\mathbf{B})_x\right] \\
G^2 &= \gamma\rho_0\left[E_y + \frac{1}{c}(\mathbf{u}\times\mathbf{B})_y\right],\ G^3 = \gamma\rho_0\left[E_z + \frac{1}{c}(\mathbf{u}\times\mathbf{B})_z\right]
\end{aligned}$$

from which (8.79) follows immediately.

15. Show that for an infinitesimal proper volume δV_0,

$$\mathbf{F} = \mathbf{G}\,(\delta V_0) \tag{8.80}$$

(Solution)

Recall that

$$\mathbf{F} = \gamma\left(\frac{q}{c}(\mathbf{E}\cdot\mathbf{u}),\ q\left[\mathbf{E} + \frac{1}{c}(\mathbf{u}\times\mathbf{B})\right]\right)$$

Assume that the infinitesimal proper volume δV_0 contains charge q, and let ρ_0 be the proper charge density. Since the charge must be invariant,

$$q = \rho_0\,(\delta V_0) = \rho\,(\delta V)$$

Consequently, we can write

$$\begin{aligned}
\mathbf{F} &= \gamma\left(\frac{\rho\,(\delta V)}{c}(\mathbf{E}\cdot\mathbf{u}),\ \rho\,(\delta V)\left[\mathbf{E}+\frac{1}{c}(\mathbf{u}\times\mathbf{B})\right]\right)\\
&= \gamma\left(\frac{\rho_0\,(\delta V_0)}{c}(\mathbf{E}\cdot\mathbf{u}),\ \rho_0\,(\delta V_0)\left[\mathbf{E}+\frac{1}{c}(\mathbf{u}\times\mathbf{B})\right]\right)\\
&= \gamma\left(\frac{\rho_0}{c}(\mathbf{E}\cdot\mathbf{u}),\ \rho_0\left[\mathbf{E}+\frac{1}{c}(\mathbf{u}\times\mathbf{B})\right]\right)(\delta V_0)\\
&= \mathbf{G}\,(\delta V_0)
\end{aligned}$$

We conclude that **G** is truly the 4-force *density*, as it is simply the 4-force per unit proper volume.

16. *(Field of Point Charge)* Show that relative to the laboratory frame $K(ct,x,y,z)$, the field of a point charge q moving with uniform velocity $\mathbf{v}=v\hat{\mathbf{x}}$ is given by

$$\begin{aligned}
\mathbf{E}_{\parallel} &= \bar{\mathbf{E}}_{\parallel},\ \mathbf{E}_{\perp}=\gamma\bar{\mathbf{E}}_{\perp}\\
\mathbf{B}_{\parallel} &= \mathbf{0},\ \mathbf{B}_{\perp}=\frac{\gamma}{c}\left(\mathbf{v}\times\bar{\mathbf{E}}\right)
\end{aligned}$$

provided that $\bar{K}(c\bar{t},\bar{x},\bar{y},\bar{z},)$ is the proper frame of the charge.

(Solution)

Let $\bar{K}(\bar{x},\bar{y},\bar{z},\bar{t})$ be the proper frame of the charge q, and assume that q is situated at the origin. Relative to the proper frame, the field is purely electrostatic, so $\bar{\mathbf{B}}=\mathbf{0}$, and

$$\bar{\mathbf{E}}=q\frac{\bar{\mathbf{x}}}{\|\bar{\mathbf{x}}\|^3}=\bar{\mathbf{E}}_{\parallel}+\bar{\mathbf{E}}_{\perp}$$

If in (8.74), we interchange barred and unbarred quantities, and replace $\mathbf{v}$ by $-\mathbf{v}$, we obtain the inverse transformation:

$$\begin{aligned}
\mathbf{E}_{\parallel} &= \bar{\mathbf{E}}_{\parallel},\ \mathbf{E}_{\perp}=\gamma\left[\bar{\mathbf{E}}_{\perp}-\frac{1}{c}\left(\mathbf{v}\times\bar{\mathbf{B}}\right)\right]\\
\mathbf{B}_{\parallel} &= \bar{\mathbf{B}}_{\parallel},\ \mathbf{B}_{\perp}=\gamma\left[\bar{\mathbf{B}}_{\perp}+\frac{1}{c}\left(\mathbf{v}\times\bar{\mathbf{E}}\right)\right]
\end{aligned}$$

But $\bar{\mathbf{B}}_{\parallel}=\bar{\mathbf{B}}_{\perp}=\mathbf{0}$, so the equations simplify to

$$\begin{aligned}
\mathbf{E}_{\parallel} &= \bar{\mathbf{E}}_{\parallel},\ \mathbf{E}_{\perp}=\gamma\bar{\mathbf{E}}_{\perp}\\
\mathbf{B}_{\parallel} &= \mathbf{0},\ \mathbf{B}_{\perp}=\frac{\gamma}{c}\left(\mathbf{v}\times\bar{\mathbf{E}}\right)
\end{aligned}$$

as claimed.

Chapter 9

Relativistic Hydrodynamics

There is no philosophy which is not founded upon the knowledge of phenomena, but to get any profit from this knowledge it is absolutely necessary to be a mathematician.

Daniel Bernoulli

9.1. Introduction

In this chapter, we will see how continuous distributions of matter can be described in mathematical terms. Our discussion will naturally lead us to the notion of the energy-momentum tensor for matter. This new tensor, which is particularly important in General Relativity, plays the role of a *source term* in Einstein's field equations.

When we discuss General Relativity, we shall see that Einstein no longer viewed gravity as a force but as a phenomemon which is linked to the curvature of spacetime. Einstein accomplished this geometrization of gravity by assuming that spacetime curvature, in turn, is directly related to the energy-momentum tensor. In other words, he assumed that this tensor somehow tells spacetime how to curve.

We begin our discussion with **classical fluids** where special relativistic effects can be safely ignored. Later on in the chapter, our equations will be suitably modified to accomodate special relativistic effects.

9.2. Newton's Second Law for Fluids

A fluid, such as a liquid or a gas, is a discrete system of particles at the microscopic scale. It turns out, however, that if the mean free path of fluid molecules is much smaller than the characteristic length of the fluid system considered, then it is permissible to ignore its discrete molecular structure, and to treat the fluid as a continuous distribution of matter. Consequently, we shall assume that prevalent physical conditions permit us to make this **continuum hypothesis**, so that all elementary fluid properties may be treated as *continuous* functions of position and time.

This means that we can characterize a fluid as continuous *field*, where we can examine any fluid property, say $F(\mathbf{x},t)$, at each point of physical space, and determine its variation.

The change in any such quantity is given by

$$dF = \frac{\partial F}{\partial t}dt + \frac{\partial F}{\partial x^a}dx^a \tag{9.1}$$

Now suppose further that we wish to determine how this property changes as we follow the motion of a small fluid parcel of *fixed identity*, then it is clear that the differentials dt and $d\mathbf{x}$ are no longer independent, but are related by

$$dx^a = u^a dt \tag{9.2}$$

where $\mathbf{u} = (u^a)$ is the 3-velocity of the fluid. Consequently, (9.1) may be rewritten as

$$\frac{dF}{dt} = \frac{\partial F}{\partial t} + \frac{dx^a}{dt}\frac{\partial F}{\partial x^a} = \frac{\partial F}{\partial t} + u^a \frac{\partial F}{\partial x^a} \tag{9.3}$$

It is important to note that the *total derivative* in (9.3) gives us the time rate of change of the field quantity F as we follow the motion of a fluid parcel, and we call (9.3) the **material derivative** of F. In standard vector notation,

$$\frac{dF}{dt} = \frac{\partial F}{\partial t} + \mathbf{u} \cdot \nabla F \tag{9.4}$$

where $\partial F/\partial t$ gives us the *local rate of change* of F at any given point, while the second term $\mathbf{u} \cdot \nabla F$ is the *convective rate of change* in F which reflects the change in F as the fluid is convected from one point to another in the direction of $\mathbf{u}$.

Since the field variable F was arbitrary, then in particular, we can use the material derivative to define the **acceleration** $\mathbf{a} = (a^k)$ of a small fluid parcel by

$$\mathbf{a} = \frac{d\mathbf{u}}{dt} = \frac{\partial \mathbf{u}}{\partial t} + \mathbf{u} \cdot \nabla \mathbf{u} \tag{9.5}$$

Now, consider a small fluid parcel with density $\rho(\mathbf{x}, t)$ and volume δV. Assume that this parcel experiences a force $\varphi(\delta V)$ where $\varphi = \varphi(\mathbf{x}, t)$ is the *total force* per unit volume. Since $\rho \delta V$ is the mass of the parcel, then by Newton's second law we must have

$$(\rho \delta V)\frac{d\mathbf{u}}{dt} = \varphi(\delta V)$$

and consequently,

$$\rho a^k = \rho \left(\frac{\partial u^k}{\partial t} + u^p \frac{\partial u^k}{\partial x^p} \right) = \varphi^k \tag{9.6}$$

which is **Newton's second law for fluids**. It is clear that the validity of (9.6) is restricted to classical or non-relativistic fluids where $\|\mathbf{u}\| \ll c$.

9.3. Continuity Equation for Fluids

Since mass is conserved in Newton's mechanics, a classical fluid must a continuity equation which is analogous to the one we found for charge conservation. Consider a fixed volume V of fluid of density ρ. Now, the mass of the fluid which is contained in V cannot change

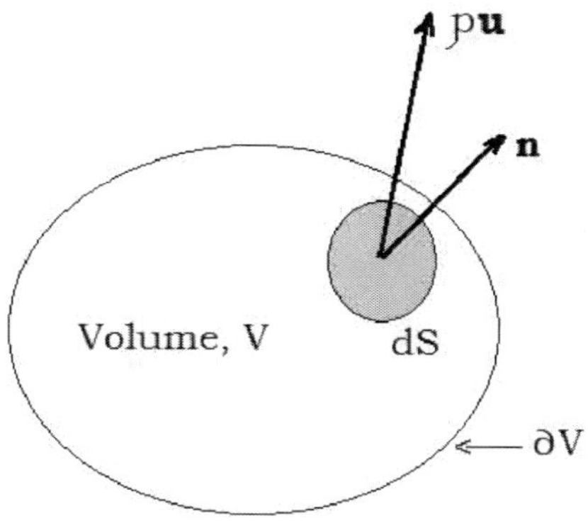

Figure 9.1. Mass leaving a closed volume V.

unless fluid enters or exits this volume through its boundary ∂V. This state of affairs is shown in Figure 9.1.

Clearly, the mass of the fluid leaving V per unit time is just the integral of the **mass-current density** $\rho\mathbf{u}$ taken over the closed surface; in other words,

$$\begin{pmatrix} \text{Mass leaving volume } V \\ \text{per unit time} \end{pmatrix} = \oint_{\partial V} \rho\mathbf{u}\cdot\mathbf{n}\,dS \tag{9.7}$$

On the other hand, the time rate of *decrease* of the total mass m in V is given by

$$-\frac{dm}{dt} = -\frac{d}{dt}\int_V \rho\,dV = -\int_V \frac{\partial\rho}{\partial t}\,dV \tag{9.9}$$

Equating (9.8) and (9.9), we have

$$-\int_V \frac{\partial\rho}{\partial t}\,dV = \oint_{\partial V} \rho\mathbf{u}\cdot\mathbf{n}\,dS$$

Employing the divergence theorem, we now convert the righthand side to a volume integral, and obtain

$$-\int_V \frac{\partial\rho}{\partial t}\,dV = \int_V (\nabla\cdot\rho\mathbf{u})\,dV \Longrightarrow \int_V \left[\frac{\partial\rho}{\partial t} + \nabla\cdot(\rho\mathbf{u})\right]\,dV = 0$$

Since the volume V is arbitrary, we must have

$$\frac{\partial\rho}{\partial t} + \nabla\cdot(\rho\mathbf{u}) = 0 \tag{9.10}$$

and this equation is known as the **continuity equation**; it expresses mass conservation for non-relativistic fluids.

9.4. The Stress Tensor

Consider a small fluid parcel which occupies a volume V. Clearly, the fluid inside this volume will exert forces on any fluid exterior to this volume. These forces must include pressure forces as well as viscous (frictional) forces on the neighboring fluid.

Let $\mathbf{f} = (f^k)$ be the force exerted by the fluid matter *inside* V on the fluid matter which is *outside* V. Now, for a **perfect fluid** which is at rest, this force solely consists of a pressure force which has the same direction as the unit normal $\mathbf{n}$ and is proportional to the area dS, so that

$$\mathbf{f} = p(\mathbf{x}, t)\mathbf{n}dS \tag{9.11}$$

where $p(\mathbf{x}, t)$ is the **fluid pressure** at the location $\mathbf{x}$, at time t.

For a general fluid, however, **shear forces**, which act *tangentially* along the boundary of V, may be present in addition to the pressure force described by (9.11). Consequently, for a general fluid, we assume that the force is still proportional to the area dS, and write

$$f^k = T^{kl} n_l dS \tag{9.12}$$

where the outward unit normal is $\mathbf{n} = (n_1, n_2, n_3)$, and the quantities T^{kl} are the components of the **stress tensor T**.

If we write the components of **T** is matrix form as

$$\mathbf{T} = (T^{ab}) = \begin{bmatrix} T^{11} & T^{12} & T^{13} \\ T^{21} & T^{22} & T^{23} \\ T^{31} & T^{32} & T^{33} \end{bmatrix} \tag{9.13}$$

then we can identify its diagonal elements T^{11}, T^{22}, and T^{33}, with *normal stresses*, while its off-diagonal elements represent*shear stresses*. In other words,

$$T^{ab} = \begin{cases} \text{Force (per unit area) in direction } a \text{ on a} \\ \text{surface } dS \text{ whose normal has direction } b. \end{cases} \tag{9.14}$$

This state of affairs is shown in Figure-9.2 below. Here, we can see the various stresses that an elemental rectangular volume δV of fluid, whose edges are parallel to the coordinate axes, exerts on the surrounding fluid. For the sake of clarity, however, some of the stresses have been omitted from the diagram.

Now, it turns out that the stress tensor is *symmetric*. To see this, consider a small fluid volume $\delta V = \delta x^1 \delta x^2 \delta x^3$ sitting in the x^1x^2-plane, as shown in Figure 9.3. Let's calculate the 3-component of the torque Γ_3 on this fluid element about the centroid $C(x, y, z)$. Recall that the torque $\mathbf{\Gamma} = \mathbf{x} \times \mathbf{f}$ where $\mathbf{x}$ is a position vector drawn from C to the point of application of the force $\mathbf{f}$. Clearly,

$$\Gamma_3 = \frac{\delta x^1}{2} f_2 - \frac{\delta x^2}{2} f_1 \tag{9.15}$$

But, the net force f_1 *acting on this element* in the x^1-direction is given by

$$\begin{aligned} f_1 &= (\delta x^1 \delta x^3) \left[\left(-T^{12}\big|_C + \frac{\delta x^2}{2} \frac{\partial T^{12}}{\partial x^2}\bigg|_C \right) + \left(-T^{12}\big|_C - \frac{\delta x^2}{2} \frac{\partial T^{12}}{\partial x^2}\bigg|_C \right) \right] \\ &= -2T^{12} \left(\delta x^1 \delta x^3 \right) \end{aligned} \tag{9.16}$$

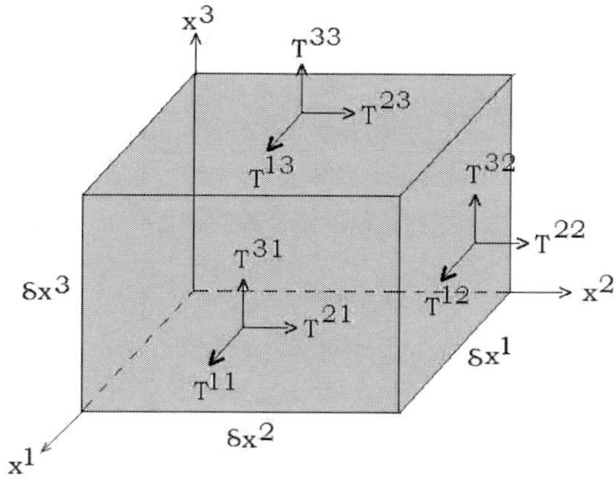

Figure 9.2. A small fluid element δV exerts stresses on the surrounding fluid. Some stresses have been omitted for clarity.

while the net force f_2 in the x^2-direction is

$$\begin{aligned} f_2 &= \left(\delta x^2 \delta x^3\right)\left[\left(-T^{21}\big|_C + \frac{\delta x^1}{2}\frac{\partial T^{21}}{\partial x^1}\bigg|_C\right) + \left(-T^{21}\big|_C - \frac{\delta x^1}{2}\frac{\partial T^{21}}{\partial x^1}\bigg|_C\right)\right] \quad (9.17) \\ &= -2T^{21}\left(\delta x^2 \delta x^3\right) \end{aligned}$$

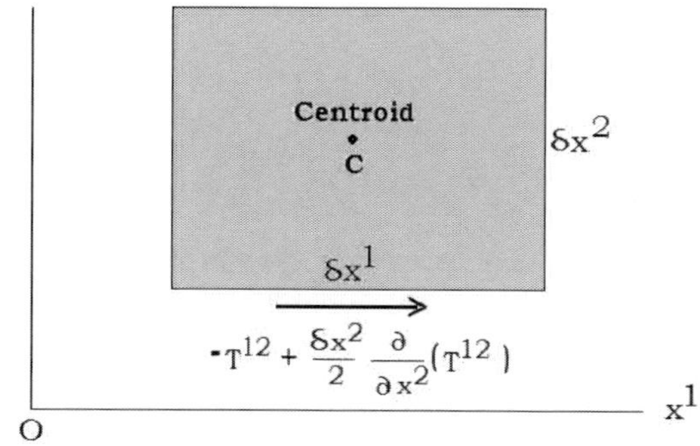

Figure 9.3. A bird's eye view of the stresses on the top and bottom faces of a small fluid element.

If we now substitute (9.16) and (9.17) into (9.15) and simplify, we obtain

$$\Gamma_3 = \frac{\delta x^1}{2}\left[-2T^{21}\left(\delta x^2 \delta x^3\right)\right] - \frac{\delta x^2}{2}\left[-2T^{12}\left(\delta x^1 \delta x^3\right)\right] = (\delta V)\left(T^{12} - T^{21}\right) \quad (9.18)$$

This equation implies that the angular momentum of the fluid element would continue to change - even in the *absence* of external forces. Consequently, we must have $T^{12} = T^{21}$.

By calculating the remaining components of the torque, we can also show that $T^{23} = T^{32}$ and $T^{31} = T^{13}$. So, the stress tensor is symmetric, and we have

$$T^{ab} = T^{ba} \tag{9.19}$$

in general.

9.5. Euler's Equation

Consider the forces acting on a small fluid element with volume V, as shown in Figure 9.4. In general, our fluid parcel will experience a **body force** $\mathbf{g} = (g^k)$ in addition to the familiar **stress force** $\mathbf{f} = (f^k)$. Body forces are present whenever a fluid parcel moves under the influence of an external field such as a gravitational field. Consequently, body forces act *at each point* of a fluid element, and they are always proportional to the *mass* of the fluid element.

Now, by (9.12), the force due to stress acting upon the fluid contained in V is given by

$$f^k = -\oint_{\partial V} T^{pk} n_p dS = -\int_V \partial_p T^{pk} dV \tag{9.20}$$

where we have used a *negative* sign, since we are computing the stress force on V due to fluid which is *outside* this small volume.

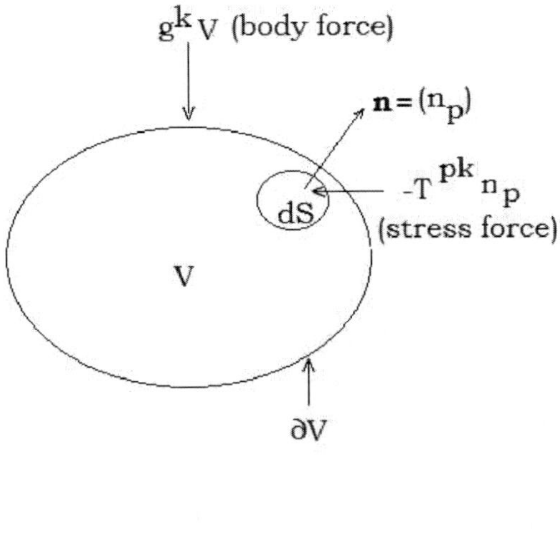

Figure 9.4. Volume and surface forces acting on a small fluid element.

Since the stress force per unit volume is simply $-\partial_p T^{pk}$, and the body force per unit volume is g^k, then the total force φ^k per unit volume is simply

$$\varphi^k = g^k - \partial_p T^{pk} \tag{9.21}$$

Consequently, by Newton's second law for fluids, we have

$$\rho\left(\partial_t u^k + u^p \partial_p u^k\right) = g^k - \partial_p T^{pk} \tag{9.22}$$

We can rewrite the lefthand side of this equation as

$$\partial_t\left(\rho u^k\right) - u^k\partial_t\rho + u^p\partial_p u^k = g^k - \partial_p T^{pk} \tag{9.23}$$

Recalling the continuity equation, we can substitute $\partial_t\rho = -\partial_p(\rho u^p)$ into the last equation to give

$$\partial_t\left(\rho u^k\right) + u^k\partial_p(\rho u^p) + u^p\partial_p u^k = g^k - \partial_p T^{pk} \tag{9.23a}$$

so that

$$\partial_t\left(\rho u^k\right) + \partial_p(\rho u^p u^k) = g^k - \partial_p T^{pk} \tag{9.24}$$

Finally, (9.24) can be rewritten as

$$\partial_t\left(\rho u^k\right) + \partial_p\left(\rho u^p u^k + T^{pk}\right) = g^k \tag{9.25}$$

which is **Euler's equation**. This equation, together with the continuity equation, controls the dynamics of any non-relativistic, continuous distribution of matter.

9.6. Relativistic Dust

We now turn our attention to finding the relativistic equations of motion for a continuous distribution of matter. First, we will take a look at the simplest distribution of matter that is possible, namely, **relativistic dust**. By "dust" we mean an incoherent collection of fluid particles that do not interact with one another in any way. Thus, pressure forces and shear forces are absent in such a field of matter. Consider the motion of a small fluid parcel with volume δV consisting of relativistic dust. Owing to the Lorentz contraction, we first observe that

$$\rho = \gamma^2\rho_0 \tag{9.26}$$

where $\gamma = \gamma(u)$, while ρ and ρ_0 are the density and proper density of the parcel, respectively, relative to an inertial frame K.

Clearly, the 3-momentum of our parcel is $(\rho\delta V)u^a$, so that

$$\frac{d}{dt}\left[(\rho\delta V)u^a\right] = g^a\,(\delta V) \tag{9.27}$$

where $\mathbf{g} = (g^a)$ is the *total 3-force* per unit volume. Note the necessity of using the material derivative operator d/dt in this equation, since we are following the motion of a particular parcel of fluid. Taking the indicated derivative in the last equation, we obtain

$$(\delta V)\frac{d}{dt}(\rho u^a) + \rho u^a\frac{d}{dt}(\delta V) = g^a\,(\delta V) \tag{9.28}$$

But it may be shown that [See the Exercises]:

$$\frac{d}{dt}(\delta V) = (\delta V)\frac{\partial u^b}{\partial x^b} \tag{9.29}$$

while by the definition of the material derivative, we have

$$\frac{d}{dt}(\rho u^a) = \frac{\partial}{\partial t}(\rho u^a) + u^b \frac{\partial}{\partial x^b}(\rho u^a) \tag{9.30}$$

Substituting (9.29) and (9.30) into (9.28), we find

$$(\delta V)\left[\frac{\partial}{\partial t}(\rho u^a) + u^b \frac{\partial}{\partial x^b}(\rho u^a)\right] + \rho u^a \frac{\partial u^b}{\partial x^b}(\delta V) = g^a (\delta V)$$

Dividing out δV, and simplifying the last equation becomes

$$\frac{\partial}{\partial t}(\rho u^a) + \frac{\partial}{\partial x^b}\left(\rho u^a u^b\right) = g^a \tag{9.31}$$

We now seek to write this equation in a manifestly covariant form. Recalling that $x^0 = ct$, $\rho = \gamma^2 \rho_0$ and $\mathbf{U} = \gamma(c, \mathbf{u})$, this equation can be rewritten as

$$\frac{\partial}{\partial x^0}(\rho_0 U^0 U^a) + \frac{\partial}{\partial x^b}\left(\rho_0 U^a U^b\right) = g^a$$

or more succinctly as

$$\frac{\partial}{\partial x^\mu}(\rho_0 U^\mu U^a) = g^a \tag{9.32}$$

We're almost there, but since $a = 1, 2, 3$ we see that (9.32) only gives us three equations governing the motion. We now seek the missing fourth equation which must involve the energy density of the matter field.

Let $w = \rho c^2$ denote the energy density of the fluid particles in δV so that

$$\frac{d}{dt}(w\delta V) = c^2 \frac{d}{dt}(\rho \delta V) = (\mathbf{g}\cdot\mathbf{u})\,\delta V \tag{9.33}$$

where, once again, $\mathbf{g}$ is the total 3-force density. The last equation may be rewritten as

$$\rho \frac{d}{dt}(\delta V) + (\delta V)\frac{d\rho}{dt} = \frac{1}{c^2}(\mathbf{g}\cdot\mathbf{u})\,\delta V$$

which, in turn, by (9.29) becomes

$$\rho\,(\delta V)\frac{\partial u^b}{\partial x^b} + (\delta V)\left[\frac{\partial \rho}{\partial t} + u^b \frac{\partial \rho}{\partial x^b}\right] = \frac{1}{c^2}(\mathbf{g}\cdot\mathbf{u})\,\delta V$$

Dividing out δV and simplifying, we obtain

$$\frac{\partial \rho}{\partial t} + \frac{\partial}{\partial x^b}(\rho u^b) = \frac{1}{c^2}(\mathbf{g}\cdot\mathbf{u}) \tag{9.34}$$

which, in the language of 4-vectors, may be written as

$$\frac{\partial}{\partial t}(\rho_0 U^0 U^0) + \frac{\partial}{\partial x^b}(\rho_0 U^0 U^b) = \frac{1}{c}(\mathbf{g}\cdot\mathbf{u}) \tag{9.35}$$

We now seek to write the righthand side of (9.35) as a the time component of bonefide 4-vector. We define the **4-force density G** as

$$\mathbf{G} = \left(\frac{1}{c}\left(\mathbf{g}\cdot\mathbf{u}\right), \mathbf{g}\right) \tag{9.36}$$

To see that **G** is truly a 4-vector, observe that by (7.53), the 4-force **F** acting on the matter inside δV may be written as

$$\mathbf{F} = \gamma\left(\frac{\mathbf{f}\cdot\mathbf{u}}{c}, \mathbf{f}\right) = \gamma\left(\frac{1}{c}\left(\mathbf{g}\cdot\mathbf{u}\right)\delta V,\ \mathbf{g}\delta V\right) = \gamma\left(\frac{1}{c}\left(\mathbf{g}\cdot\mathbf{u}\right),\ \mathbf{g}\right)\delta V$$

Due to the Lorentz contraction, however, $\delta V = \gamma^{-1}\delta V_0$ where δV_0 is the proper volume. Consequently, the last equation becomes

$$\mathbf{F} = \left(\frac{1}{c}\left(\mathbf{g}\cdot\mathbf{u}\right),\ \mathbf{g}\right)\delta V_0 \tag{9.37}$$

Since **F** is a 4-vector, and the proper volume δV_0 is a scalar invariant, then we see that $\mathbf{G} = (G^\mu)$ is a genuine 4-vector. This means that (9.35) may be nicely rewritten as

$$\frac{\partial}{\partial t}\left(\rho_0 U^0 U^0\right) + \frac{\partial}{\partial x^b}\left(\rho_0 U^0 U^b\right) = G^0 \tag{9.38}$$

and based upon the above analysis, we can interpret the components of the 4-force density $\mathbf{G} = (G^\mu)$ physically as

$$\begin{cases} \gamma G^0 \delta V = \text{Work done on matter inside } \delta V \text{ per unit time.} \\ \gamma G^a \delta V = a\text{-th component of the total 3-force } \mathbf{g} \text{ exerted on } \delta V. \end{cases} \tag{9.39}$$

Finally, we can combine (9.32) and (9.38) into a single manifestly covariant equation which governs the motion of relativistic dust:

$$\frac{\partial}{\partial x^\mu}\left(T^{\mu\nu}\right) = G^\nu \tag{9.40}$$

where

Relativistic Dust

$$T^{\mu\nu} = \rho_0 U^\mu U^\nu \tag{9.41}$$

is called the **energy-momentum tensor** of relativistic dust. We note in passing that, the like the stress-energy tensor for a non-relativistic fluid, our new tensor $\mathbf{T} = (T^{\mu\nu})$ is *symmetric*. It's easy to show that the components of this tensor may be written in matrix form as

$$\mathbf{T} = \left(T^{\mu\nu}\right) = \rho \begin{bmatrix} c^2 & cu_x & cu_y & cu_z \\ cu_x & u_x^2 & u_x u_y & u_x u_z \\ cu_y & u_y u_x & u_y^2 & u_y u_z \\ cu_z & u_z u_x & u_z u_y & u_z^2 \end{bmatrix} \tag{9.42}$$

Secondly, in the absence of external forces, (9.40) implies that

$$\frac{\partial}{\partial x^{\mu}}\left(T^{\mu\nu}\right)=0 \tag{9.43}$$

This is a *conservation equation*; it expresses the conservation of energy and momentum.

9.7. Relativistic Perfect Fluids

Now, we attempt to generalize the results of our previous work on relativistic dust to perfect fluids. In a **perfect fluid**, we allow for collisions amongst its constituent particles to create pressure, but will still ignore the possibility of shearing forces between adjacent fluid elements. Thus, for a perfect fluid which is at rest, we assume that the provisional stress tensor $\theta=(\theta^{ab})$ assumes the form

$$\theta^{ab}=p(\mathbf{x},t)\delta^{ab} \tag{9.44}$$

where $p(\mathbf{x},t)$ is the pressure. Now, in the limit $\theta^{ab}\longrightarrow 0$, we should recover relativistic dust, so we assume that the energy-momentum tensor for a perfect fluid looks like

$$T^{\mu\nu}=T^{\mu\nu}_{\text{dust}}+\theta^{\mu\nu}=\rho_0 U^{\mu}U^{\nu}+\theta^{\mu\nu} \tag{9.45}$$

where

$$\theta=(\theta^{\mu\nu})=\begin{bmatrix} 0 & 0 & 0 & 0 \\ 0 & \theta^{11} & \theta^{12} & \theta^{13} \\ 0 & \theta^{21} & \theta^{22} & \theta^{23} \\ 0 & \theta^{31} & \theta^{32} & \theta^{33} \end{bmatrix} \tag{9.46}$$

Furthermore, in order to guarantee the covariance of (9.45), we *require* that under a Lorentz transformation, the components of our new tensor transform as a genuine 4-tensor:

$$\bar{\theta}^{\mu\nu}=\Lambda^{\mu}_{\sigma}\,\Lambda^{\nu}_{\tau}\,\theta^{\sigma\tau} \tag{9.47}$$

Now, let the inertial frame $\bar{K}(c\bar{t},\bar{\mathbf{x}})$ represent the instantaneous rest frame of a parcel of fluid. So, relative to this frame, we have

$$\bar{\theta}=\left(\bar{\theta}^{\mu\nu}\right)=\begin{bmatrix} 0 & 0 & 0 & 0 \\ 0 & p & 0 & 0 \\ 0 & 0 & p & 0 \\ 0 & 0 & 0 & p \end{bmatrix} \tag{9.48}$$

Under a general Lorentz boost $\mathbf{u}$, we can write the inverse transformation $\bar{x}^{\mu}\longrightarrow x^{\mu}$ as

$$x^a = \left[\delta^{ab}+\frac{(\gamma-1)}{\|\mathbf{u}\|^2}u^a u^b\right]\bar{x}^b+\gamma u^a t \tag{9.49a}$$

$$x^0 = \gamma\left(\bar{x}^0+\frac{1}{c}u^a x^a\right) \tag{9.49b}$$

where $a, b = 1, 2, 3$. Consequently, we can use equations (9.47) - (9.49) to find the components of θ relative to our laboratory frame $K(ct, \mathbf{x})$. To see this, we first invert (9.47) to obtain

$$\theta^{\mu\nu} = \left(\Lambda^{-1}\right)^{\mu}_{\ \sigma} \left(\Lambda^{-1}\right)^{\nu}_{\ \tau} \bar{\theta}^{\sigma\tau} \tag{9.50}$$

So, for example, if we let $\mu = \nu = 0$, then we get

$$\theta^{00} = \left(\Lambda^{-1}\right)^{0}_{\ \sigma} \left(\Lambda^{-1}\right)^{0}_{\ \tau} \bar{\theta}^{\sigma\tau} = p\left(\gamma^2 - 1\right) \tag{9.51}$$

Similarly, after some tedious algebra, we find that

$$\begin{aligned}
\theta^{0b} &= p\gamma U^b \\
\theta^{ab} &= \frac{1}{c^2} p U^a U^b \qquad \text{(for } a \neq b\text{)} \\
\theta^{aa} &= p\left(1 + \frac{1}{c^2} U^a U^a\right) \quad \text{(no summation on } a\text{)}
\end{aligned} \tag{9.52}$$

We can summarize (9.51) and (9.52) by writing

$$\theta^{\mu\nu} = p\left(-\eta^{\mu\nu} + \frac{1}{c^2} U^\mu U^\nu\right) \tag{9.53}$$

Now, if we substitute (9.53) into (9.45), and rearrange terms, then we finally obtain

Relativistic Perfect Fluid

$$T^{\mu\nu} = -p\eta^{\mu\nu} + \left(\rho_0 + \frac{1}{c^2} p\right) U^\mu U^\nu \tag{9.54}$$

as the energy-momentum tensor for a perfect fluid. Observe that this tensor is *symmetric*.

Furthermore, in the absence of external forces, a perfect fluid satisfies

$$\frac{\partial}{\partial x^\mu}\left(T^{\mu\nu}\right) = 0 \tag{9.55}$$

which expresses the conservation of energy and momentum. In fact, it may be shown [See the Exercises] that (9.55) implies the **relativistic continuity equation**

$$\frac{\partial}{\partial x^\mu}\left[\left(\rho_0 + \frac{1}{c^2} p\right) U^\mu\right] = \frac{1}{c^2} U^\mu \frac{\partial p}{\partial x^\mu} \tag{9.56}$$

and the **relativistic Euler equation** for a perfect fluid,

$$\left(\rho_0 + \frac{1}{c^2} p\right) U^\mu \frac{\partial U^\nu}{\partial x^\mu} = \left(\eta^{\mu\nu} - \frac{1}{c^2} U^\mu U^\nu\right) \frac{\partial p}{\partial x^\mu} \tag{9.57}$$

These equations, in turn, suggest that the quantity

$$\sigma = \left(\rho_0 + \frac{1}{c^2} p\right) \tag{9.58}$$

plays the role of the *inertial mass* (per unit volume), and it often appears in the equations of relativistic hydrodynamics. It is sometimes called the **specific relativistic enthalpy**.

9.8. Exercises

1. Derive equation (9.29); that is, show that

$$\frac{d}{dt}(\delta V) = (\delta V)\frac{\partial u^a}{\partial x^a}$$

for a small fluid parcel occupying an infinitesimal volume δV and with 3-velocity $\mathbf{u} = (u^a)$.

(Solution)

If we follow the motion of a small volume δV of fluid, bounded by the closed surface Σ, the change in δV during a time interval dt is

$$d(\delta V) = (dt)\oint_{\Sigma} \mathbf{u}\cdot d\mathbf{S} = \int_{\delta V} (\nabla\cdot\mathbf{u})\, dV$$

Now, since δV is *infinitesimal*, the last equation gives

$$d(\delta V) = (\nabla\cdot\mathbf{u})\,\delta V\ (dt)$$

which implies

$$\frac{d}{dt}(\delta V) = (\nabla\cdot\mathbf{u})\,\delta V = (\delta V)\frac{\partial u^a}{\partial x^a}.$$

2. *(Incompressible Fluid)* A non-relativistic fluid motion is said to be **incompressible** if $\nabla\cdot\mathbf{u} = 0$. Show that for an incompressible flow, the density of a small fluid parcel cannot change with time:

$$\frac{d\rho}{dt} = 0 \tag{9.59}$$

(Solution)

We have

$$\begin{aligned}\frac{d\rho}{dt} &= \frac{\partial\rho}{\partial t} + \mathbf{u}\cdot\nabla\rho = \left[\frac{\partial\rho}{\partial t} + \nabla\cdot(\rho\mathbf{u})\right] - \rho\,(\nabla\cdot\mathbf{u}) \\ &= 0 - \rho\,(\nabla\cdot\mathbf{u}) \quad \text{[By the continuity equation.]} \\ &= 0 \qquad \text{[Since the fluid is incompressible.]}\end{aligned}$$

3. *(Euler's Equation)* Show that for a non-relativistic, perfect fluid, Euler's equation (9.23a) may be written as

$$\frac{\partial\mathbf{u}}{\partial t} + \mathbf{u}\cdot\nabla\mathbf{u} = -\frac{1}{\rho}\nabla p\,, \tag{9.60}$$

assuming an incompressible flow.

(Solution)

If we let $c \longrightarrow \infty$ in (9.54), we see that $\rho \longrightarrow \rho_0$, and the relativistic energy-momentum tensor reduces to the stress tensor for a non-relativistic, perfect fluid:

Non-relativistic Perfect Fluid

$$T^{ab} = p\delta^{ab} + \rho u^a u^b \tag{9.61}$$

By (9.24), however, we have

$$\partial_t(\rho u^a) + \partial_b(\rho u^a u^b) = -\,\partial_b T^{ab} + g^a$$

On the other hand,

$$\begin{aligned}\partial_b T^{ab} &= \partial_b\left(p\delta^{ab} + \rho u^a u^b\right) = \partial_a p\ + \partial_b\left(\rho u^a u^b\right)\\ &= \partial_a p\ + \rho u^a\left(\partial_b u^b\right) + u^b \partial_b(\rho u^a)\\ &= \partial_a p\ + u^b\partial_b(\rho u^a) \quad [\text{Since } \partial_b u^b = 0.]\end{aligned}$$

Substituting this result into the previous equation, we get

$$\partial_t(\rho u^a) + \partial_b(\rho u^a u^b) + u^b \partial_b(\rho u^a) = -\,\partial_a p$$

Taking the indicated derivatives, and rearranging terms, we obtain

$$\rho\left[\partial_t u^a + u_b\partial_b u^a\right] + u^a\left[\partial_t\rho + \partial_b(\rho u^b)\right] = -\,\partial_a p$$

But the second term in the brackets is zero by virtue of the continuity equation, so the last equation reduces to

$$\rho\left[\partial_t u^a + u_b\partial_b u^a\right] = -\,\partial_a p$$

which may be written in vector form as (9.60).

4. *(Relativistic Dust)* Show that relativistic dust satisfies the continuity equation

$$\frac{\partial\rho}{\partial t} + \nabla\cdot(\rho\mathbf{u}) = 0$$

(Solution)

Setting $\nu = 0$ in (9.43) gives

$$\frac{\partial}{\partial x^\mu}\left(T^{\mu 0}\right) = \frac{\partial}{\partial x^\mu}\left(\rho_0 U^\mu U^0\right) = 0$$

Recalling that $\mathbf{U} = \gamma(c, \mathbf{u})$, where $\mathbf{u}$ is the 3-velocity, and $x^0 = ct$, the last equation becomes

$$\frac{\partial}{\partial x^\mu}\left(\rho_0 U^\mu U^0\right) = \frac{\partial}{c\partial t}\left(\rho_0\gamma^2 c^2\right) + \frac{\partial}{\partial x^a}\left(\rho_0\gamma^2 u^a c\right) = 0$$

Recalling that $\rho = \gamma^2 \rho_0$, the last equation simplifies to

$$\frac{\partial \rho}{\partial t} + \frac{\partial}{\partial x^a}(\rho u^a) = 0 \implies \frac{\partial \rho}{\partial t} + \nabla \cdot (\rho \mathbf{u}) = 0$$

5. *(Relativistic Dust)* In the absence of external forces, show that relativistic dust must satisfy the equation:

$$\rho \frac{d\mathbf{u}}{dt} = \rho \left(\frac{\partial \mathbf{u}}{\partial t} + \mathbf{u} \cdot \nabla \mathbf{u} \right) = \mathbf{0} \tag{9.62}$$

(Solution)

Setting $\nu = a$ in (9.43) gives

$$\frac{\partial}{\partial x^\mu}(T^{\mu a}) = \frac{\partial}{\partial x^0}(\rho_0 U^0 U^a) + \frac{\partial}{\partial x^b}\left(\rho_0 U^a U^b\right) = 0$$

Once again, recalling that $\mathbf{U} = \gamma(c, \mathbf{u})$ and $x^0 = ct$, the last equation simplifies to

$$\frac{\partial}{\partial t}(\gamma^2 \rho_0 u^a) + \frac{\partial}{\partial x^b}\left(\gamma^2 \rho_0 u^a u^b\right) = 0$$

Recalling that $\rho = \gamma^2 \rho_0$, the last equation becomes

$$\frac{\partial}{\partial t}(\rho u^a) + \frac{\partial}{\partial x^b}\left(\rho u^a u^b\right) = 0,$$

or equivalently,

$$\rho \left[\frac{\partial u^a}{\partial t} + u^b \partial_b u^a \right] + u^a \left[\frac{\partial \rho}{\partial t} + \frac{\partial}{\partial x^b}(\rho u^b) \right] = 0$$

Invoking the continuity equation, the last equation becomes

$$\rho \left[\frac{\partial u^a}{\partial t} + u^b \partial_b u^a \right] = 0 \Longrightarrow \rho \left(\frac{\partial \mathbf{u}}{\partial t} + \mathbf{u} \cdot \nabla \mathbf{u} \right) = \rho \frac{d\mathbf{u}}{dt} = \mathbf{0}$$

as required. Observe that (9.62) is just Euler's equation for the case where both pressure and external forces are absent.

6. *(Relativistic Dust)* Show that in the absence of external forces, the momentum of a parcel of relativistic dust is *conserved*.

(Solution)

Clearly, the 3-momentum of a small parcel δV of relativistic dust is simply

$$\mathbf{p} = \rho \, (\delta V) \, \mathbf{u}$$

Consequently, the time rate of change of momentum is

$$\begin{aligned}
\frac{d\mathbf{p}}{dt} &= \rho\,(\delta V)\,\frac{d\mathbf{u}}{dt}+\rho\mathbf{u}\frac{d}{dt}\,(\delta V)+\mathbf{u}(\delta V)\,\frac{d\rho}{dt} \\
&= \rho\,(\delta V)\,\frac{d\mathbf{u}}{dt}+\rho\mathbf{u}\,[\delta V\,(\nabla\cdot\mathbf{u})]+\mathbf{u}(\delta V)\,\frac{d\rho}{dt} \quad \text{[By (9.29)]} \\
&= (\delta V)\left\{\rho\frac{d\mathbf{u}}{dt}+\rho\mathbf{u}\,(\nabla\cdot\mathbf{u})+\mathbf{u}\frac{d\rho}{dt}\right\} \\
&= (\delta V)\left\{\rho\mathbf{u}\,(\nabla\cdot\mathbf{u})+\mathbf{u}\frac{d\rho}{dt}\right\} \quad \left[\text{Since } \rho\frac{d\mathbf{u}}{dt}=\mathbf{0} \text{ by (9.62).}\right] \\
&= (\delta V)\left\{\rho\mathbf{u}\,(\nabla\cdot\mathbf{u})+\mathbf{u}\left(\frac{\partial\rho}{\partial t}+\mathbf{u}\cdot\nabla\rho\right)\right\} \\
&= \mathbf{u}\,(\delta V)\left\{\frac{\partial\rho}{\partial t}+\nabla\cdot(\rho\mathbf{u})\right\}=\mathbf{0} \quad \text{[By Exercise 4.]}
\end{aligned}$$

In other words, $d\mathbf{p}/dt=\mathbf{0}$ so momentum is conserved.

7. *(Relativistic Perfect Fluid)* Show that in the absence of external forces, a relativistic perfect fluid satisfies the relativistic continuity equation:

$$\frac{\partial}{\partial x^\mu}\left[\left(\rho_0+\frac{1}{c^2}p\right)U^\mu\right]=\frac{1}{c^2}U^\mu\frac{\partial p}{\partial x^\mu} \tag{9.56}$$

(Solution)

Recalling (9.54), we can write

$$T^{\mu\nu}=-p\eta^{\mu\nu}+\sigma U^\mu U^\nu,\ \text{ where } \sigma=\left(\rho_0+\frac{1}{c^2}p\right)$$

Consequently, since $\partial_\mu T^{\mu\nu}=0$, we have

$$\partial_\mu T^{\mu\nu}=-\eta^{\mu\nu}\partial_\mu p+\sigma U^\mu\partial_\mu U^\nu+\sigma U^\nu\partial_\mu U^\mu+U^\mu U^\nu\partial_\mu\sigma=0 \tag{9.63}$$

Next, we contract the last equation with contract U_ν, to obtain

$$U_\nu\partial_\mu T^{\mu\nu}=-\eta^{\mu\nu}U_\nu\partial_\mu p+\sigma U^\mu\left(U_\nu\partial_\mu U^\nu\right)+\sigma\left(U^\nu U_\nu\right)\partial_\mu U^\mu+U^\mu\left(U_\nu U^\nu\right)\partial_\mu\sigma=0$$

Now, if we observe that

$$\begin{aligned}
U^\nu U_\nu &= \|\mathbf{U}\|^2=c^2 \\
U_\nu\partial_\mu U^\nu &= \frac{1}{2}\partial_\mu\left(U_\nu U^\nu\right)=\frac{1}{2}\partial_\mu\left(c^2\right)=0\,,
\end{aligned}$$

then the last equation simplifies to

$$U_\nu\partial_\mu T^{\mu\nu}=-U^\mu\partial_\mu p+\sigma c^2\partial_\mu U^\mu+c^2U^\mu\partial_\mu\sigma=0$$

Rearranging terms, we obtain

$$\sigma\partial_\mu U^\mu+U^\mu\partial_\mu\sigma=\frac{1}{c^2}U^\mu\partial_\mu p\Longrightarrow\partial_\mu\left(\sigma U^\mu\right)=\frac{1}{c^2}U^\mu\partial_\mu p$$

Finally, recalling the definition of σ, we obtain

$$\frac{\partial}{\partial x^\mu}\left[\left(\rho_0+\frac{1}{c^2}p\right)U^\mu\right]=\frac{1}{c^2}U^\mu\frac{\partial p}{\partial x^\mu}$$

as required.

8. *(Relativistic Perfect Fluid)* Show that (9.56), the relativistic continuity equation, reduces to the classical continuity equation in the limit of small velocities, $\|\mathbf{u}\| << c$.

(Solution)

For $\|\mathbf{u}\| << c$, we have $\gamma \approx 1$, so that

$$\frac{p}{c^2}\approx 0,\ \mathbf{U}\approx(c,\mathbf{u}),\text{ and }\rho\approx\rho_0$$

Note that the approximation $p/c^2 \approx 0$ makes sense since fluid pressure is caused by collisions of fluid particles, and we are assuming small velocities. Consequently, for small velocities, (9.56) becomes

$$\frac{\partial}{\partial x^\mu}(\rho U^\mu)=\frac{1}{c^2}U^\mu\frac{\partial p}{\partial x^\mu}$$

Writing out the implied sums, we obtain

$$\begin{aligned}\frac{\partial}{\partial x^0}(\rho U^0)+\frac{\partial}{\partial x^a}(\rho U^a) &= \frac{1}{c^2}\left(U^0\frac{\partial p}{\partial x^0}+U^a\frac{\partial p}{\partial x^a}\right)\\ &\Longrightarrow \frac{\partial\rho}{\partial t}+\frac{\partial}{\partial x^a}(\rho u^a)=\frac{1}{c^2}\left(\frac{\partial p}{\partial t}+u^a\frac{\partial p}{\partial x^a}\right)\\ &\Longrightarrow \frac{\partial\rho}{\partial t}+\frac{\partial}{\partial x^a}(\rho u^a)=\frac{1}{c^2}\frac{dp}{dt}\end{aligned}$$

But since $p/c^2 \approx 0$, we have $(1/c^2)dp/dt \approx 0$ to the same order of approximation; so the last equation reduces to the classical continuity equation.

9. *(Relativistic Perfect Fluid)* Starting with equation (9.55), derive the relativistic Euler equation:

$$\left(\rho_0+\frac{1}{c^2}p\right)U^\mu\frac{\partial U^\nu}{\partial x^\mu}=\left(\eta^{\mu\nu}-\frac{1}{c^2}U^\mu U^\nu\right)\frac{\partial p}{\partial x^\mu} \tag{9.57}$$

(Solution)

Beginning with $\partial_\mu T^{\mu\nu}=0$, we again arrive at (9.63):

$$\partial_\mu T^{\mu\nu}=-\eta^{\mu\nu}\partial_\mu p+\sigma U^\mu\partial_\mu U^\nu+\sigma U^\nu\partial_\mu U^\mu+U^\mu U^\nu\partial_\mu\sigma=0 \tag{9.63}$$

where σ is the specific relativistic entropy. By equation (9.56), however, we can write

$$\begin{aligned}\partial_\mu(\sigma U^\mu) &= \sigma\partial_\mu U^\mu+U^\mu\partial_\mu\sigma=\frac{1}{c^2}U^\mu\partial_\mu p\\ &\Longrightarrow U^\mu\partial_\mu\sigma=\frac{1}{c^2}U^\mu\partial_\mu p-\sigma\partial_\mu U^\mu\end{aligned} \tag{9.56a}$$

Substituting (9.56a) into (9.63), and simplifying, we obtain

$$\sigma U^{\mu}\partial_{\mu}U^{\nu} = \left(\eta^{\mu\nu} - \frac{1}{c^2}U^{\mu}U^{\nu}\right)\partial_{\mu}p$$

which is equivalent to (9.57).

10. *(Relativistic Perfect Fluid)* Show that in the limit of small velocities, $\|\mathbf{u}\| << c$, equation (9.57) reduces to Euler's equation for a classical fluid.

(Solution)

For $\|\mathbf{u}\| << c$, we have $\gamma \approx 1$, so that

$$\frac{p}{c^2} \approx 0,\ \mathbf{U} \approx (c, \mathbf{u}),\ \rho \approx \rho_0$$

In view of the above approximations, if we set $\nu = b$ in (9.57), we obtain

$$\rho U^{\mu}\frac{\partial U^b}{\partial x^{\mu}} = -\frac{\partial p}{\partial x^b} - \frac{\partial p}{\partial x^{\mu}}\left(\frac{1}{c^2}U^{\mu}U^b\right)$$

Expanding out the implied sums gives

$$\rho\frac{\partial u^b}{\partial t} + \rho u^a\frac{\partial u^b}{\partial x^a} = -\frac{\partial p}{\partial x^b} - \frac{1}{c^2}\frac{\partial p}{\partial t} - \frac{\partial p}{\partial x^a}\left(\frac{1}{c^2}u^a u^b\right)$$

Now, to the same order of approximation, we have

$$\frac{1}{c^2}\frac{\partial p}{\partial t} \approx 0 \text{ and } \left(\frac{1}{c^2}u^a u^b\right) \approx 0$$

so that the last equation simplifies to

$$\rho\left(\frac{\partial u^b}{\partial t} + u^a\frac{\partial u^b}{\partial x^a}\right) = -\frac{\partial p}{\partial x^b}$$

or equivalently, in vector form, it becomes

$$\frac{\partial \mathbf{u}}{\partial t} + \mathbf{u}\cdot\nabla\mathbf{u} = -\frac{1}{\rho}\nabla p,$$

which is Euler's equation for a classical fluid.

Chapter 10

Foundations of General Relativity

The supreme task of the physicist is to arrive at those universal elementary laws from which the cosmos can be built up by pure deduction. There is no logical path to these laws; only intuition, resting on sympathetic understanding of experience, can reach them.

Albert Einstein

10.1. Introduction

We now turn our attention to the physical foundations of general relativity, and to a discussion of those philosophical and physical principles that guided Einstein in his effort to understand the phenomenon of gravitation.

Although such a discussion may seem unnecessary, this approach forces us to reexamine the soundness of our basic concepts and naturally leads us to a deeper understanding of the underlying theory. If we were not to do so, we would be like novice tourists, who may marvel at the majesty of a beautiful building, but at the same time, be totally oblivious to its significance in history and ignorant of the scaffolding that was used in its construction.

There are several physical principles that Einstein used to arrive at generalization of his special theory. They include Galileo's equivalence principle, Einstein's equivalence principle, the principle of general covariance and the principle of minimal gravitational coupling.

Guided by these basic principles, we will provide a logical and heuristic motivation for the validity of Einstein's field equations. Before doing so, however, we review the basic elements of Newtonian gravitation.

10.2. Newtonian Gravitation

Newton's gravitational theory is an **action-at-a-distance** theory. It is assumed that the gravitational force $\mathbf{F}(\mathbf{r})$ exerted by a point mass M on another point mass m is given by

$$\mathbf{F}(\mathbf{r}) = -G\frac{Mm}{r^2}\,\hat{\mathbf{r}} \tag{10.1}$$

where, as shown in Figure 10.1, the point mass M is situated at the origin of a coordinate system, $\mathbf{r}$ is the position vector corresponding to m, and $r = \|\mathbf{r}\|$. Note that the minus sign in this equation asserts that gravitational force is always *attractive* in nature.

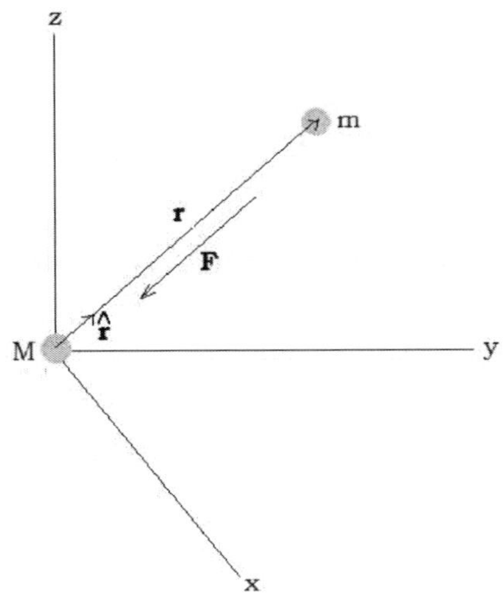

Figure 10.1. Conceptual setting for Newtonian gravitation.

In developing the Newtonian theory, we will proceed in a manner analogous to electrostatics. First, we define the **gravitational field intensity** $\mathbf{g}(\mathbf{r})$ as the gravitational force per unit test mass m so that

$$\mathbf{F}(\mathbf{r}) = m\mathbf{g}(\mathbf{r}) \tag{10.2}$$

and the **gravitational flux** Ψ_G through a surface Σ with unit normal $\mathbf{n}$ as the surface integral

$$\Psi_G = \int_{\Sigma} \mathbf{g}(\mathbf{r}) \cdot \mathbf{n}\, dS \tag{10.3}$$

Thus, Ψ_G can be conceptually viewed as a quantity that is directly proportional to the number of lines of force that pass through the surface Σ. From this basic definition, it's easy to show that for any *closed* region Ω containing a total mass M of matter, we can write

$$\oint_{\partial\Omega} \mathbf{g}(\mathbf{r}) \cdot \mathbf{n}\, dS = -4\pi GM \tag{10.4}$$

which is simply **Gauss's law** for the gravitational field. Like its electrostatic counterpart, this law may be profitably applied to determine $\mathbf{g}(\mathbf{r})$ in problems where a high degree of symmetry is present.

Secondly, since $\mathbf{F}(\mathbf{r})$ is a central force field, its easy to show that it is *conservative*. Consequently, there exists a scalar field $\varphi(\mathbf{r})$ called the **gravitational potential**, such that

$$\mathbf{g}(\mathbf{r}) = -\nabla\varphi(\mathbf{r})\,, \tag{10.5}$$

where

$$\varphi(\mathbf{r}) = -\int_{\mathbf{r}_0}^{\mathbf{r}} \mathbf{g}(\mathbf{r}') \cdot d\mathbf{r}' \tag{10.5a}$$

For any finite distribution of matter, we usually take the reference point $\mathbf{r}_0$ at infinity, so that $\varphi(r=\infty)=0$. Thus, in particular, the potential due to the point mass M in Figure 10.1 is simply

$$\varphi(\mathbf{r}) = -\int_{\infty}^{r} -G\frac{M}{r'^2}\hat{\mathbf{r}}\cdot(\hat{\mathbf{r}}dr') = \int_{\infty}^{r} G\frac{M}{r'^2}dr' = -G\frac{M}{r} \tag{10.6}$$

For the case of a continuous distribution of matter with mass density $\rho(\mathbf{r})$, filling a region of space Ω, we use the superposition principle to write the potential as the volume integral

$$\varphi(\mathbf{r}) = -\int_{\Omega} G\frac{\rho(\mathbf{r}')}{\|\mathbf{r}-\mathbf{r}'\|}d^3\mathbf{r}' \tag{10.7}$$

Using the above constructs, it's easy to show that the potential due to a continuous distribution of matter with density $\rho(\mathbf{r})$ must satisfy Poisson's equation:

$$\nabla^2\varphi(\mathbf{r}) = 4\pi G\rho(\mathbf{r}) \tag{10.8}$$

We note in passing, that (10.7) is the formal solution of this equation. In free space, or in space which is absent of matter, $\rho(\mathbf{r})=0$, so Poisson's equation reduces to Laplace's equation,

$$\nabla^2\varphi(\mathbf{r}) = 0\,. \tag{10.9}$$

From Newton's second law, and equations (10.2) and (10.4), we can write

$$m\frac{d^2\mathbf{r}}{dt^2} = m\mathbf{g}(\mathbf{r}) = -m\,\nabla\varphi(\mathbf{r})$$

so we obtain the **equation of motion** for a Newtonian field as simply,

$$\frac{d^2\mathbf{r}}{dt^2} = -\nabla\varphi(\mathbf{r}) \tag{10.10}$$

Observe that the Newtonian equation of motion implies that the acceleration of a test particle and consequently, the path its follows in space, are both independent of its mass, or any other intrinsic, physical property.

10.3. Inertial and Gravitational Mass

In Newtonian mechanics, the mass of a body can play different roles, depending upon how it enters into a given equation. For example, the **inertial mass** $m^{(I)}$ of a body is simply the mass that appears in Newton's second law

$$\mathbf{F} = m^{(I)}\mathbf{a} \tag{10.11}$$

It is a measure of the **inertia** of a body, or its resistance to a change in the state of its motion.

On the other hand, **gravitational mass** is the quantity that makes it appearance in Newton's law of gravitation. Consider two bodies as shown in Figure 10.2, separated by a distance r.

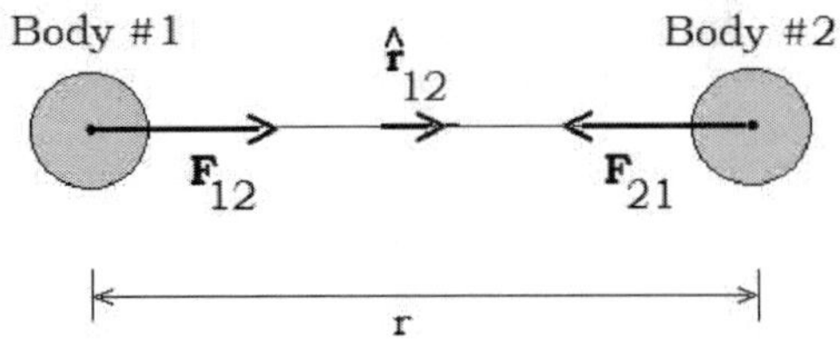

Figure 10.2. Gravitational interaction of two bodies.

According to Newton's theory, the gravitational force that the first body exerts on the second body is

$$\mathbf{F}_{12} = -\,G\frac{m_1^{(A)}m_2^{(P)}}{r^2}\,\hat{\mathbf{r}}_{12} \tag{10.12}$$

where $\hat{\mathbf{r}}_{12}$ is a unit vector in the indicated direction. Also, $m_1^{(A)}$ is called the **active gravitational mass** of the first body, since this body is viewed as the *source* of the field, and $m_2^{(P)}$ is called the **passive gravitational mass** of the second body, since it measures the *response* of the second body to the source. Similarly, the force $\mathbf{F}_{21}$ that the second body exerts on the first body is

$$\mathbf{F}_{21} = -\,G\frac{m_1^{(P)}m_2^{(A)}}{r^2}\,\hat{\mathbf{r}}_{21} \tag{10.13}$$

By Newton's third law, we know that $\mathbf{F}_{12} = -\,\mathbf{F}_{21}$, so we can write

$$\|\mathbf{F}_{12}\| = \|\mathbf{F}_{21}\| \Longrightarrow \frac{m_1^{(A)}}{m_1^{(P)}} = \frac{m_2^{(A)}}{m_2^{(P)}} = \text{constant} \qquad ((10.14))$$

We conclude that the ratio of any body's active gravitational mass to its passive gravitational mass is always the *same* constant, and so, by an appropriate choice of units, we can take this constant equal to unity. Consequently, for any body, we have

$$m^{(A)} = m^{(P)}, \tag{10.15}$$

so its active and passive gravitational masses are identical.

Now, let's recall Galileo's famous experimental result that states when two bodies, initially at rest, are released from the same height above the earth's surface, they will experience the *same* acceleration, and will strike the ground at the precisely the same time. Furthermore, the acceleration of a freely falling test particle is independent of its mass and composition. This experimental result is sometimes called the Galilean Equivalence Principle:

Galilean Equivalence Principle

The path of a test particle in a gravitational field is independent of its internal composition, and depends only upon its initial position and velocity. In addition, under identical initial conditions, all test particles undergo the same acceleration.

Galileo's equivalence principle implies that inertial and gravitational mass are identical. To see this, consider two bodies that are released from the same height H above the earth's surface, as shown in Figure 10.3 below. We have both

$$m_1^{(I)} a_1 = G \frac{m_1^{(P)} M^{(A)}}{(R_0 + H)^2} \text{ and } m_2^{(I)} a_2 = G \frac{m_2^{(P)} M^{(A)}}{(R_0 + H)^2}, \tag{10.16}$$

where R_0 and $M^{(A)}$ are the radius and active gravitational mass of the earth, respectively. According to Galileo's equivalence principle, however, if we neglect air resistance, then their accelerations are identical, $a_1 = a_2$, and will they strike the ground at the same time. We conclude that

$$a_1 = a_2 \Longrightarrow \frac{m_1^{(P)}}{m_1^{(I)}} = \frac{m_2^{(P)}}{m_2^{(I)}} = \text{constant} \tag{10.17}$$

so by an appropriate choice of units, we see that by (10.17), and our previous argument, the inertial mass of any body is equal to both its active and passive gravitational mass:

$$m^{(I)} = m^{(P)} = m^{(A)} \tag{10.18}$$

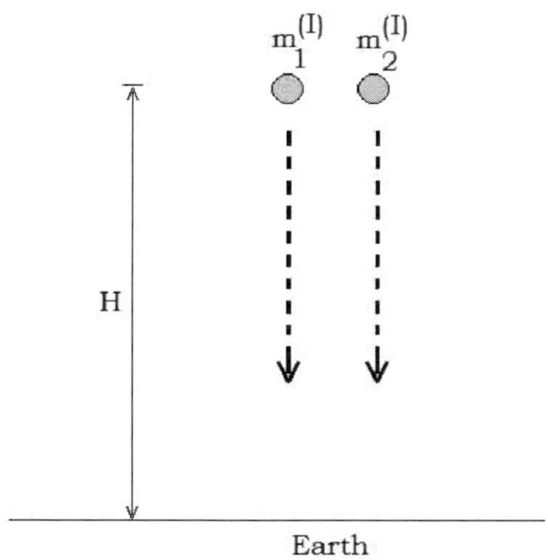

Figure 10.3. Two bodies released from rest at the same time.

The equality of inertial and gravitational mass has been tested with great accuracy. In fact, R.H. Dicke and his coworkers at Princeton have found this equality to hold good to within an error of less than one part in 10^{11}.

10.4. The Principle of Equivalence

Consider an observer *at rest*, in a small laboratory, who views the motion of a test particle in a uniform gravitational field, as shown in Figure 10.4(a). He will simply observe that the test particle is accelerated towards the floor of his laboratory with a constant downward acceleration g.

Now, suppose we transport his laboratory into deep space, to a location that is free from all gravitational effects, and then accelerate the laboratory upward with an acceleration g, as shown in part (b) of the Figure 10.4. Since the floor of the laboratory is accelerated vertically upward (while the test particle remains stationary), the observer would see the test particle move downward with an acceleration of g. In the absence of any additional evidence, he would naturally conclude his laboratory had been placed in a uniform gravitational field.

From this simple analysis, we conclude that given a reference frame at rest in a uniform gravitational field, there exists a uniformly accelerated reference frame in free space, such that the motion of a test particle in the second frame is identical to the first. This simple thought experiment led Einstein to conclude that there is a *physical equivalence* between a reference frame at rest in a gravitational field, and the field of acceleration produced by the corresponding accelerated reference frame.

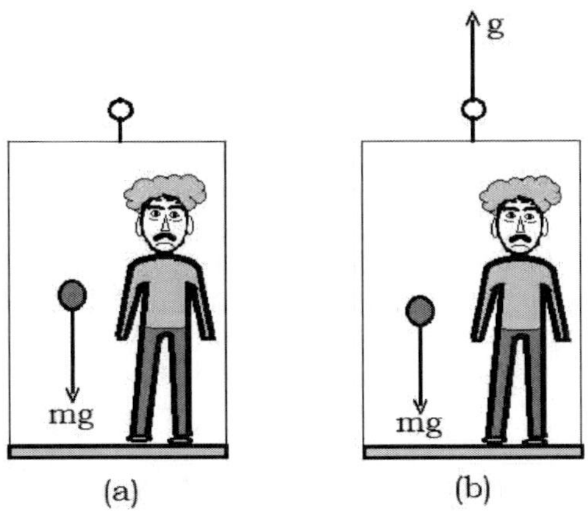

Figure 10.4. (a) Observer is at rest in a uniform gravitational field; (b) Observer's lab is pulled upward with an acceleration of g.

This physical equivalence is a local phenomenon since genuine gravitational fields may be considered as uniform only over small regions of space. When we consider sufficiently large spatial regions, an observer would notice that the paths of two such test particles would converge towards the center of the gravitational force; this is called a **tidal effect** of a gravitational field. Keeping this basic caveat in mind, we state the first version of the equivalence principle:

Equivalence Principle (Version 1)

An observer cannot conduct any *local* experiment which will allow him distinguish between the presence of a gravitational field, and a corresponding, appropriately accelerated reference frame which moves in the absence of gravity.

This principle, in turn, implies that a gravitational field may be viewed as an *inertial force* which directly results from using an *accelerated* frame of reference.

On the other hand, we can, at least locally, transform away the effects of gravitation by introducing a non-rotating, free-falling reference frame. As shown in Figure 10.5, if our laboratory (which had been placed in a uniform gravitational field that produces a downward acceleration of g), is now allowed to fall freely with the same acceleration, it is clear that both the observer and the test particle would no longer experience any gravitational forces.

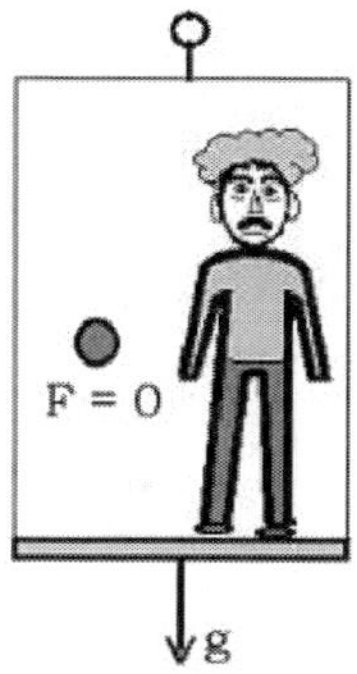

Figure 10.5. Observer and laboratory fall freely in a uniform field.

It is crucial to note that from a purely *operational* point of view, such a freely falling laboratory is also a **local inertial frame**. This is because a test particle which is initially at rest relative to such a frame, remains at rest. Secondly, in a freely falling frame, any test particle which is initially in uniform motion, will remain in uniform motion. This observation leads to the our second version of the equivalence principle:

Equivalence Principle (Version 2)

A freely falling (non-rotating) frame in a gravitational field is a **local inertial frame**. In other words, an observer in such a frame cannot conduct any local experiment to distinguish between uniform motion in the *absence* of a gravitational field, and (non-rotating) free fall in the *presence* of a gravitational field.

It is crucial to note that both versions of the equivalence principle rest upon the *exact* equality of inertial and gravitational mass. In Newtonian mechanics, this equality appears to be accidental, but in general relativity, it is taken as a fundamental postulate.

10.5. Non-Inertial Frames in Minkowski Space

Since the second version of the Equivalence Principle states that we can transform away the local effects of gravity by switching to an appropriate accelerated reference frame, let's see what this implies in Minkowski space.

Assume that the reference frames $K(ct,x,y,z)$ and $\bar{K}(c\bar{t},\bar{x},\bar{y},\bar{z})$ share a common z-axis, while $\bar{K}$ rotates counterclockwise at a constant an- gular speed ω relative to K, as shown in Figure 10.6. We also assume both frames coincide at time $t = 0$.

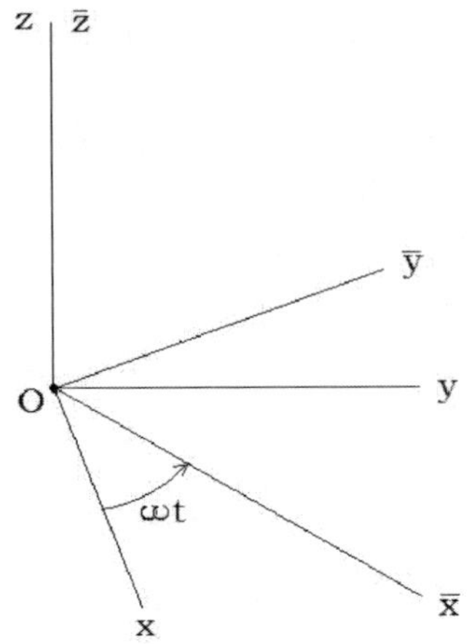

Figure 10.6. The frame $\bar{K}$ rotates at a constant angular speed ω relative to the inertial frame K.

If we assume that a Galilean transformation is permissible in this situation, and switch to cylindrical coordinates, then the coordinates in both frames are related by

$$\bar{t} = t, \ \ \bar{r} = r, \ \ \bar{\theta} = \theta - \omega t, \ \ \bar{z} = z \tag{10.19}$$

Now while the line element in $K(ct, r, \theta, z)$ is given by the familiar expression

$$ds^2 = c^2 dt^2 - dr^2 - r^2 d\theta^2 - dz^2 \tag{10.20}$$

the corresponding line element in $\bar{K}(c\bar{t}, \bar{r}, \bar{\theta}, \bar{z})$ will be given by

$$\begin{aligned} d\bar{s}^2 &= c^2 d\bar{t}^2 - d\bar{r}^2 - \bar{r}^2 \left(d\bar{\theta} + \omega d\bar{t}\right)^2 - dz^2 \\ &= c^2 \left(1 - \frac{\bar{r}^2 \omega^2}{c^2}\right) d\bar{t}^2 - d\bar{r}^2 - 2\bar{r}^2 \omega d\bar{\theta} d\bar{t} - dz^2 \end{aligned}$$

For simplicity, we remove the bars from the last expression and simply write

$$ds^2 = c^2 \left(1 - \frac{r^2 \omega^2}{c^2}\right) dt^2 - dr^2 - 2r^2 \omega d\theta dt - dz^2 \tag{10.21}$$

Setting $(x^0, x^1, x^2, x^3) = (ct, r, \theta, z)$, we see that the metric tensor for the non-inertial frame $\bar{K}$ assumes the form

$$(g_{\mu\nu}) = \begin{bmatrix} \left(1 - \frac{r^2\omega^2}{c^2}\right) & 0 & -\frac{r^2\omega}{c} & 0 \\ 0 & -1 & 0 & 0 \\ -\frac{r^2\omega}{c} & 0 & -r^2 & 0 \\ 0 & 0 & 0 & -1 \end{bmatrix} \tag{10.22}$$

Note that the rotation results in the appearance of the non-zero, off-diagonal terms g_{02} and g_{20}, and causes the g_{00} term to be different from unity. We now investigate the equations of motion of a particle in this non-inertial frame.

Assume that a particle moves along the world line $x^\mu = x^\mu(\tau)$ where the parameter τ is the proper time. If we set

$$\varphi = g_{\mu\nu}\frac{dx^\mu}{d\tau}\frac{dx^\nu}{d\tau} = \left(\frac{ds}{d\tau}\right)^2 \tag{10.23}$$

then the Euler-Lagrange equations can be written in the form

$$\frac{d}{d\tau}\left[\frac{\partial\varphi}{\partial\dot{x}^\mu}\right] - \frac{\partial\varphi}{\partial x^\mu} = 0 \tag{10.24}$$

From (10.21), we find

$$\varphi = \left(\frac{ds}{d\tau}\right)^2 = c^2\left(1 - \frac{r^2\omega^2}{c^2}\right)\dot{t}^2 - d\dot{r}^2 - 2r^2\omega\dot{\theta}\dot{t} - \dot{z}^2 \tag{10.25}$$

where the dots denote differentiation with respect to τ. Substituting (10.25) into (10.24) and simplifying, we obtain a coupled system of four differential equations:

$$c^2\left(1 - \frac{r^2\omega^2}{c^2}\right)\ddot{t} - 2r\omega^2\dot{r}\dot{t} - 2r\omega\dot{r}\dot{\theta} - r^2\omega\ddot{\theta} = 0 \tag{10.26a}$$

$$\ddot{r} - r\omega^2\dot{t}^2 - 2r\omega\dot{\theta}\dot{t} - r\dot{\theta}^2 = 0 \tag{10.26b}$$

$$\ddot{\theta} + \omega\ddot{t} + \frac{2\omega}{r}\dot{r}\dot{t} + \frac{2}{r}\dot{r}\dot{\theta} = 0 \tag{10.26c}$$

$$\ddot{z} = 0 \tag{10.26d}$$

Although this system is quite messy, if we substitute (10.26c) into (10.26a) and simplify, we find that $\ddot{t} = 0$. So, the original system of equations simplifies to

$$\begin{aligned} \ddot{t} &= 0 \\ \ddot{r} - r\omega^2\dot{t}^2 - 2r\omega\dot{\theta}\dot{t} - r\dot{\theta}^2 &= 0 \\ \ddot{\theta} + \frac{2\omega}{r}\dot{r}\dot{t} + \frac{2}{r}\dot{r}\dot{\theta} &= 0 \\ \ddot{z} &= 0 \end{aligned} \tag{10.27}$$

Now, we can read off the Christoffel symbols from (10.27) to obtain

$$\begin{aligned} \Gamma^1_{00} &= -r\omega^2,\ \Gamma^1_{02} = -r\omega,\ \ \Gamma^1_{22} = -r, \\ \Gamma^2_{01} &= \frac{\omega}{r},\ \ \Gamma^2_{12} = \frac{1}{r} \end{aligned} \tag{10.28}$$

and all of the remaining such symbols are identically zero.

The appearance of the variable Christoffel symbols often signals the presence of inertial forces in the non-inertial frame. To see this, let's find the Newtonian approximation to equations (10.27). Let's assume small velocities so that $r^2\omega/c^2 << 1$. Now, the proper time differential $d\tau$ is related to the coordinate time differential dt by

$$d\tau = \sqrt{1 - \frac{u^2}{c^2}}dt = \sqrt{1 - \frac{r^2\omega^2}{c^2}}dt$$

so that in the limit of small velocities, we have $d\tau \approx dt$, and consequently, $i \approx 1$. We find that equations (10.27) simplify to

$$\begin{aligned}
\frac{d^2r}{dt^2} &= r\left(\omega + r\frac{d\theta}{dt}\right)^2 \\
\frac{d^2\theta}{dt^2} &= -\frac{2}{r}\frac{dr}{dt}\left(\omega + \frac{d\theta}{dt}\right) \\
\frac{d^2z}{dt^2} &= 0
\end{aligned} \tag{10.29}$$

The righthand side of the first equation is simply the *centripetal acceleration* experienced by a particle of unit mass which has a total angular velocity of $(\omega + d\theta/dt)$. Similarly, we may interpret the righthand side of the second equation as a *Coriolis acceleration*. The third equation tells us that there is no acceleration in the z-direction.

We now examine how the measurement of *spatial* distances compare in both frames. One way of doing this is to find a *purely spatial* metric h_{ab} for the non-inertial frame. We first recall that if we are given a coordinate basis $\{\mathbf{e}_\mu\}$ for the tangent space at some point p, these basis vectors are not necessarily orthogonal. We can, however, still construct a basis such that $\mathbf{e}_1$,$\mathbf{e}_2$, and $\mathbf{e}_3$ are orthogonal to $\mathbf{e}_0$ at p.

First, recall that the 4-velocity $\mathbf{U}$ is always tangent to the world line of a particle which passes through the point p. In the instantaneous rest frame of the particle, we must have $\bar{\mathbf{U}} = (c, \mathbf{0})$, so the 4-velocity points in the same direction as the $\bar{x}^0$-axis of the comoving frame. Consequently, we can conveniently set $\mathbf{e}_0 = \bar{\mathbf{U}}$, and then find the parallel and orthogonal projections of each vector $\mathbf{e}_a$ by

$$\begin{aligned}
\mathbf{e}_a^{(\parallel)} &= \left(\frac{\mathbf{e}_a \cdot \mathbf{e}_0}{\mathbf{e}_0 \cdot \mathbf{e}_0}\right)\mathbf{e}_0 = \left(\frac{g_{a0}}{g_{00}}\right)\mathbf{e}_0 \\
\mathbf{e}_a^{(\perp)} &= \mathbf{e}_a - \mathbf{e}_a^{(\parallel)}
\end{aligned} \tag{10.30}$$

Next, we find the **spatial metric** h_{ab}; we define this as the metric which corresponds to the *three dimensional* vector subspace (of the tangent space at p) and which is orthogonal to the 4-velocity at p. We set $h_{0a} = h_{a0} = 0$ and define

$$h_{ab} = -\left(\mathbf{e}_a^{(\perp)} \cdot \mathbf{e}_b^{(\perp)}\right) \tag{10.31}$$

We find (exercise) that the spatial metric is

$$h_{ab} = -g_{ab} + \frac{g_{a0}g_{b0}}{g_{00}} \tag{10.32}$$

and so, the **spatial distance** dl is given by

$$dl^2 = h_{ab}dx^a dx^b = \left(-g_{ab} + \frac{g_{a0}g_{b0}}{g_{00}}\right)dx^a dx^b \tag{10.33}$$

Here, dl is the *physical distance* (as measured by a measuring stick) between any two simultaneous events in the *instantaneous rest frame* of the point p, and where $g_{\mu\nu}$ is the spacetime metric associated with this frame.

For our rotating reference frame, we find

$$h_{11} = -g_{11} = 1,\ h_{33} = -g_{33} = 1$$

while

$$h_{22} = -g_{22} + \frac{g_{20}g_{20}}{g_{00}} = \frac{r^2}{\left(1 - \frac{r^2\omega^2}{c^2}\right)}$$

So the spatial distance dl is simply

$$dl^2 = dr^2 + \frac{r^2}{\left(1 - \frac{r^2\omega^2}{c^2}\right)} d\theta^2 + dz^2 \tag{10.34}$$

Now, consider a circle of radius $r = a$ sitting in the plane of the frame $\bar{K}$. Since $dr = dz = 0$, the circumference of the circle is given by

$$C = \int_0^{2\pi} dl = \int_0^{2\pi} \frac{a}{\sqrt{1 - \frac{a^2\omega^2}{c^2}}} d\theta = \frac{2\pi a}{\sqrt{1 - \frac{a^2\omega^2}{c^2}}} \tag{10.35}$$

So, as long as $\omega \neq 0$, we will have $C > 2\pi a$; this implies that the geometry of the *space* is non-Euclidean; however, the associated *spacetime* when taken as a whole, is still *flat*. To see this, recall that for the Minkowski metric given in (10.20) we have

$$R^{\mu}_{\nu\rho\sigma} \equiv 0$$

Since tensor equations retain the *same form* under any coordinate transformation, we must have

$$\bar{R}^{\mu}_{\nu\rho\sigma} \equiv 0$$

for our rotating frame as well.

10.6. The Role of the Spacetime Metric

In the previous section, we saw that the use of a non-inertial reference frame in Minkowski space leads to the appearance of inertial forces. In each instance, the *intensity* of a particular inertial force was directly related to the metric connection Γ.

Now, according to the Equivalence Principle, inertial forces are locally equivalent to a gravitational forces, so we suspect that the intensity of a gravitational field should be related to the metric connection as well. Furthermore, as the metric connection contains the first derivatives of the metric components, we also suspect that for a genuine graviational field, the quantities $g_{\mu\nu}$ must be related to the *gravitational potential*.

Unfortunately, the *spacetime* associated with the rotating coordinate system is of limited instructional value, as it is still pseudo-Euclidean. This is to be expected since we began our discussion in Minkowski space which is flat, and any change of coordinates, no matter how clever, will alter this fact. To remedy this situation, we shall look at a spacetime that is slightly non-Euclidean.

Let's investigate the spacetime associated with spherically symmetric gravitational field, such as that outside a spherical mass M. We shall employ spherical coordinates, and we assume that this spacetime is *approximately* Euclidean, so our metric has the form

$$g_{\mu\nu} = \eta_{\mu\nu} + \varepsilon_{\mu\nu} \tag{10.36}$$

where $|\varepsilon_{\mu\nu}| \ll 1$ is a small perturbation[1]. By symmetry, we assume that the quantities $\varepsilon_{\mu\nu}$ comprise a *diagonal* matrix. Also, we assume they depend only upon the radial distance so that

$$\varepsilon_{\mu\nu} = \varepsilon_{\mu\nu}(r) \tag{10.37}$$

We are making what is called the **static weak-field approximation** to the field outside a gravitating mass. We use the term *static*, since we are assuming that

$$\frac{\partial}{\partial t} g_{\mu\nu} = 0, \tag{10.38}$$

so *the metric components have no explicit time dependence*. If we now make the identification

$$(x^0, x^1, x^2, x^3) = (ct, r, \theta, \phi),$$

then we can write

$$ds^2 = (1+\varepsilon_{00})\, c^2 dt^2 - (1+\varepsilon_{11})\, dr^2 - (1+\varepsilon_{22})\, r^2 d\theta^2 - (1+\varepsilon_{33})\, r^2 \sin^2\theta d\phi^2 \tag{10.39}$$

so that

$$(g_{\mu\nu}) = \begin{bmatrix} (1+\varepsilon_{00}) & 0 & 0 & 0 \\ 0 & -(1+\varepsilon_{11}) & 0 & 0 \\ 0 & 0 & -(1+\varepsilon_{22})\, r^2 & 0 \\ 0 & 0 & 0 & -(1+\varepsilon_{33})\, r^2 \sin^2\theta \end{bmatrix} \tag{10.40}$$

Since $(g_{\mu\nu})$ is a diagonal matrix, the components of the reciprocal metric tensor $\mathbf{g}^{-1} = (g^{\mu\nu})$ are easily found to be

$$(g^{\mu\nu}) = \begin{bmatrix} \frac{1}{1+\varepsilon_{00}} & 0 & 0 & 0 \\ 0 & -\frac{1}{1+\varepsilon_{11}} & 0 & 0 \\ 0 & 0 & -\frac{1}{r^2(1+\varepsilon_{22})} & 0 \\ 0 & 0 & 0 & -\frac{1}{r^2 \sin^2\theta (1+\varepsilon_{33})} \end{bmatrix} \tag{10.41}$$

We now investigate the *slow radial motion* of a typical test particle. The geodesic equation (6.149) becomes

$$\frac{d^2 r}{d\tau^2} + \Gamma^1_{\mu\nu} \frac{dx^\mu}{d\tau} \frac{dx^\nu}{d\tau} = 0 \tag{10.42}$$

where the parameter τ is the proper time. Since the motion is *radial*, we have $d\theta/d\tau = d\phi/d\tau = 0$, so (10.42) reduces to

$$\frac{d^2 r}{d\tau^2} + \Gamma^1_{00} c^2 \left(\frac{dt}{d\tau}\right)^2 + 2c\Gamma^1_{01} \frac{dt}{d\tau}\frac{dr}{d\tau} + \Gamma^1_{11} \left(\frac{dr}{d\tau}\right)^2 = 0 \tag{10.43}$$

The Christoffel symbols are given by

$$\begin{aligned}
\Gamma^1_{00} &= \frac{1}{2}g^{11}\left(-g_{00,1}+g_{10,0}+g_{01,0}\right) = -\frac{1}{2}g^{11}g_{00,1} = -\frac{1}{2\left(1+\varepsilon_{11}\right)}\frac{d\varepsilon_{00}}{dr} \\
\Gamma^1_{01} &= \frac{1}{2}g^{11}\left(-g_{01,1}+g_{10,1}+g_{11,0}\right) = \frac{1}{2}g^{11}g_{11,0} = 0 \\
\Gamma^1_{11} &= \frac{1}{2}g^{1\sigma}\left(-g_{11,1}+g_{11,1}+g_{11,1}\right) = \frac{1}{2}g^{11}g_{11,1} = \frac{1}{2\left(1+\varepsilon_{11}\right)}\frac{d\varepsilon_{11}}{dr}
\end{aligned} \tag{10.44}$$

Since the quantities $\left|\varepsilon_{\mu\nu}\right| \ll 1$, we can expand the coefficients of the non-zero Christoffel symbols in a power series in ε_{11}, and retain only first order terms to obtain

$$\begin{aligned}
\Gamma^1_{00} &= -\left(-\frac{1}{2}+\frac{1}{2}\varepsilon_{11}+\ldots\right)\frac{d\varepsilon_{00}}{dr} \approx \frac{1}{2}\frac{d\varepsilon_{00}}{dr} \\
\Gamma^1_{11} &= \left(-\frac{1}{2}+\frac{1}{2}\varepsilon_{11}+\ldots\right)\frac{d\varepsilon_{11}}{dr} \approx -\frac{1}{2}\frac{d\varepsilon_{11}}{dr}
\end{aligned} \tag{10.45}$$

Substituting (10.45) into (10.43), we obtain

$$\frac{d^2r}{d\tau^2}+\frac{c^2}{2}\left[\left(\frac{d\varepsilon_{00}}{dr}\right)\left(\frac{dt}{d\tau}\right)^2-\frac{1}{c^2}\left(\frac{d\varepsilon_{11}}{dr}\right)\left(\frac{dr}{d\tau}\right)^2\right]=0 \tag{10.46}$$

Since we have assumed slow, radial motion, then the radial velocity component $u_r = dr/dt$ must satisfy the condition $(u_r/c)^2 \ll 1$, and

$$d\tau = \sqrt{1-\frac{{u_r}^2}{c^2}}dt \approx dt$$

So we find $dt/d\tau \approx 1$, and the last equation becomes

$$\frac{d^2r}{dt^2} = -\frac{c^2}{2}\left[\frac{d\varepsilon_{00}}{dr}-\left(\frac{u_r}{c}\right)^2\frac{d\varepsilon_{11}}{dr}\right],$$

which, in turn, simplifies to

$$\frac{d^2r}{dt^2} = -\frac{c^2}{2}\frac{d\varepsilon_{00}}{dr} \tag{10.47}$$

since $(u_r/c)^2 \ll 1$.

Now, from (10.10), we see that according to Newton, the radial equation for the test particle should be

$$\frac{d^2r}{dt^2} = -\frac{d\varphi}{dr} \tag{10.48}$$

Comparing the righthand sides of the last two equations, we find that

$$\varphi = \frac{c^2}{2}\varepsilon_{00}+K$$

where K is a constant. Far away from the gravitating mass ($r \longrightarrow \infty$) we expect that spacetime is nearly flat so that $\varepsilon_{00} \longrightarrow 0$, while we must have $\varphi \longrightarrow 0$. So, we conclude that $K = 0$, and

$$g_{00} = 1 + \frac{2\varphi}{c^2} \tag{10.49}$$

So, g_{00} is directly related to the Newtonian gravitational potential as we had suspected in the first place!

10.7. Geodesic Deviation

We know that the physical equivalence between genuine gravitational fields and fields of acceleration is a *local* phenomenon since bonafide gravitational fields may be considered as uniform only over small regions of spacetime.

When we consider sufficiently large regions of spacetime, an observer would notice that the paths of any two test particles would converge towards the center of gravitational force. In this section, we shall show that the deviation in the motion of test particles is directly related to the curvature of spacetime.

Consider a 2-surface Σ in spacetime. We imagine that this surface is filled by a family of timelike geodesics such that one and only one geodesic passes through each point $p \in \Sigma$, as shown in Figure 10.7. Such a family of space filling curves is called a **congruence**.

If each geodesic is parametrized by the proper time τ, and we label each geodesic in the congruence of curves by λ, then the surface Σ can be given the parametric representation

$$x^{\sigma} = x^{\sigma}(\tau, \lambda) \tag{10.50}$$

where different geodesics correspond to distinct values of the label λ.

We define the vector fields $\mathbf{U} = (U^{\alpha})$ and $\mathbf{V} = (V^{\alpha})$ by

$$U^{\alpha} = \frac{\partial x^{\alpha}}{\partial \tau} \quad \text{and} \quad V^{\alpha} = \frac{\partial x^{\alpha}}{\partial \lambda} \tag{10.51}$$

so that for a typical geodesic which passes through the point $p \in \Sigma$, the vector $\mathbf{U}$ is tangent to the geodesic, and the **connecting vector V** serves to join adjacent geodesics.

As shown in Figure 10.8, consider the motion of two free test particles that move along adjacent timelike geodesics, C_1 and C_2. We want to find an equation that tells us how their separation or **deviation** V^{α} changes as they move.

Now, it may be shown (exercise) that at any point $p \in \Sigma$, the Lie bracket of $\mathbf{U}$ and $\mathbf{V}$ vanishes, so that

$$[\mathbf{U}, \mathbf{V}] = \mathbf{0} \tag{10.52}$$

From equation (6.40), we recall that the torsion tensor $\mathbf{T}$ is defined by

$$\mathbf{T}(\mathbf{U}, \mathbf{V}) = \nabla_{\mathbf{U}}\mathbf{V} - \nabla_{\mathbf{V}}\mathbf{U} - [\mathbf{U}, \mathbf{V}]$$

If we assume that physical spacetime is *torsion-free* ($\mathbf{T} = \mathbf{0}$), the last two equations lead us to conclude that

$$\nabla_{\mathbf{U}}\mathbf{V} = \nabla_{\mathbf{V}}\mathbf{U} \tag{10.53}$$

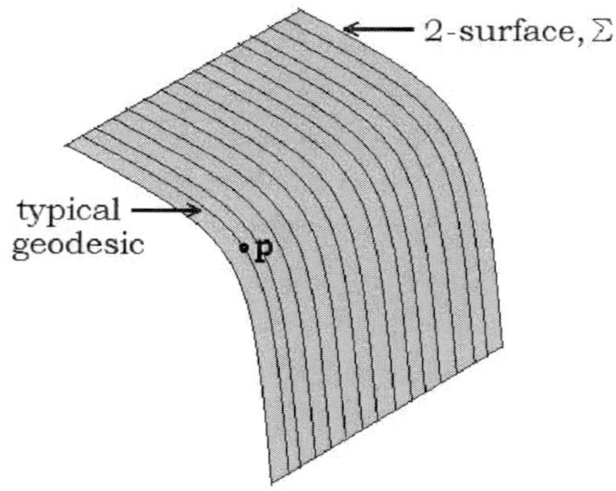

Figure 10.7. A congruence of geodesics filling the 2-surface, Σ.

Now recall that for *any* smooth vector fields $\mathbf{X}, \mathbf{Y}$,and $\mathbf{Z}$ the Riemann curvature tensor is given by

$$\mathbf{R}(\mathbf{X},\mathbf{Y})\mathbf{Z} = \nabla_{\mathbf{X}}(\nabla_{\mathbf{Y}}\mathbf{Z}) - \nabla_{\mathbf{Y}}(\nabla_{\mathbf{X}}\mathbf{Z}) - \nabla_{[\mathbf{X},\mathbf{Y}]}\mathbf{Z} \tag{10.54}$$

If we set $\mathbf{X} = \mathbf{Z} = \mathbf{U}$, and $\mathbf{Y} = \mathbf{V}$, this equation becomes

$$\mathbf{R}(\mathbf{U},\mathbf{V})\mathbf{U} = \nabla_{\mathbf{U}}(\nabla_{\mathbf{V}}\mathbf{U}) - \nabla_{\mathbf{V}}(\nabla_{\mathbf{U}}\mathbf{U}) - \nabla_{[\mathbf{U},\mathbf{V}]}\mathbf{U} \tag{10.55}$$

But $[\mathbf{U},\mathbf{V}] = \mathbf{0}$, so if we substitute (10.53) into (10.55) we obtain

$$\mathbf{R}(\mathbf{U},\mathbf{V})\mathbf{U} = \nabla_{\mathbf{U}}(\nabla_{\mathbf{U}}\mathbf{V}) - \nabla_{\mathbf{V}}(\nabla_{\mathbf{U}}\mathbf{U}) \tag{10.55a}$$

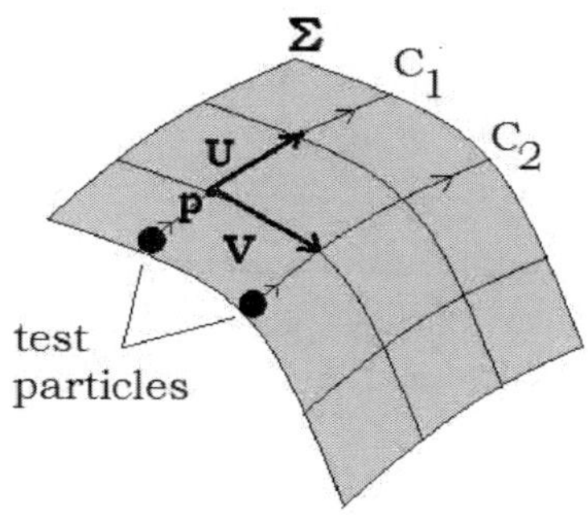

Figure 10.8. Test particles moving on timelike geodesics, C_1 and C_2.

Since $\mathbf{U}$ is tangent to a *geodesic*, we must have

$$\frac{D\mathbf{U}}{D\tau} = \nabla_{\mathbf{U}}\mathbf{U} = \mathbf{0}$$

so the last equation becomes

$$\nabla_{\mathbf{U}}(\nabla_{\mathbf{U}}\mathbf{V}) = \mathbf{R}(\mathbf{U},\mathbf{V})\mathbf{U} \tag{10.56}$$

which is called **Jacobi's equation**, or the **equation of geodesic deviation**. We can write this equation in component form as

$$\frac{D^2V^\alpha}{D\tau^2} = R^\alpha_{\beta\gamma\delta}U^\beta U^\gamma V^\delta \tag{10.57}$$

This equation provides us with a measure of the *relative acceleration* of any two test particles that move along adjacent timelike geodesics. If we assume that a gravitational field is associated with spacetime curvature, then (10.57) tells us that this curvature is responsible for the appearance of tidal forces we mentioned previously.

Clearly, for Minkowski space, the righthand side of (10.57) vanishes so that

$$\frac{d^2V^\alpha}{d\tau^2} = 0 \Longrightarrow V^\alpha(\tau) = A^\alpha\tau + B^\alpha \tag{10.58}$$

where A^α, and B^α are constants. Since the geodesics in this case are simply straight lines, if we assume that the initial rate of deviation $\dot{V}^\alpha = 0$, then V^α will remain constant for all (proper) time. This is exactly what we would expect to happen in the accelerated laboratory of Figure 10.4(b).

10.8. Einstein's Field Equations

The first version of the equivalence principle implies that gravitational forces are equivalent to *inertial forces* that result from employing *accelerated* frames of reference. Einstein used this principle to make a bold hypothesis; he assumed that the motion of a test particle in a gravitational field must be *equivalent* to *force-free* motion in a *curved spacetime*. Furthermore, any such test particle must travel along a *timelike* geodesic.

But what determines the curvature of spacetime? In his effort to answer this question, Einstein was strongly influenced by **Mach's principle**. This principle, set forth by the German physicist and philosopher Ernst Mach, asserts that the inertia of any given body is the result of its *interaction* with all of the matter in the rest of the universe. That is, the inertia of a test mass is felt when it is accelerated *relative* to the enormous masses (such as the fixed stars) that are distributed throughout the universe.

Although this principle is purely philosophical, and has proved to be the source of much controversy, it certainly guided Einstein in his thinking. Mach's principle implies that a test body in an otherwise empty universe would not have any inertial properties. It also implies that inertia itself, has a gravitational origin, and is due to the *distribution* of distant matter in the universe.

In view of the equivalence of mass and energy, Einstein extended Mach's principle and assumed that distribution both matter and energy determines spacetime curvature. Thus, Einstein's new theory is truly a *field theory* where spacetime itself plays the role of the underlying field.

Furthermore, since the second version of the equivalence principle states that the *local effects* of gravity can be transformed away by choosing a suitable freely falling reference frame, Einstein was led to conclude that *spacetime must look locally like Minkowski space*. Consequently, for sufficiently small regions of spacetime, the equations of Special Relativity must hold good.

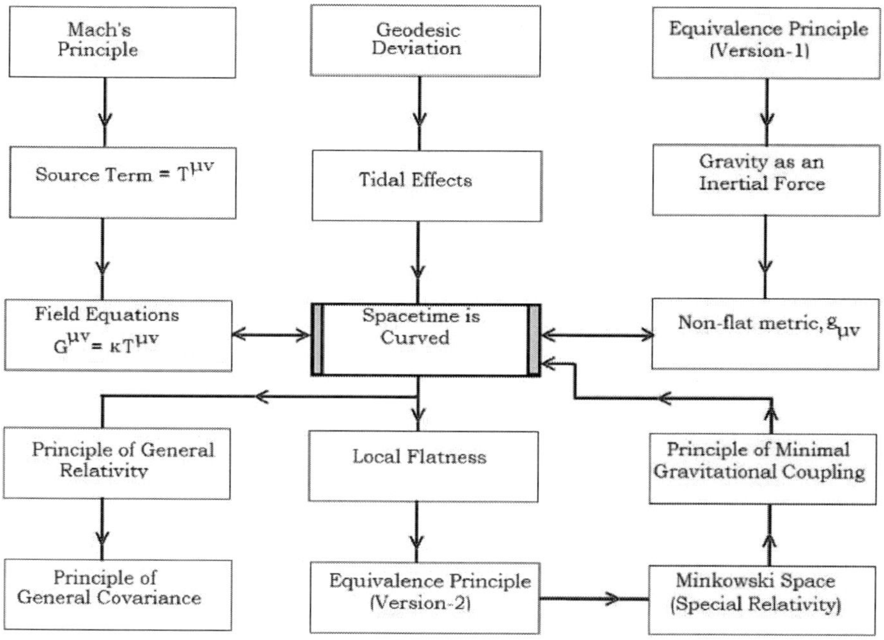

Figure 10.9. Conceptual map of key concepts in General Relativity.

Recall that the Principle of Special Relativity asserts that the laws of physics must assume the same mathematical form for all *inertial* observers. In order to accomodate the phenomenon of gravity, Einstein extended this principle by assuming that the laws of physics must remain the same for all observers, whether they are situated in an inertial frames or not. This is called the **Principle of General Relativity**. In other words, all observers (and their associated reference frames) are equally valid when determining the laws of physics.

The Principle of General Relativity, in turn, implies the **Principle of General Covariance**: the equations of physics must be *covariant* under general coordinate transformations that relate different coordinate frames, including non-inertial frames. Since tensor equations remain covariant under general transformations, Einstein assumed that the equations of physics should be formulated in *tensorial form*.

When making the transition from special to general relativity, how do we generalize an equation which is known to be valid in the special theory? Here, Einstein implicitly used the **Principle of Minimal Gravitational Coupling** which states that we simply replace any occurrence of a partial derivative (in an equation of Special Relativity) by a *covariant derivative* - without adding in any extra terms involving the curvature tensor or any of its contractions:

$$\frac{\partial}{\partial x^{\mu}} \Longrightarrow \nabla_{\mu}$$

For example, in Minkowski space, we have shown that relativistic dust satisfies the conservation equation

$$\frac{\partial}{\partial x^{\mu}}(T^{\mu\nu}) = 0$$

So, in general relativity, we assume that relativistic dust will satisfy the *tensor equation*

$$\nabla_{\mu} T^{\mu\nu} = 0.$$

Note that this is a *simplicity principle*; it is the simplest possible assumption that we can make to arrive at the general equations.

We are now ready to write down the field equations of general relativity. After ten years of research, Einstein had formed the basic idea that curvature of spacetime is directly related to the distribution of matter and energy. This suggested that the energy-momentum tensor $T^{\mu\nu}$ should be used as a *source term* in his field equations so that

$$\text{Spacetime Curvature} \propto T^{\mu\nu}$$

On the other hand, we know that the curvature of spacetime is given by the Riemann curvature tensor. Unfortunately, this is a fourth order tensor. To remedy this problem, we contract it, to obtain the Ricci tensor, which is of second order. So, our first guess might be

$$R^{\mu\nu} \propto T^{\mu\nu} \ ?$$

But there is still a problem, since the righthand side of the last equation is *divergence free*, while the lefthand side is not. However, from (6.123) we do know that the covariant derivative of the Einstein tensor vanishes identically, so we can replace our first guess by

$$G^{\mu\nu} = \kappa T^{\mu\nu} \tag{10.59}$$

where κ is a **coupling constant** (to be determined), and where

$$G^{\mu\nu} = R^{\mu\nu} - \frac{R}{2}\delta^{\mu\nu} \tag{10.60}$$

Equations (10.59) are the equations we have been seeking, they are called the **full field equations** of general relativity.

For a region of spacetime which is outside any gravitating matter or energy, the right-hand side of (10.59) vanishes, and we obtain

$$G^{\mu\nu} = 0\ . \tag{10.61}$$

These are the **vacuum field equations** of general relativity.

10.9. The Newtonian Limit

In order to find the coupling constant, we shall find the **Newtonian limit** of the full field equations. We assume that the source of gravity is a distribution of dust particles which persists over a fairly large region, and we consider the field at any point *inside* this region.

We assume that:

1. There exists a coordinate system (x^α) such that the velocities of particles in the dust cloud are small compared to the velocity of light.

2. The metric has the form

$$g_{\mu\nu} = \eta_{\mu\nu} + \varepsilon_{\mu\nu} \tag{10.62}$$

where $|\varepsilon_{\mu\nu}| \ll 1$, and the quantities $\varepsilon_{\mu\nu}$ are small perturbations to the Minkowski metric.

3. The above metric is *static*, so that derivatives of the metric with respect to x^0 may be ignored.

4. We can ignore any *second order* terms in the $\varepsilon_{\mu\nu}$ so the full field equations may be *linearized*.

5. The **Correspondence Principle** holds; that is, the linearized version of the full field equations should be equivalent to Poisson's equation (10.8), in the Newtonian limit.

From the (10.62) and the relation $g^{\mu\sigma}g_{\sigma\nu} = \delta^\mu_\nu$, we see that, up to first order terms, the reciprocal metric is given by

$$g^{\mu\nu} = \eta^{\mu\nu} - \varepsilon^{\mu\nu} \quad \text{where} \quad \varepsilon^{\mu\nu} = \eta^{\mu\sigma}\eta^{\nu\tau}\varepsilon_{\sigma\tau} \tag{10.63}$$

As we will be retaining only first order terms, we shall use $\eta^{\mu\nu}$ and $\eta_{\mu\nu}$ to raise and lower indices in $\varepsilon_{\mu\nu}$, respectively.

Before we attempt to introduce these ideas into the full fields equations, we will first cast these equations into a more convenient form. We first note that the full field equations may be written in the form

$$G^{\mu\sigma} = R^{\mu\sigma} - \frac{R}{2}\delta^{\mu\sigma} = \kappa T^{\mu\sigma} \tag{10.64}$$

If we multiply both sides of this equation by $g_{\nu\sigma}$ we obtain

$$G^\mu_\nu = R^\mu_\nu - \frac{R}{2}\delta^\mu_\nu = \kappa T^\mu_\nu \tag{10.65}$$

Contracting on μ and ν, we find

$$R - \frac{R}{2}(4) = \kappa T \quad \text{where } T = Tr(\mathbf{T}) = T^\mu_\mu \tag{10.66}$$

and R is the Ricci curvature scalar. The last equation gives

$$R = -\kappa T \tag{10.67}$$

Finally, we substitute (10.67) into (10.65), and lower indices, to cast the field equations into the convenient form:

$$R_{\mu\nu} = \kappa\left(T_{\mu\nu} - \frac{1}{2}T\, g_{\mu\nu}\right) \tag{10.68}$$

Now, recall that the energy-momentum tensor for a dust cloud is

$$T^{\alpha\beta} = \rho_0 U^\alpha U^\beta$$

so that its covariant counterpart is

$$T_{\mu\nu} = \rho_0 g_{\mu\alpha} g_{\nu\beta} U^\alpha U^\beta$$

Since the particles in the dust cloud are nearly stationary, we see that $\mathbf{U} \approx (c, \mathbf{0})$ and consequently, only the 00-component of this tensor survives, to give

$$\begin{aligned} T_{00} &= \rho_0 g_{0\alpha} g_{0\beta} U^\alpha U^\beta \approx \rho_0 c^2 (g_{00})^2 \\ &\Longrightarrow T_{00} \approx \rho_0 c^2 \end{aligned} \tag{10.69}$$

Similarly,

$$\begin{aligned} T &= T^\sigma_\sigma = g^{\sigma\mu} T_{\mu\nu} \approx g^{00} T_{00} \approx T_{00} \\ &\Longrightarrow T \approx \rho_0 c^2 \end{aligned} \tag{10.70}$$

Substituting the last two approximations into the righthand side of (10.68), we find

$$\kappa \left(T_{00} - \frac{1}{2} T\, g_{00} \right) \approx \frac{1}{2} \kappa \rho c^2 \tag{10.71}$$

where ρ_0 has been replaced by ρ for obvious reasons.

We now work on the lefthand side of the field equations. Recall that the Ricci tensor is given by (6.118) as

$$R_{\mu\nu} = \frac{\partial^2}{\partial x^\mu \partial x^\nu} (\log \sqrt{|g|}) - \frac{1}{\sqrt{|g|}} \frac{\partial}{\partial x^\alpha} \left(\Gamma^\alpha_{\mu\nu} \sqrt{|g|} \right) + \Gamma^\alpha_{\mu\beta} \Gamma^\beta_{\nu\alpha} \tag{10.72}$$

so that

$$R_{00} = \frac{\partial^2}{\partial x^0 \partial x^0} (\log \sqrt{|g|}) - \frac{1}{\sqrt{|g|}} \frac{\partial}{\partial x^\alpha} \left(\Gamma^\alpha_{00} \sqrt{|g|} \right) + \Gamma^\alpha_{0\beta} \Gamma^\beta_{0\alpha} \tag{10.73}$$

If we ignore the last term which is quadratic in the Christoffel symbols, and observe that $\sqrt{|g|} \approx 1$, the last equation simplifies to (exercise):

$$R_{00} = -\frac{\partial}{\partial x^\alpha} (\Gamma^\alpha_{00}) \approx -\frac{1}{2} \nabla^2 g_{00} \tag{10.74}$$

Equating (10.74) and (10.71), we obtain

$$\nabla^2 g_{00} = \left(-\kappa c^2 \right) \rho$$

Now, from (11.49), we know that for a static weak-field approximation, $g_{00} = 1 + 2\varphi/c^2$, so the last equation may be rewritten as

$$\nabla^2 \varphi = \left(-\frac{\kappa}{2} c^4 \right) \rho \tag{10.75}$$

Comparing this equation with Poisson's equation $\nabla^2\varphi = 4\pi G\rho$ shows that the coupling constant must be

$$\kappa = -\frac{8\pi G}{c^4} \tag{10.76}$$

We conclude that the full field equations can now be written as

$$R_{\mu\nu} = -\frac{8\pi G}{c^4}\left(T_{\mu\nu} - \frac{1}{2}T\, g_{\mu\nu}\right) \tag{10.77}$$

10.10. Geodesic Motion from the Field Equations

We now turn our attention to the so-called *geodesic postulate* which asserts that *free particles* - particles moving only under the influence of gravity - travel along geodesics in spacetime. We shall show that this postulate follows directly from the full field equations and is therefore, unnecessary.

Consider a continuous distribution matter such as relativistic dust. Recall that the energy-momentum tensor of dust is

$$T^{\mu\nu} = \rho_0 U^\mu U^\nu$$

Now, the full field equations are $G^{\mu\nu} = \kappa T^{\mu\nu}$, so the fact that the Einstein tensor is *divergence free* requires that

$$\nabla_\nu T^{\mu\nu} = 0 \tag{10.78}$$

as well. We shall show that equations (10.78) represent the equations of motion of the dust particles.

We find

$$\nabla_\nu T^{\mu\nu} = \nabla_\nu\left[(\rho_0 U^\nu) U^\mu\right] = \left[\nabla_\nu(\rho_0 U^\nu)\right] U^\mu + \rho_0 U^\nu(\nabla_\nu U^\mu) = 0 \tag{10.79}$$

Multiplying this equation by U_ν and recalling that $U_\nu U^\nu = c^2$, we obtain

$$c^2\left[\nabla_\nu(\rho_0 U^\nu)\right] U^\mu + \rho_0 U^\nu(U_\mu \nabla_\nu U^\mu) = 0 \tag{10.80}$$

Observe that,

$$U_\mu \nabla_\nu U^\mu = \frac{1}{2}\nabla_\nu(U_\mu U^\mu) = \frac{1}{2}\nabla_\nu(c^2) = 0 \tag{10.81}$$

so substituting (10.81) into (10.80), we find

$$\left[\nabla_\nu(\rho_0 U^\nu)\right] U^\mu = 0 \tag{10.82}$$

Finally, substituting the last result into (10.79) implies that the equations of motion are given by

$$U^\nu \nabla_\nu U^\mu = 0 \tag{10.83}$$

We can rewrite these equations in a familiar form. Since $U^\nu = dx^\nu/d\tau$, we have

$$\begin{aligned}\frac{dx^\nu}{d\tau}\nabla_\nu\left(\frac{dx^\mu}{d\tau}\right) &= 0\\ &\Longrightarrow \frac{dx^\nu}{d\tau}\left\{\frac{\partial}{\partial x^\nu}\left(\frac{dx^\mu}{d\tau}\right)+\Gamma^\mu_{\nu\sigma}\frac{dx^\sigma}{d\tau}\right\}=0\\ &\Longrightarrow \frac{dx^\nu}{d\tau}\frac{\partial}{\partial x^\nu}\left(\frac{dx^\mu}{d\tau}\right)+\Gamma^\mu_{\nu\sigma}\frac{dx^\nu}{d\tau}\frac{dx^\sigma}{d\tau}=0\end{aligned}$$

Using the chain rule on the first term of the last equation, we obtain the familiar geodesic equation

$$\frac{d^2x^\mu}{d\tau^2}+\Gamma^\mu_{\nu\sigma}\frac{dx^\nu}{d\tau}\frac{dx^\sigma}{d\tau}=0$$

We conclude that the dust particles must move along geodesic worldlines $x^\mu = x^\mu(\tau)$ in spacetime.

Remarkably, Einstein's full field equations *alone* provide us with this conclusion. This state of affairs should be contrasted to Maxwell's theory of the electromagnetic field where the field equations fail to provide an equation of motion for charged particles. Here, the Lorentz force law must either be *postulated* or derived from a separate argument where we invoke the requirement of *manifest covariance*, as we did in Section 8.3.

This discrepancy may be understood by noting that Maxwell's theory is a *linear theory* so that an electromagnetic field cannot inherently act as a source of itself; it can't pull itself up by its own bootstraps. On the otherhand, however, the Einstein's field equations are *non-linear*, and this non-linearity is a result of the fact that a gravitational field can act as its own source.

10.11. Formal Postulates of General Relativity

In this section, let's summarize the story so far, and spell out some of the assumptions we have been making all along. These assumptions may be characterized in terms of formal mathematical postulates and physical postulates.

First, we formally identify **spacetime**, which comprises the set of all possible events, with the *field* of general relativity. We assume further, that spacetime is a four-dimensional, semi-Riemannian space $(S^4, \mathbf{g})$ such that S^4 is a smooth manifold with the Hausdorf property.[2]

We have also assume that S^4 it is equipped with both an indefinite metric $\mathbf{g}$ that has signature -2 at each point $p \in S^4$, and the Levi-Civita connection ∇. We take it for granted that the manifold S^4 is **connected**; that is, given any two distinct points $p_1, p_2 \in S^4$ there exists a smooth curve which joins these two points.

We assume that *locally*, S^4 is the flat Minkowski spacetime M^4. Stated differently, at each point $p \in S^4$ the tangent space $T_p(S^4)$ is a Minkowski space so that the laws of special relativity hold locally. This assumption is inspired by the second version of the equivalence principle.

The assumption of the *local flatness* of S^4 is perfectly reasonable, as it is a direct result of the fact that S^4 is a smooth manifold. Recall that by the introduction of a geodesic coordinate transformation, we can induce the derivatives of the metric tensor and the affine connection coefficients to vanish at any given point. But, as the *derivatives* of the connection coefficients fail to vanish at such a point, however, we can't transform away the intrinsic curvature of S^4, so we must be careful to use the term 'local.' This is the mathematical content of the second version of the equivalence principle.

We adopt the **geodesic postulate** by assuming that freely falling test particles move along timelike geodesics in S^4. Here, a freely falling particle is one that moves only under the influence of gravity. As shown in the previous section, this postulate is, to some extent, unnecessary in that it follows directly from the full field equations.

We take the principle of general covariance as a basic premise so we require that the equations of physics must be *covariant* under *general* coordinate transformations that relate different coordinate frames, including non-inertial frames.

We postulate that the principle of minimal gravitational coupling is applicable so that any physical laws that hold in Minkowski spacetime will continue to work in the general spacetime S^4, provided that we replace any partial derivative (in an equation of special relativity) by the corresponding *covariant derivative*, and we are careful not to add any extra terms involving the curvature tensor.

Finally, we postulate that the *full field equations*

$$R_{\mu\nu} = \kappa\left(T_{\mu\nu} - \frac{1}{2}T\, g_{\mu\nu}\right)$$

are valid at any point of a region containing a source, and outside such a region, the *vacuum field equations*

$$R_{\mu\nu} = 0$$

hold good. Finally, note that the field equations must be postulated; Einstein could not *formally* derive these equations any more than Schrodinger could derive his celebrated wave equation.

10.12. Exercises

1. *(Newtonian Potential)* Show that the Newtonian gravitational potential must satisfy equation (10.8):

$$\nabla^2\varphi(\mathbf{r}) = 4\pi G\rho(\mathbf{r})$$

(Solution)

Assume a continuous distribution of matter with mass density $\rho(\mathbf{r})$, which fills an *arbitrary* region of space Ω. From (10.4), we have

$$\oint_{\partial\Omega} \mathbf{g}(\mathbf{r})\cdot\mathbf{n}\, dS = -4\pi GM = -4\pi G\int_{\Omega}\rho(\mathbf{r})dV$$

where $\rho(\mathbf{r})$ is the mass density. On the other hand, by the divergence theorem,

$$\oint_{\partial\Omega} \mathbf{g}(\mathbf{r}) \cdot \mathbf{n}\, dS = \int_{\Omega} [\nabla \cdot \mathbf{g}(\mathbf{r})]\, dV = \int_{\Omega} [\nabla \cdot -\nabla\varphi(\mathbf{r})]\, dV = -\int_{\Omega} \left[\nabla^2\varphi(\mathbf{r})\right] dV$$

Equating the last two results, and rearranging terms gives

$$\int_{\Omega} \left[\nabla^2\varphi(\mathbf{r}) - 4\pi G\rho(\mathbf{r})\right] dV = 0$$

Since this equation must hold for arbitrary regions Ω, its integrand must vanish, and the result follows immediately.

2. *(Newtonian Gravity)* As shown in Figure 10.10, consider a spherical distribution of matter of radius a. Assuming the mass density ρ is uniform, show that at any point p, the Newtonian field intensity $\mathbf{g}$ is given by

$$\mathbf{g} = \begin{cases} -\frac{4\pi G\rho r}{3}\hat{\mathbf{r}}, & \text{if } r \leq a \text{ (interior solution)} \\ -\frac{4\pi a^3 G\rho}{3r^2}\hat{\mathbf{r}}, & \text{if } r > a \text{ (exterior solution)} \end{cases}$$

where r is the distance from the center. and $\hat{\mathbf{r}}$ is a unit vector in the radial direction.

(Solution)

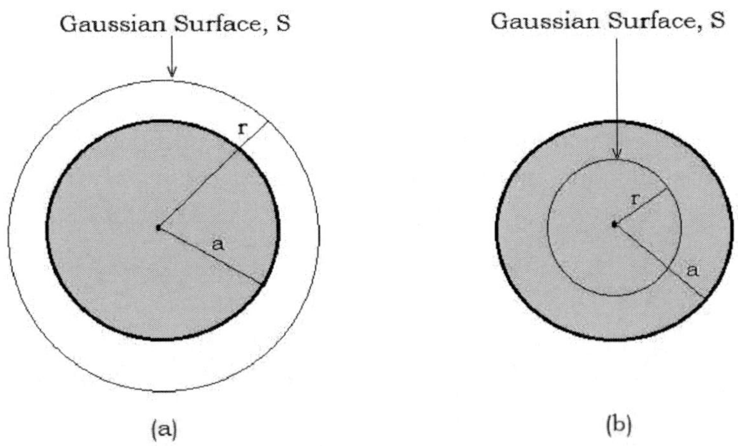

Figure 10.10.

We shall find the exterior solution. As shown in Figure 10.10(a) we construct a Gaussian surface outside the spherical mass. According to Gauss's law for gravitational fields, we have

$$\oint_{\partial\Omega} \mathbf{g}(\mathbf{r}) \cdot \mathbf{n}\, dS = -4\pi GM$$

By symmetry, $\mathbf{g}(\mathbf{r}) = -g\hat{\mathbf{r}}$ and the unit normal $\mathbf{n} = \hat{\mathbf{r}}$. We find that the lefthand side of the last equation is

$$\oint_{\partial\Omega} \mathbf{g}(\mathbf{r}) \cdot \mathbf{n}\, dS = -\oint_{\partial\Omega} g\,(\hat{\mathbf{r}} \cdot \hat{\mathbf{r}})\; dS = -g \oint_{\partial\Omega} dS = -4\pi r^2 g$$

Consequently,

$$\begin{aligned} -4\pi r^2 g &= -4\pi GM \\ &\Longrightarrow \; g = G\frac{M}{r^2} \\ &\Longrightarrow \; \mathbf{g} = -\frac{4\pi a^3 G\rho}{3r^2}\hat{\mathbf{r}} \end{aligned}$$

as required. In order to find the interior solution, we construct a Gaussian surface S inside the spherical mass (as shown in part (b)) and proceed in a similar manner.

3. *(Newtonian Gravity)* As shown in Figure 10.11, assume that planetary engineers drill a hole connecting the north pole to the south pole of a planet, and install a frictionless elevator. Assuming the elevator falls freely, show that it undergoes periodic motion with period

$$P = \sqrt{\frac{3\pi}{G\rho}}$$

where ρ is the mean mass density of the planet.

(Solution)

Let r denote the distance of the elevator from the center of the planet. Employing the *interior solution* of Exercise 2, we see that we must solve the initial value problem

$$\begin{aligned} \frac{d^2r}{dt^2} &= -\left(\frac{4\pi G\rho}{3}\right) r \\ r(0) &= a \\ \dot{r}(0) &= 0 \end{aligned}$$

where, without loss of generality, we have assumed that the elevator is initially at rest, and sitting on the surface of the planet.

The solution of the above initial value problem is clearly

$$r(t) = a\cos\left[\left(\sqrt{\frac{4\pi G\rho}{3}}\right) t\right]$$

so the motion is periodic with amplitude a, and period

$$P = \frac{2\pi}{\sqrt{\frac{4\pi G\rho}{3}}} = \sqrt{\frac{3\pi}{G\rho}}$$

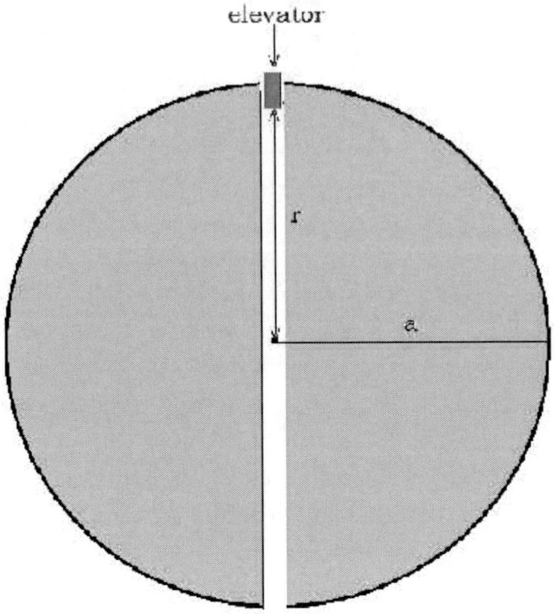

Figure 10.11.

as claimed.

4. *(Spatial Metric)* Fill in the missing steps that lead to equation (10.32) for the spatial metric.

(Solution)

From (10.31), we have

$$h_{ab} = -\left(\mathbf{e}_a^{(\perp)} \cdot \mathbf{e}_b^{(\perp)}\right)$$

Using equation (10.30), we can write

$$\begin{aligned} h_{ab} &= -\left(\mathbf{e}_a - \mathbf{e}_a^{(\|)}\right) \cdot \left(\mathbf{e}_b - \mathbf{e}_b^{(\|)}\right) = -\mathbf{e}_a \cdot \mathbf{e}_b + \mathbf{e}_a \cdot \mathbf{e}_b^{(\|)} + \mathbf{e}_a^{(\|)} \cdot \mathbf{e}_b - \mathbf{e}_a^{(\|)} \cdot \mathbf{e}_b^{(\|)} \\ &= -g_{ab} + \mathbf{e}_a \cdot \left(\frac{g_{b0}}{g_{00}} \mathbf{e}_0\right) + \left(\frac{g_{a0}}{g_{00}} \mathbf{e}_0\right) \cdot \mathbf{e}_b - \left(\frac{g_{a0}}{g_{00}} \mathbf{e}_0\right) \cdot \left(\frac{g_{b0}}{g_{00}} \mathbf{e}_0\right) \\ &= -g_{ab} + \frac{g_{a0} g_{b0}}{g_{00}} + \frac{g_{a0} g_{0b}}{g_{00}} - \frac{g_{a0} g_{b0}}{g_{00}} \\ &= -g_{ab} + \frac{g_{a0} g_{b0}}{g_{00}} \end{aligned}$$

Consequently,

$$h_{ab} = -g_{ab} + \frac{g_{a0} g_{b0}}{g_{00}}$$

5. *(Spatial Metric)* Given a spacetime metric $g_{\mu\nu}$ which has the *diagonal* form

$$ds^2 = g_{00}c^2dt^2 - g_{11}(dx^1)^2 - g_{22}(dx^2)^2 - g_{33}(dx^3)^2$$

show that the corresponding *spatial distance* dl^2 is given by

$$dl^2 = -\left[g_{11}(dx^1)^2 + g_{22}(dx^2)^2 + g_{33}(dx^3)^2\right] \tag{10.84}$$

(Solution)

From (10.32), the spatial metric h_{ab} is

$$h_{ab} = -g_{ab} + \frac{g_{a0}g_{b0}}{g_{00}} = -g_{ab} \text{ [Since } g_{a0} = g_{b0} = 0.]$$

Consequently, from (10.33), we find that

$$dl^2 = h_{ab}dx^a dx^b = -\left[g_{11}(dx^1)^2 + g_{22}(dx^2)^2 + g_{33}(dx^3)^2\right]$$

as required. So, the moral of the story is that whenever we are presented with a *diagonal* spacetime metric, the spatial distance dl^2 is simply given by the *spatial part* of the spacetime line element.

6. Derive equation (10.52).

(Solution)

From the definition of the Lie bracket, we can write

$$\begin{aligned}
[\mathbf{U},\mathbf{V}]^\alpha &= U^\sigma\partial_\sigma V^\alpha - V^\sigma\partial_\sigma U^\alpha \\
&= \frac{\partial x^\sigma}{\partial\tau}\frac{\partial V^\alpha}{\partial x^\sigma} - \frac{\partial x^\sigma}{\partial\lambda}\frac{\partial U^\alpha}{\partial x^\sigma} \quad \text{[By (10.51)]} \\
&= \frac{\partial V^\alpha}{\partial\tau} - \frac{\partial U^\alpha}{\partial\lambda} \quad \text{[Chain Rule]} \\
&= \frac{\partial^2 x^\alpha}{\partial\tau\partial\lambda} - \frac{\partial^2 x^\alpha}{\partial\lambda\partial\tau} = 0,
\end{aligned}$$

so $[\mathbf{U},\mathbf{V}] = \mathbf{0}$.

7. Show that up to first order terms, (10.63) is the reciprocal metric tensor corresponding to (10.62).

(Solution)

From (10.62) and (10.61), we have

$$\begin{aligned}
g^{\mu\sigma}g_{\sigma\nu} &= (\eta^{\mu\sigma} - \varepsilon^{\mu\sigma})(\eta_{\sigma\nu} + \varepsilon_{\sigma\nu}) = \eta^{\sigma\mu}\eta_{\sigma\nu} + \eta^{\sigma\mu}\varepsilon_{\sigma\nu} - \varepsilon^{\sigma\mu}\eta_{\sigma\nu} - \varepsilon^{\sigma\mu}\varepsilon_{\sigma\nu} \\
&\approx \delta^\mu_\nu + \varepsilon^\mu_\nu - \varepsilon^\mu_\nu \approx \delta^\mu_\nu
\end{aligned}$$

as required.

8. Justify the approximation given in (10.74).

(Solution)

We have

$$\begin{aligned} R_{00} &= -\frac{\partial}{\partial x^{\alpha}}\left(\Gamma^{\alpha}_{00}\right) \approx -\frac{1}{2}\eta^{\alpha\sigma}\left(-g_{00,\sigma}+g_{\sigma 0,0}+g_{0\sigma,0}\right) \\ &\approx \frac{1}{2}\eta^{ab}g_{00,b} \text{ [Since } g_{\sigma 0,0}=g_{0\sigma,0}=0 \text{ as metric is static.]} \\ &\approx -\frac{1}{2}\nabla^2 g_{00} \end{aligned}$$

9. Show that the vanishing of the Einstein tensor implies the vanishing of the Ricci tensor, so that the vacuum field equations may be written in the form:

$$R_{\mu\nu}=0 \tag{10.85}$$

(Solution)

Assume that $G^{\mu}_{\nu}=0$. From (10.65), we have

$$R^{\mu}_{\nu}-\frac{R}{2}\delta^{\mu}_{\nu}=0$$

Contracting on μ and ν, we get

$$R^{\mu}_{\mu}-\frac{R}{2}\delta^{\mu}_{\mu}=R-\frac{R}{2}(4)=0 \Longrightarrow R=0$$

so that $R^{\mu}_{\nu}=0$ as well. If we lower the upper index, then the result follows immediately.

10. *(Newtonian Deviation)* Consider two particles moving only under the influence of a Newtonian field. Assume that one particle moves along the path $x^a=x^a(t)$, while an adjacent particle moves along the trajectory $y^a(t)=x^a(t)+V^a(t)$ where $V=(V^a)$ is an *infinitesimal* connecting vector. Show that

$$\frac{d^2V^a}{dt^2}=-\left(\frac{\partial^2\varphi}{\partial x^a\partial x^b}\right)V^b \tag{10.86}$$

where $\varphi=\varphi(\mathbf{x})$ is the Newtonian potential. This called the **equation of Newtonian deviation**.

(Solution)

The equations of motion of the particles are:

$$\frac{d^2x^a}{dt^2} = -\frac{\partial\varphi(\mathbf{x})}{\partial x^a} \tag{10.87a}$$

$$\frac{d^2y^a}{dt^2} = \frac{d^2x^a}{dt^2}+\frac{d^2V^a}{dt^2}=-\frac{\partial\varphi(\mathbf{x}+\mathbf{V})}{\partial x^a} \tag{10.87b}$$

Now, we Taylor expand $\varphi(\mathbf{x}+\mathbf{V})$ as

$$\varphi(\mathbf{x}+\mathbf{V}) = \varphi(\mathbf{x}) + V^b \frac{\partial}{\partial x^b}\varphi(\mathbf{x}) + \dots$$

so that

$$-\frac{\partial \varphi(\mathbf{x}+\mathbf{V})}{\partial x^a} = -\frac{\partial \varphi(\mathbf{x})}{\partial x^a} - \left(\frac{\partial^2 \varphi(\mathbf{x})}{\partial x^a \partial x^b}\right) V^b$$

and consequently, (10.87b) becomes

$$\frac{d^2 y^a}{dt^2} = \frac{d^2 x^a}{dt^2} + \frac{d^2 V^a}{dt^2} = -\frac{\partial \varphi(\mathbf{x})}{\partial x^a} - \left(\frac{\partial^2 \varphi(\mathbf{x})}{\partial x^a \partial x^b}\right) V^b \tag{10.88}$$

Finally, subtracting (10.87a) from the last equation, we obtain

$$\frac{d^2 V^a}{dt^2} = -\left(\frac{\partial^2 \varphi}{\partial x^a \partial x^b}\right) V^b$$

11. *(Newtonian limit of Geodesic Deviation)* Show that the relativistic equation of geodesic deviation (10.57) reduces to (10.86) in the Newtonian limit.

(Solution)

Recall that

$$\frac{dV^\alpha}{d\tau^2} = R^\alpha_{\beta\gamma\delta} U^\beta U^\gamma V^\delta$$

We assume that the assumptions of Section 10.9 are in force, so that in the Newtonian limit of small velocities, $\mathbf{U} \approx (c, \mathbf{0})$ and $d\tau \approx dt$. Thus, this equation reduces to

$$\frac{dV^a}{dt^2} = c^2 R^a_{00d} V^d \tag{10.89}$$

On the other hand, recall that

$$R^a_{bcd} = \frac{\partial}{\partial x^c}\Gamma^a_{db} - \frac{\partial}{\partial x^d}\Gamma^a_{cb} + \Gamma^a_{cp}\Gamma^p_{db} - \Gamma^a_{dp}\Gamma^p_{cb}$$

so that ignoring products of the connection coefficients,

$$R^a_{00d} \approx \frac{\partial}{\partial x^0}\Gamma^a_{d0} - \frac{\partial}{\partial x^d}\Gamma^a_{00}$$

Now, it's easy to show that $\Gamma^a_{d0} = 0$ while

$$\begin{aligned}
\Gamma^a_{00} &\approx \frac{1}{2}\eta^{ap}\left(-g_{00,p} + g_{p0,0} + g_{0p,0}\right) = \frac{1}{2}\frac{\partial}{\partial x^a}(g_{00}) \\
&= \frac{1}{2}\frac{\partial}{\partial x^a}\left(1 + \frac{2\varphi}{c^2}\right) = \frac{1}{c^2}\frac{\partial \varphi}{\partial x^p}
\end{aligned}$$

Consequently, in the Newtonian limit, we have

$$R^a_{00d} \approx -\frac{\partial}{\partial x^d}\left(\frac{1}{c^2}\frac{\partial \varphi}{\partial x^a}\right) = -\frac{1}{c^2}\left(\frac{\partial^2 \varphi}{\partial x^a \partial x^d}\right) \tag{11.90}$$

Finally, substituting (10.90) into (10.89), we obtain

$$\frac{dV^a}{dt^2} = -\left(\frac{\partial^2 \varphi}{\partial x^a \partial x^d}\right) V^d$$

which is the equation of Newtonian deviation.

12. *(Spherical Star)* Consider a spherical star which is comprised of plasma, that can be modelled as a perfect fluid. Show that in the Newtonian limit of Section 10.9, the Newtonian potential is governed by the equation

$$\nabla^2 \varphi = 4\pi G \left(\rho_0 + \frac{3p}{c^2}\right) \tag{10.91}$$

(Solution)

Recall that from (9.54), the energy-momentum tensor for a perfect fluid is

$$T^{\mu\nu} = -p\eta^{\mu\nu} + \left(\rho_0 + \frac{1}{c^2}p\right) U^\mu U^\nu$$

From (10.68), the Einstein field equations are

$$R_{\mu\nu} = \kappa\left(T_{\mu\nu} - \frac{1}{2}T\, g_{\mu\nu}\right)$$

which may be rewritten in the form

$$R^{\mu\nu} = \kappa\left(T^{\mu\nu} - \frac{1}{2}T\, \delta^{\mu\nu}\right)$$

Now, in the Newtonian limit, only the 00-component of each tensor is of any importance. So, the Einstein field equations reduce to

$$\begin{aligned} R^{00} &= \kappa\left(T^{00} - \frac{1}{2}T\, \delta^{00}\right) = \kappa\left[T^{00} - \frac{1}{2}\left(T^{00} + T^{11} + T^{22} + T^{33}\right)\right] \\ \Longrightarrow \quad & R^{00} = \frac{\kappa}{2}\left(T^{00} - T^{11} - T^{22} - T^{33}\right) \end{aligned}$$

Assuming small plasma velocities so that $\mathbf{U} \approx (c, \mathbf{0})$, we find that

$$\begin{aligned} T^{00} &= -p + \left(\rho_0 + \frac{1}{c^2}p\right) c^2 = \rho_0 c^2 \\ T^{11} &= T^{22} = T^{33} = -p \end{aligned}$$

so that

$$R^{00} = \frac{\kappa}{2}\left(\rho_0 c^2 + 3p\right)$$

On the other hand, from (10.74), we have

$$R^{00} = \eta^{0\alpha}\eta^{0\beta}R_{\alpha\beta} = R_{00} = -\frac{1}{2}\nabla^2 g_{00} = -\frac{1}{2}\nabla^2\left(1 + \frac{2\varphi}{c^2}\right) = -\frac{1}{c^2}\nabla^2\varphi$$

Combining the last two equations, we obtain

$$\begin{aligned}
\nabla^2\varphi &= -\frac{\kappa c^2}{2}\left(\rho_0 c^2 + 3p\right) \\
&= -\frac{c^2}{2}\left(-\frac{8\pi G}{c^4}\right)\left(\rho_0 c^2 + 3p\right) \quad \text{[From (10.76)]} \\
&= 4\pi G\left(\rho_0 + 3\frac{p}{c^2}\right)
\end{aligned}$$

as required. From this equation, we see that both the proper density ρ_0, and the *pressure p* act as a *sources* of gravity.

Chapter 11

Predictions of General Relativity

Our ultimate analysis of space leads us not to a 'here' and a 'there', but to an extension such as that which relates 'here' and 'there'. To put the conclusion rather crudely, space is not a lot of points close together; it is a lot of distances interlocked.

Sir Arthur Stanley Eddington

11.1. Introduction

Since the Einstein tensor is symmetric, the Einstein field equations (10.59) provide us with $4(4+1)/2 = 10$ independent relations that must be satisfied by the 10 independent components of the metric tensor $\mathbf{g}$. At the same time, however, the *Bianchi identities* act as four *differential constraints*, so that leaves us with a total of only six independent field equations.

This means that we have a certain amount of *coordinate freedom* in that we can assign four out of the ten components of the metric tensor by judiciously choosing our coordinates. This is possible because we have assumed that spacetime is a differentiable manifold, so smooth but otherwise arbitrary coordinate transformations are permissible. The remaining six components of the metric tensor can then be determined from the six independent field equations.

Our problem is considerably simplified in those cases where the gravitational field possesses some sort of symmetry. In such cases, the metric tensor will be *form invariant* under some group of diffeomorphisms of spacetime into itself. In this regard, an important class of gravitational fields are those that possess *spherical symmetry*.

Intuitively, we expect that such fields arise from spherical distributions of matter such as stars and planets. If we make the identification

$$(x^0, x^1, x^2, x^3) = (ct, x, y, z) \tag{11.1}$$

then the metric tensor must be form-invariant with respect to the group $SO(3)$ of three dimensional rotations for each *fixed* value of the time t. Although the underlying three-space is non-Euclidean, we already know that the scalar product of any two three-vectors is invariant under rotations, so that the metric $\mathbf{g}(\mathbf{x}, t)$ can only depend upon products of the form

$$\begin{aligned} \mathbf{x}\cdot\mathbf{x} &= r^2 \\ d\mathbf{x}\cdot d\mathbf{x} &= dr^2 + r^2 d\theta^2 + r^2\sin^2\theta d\phi^2 \\ \mathbf{x}\cdot d\mathbf{x} &= r dr \end{aligned} \tag{11.2}$$

when written in terms of spherical polar coordinates. Assuming such spherical symmetry, we see that the most general expression for the line element ds^2 is

$$ds^2 = Ac^2dt^2 - B(d\mathbf{x}\cdot d\mathbf{x}) - Cdt\,(\mathbf{x}\cdot d\mathbf{x}) - D\,(\mathbf{x}\cdot d\mathbf{x})^2 \tag{11.3}$$

where A,B,C, and D are functions of $r = \|\mathbf{x}\|$ and t. We assume further that the coefficients A,B,C, and D are *positive* for all values of their arguments, so the line element always has signature -2.

If we substitute (11.2) into (11.3), and rearrange terms, we obtain

$$ds^2 = Ac^2dt^2 - \left(B + Dr^2\right)dr^2 - Br^2\left(d\theta^2 + \sin^2\theta d\phi^2\right) - Crdtdr \tag{11.4}$$

Now, let $B' = B + Dr^2$ so the line element simplifies to

$$ds^2 = Ac^2dt^2 - B'dr^2 - Br^2\left(d\theta^2 + \sin^2\theta d\phi^2\right) - Crdtdr \tag{11.5}$$

We now seek to eliminate the cross-product term $Crdtdr$. We introduce a new time variable

$$t' = t + R(r) \tag{11.6}$$

where R is a function of r alone. In making this substitution, only the first and last terms of (11.5) will be affected so that

$$Ac^2dt^2 - Crdtdr = Ac^2dt'^2 + (Ac^2R'^2 + CrR')dr^2 - (Cr + 2c^2AR')drdt' \tag{11.7}$$

From the last expression, we see that if we require R to be a solution of the differential equation

$$\frac{dR}{dr} = -\frac{C}{2c^2A}r \tag{11.8}$$

then the cross-product term will vanish. Assuming this has been done, the line element simplifies to

$$ds^2 = Ac^2dt'^2 - (Ac^2R'^2 + B' + CrR')dr^2 - Br^2\left(d\theta^2 + \sin^2\theta d\phi^2\right) \tag{11.9}$$

Next, we transform the radial coordinate by taking

$$r' = \sqrt{B}\,r \tag{11.10}$$

so that $dr' = \sqrt{B}dr$, and the last expression becomes

$$ds^2 = Ac^2dt'^2 - \left(\frac{Ac^2R'^2 + B' + CrR'}{B}\right)dr'^2 - r'^2\left(d\theta^2 + \sin^2\theta d\phi^2\right)$$

Consequently, relabeling the second coefficient and dropping primes, we conclude that our line element becomes:

$$ds^2 = A(r,t)c^2dt - B(r,t)dr^2 - r^2\left(d\theta^2 + \sin^2\theta d\phi^2\right) \tag{11.11}$$

Finally, as is customary, we introduce the functions $\alpha(r,t)$ and $\beta(r,t)$ defined by the relations

$$A(r,t) = e^{\alpha(r,t)} \text{ and } B(r,t) = e^{\beta(r,t)} \tag{11.11a}$$

so that the line element assumes the convenient form

$$ds^2 = e^{\alpha(r,t)}c^2dt^2 - e^{\beta(r,t)}dr^2 - r^2\left(d\theta^2 + \sin^2\theta d\phi^2\right) \tag{11.12}$$

This expression is known as the **Schwarzchild line element** in honor of the German mathematician Karl Schwarzchild, who first investigated the ramifications of its static counterpart in 1916.

11.2. Schwarzchild's Exterior Solution

We now find the solution of Einstein's vacuum field equations for the region outside a spherically symmetric source; this is known as **Schwarzchild's exterior solution**. This amounts to determining the components of the metric function in the Schwarzchild line element (11.12). Once we have done this, the resulting spacetime geometry as well as the paths of test particles in an exterior region are completely determined.

We must begin our task by calculating the connection coefficients corresponding to the Schwarzchild metric. We note that

$$\begin{aligned}\mathbf{g} &= (g_{\mu\nu}) = diag\left[e^{\alpha}, -e^{\beta}, -r^2, -r^2\sin^2\theta\right] \\ \mathbf{g}^{-1} &= (g^{\mu\nu}) = diag\left[e^{-\alpha}, -e^{-\beta}, -\frac{1}{r^2}, -\frac{1}{r^2\sin^2\theta}\right]\end{aligned} \tag{11.13}$$

From (11.13), we find (exercise) that the non-zero Christoffel symbols are:

$$\begin{aligned}\Gamma^0_{00} &= \frac{1}{2c}\dot{\alpha},\ \Gamma^0_{01} = \frac{1}{2}\alpha',\ \Gamma^0_{11} = \frac{1}{2c}e^{\beta-\alpha}\dot{\beta},\ \Gamma^1_{00} = \frac{1}{2}e^{\alpha-\beta}\alpha' \\ \Gamma^1_{01} &= \frac{1}{2c}\dot{\beta},\ \Gamma^1_{11} = \frac{1}{2}\beta',\ \Gamma^1_{22} = -re^{-\beta},\ \Gamma^1_{33} = -re^{-\beta}\sin^2\theta \\ \Gamma^2_{12} &= \frac{1}{r},\ \Gamma^2_{33} = -\sin\theta\cos\theta,\ \Gamma^3_{13} = \frac{1}{r},\ \Gamma^3_{23} = \cot\theta\end{aligned} \tag{11.14}$$

where the primes denote differentiation with respect to radial coordinate r, while the dots represent differentiation with respect to t.

Recalling the definition of the Ricci tensor given by:

$$R_{\mu\nu} = \frac{\partial}{\partial x^{\nu}}\left(\Gamma^{\sigma}_{\mu\sigma}\right) - \frac{\partial}{\partial x^{\sigma}}\left(\Gamma^{\sigma}_{\mu\nu}\right) - \Gamma^{\sigma}_{\mu\nu}\Gamma^{\tau}_{\sigma\tau} + \Gamma^{\sigma}_{\mu\tau}\Gamma^{\tau}_{\sigma\nu} \tag{11.15}$$

we find (exercise) that the vacuum field equations become:

$$R_{00} = -e^{\alpha-\beta}\left(\frac{\alpha''}{2}-\frac{\alpha'\beta'}{4}+\frac{\alpha'}{r}+\frac{\alpha'^2}{4}\right)+\frac{1}{c^2}\left(\frac{\ddot{\beta}}{2}-\frac{\dot{\alpha}\dot{\beta}}{4}+\frac{\dot{\beta}^2}{4}\right)=0 \quad (11.16)$$

$$R_{01} = -\frac{1}{cr}\dot{\beta}=0 \quad (11.17)$$

$$R_{11} = \left(\frac{\alpha''}{2}+\frac{\alpha'^2}{4}-\frac{\alpha'\beta'}{4}-\frac{\beta'}{r}\right)-\frac{1}{c^2}e^{\beta-\alpha}\left(\frac{\ddot{\beta}}{2}-\frac{\dot{\alpha}\dot{\beta}}{4}+\frac{\dot{\beta}^2}{4}\right)=0 \quad (11.18)$$

$$R_{22} = e^{-\beta}\left[\frac{r}{2}(\alpha'-\beta')+1\right]-1=0 \quad (11.19)$$

$$R_{33} = R_{22}\sin^2\theta=0 \quad (11.20)$$

We set ourselves upon the task of solving these equations for the unknown functions $\alpha(r,t)$ and $\beta(r,t)$. We first note from (11.17) that

$$R_{01} = -\frac{1}{cr}\dot{\beta}=0 \Longrightarrow \beta = F(r) \quad (11.21)$$

so β depends only upon r. Substituting $\beta = F(r)$ into (11.19), we find that α' can depend only upon r as well. Thus, there must exist functions $G(r)$ and $H(t)$ such that

$$\alpha = G(r)+H(t) \quad (11.22)$$

But we can eliminate the appearance of $H(t)$ in the line element by introducing a new time variable t' which satisfies

$$dt' = e^{H(t)/2}dt \quad (11.23)$$

so that $e^{\alpha(r,t)}c^2dt^2 \longrightarrow e^{G(r)}c^2dt'^2$. Assuming this has been done, we see that the line element is totally independent of the time. We have proved *Birkhoff's Theorem*:

Birkhoff's Theorem

Every *spherically symmetric* spacetime metric is independent of t, and so, is necessarily *static*.

Exploiting this time independence, the vacuum equations reduce to

$$R_{00} = \frac{\alpha''}{2}-\frac{\alpha'\beta'}{4}+\frac{\alpha'}{r}+\frac{\alpha'^2}{4}=0 \quad (11.16a)$$

$$R_{01} = 0 \quad (11.17a)$$

$$R_{11} = \frac{\alpha''}{2}+\frac{\alpha'^2}{4}-\frac{\alpha'\beta'}{4}-\frac{\beta'}{r}=0 \quad (11.18a)$$

$$R_{22} = e^{-\beta}\left[\frac{r}{2}(\alpha'-\beta')+1\right]-1=0 \quad (11.19a)$$

$$R_{33} = R_{22}\sin^2\theta=0 \quad (12.20a)$$

Equation (11.17a) tells us nothing, while (11.20a) is equivalent to (11.19a). Adding (11.16a) and (11.18a) gives

$$\alpha'+\beta'=0 \Longrightarrow \beta'=-\alpha' \quad (11.24)$$

If we substitute this result into (11.19a), we obtain

$$re^{-\beta}\beta' + e^{-\beta} = 1 \tag{11.25}$$

Making a the substitution $w = e^{-\beta}$, the last equation becomes

$$rw' + w = \frac{d}{dr}(rw) = 1 \Longrightarrow rw = r + K \tag{11.26}$$

where K is a constant. Consequently,

$$e^{-\beta} = 1 + \frac{K}{r} \text{ and } e^{\alpha} = 1 + \frac{K}{r} \tag{11.27}$$

so that our metric now assumes the form

$$ds^2 = \left(1 + \frac{K}{r}\right) c^2 dt^2 - \left(1 + \frac{K}{r}\right)^{-1} dr^2 - r^2 \left(d\theta^2 + \sin^2\theta d\phi^2\right) \tag{11.28}$$

We note that as $r \longrightarrow \infty$, both $g_{00} \longrightarrow 1$ and $g_{11} \longrightarrow 1$ so that (11.28) just reduces to the Minkowski metric in spherical coordinates.

We now seek to determine the constant K. We recall that in the weak field limit,

$$g_{00} \approx 1 + \frac{2\varphi}{c^2} \tag{11.29}$$

where the gravitational potential φ is is given by

$$\varphi = -\frac{GM}{R} \tag{11.30}$$

where R is the distance between the field point (where the field is felt) and the center of the source, and M is the total mass of the source. We have been careful to use the variable R to denote the radial distance since the r that appears in (11.28) is not a true radial distance. In fact, from (11.28), however, we see that for purely radial displacements, the *spatial radial distance* dR^2 is

$$dR^2 = \left(1 + \frac{K}{r}\right)^{-1} dr^2 \Longrightarrow dR = \frac{1}{\sqrt{1 + \frac{K}{r}}} dr \tag{11.31}$$

so that at large distances ($r \longrightarrow \infty$) we have $dr \approx dR$, and consequently, $r \approx R$. Comparing (11.29) and (11.30), we conclude that

$$K = -\frac{2GM}{c^2} \tag{11.32}$$

and the Schwarzchild metric finally becomes:

$$ds^2 = \left(1 - \frac{2GM}{c^2 r}\right) c^2 dt^2 - \frac{1}{\left(1 - \frac{2GM}{c^2 r}\right)} dr^2 - r^2 \left(d\theta^2 + \sin^2\theta d\phi^2\right) \tag{11.33}$$

11.3. Classical Kepler Motion

In this section, we will review the basic features of the classical Kepler problem; that is the motion of a particle in the central force field, such as the gravitational field of a planet. We assume that a particle of mass m moves under the influence of a gravitational field so its experiences a force given by

$$\mathbf{f} = -G\frac{Mm}{r^2}\hat{\mathbf{r}} \tag{11.34}$$

Consequently, according the Newton's second law, the equation of motion is

$$\frac{d^2\mathbf{r}}{dt^2} = -G\frac{M}{r^2}\hat{\mathbf{r}} \tag{11.35}$$

Now, it turns out the motion of the particle is confined to a plane. To see this, recall that the angular momentum $\boldsymbol{\Omega}$ of the mass m is given by

$$\boldsymbol{\Omega} = \mathbf{r} \times m\dot{\mathbf{r}} \tag{11.36}$$

and observe that

$$\frac{d\boldsymbol{\Omega}}{dt} = \frac{d}{dt}(\mathbf{r} \times m\dot{\mathbf{r}}) = m(\dot{\mathbf{r}} \times \dot{\mathbf{r}}) + (\mathbf{r} \times m\ddot{\mathbf{r}}) = -G\frac{Mm}{r^2}(\mathbf{r} \times \hat{\mathbf{r}}) = \mathbf{0} \tag{11.37}$$

This equation tells us that the angular momentum $\boldsymbol{\Omega}$ is conserved, so we can write

$$\boldsymbol{\Omega} = m\mathbf{h} \tag{11.38}$$

where the constant vector $\mathbf{h}$ is simply the angular momentum per unit mass. This means that the ordered triple of vectors $(\mathbf{r}, \dot{\mathbf{r}}, m\mathbf{h})$ forms a righthanded system, so that the motion of the mass m is confined to the *plane* determined by the vectors $\mathbf{r}$ and $\dot{\mathbf{r}}$. This situation is depicted in Figure 11.1.

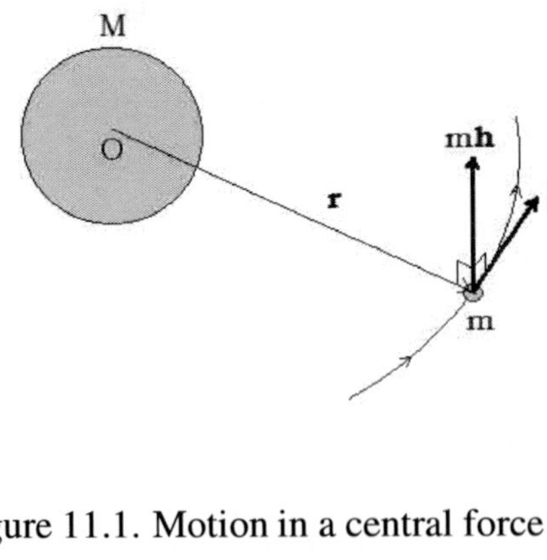

Figure 11.1. Motion in a central force field.

Since the motion is confined to a plane, we can introduce plane polar coordinates (R, θ). It's easy to show that the velocity $\mathbf{v}$ and acceleration $\mathbf{a}$ of the particle are given by

$$\mathbf{v} = (\dot{R})\hat{\mathbf{R}} + (R\dot{\theta})\hat{\theta} \tag{11.39a}$$

$$\mathbf{a} = \ddot{\mathbf{R}} = \left(\ddot{R} - R\dot{\theta}^2\right)\hat{\mathbf{R}} + \left(R\ddot{\theta} + 2\,\dot{R}\dot{\theta}\right)\hat{\theta} \tag{11.39b}$$

so that (11.35) becomes

$$\left(\ddot{R}-R\dot{\theta}^2\right)\hat{\mathbf{R}}+\left(R\ddot{\theta}+2\,\dot{R}\dot{\theta}\right)\hat{\theta}=-\frac{GM}{R^2}\hat{\mathbf{R}} \tag{11.40}$$

Equating radial and angular components, we obtain

$$\ddot{R}-R\dot{\theta}^2 = -\frac{GM}{R^2} \tag{11.41a}$$

$$R\ddot{\theta}+2\,\dot{R}\dot{\theta} = \frac{d}{dt}\left(R^2\dot{\theta}\right)=0 \tag{11.41b}$$

The second equation is just the conservation of angular momentum expressed in terms of polar coordinates, so that

$$R^2\dot{\theta}=h$$

where $h=\|\mathbf{h}\|$.

We now direct our attention to (11.41a). We assume the path of the particle is of the form $R=R(\theta)$, and we introduce a new variable u:

$$u=\frac{1}{R(\theta)} \tag{11.42}$$

We seek to rewrite (11.41a) in terms of this new variable. We find that (exercise)

$$\ddot{R}=-h^2u^2\frac{d^2u}{d\theta^2} \tag{11.43}$$

so that (11.41a) simplifies to

$$\frac{d^2u}{d\theta^2}+u=\frac{GM}{h^2} \tag{11.44}$$

which is called **Binet's equation**. This is just the equation of a simple harmonic oscillator with a constant forcing term. Its solution can be written in the form (exercise):

$$u=\frac{GM}{h^2}+C\cos\left(\theta-\theta_0\right) \tag{11.45}$$

where C is a constant, and θ_0 is a phase angle. We can choose our coordinate axes so that $\theta_0=0$, and the previous equation becomes

$$R=\frac{1}{\frac{GM}{h^2}+C\cos\theta} \tag{11.46}$$

Now, recall that when expressed in polar coordinates, the general equation of a conic section is

$$R=\frac{p}{1+e\cos\theta} \tag{11.47}$$

where p is the *semi-latus rectum*, and e is the *eccentricity*. So, if we compare the last two equations, we find that

$$p=\frac{h^2}{GM}\quad\text{and}\quad e=\frac{GM}{h^2}, \tag{11.48}$$

so finally, our solution may be written as

$$R = \frac{h^2/GM}{1+e\cos\theta} \tag{11.49}$$

It's easy to show that R achieves a maximum value when θ is an odd multiple of π, and a minimum value when θ is an even multiple of π. In fact, we find that

$$R_{\text{max}} = \frac{h^2/GM}{1-e} \quad \text{and} \quad R_{\text{min}} = \frac{h^2/GM}{1-e} \ . \tag{11.50}$$

The distance R_{max} is called the **aphelion** while the distance of closest approach R_{min} is called the **perihelion**. Now, let a be the semi-major axis of the elliptical orbit, as shown in Figure 11.2. Since $2a = R_{\text{max}} + R_{\text{min}}$ then from (11.50), we find

$$a = \frac{h^2}{GM}(1-e^2) \tag{11.51}$$

and consequently,

$$R_{\text{max}} = a(1+e) \quad \text{and} \quad R_{\text{min}} = a(1-e) \tag{11.52}$$

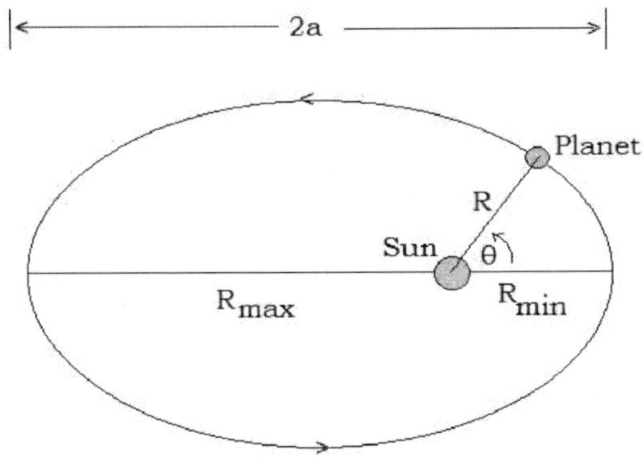

Figure 11.2. Motion of a planet around the Sun.

11.4. The Precession of Planetary Orbits

In this section, we apply the Schwarzchild metric to the problem of planetary motion. We seek the relativistic equations of motion of a planet which moves under the gravitational influence of a central body of mass M. We first write (11.33) in compact form as

$$ds^2 = \left(1 - \frac{2\sigma}{r}\right)c^2dt^2 - \left(1 - \frac{2\sigma}{r}\right)^{-1} dr^2 - r^2\left(d\theta^2 + \sin^2\theta d\phi^2\right) \tag{11.53}$$

where we have put

$$\sigma = \frac{GM}{c^2} \tag{11.54}$$

We parametrize the trajectory $x^\alpha = x^\alpha(\tau)$ of a planet in terms of the proper time τ. We can obtain the geodesic equations by finding the extremals of the Lagrangian $L = g_{\alpha\beta}\dot{x}^\alpha\dot{x}^\beta$ given by

$$L = \left(\frac{ds}{d\tau}\right)^2 = \left(1-\frac{2\sigma}{r}\right)c^2\dot{t}^2 - \left(1-\frac{2\sigma}{r}\right)^{-1}\dot{r}^2 - r^2\left(\dot{\theta}^2 + \sin^2\theta\,\dot{\phi}^2\right) = c^2 \quad (11.55)$$

The geodesic equations are then found by substituting (11.55) into the Euler-Lagrange equations

$$\frac{d}{d\tau}\left(\frac{\partial L}{\partial \dot{x}^\alpha}\right) - \frac{\partial L}{\partial x^\alpha} = 0 \quad (11.56)$$

The geodesic equations corresponding to $\alpha = 0, 1, 2, 3$ are

$$\frac{d}{d\tau}\left[\left(1-\frac{2\sigma}{r}\right)\dot{t}\right] = 0 \quad (11.57)$$

$$\frac{d}{d\tau}\left[\left(1-\frac{2\sigma}{r}\right)\dot{r}\right] - \frac{\sigma}{r^2}c^2\dot{t}^2 - \frac{\sigma}{r^2}\left(1-\frac{2\sigma}{r}\right)^{-2}\dot{r}^2 - r\dot{\theta}^2 - \left(r\sin^2\theta\right)\dot{\phi}^2 = 0 \quad (11.58)$$

$$\frac{d}{d\tau}\left(r^2\dot{\theta}\right) - \left(r^2\sin\theta\cos\theta\right)\dot{\phi}^2 = 0 \quad (11.59)$$

$$\frac{d}{d\tau}\left[\left(r^2\sin^2\theta\right)\dot{\phi}\right] = 0 \quad (11.60)$$

Now, we will disregard equation (11.58) as it is quite messy. We can do this because (11.55) already provides us with a first integral of the motion, and we only need four equations to determine four unknowns.

In order to simplify things further, we shall only consider motion in the 'equatorial plane' where $\theta = \pi/2$. Our set of working equations then reduce to

$$\left(1-\frac{2\sigma}{r}\right)c^2\dot{t}^2 - \left(1-\frac{2\sigma}{r}\right)^{-1}\dot{r}^2 - r^2\dot{\phi}^2 = c^2 \quad (11.55a)$$

$$\left(1-\frac{2\sigma}{r}\right)\dot{t} = k \quad (11.57a)$$

$$r^2\dot{\phi} = h \quad (11.59a)$$

where k and h are constants. Substituting (11.57a) and (11.59a) into (11.55a), we obtain

$$\dot{r}^2 + \frac{h^2}{r^2}\left(1-\frac{2\sigma}{r}\right) = c^2\left[k^2 - \left(1-\frac{2\sigma}{r}\right)\right] \quad (11.61)$$

From our previous experience with Kepler motion, we set $u = 1/r$ so that

$$\dot{r} = \frac{dr}{d\tau} = -h\frac{du}{d\phi} \quad (11.62)$$

and the last equation becomes

$$\left(\frac{du}{d\phi}\right)^2 + (1-2\sigma u)u^2 = \frac{c^2}{h^2}\left[k^2 - (1-2\sigma u)\right] \quad (11.63)$$

Differentiating both sides of this equation with respect to ϕ and simplifying, we obtain

$$\frac{d^2u}{d\phi^2} + u = \frac{GM}{h^2} + \frac{3GM}{c^2}u^2 \tag{11.64}$$

This is the relativistic counterpart of Binet's equation, and it is non-linear due to the presence of the term containing u^2.

We now seek an approximate solution to (11.64). First, we introduce the small parameter

$$\varepsilon = \frac{3GM}{c^2} \tag{11.65}$$

so that (11.64) becomes

$$\frac{d^2u}{d\phi^2} + u = \frac{GM}{h^2} + \varepsilon u^2 \tag{11.66}$$

and we seek a solution to (11.66) as a perturbation series of the form

$$u = u_0 + \varepsilon u_1 + O(\varepsilon^2) \tag{11.67}$$

Substituting (11.67) into (11.66), equating equal powers of ε, and retaining only first order terms, we obtain

$$u_0'' + u_0 = \frac{GM}{h^2} \tag{11.68}$$

$$u_1'' + u_1 = u_0^2 \tag{11.69}$$

Observe that (11.68) is just Binet's equation, while (11.69) gives us the first order correction. We can, however, immediately write down the solution to Binet's equation as

$$u_0 = \frac{GM}{h^2}(1 + e\cos\phi) \tag{11.70}$$

Substituting this result into (11.69), we obtain

$$u_1'' + u_1 = \alpha\left(1 + 2e\cos\phi + e^2\cos^2\phi\right) \tag{11.71}$$

where we have set

$$\alpha = \frac{G^2M^2}{h^4} \tag{11.72}$$

We now use the method of undetermined coefficients to find a particular solution of (11.71). If we assume a particular solution of the form

$$u_1 = A + B\phi\sin\phi + C\cos 2\phi \tag{11.73}$$

and substitute this expression into (11.71), then we obtain

$$A = \alpha\left(1 + \frac{e^2}{2}\right),\ B = \alpha e,\ C = -\frac{1}{6}\alpha e^2 \tag{11.74}$$

and consequently,

$$u_1 = \alpha\left[1 + e^2\left(\frac{1}{2} - \frac{1}{6}\cos 2\phi\right) + e\phi\sin\phi\right] \tag{11.75}$$

Since $\alpha \ll 1$, we can neglect the first two terms in (11.75). The third term, however, cannot be ignored as it increases with each revolution of the planet. So, we simply set

$$u_1 = \alpha e \phi \sin\phi \tag{11.75a}$$

Substituting this result into the original perturbation series, we find

$$u = \frac{GM}{h^2}\left[1 + e\left(\cos\phi + \phi\delta\sin\phi\right)\right] \tag{11.76}$$

where

$$\delta = \frac{3G^2M^2}{c^2h^2} \tag{11.77}$$

Clearly, $\delta \ll 1$ and we can exploit this fact to simplify (11.76) even further. Observe that

$$\begin{aligned} \cos\left[\phi\left(1-\delta\right)\right] &= \cos\phi\cos\left(\phi\delta\right) + \sin\phi\sin\left(\phi\delta\right) \approx \cos\phi\cdot(1) + \sin\phi\cdot(\phi\delta) \\ &\approx \cos\phi + (\phi\delta)\sin\phi \end{aligned}$$

so our final solution becomes

$$u = \frac{GM}{h^2} 1 + e\cos\left[(1-\delta)\,\phi\right] \tag{11.78}$$

From this equation, we see that the orbit of the planet is periodic with the period

$$P = \frac{2\pi}{1-\delta} \approx 2\pi + 2\pi\delta \tag{11.79}$$

which is *greater than* the usual period of 2π we normally associate with classical Kepler motion. This means that each revolution, the major axis of the ellipse rotates counterclockwise by an *additional angle* of $2\pi\delta$ from its position at the previous orbit. Stated differently, the ellipse *precesses* by a small amount $2\pi\delta$ each revolution. Since $2\pi\delta > 0$, the precession has the same direction as the motion of the planet.

Furthermore, from (11.51) and (11.77), we can express the amount of precession *per revolution* as

$$\delta = \frac{6\pi GM}{ac^2(1-e^2)} \tag{11.80}$$

It is instructive to apply this equation to Mercury's orbit in order to calculate the expected precession after one century. Since the period of Mercury's orbit is about 88 days, it will undergo about 415 revolutions per century. Using the known values:

$$\begin{aligned} G &= 6.67\times10^{-11}\ \text{Nt-m/kg}^2 \\ M &= 2\times10^{30}\ \text{kg} \\ a &= 5.8\times10^{10}\ \text{m} \\ e &= 0.2 \\ c &= 3\times10^{8}\ \text{m/sec} \end{aligned}$$

we find that

$$\begin{aligned}\delta &\approx 415\times\frac{6\pi\times(6.67\times10^{-11})(2\times10^{30})}{(5.8\times10^{10})\,(3\times10^{8})^2\,(0.96)} = 2.08\times10^{-4}\text{ radians/century}\\ &= 43''\text{ per century}\end{aligned}$$

Now, the *observed* total precession for Mercury is

$$(5599.7\pm0.4)''\text{ per century.}$$

However, after one accounts for the precession due to the influence of nearby planets, the observed *residual* precession is $(43.1\pm0.5)''$.

We see that residual precession predicted by Einstein's theory is in excellent agreement with experiment.

It is important to note that this residual precession could not be explained by standard Newtonian theory. Historically, Urbain Jean Joseph Le Verrier had pointed out such an anomaly in Mercury's orbit as early as 1859, and in 1882, Simon Newcomb obtained the generally accepted value of about 43″per century. In the course of time, numerous theoretical proposals were made to explain this anomaly. These attempts involved such hypotheses as the existence of an unseen intramercurial planet, the possible existence of a ring of intermercurial asteroids, and proposing modifications to Newton's inverse square law.[1]

11.5. The Deflection of Light Rays

We consider the trajectory of a photon in Schwarzchild spacetime, in the vicinity of a gravitating mass M. We shall proceed as in the previous section, however, as photons travel along *null geodesics*, we must be careful not to use the proper time τ as a parameter. Instead, we assume a trajectory of the photon assumes the form

$$x^\alpha = x^\alpha(\lambda) \tag{11.81}$$

for some affine parameter λ, and we require that

$$\left(\frac{ds}{d\lambda}\right)^2 = g_{\mu\nu}\dot{x}^\mu\dot{x}^\nu = 0 \quad \text{where } \dot{x}^\mu = \frac{dx^\alpha}{d\lambda} \tag{11.82}$$

As in the previous section, we find the extremals of the Lagrangian $L = g_{\mu\nu}\dot{x}^\mu\dot{x}^\nu$, and restrict our attention to motion in the 'equatorial plane' $\theta = \pi/2$. After doing so, the relevant equations become

$$\left(1-\frac{2\sigma}{r}\right)c^2\dot{t}^2 - \left(1-\frac{2\sigma}{r}\right)^{-1}\dot{r}^2 - r^2\dot{\theta}^2 = 0 \tag{11.83}$$

$$\left(1-\frac{2\sigma}{r}\right)\dot{t} = k \tag{11.84}$$

$$r^2\dot{\phi} = h \tag{11.85}$$

We substitute the last two equations into the first equation, and simplify, to obtain

$$\dot{r}^2 + \frac{h^2}{r^2}\left(1 - \frac{2\sigma}{r}\right) = c^2k^2 \tag{11.86}$$

As in the past, we set $u = 1/r$, and observe that

$$\dot{r} = \frac{dr}{d\lambda} = \frac{dr}{d\phi}\frac{d\phi}{d\lambda} = \left(-\frac{1}{u^2}\frac{du}{d\phi}\right)\left(hu^2\right) = -h\frac{du}{d\phi} \tag{11.87}$$

Next, we substitute (11.86) into (11.85), and simplify to get

$$\left(\frac{du}{d\phi}\right)^2 + (1-2\sigma u)u^2 = \frac{c^2k^2}{h^2}$$

We differentiate both sides of this equation to obtain

$$\frac{d^2u}{d\phi^2} + u = 3\sigma u^2 \tag{11.88}$$

It's instructive to find the classical limit of (11.88). Recall that $\sigma = GM/c^2$ so in the absence of a gravitating body $\sigma = 0$, and this equation reduces to

$$\frac{d^2u}{d\phi^2} + u = 0 \tag{11.89}$$

Clearly, (11.89) governs the motion of a photon in the flat space of special relativity. Its solution may be written as

$$u = C\sin(\phi + \phi_0)$$

where C is a constant, and ϕ_0 is a constant phase angle. By a con- venient choice of axes, we can take $\phi_0 = 0$. If we assume further that $u = 1/r_0$ when $\phi = 0$, the solution simply becomes

$$u = \frac{1}{r} = \frac{1}{r_0}\sin\phi \tag{11.90}$$

As expected, in the absence of gravitational effects, the photon moves along a straight line. This state of affairs is shown in Figure 11.3.

We now seek an approximate solution to the non-linear equation (11.88). First, we introduce the small parameter $\varepsilon = 3\sigma$ so that the equation becomes

$$\frac{d^2u}{d\phi^2} + u = \varepsilon u^2 \tag{11.91}$$

and we look for a solution as a perturbation series of the form

$$u = u_0 + \varepsilon u_1 + O(\varepsilon^2) \tag{11.92}$$

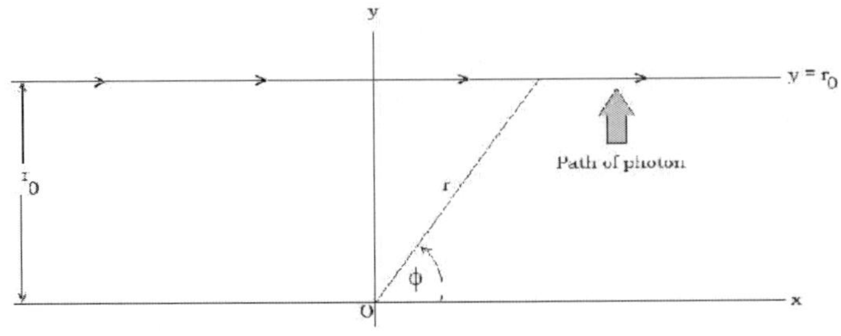

Figure 11.3. Path of a photon in Minkowski space ($\sigma = 0$).

If we substitute (11.92) into (11.91), equate equal powers of ε, and retain only first order terms, we obtain

$$u_0'' + u_0 = 0 \tag{11.93}$$
$$u_1'' + u_1 = u_0^2 \tag{11.94}$$

Observe that the solution of (11.93) is given by (11.90), while the second equation gives us the first order correction. Substituting (11.90) into (11.94), we obtain

$$u_1'' + u_1 = \frac{1}{r_0^2}\sin^2\phi \tag{11.95}$$

We now use the method of undetermined coefficients to find a particular solution of (11.95). We assume a particular solution of the form

$$u_1 = A + B\phi\cos\phi + C\cos 2\phi \tag{11.96}$$

and substitute this expression into (11.95), to obtain

$$A = \frac{1}{2r_0^2},\ B = 0,\ C = \frac{1}{6r_0^2} \tag{11.97}$$

so that,

$$u_1 = \frac{1}{r_0^2}\left(\frac{1}{2} + \frac{1}{6}\cos 2\phi\right) \tag{11.98}$$

Finally, substituting $u_0 = (1/r_0)\sin\phi$ and (11.98) into the perturbation series, we obtain

$$u = \frac{1}{r} = \frac{1}{r_0}\sin\phi + \frac{3GM}{2r_0^2c^2}\left(\frac{1}{2} + \frac{1}{6}\cos 2\phi\right) \tag{11.99}$$

Now, let us consider the behaviour of this solution as $r \longrightarrow \infty$, or equivalently, as $u \longrightarrow 0$. We seek to determine the total angle of deflection δ of the path of a light ray in the neighborhood of a gravitating body such as the Sun.

If we assume only a small deflection, we can take $\sin\phi \approx \phi$ and $\cos 2\phi \approx 1$ at infinity. As shown in Figure 11.4, assume that the lefthand and righthand asymptotes make angles

$\pi + \delta_1$ and δ_2 respectively, with respect to the x-axis. Since $u \longrightarrow 0$ as $r \longrightarrow \infty$, we must have

$$\begin{aligned} -\frac{\delta_1}{r_0} + \frac{2GM}{r_0^2 c^2} &= 0 \quad (\text{when } \phi = \pi + \delta_1) \\ -\frac{\delta_2}{r_0} + \frac{2GM}{r_0^2 c^2} &= 0 \quad (\text{when } \phi = -\delta_2) \end{aligned}$$

and consequently,

$$\delta_1 = \delta_2 = \frac{2GM}{r_0 c^2} \tag{11.100}$$

Consequently, the total deflection is $\delta = \delta_1 + \delta_2$ or

$$\delta = \frac{4GM}{r_0 c^2} \tag{11.101}$$

This is the famous deflection formula obtained by Einstein in 1915.

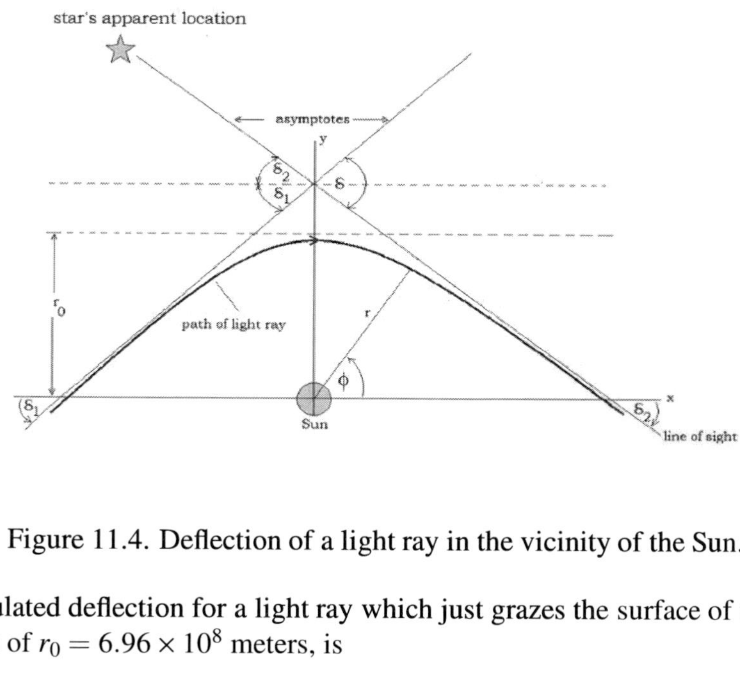

Figure 11.4. Deflection of a light ray in the vicinity of the Sun.

The calculated deflection for a light ray which just grazes the surface of the Sun, using a solar radius of $r_0 = 6.96 \times 10^8$ meters, is

$$\delta = \frac{4\left(6.67 \times 10^{-11}\right)\left(2 \times 10^{30}\right)}{\left(6.96 \times 10^8\right)\left(3 \times 10^8\right)^2} = 8.5 \times 10^{-6} \text{ radians} \approx 1.75''$$

or 1.75 arc seconds. This prediction excited the imaginations of several astronomers, especially Sir Arthur Eddington, who was at the time, a recent convert to relativity theory. Astronomers suggested that the predicted deflection might most easily be measured by photographing the field of stars in the vicinity of the Sun during a total eclipse of the Sun by the moon, and then comparing the photographs of the same stars taken at a later time when the Sun was not present.

As you can see from Figure 11.4, due to the deflection of light, the apparent positions of the stars should move outward, away from their true positions when the influence of the Sun's gravitational field is absent. Consequently, in 1919, Eddington led an expedition to Principe island as a solar eclipse was predicted to occur on May 29th of that year.

In spite of cloudy skies, Eddington and his team managed to obtain several decent exposures. Upon examining the photographic plates, Eddington reported a bending of starlight of $1.61'' \pm 0.30''$, providing good agreement with the prediction of general relativity.[3] In recent times, several researchers have used more precise methods to measure the predicted deflection, and all of their results are in good qualitative agreement with Einstein's theoretical value.

11.6. Gravitational Red Shift

We now turn our attention to another prediction of Einstein's theory which entails the retardation of clocks in a gravitational field. Consider two clocks *at rest* in a Schwarzchild spacetime. We assume, as shown in Figure 11.5, that the first clock is placed at the point $P(ct, r, \theta, \phi)$, while a second identical clock is placed at the point $P'(ct, r', \theta', \phi')$, where $r' > r$.

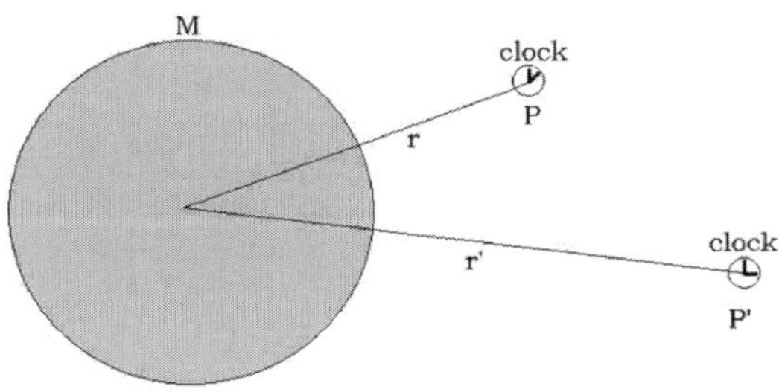

Figure 11.5. Two identical clocks at rest in a Schwarzchild field.

The proper time $d\tau$ recorded by the first clock is simply

$$d\tau = \sqrt{g_{00}(P)}\, cdt \,, \tag{11.102}$$

while the proper time $d\tau'$ registered by the second clock is

$$d\tau' = \sqrt{g_{00}(P')}\, cdt \tag{11.103}$$

From the above equations, we can see that for *equal values* of the coordinate time t, the proper times recorded by the two clocks will generally differ. In fact, these equations imply that

$$\frac{d\tau}{d\tau'} = \sqrt{\frac{g_{00}(P)}{g_{00}(P')}} = \sqrt{\frac{1 - 2GM/rc^2}{1 - 2GM/r'c^2}} \tag{11.104}$$

Since $r' > r$, then $d\tau < d\tau'$. So, the proper time interval recorded by the first clock (where the gravitational field is stronger) will be *less than* that recorded by its counterpart (where the gravitational is weaker).

Now, let us imagine that an observer situated at the point P trans- mits a short pulse of radio waves of frequency f and duration $d\tau$, to the point P'. During the proper time interval $d\tau$ a total of $fd\tau$ waves will leave the point P. An observer at P' will receive the *same number* of waves during a proper time interval $d\tau'$, so he will calculate the frequency f' of the received signal as

$$f' = \frac{\text{number of waves}}{\text{proper time interval at } P'} = \frac{fd\tau}{d\tau'} = \sqrt{\frac{g_{00}(P)}{g_{00}(P')}}$$

Consequently, the ratio of the frequency of the received signal to that of the transmitted signal is

$$\frac{f'}{f} = \sqrt{\frac{g_{00}(P)}{g_{00}(P')}} \tag{11.105}$$

Since the righthand side of this equation is less than unity, the received frequency f' will be *less than* the transmitted frequency f. Stated differently, the original signal is shifted towards the red end of the spectrum. This phenomenon is known as the **gravitational red shift**.

In the weak field limit, we know that $g_{00} = 1 + 2\varphi/c^2$, so the last equation (exercise) implies that the **frequency shift** is

$$\frac{\Delta f}{f} = \frac{f' - f}{f} \approx \frac{\varphi - \varphi'}{c^2} \tag{11.106}$$

Assuming a Schwarzchild spacetime, the frequency shift can be rewritten as

$$\frac{\Delta f}{f} \approx \frac{1}{c^2}\left(\frac{GM}{r'} - \frac{GM}{r}\right) \tag{11.107}$$

Since $r' > r$, we see once again that the frequency shift is *negative*.

It is interesting to note that equation (11.106) can be also obtained by a simple argument involving involving energy conservation. To see this, consider the motion of a single photon which moves radially outward from P to P' in a Schwarzchild field, as shown in Figure 11.6.

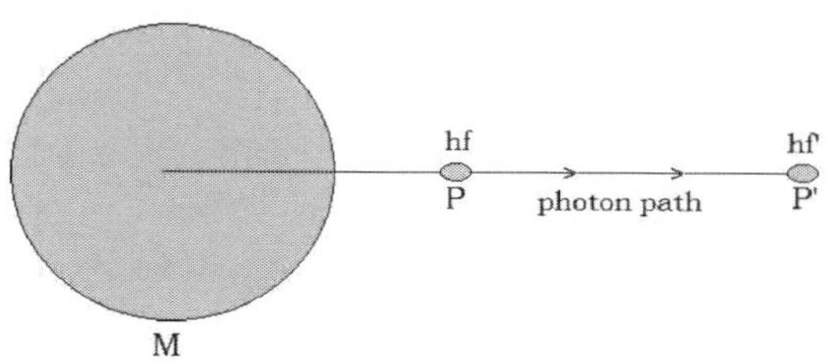

Figure 11.6. A photon moving radially outward from P to P'.

Owing to the equivalence of mass and energy, the photon has an inertial mass of hf/c^2 at the point P. As it moves from P to P' it must do an amount of work against the gravitational field given by

$$\text{work } (P \Longrightarrow P') \approx \frac{hf}{c^2}(\varphi - \varphi') \tag{11.108}$$

We write this as an *approximate* equality since the photon's frequency actually changes all along this path. Now, the work done against the field represents *energy lost* to the photon, so we must have

$$hf - \frac{hf}{c^2}(\varphi - \varphi') \approx hf' \tag{11.109}$$

The last equation again implies equation (11.106):

$$\frac{f - f'}{f} \approx \frac{\varphi - \varphi'}{c^2}$$

We can estimate the frequency shift for the special case of a beam of light emitted near the surface of the Sun. If we ignore the gravitational potential of the Earth (by setting $\varphi' = 0$), and take the radius of the Sun as $r = 6.96 \times 10^8$ meters, we find an expected frequency shift of

$$\frac{\Delta f}{f} \approx -\frac{(6.67 \times 10^{-11})(2.0 \times 10^{30})}{(6.96 \times 10^8)(3 \times 10^8)^2} \approx -2.1 \times 10^{-6}$$

Observations of solar spectrum emitted from the surface of the Sun give results in agreement with this prediction, up to the order of magnitude. But such observations are met with numerous technical difficulties, and are not totally reliable.

In 1960, however, Robert Pound and his assistant Glen Rebka devised a *terrestrial* experiment to measure the expected red shift. Using an extremely clever experimental setup, they measured the frequency shift in gamma ray photons that travelled a distance of about 22.5 meters in a shaft of the Jefferson Physics building. They found a result which was within about 10% of the predicted value.

In 1980, a R.F.C. Vessot, *etal* measured this predicted effect with great accuracy by observing the frequency shift of a hydrogen maser placed in rocket traveling radially outward from the Earth at about 8.5 kilometers per second. They monitored the frequency change until the rocket achieved a maximum altitude of about 10,000 kilometers. The investigators found agreement with the predicted frequency shift to within about 0.01%.

11.7. Gravitational Waves

In this section, we shall see how the *linearization* of Einstein's field equations leads to the prediction of gravitational waves. We assume that a weak gravitational field prevades a substantial region of spacetime, and that within this region, there exists a coordinate system (x^α) such that the metric can be written in the form

$$g_{\mu\nu} = \eta_{\mu\nu} + h_{\mu\nu} \quad \text{where } |h_{\mu\nu}| \ll 1 \tag{11.110}$$

Here, the quantities $h_{\mu\nu}$ can be viewed as *small perturbations* on an otherwise flat Minkowski space. Consequently, such a coordinate system (x^α) is often called **quasi-Minkowskian**. Thus, our approach is similar to the one we took in finding the Newtonian limit of the field equations, with the notable exception that the quantities $h_{\mu\nu}$ may depend upon the coordinate time x^0.

As part of the linearization process, we assume that:

1. We can lower and raise indices by using the Minkowski metric $\eta_{\mu\nu}$ and its reciprocal $\eta^{\mu\nu}$, instead of using $g_{\mu\nu}$ and $g^{\mu\nu}$.

2. All of the partial derivatives of $h_{\mu\nu}$ are of the same order of smallness as $h_{\mu\nu}$, so we can ignore products involving the h's with themselves, or with any of their derivatives.

Following the rules for linearization, it turns out that the reciprocal metric is given by

$$g^{\mu\nu} = \eta^{\mu\nu} - h^{\mu\nu} \tag{11.111}$$

To see this, observe that

$$\begin{aligned} g_{\mu\sigma}\left(\eta^{\sigma\nu} - h^{\sigma\nu}\right) &= \left(\eta_{\mu\sigma} + h_{\mu\sigma}\right)\left(\eta^{\sigma\nu} - h^{\sigma\nu}\right) \\ &= \eta^{\sigma\nu}\eta_{\sigma\mu} - \eta_{\mu\sigma}h^{\sigma\nu} + \eta^{\sigma\nu}h_{\mu\sigma} - h_{\mu\sigma}h^{\sigma\nu} \\ &\approx \delta_\mu^\nu - h_\mu^\nu + h_\mu^\nu \\ &\approx \delta_\mu^\nu \end{aligned}$$

where we have ignored the second order term $h_{\mu\sigma}h^{\sigma\nu}$.

With the reciprocal metric in hand, we now calculate the connection coefficients. We find

$$\begin{aligned} \Gamma^\alpha_{\beta\gamma} &= \frac{1}{2}g^{\alpha\sigma}\left(-\partial_\sigma g_{\beta\gamma} + \partial_\gamma g_{\sigma\beta,\gamma} + \partial_\beta g_{\gamma\sigma}\right) \\ &= \frac{1}{2}\left(\eta^{\alpha\sigma} - h^{\alpha\sigma}\right)\left(-\partial_\sigma h_{\beta\gamma} + \partial_\gamma h_{\sigma\beta,\gamma} + \partial_\beta h_{\gamma\sigma}\right) \\ &= \frac{1}{2}\left(-\partial^\alpha h_{\beta\gamma} + \partial_\gamma h^\alpha_\beta + \partial_\beta h^\alpha_\gamma\right) + (\text{higher order terms}) \end{aligned}$$

so that

$$\Gamma^\alpha_{\beta\gamma} = \frac{1}{2}\left(-\partial^\alpha h_{\beta\gamma} + \partial_\gamma h^\alpha_\beta + \partial_\beta h^\alpha_\gamma\right) \tag{11.112}$$

Next, we find the linearized curvature tensor. We first recall that

$$R^\alpha_{\beta\gamma\delta} = \partial_\gamma\left(\Gamma^\alpha_{\beta\delta}\right) - \partial_\delta\left(\Gamma^\alpha_{\beta\gamma}\right) + \Gamma^\alpha_{\gamma\sigma}\Gamma^\sigma_{\delta\beta} - \Gamma^\alpha_{\delta\sigma}\Gamma^\sigma_{\beta\gamma}$$

From this expression, we see that we can ignore the last two terms, as products of the connection coefficients will be of second order in the perturbation $h_{\mu\nu}$. We then obtain

$$\begin{aligned} R^\alpha_{\beta\gamma\delta} &= \partial_\gamma\left(\Gamma^\alpha_{\beta\delta}\right) - \partial_\delta\left(\Gamma^\alpha_{\beta\gamma}\right) \\ &= \frac{1}{2}\partial_\gamma\left(-\partial^\alpha h_{\beta\delta} + \partial_\delta h^\alpha_\beta + \partial_\beta h^\alpha_\delta\right) - \frac{1}{2}\partial_\delta\left(-\partial^\alpha h_{\beta\gamma} + \partial_\gamma h^\alpha_\beta + \partial_\beta h^\alpha_\gamma\right) \end{aligned}$$

so that

$$R^\alpha_{\beta\gamma\delta} = \frac{1}{2}\left(\partial^\alpha\partial_\delta h_{\beta\gamma} - \partial^\alpha\partial_\gamma h_{\beta\delta} + \partial_\beta\partial_\gamma h^\alpha_\delta - \partial_\beta\partial_\delta h^\alpha_\gamma\right) \tag{11.113}$$

Consequently, by contracting the indices α and γ in (11.113), we find that the linearized Ricci tensor is given by

$$\begin{aligned} R_{\beta\delta} &= R^{\alpha}_{\beta\alpha\delta} = \frac{1}{2}\left(\partial^{\alpha}\partial_{\delta}h_{\beta\alpha} - \partial^{\alpha}\partial_{\alpha}h_{\beta\delta} + \partial_{\beta}\partial_{\alpha}h^{\alpha}_{\delta} - \partial_{\beta}\partial_{\delta}h^{\alpha}_{\alpha}\right) \\ &= \frac{1}{2}\left(\partial^{\alpha}\partial_{\delta}h_{\alpha\beta} - \Box^2 h_{\beta\delta} + \partial_{\alpha}\partial_{\beta}h^{\alpha}_{\delta} - \partial_{\beta}\partial_{\delta}h\right) \end{aligned}$$

where in the last line, we have put

$$\begin{aligned} \Box^2 &= \partial_{\alpha}\partial^{\alpha} = \frac{\partial^2}{c^2\partial t^2} - \nabla^2 \quad \text{(d' Alembertian operator)} \\ h &= h^{\alpha}_{\alpha} = Tr(h^{\alpha}_{\beta}) \end{aligned}$$

Lowering the index on the operator ∂^{α}, and relabeling in indices, we have

$$R_{\alpha\beta} = \frac{1}{2}\left(\partial_{\alpha}\partial_{\sigma}h^{\sigma}_{\beta} + \partial_{\beta}\partial_{\sigma}h^{\sigma}_{\alpha} - \Box^2 h_{\alpha\beta} - \partial_{\alpha}\partial_{\beta}h\right) \tag{11.114}$$

Consequently, the linearized Ricci scalar becomes

$$R = \partial_{\sigma}\partial_{\tau}h^{\sigma\tau} - \Box^2 h \tag{11.115}$$

We can now find the linearized field equations themselves. Recall that the full equations can be written as

$$R_{\alpha\beta} - \frac{R}{2}g_{\alpha\beta} = \kappa T_{\alpha\beta}$$

Substituting (11.114) and (11.115) into this equation, we see that the linearized version of its lefthand side becomes

$$R_{\alpha\beta} - \frac{R}{2}g_{\alpha\beta} = \frac{1}{2}\left[\partial_{\alpha}\partial_{\sigma}h^{\sigma}_{\beta} + \partial_{\beta}\partial_{\sigma}h^{\sigma}_{\alpha} - \Box^2 h_{\alpha\beta} - \partial_{\alpha}\partial_{\beta}h - \eta_{\alpha\beta}\left(\partial_{\sigma}\partial_{\tau}h^{\sigma\tau} - \Box^2 h\right)\right]$$

So, the **linearized field equations** are

$$\partial_{\alpha}\partial_{\sigma}h^{\sigma}_{\beta} + \partial_{\beta}\partial_{\sigma}h^{\sigma}_{\alpha} - \Box^2 h_{\alpha\beta} - \partial_{\alpha}\partial_{\beta}h - \eta_{\alpha\beta}\left(\partial_{\sigma}\partial_{\tau}h^{\sigma\tau} - \Box^2 h\right) = 2\kappa T_{\alpha\beta} \tag{11.116}$$

and consequently, the **linearized vacuum equations** become

$$\partial_{\alpha}\partial_{\sigma}h^{\sigma}_{\beta} + \partial_{\beta}\partial_{\sigma}h^{\sigma}_{\alpha} - \Box^2 h_{\alpha\beta} - \partial_{\alpha}\partial_{\beta}h - \eta_{\alpha\beta}\left(\partial_{\sigma}\partial_{\tau}h^{\sigma\tau} - \Box^2 h\right) = 0 \tag{11.117}$$

The linearized field equations are quite messy. To remedy this problem, we introduce the **trace reverse** of $h_{\alpha\beta}$ which is defined by

$$\bar{h}_{\alpha\beta} = h_{\alpha\beta} - \frac{1}{2}\eta_{\alpha\beta}h \tag{11.118}$$

It's easy to show that the trace reverse has the properties

$$Tr(\bar{h}_{\alpha\beta}) \equiv \bar{h} = -h \quad \text{and} \quad h_{\alpha\beta} = \bar{h}_{\alpha\beta} - \frac{1}{2}\eta_{\alpha\beta}\bar{h}\,, \tag{11.119}$$

Clearly, the term 'trace reverse' is derived from the first property. When written in terms of the trace reverse, the linearized field equations become (exercise):

$$\Box^2\bar{h}_{\alpha\beta} + \left(\eta_{\alpha\beta}\partial_\sigma\partial_\tau\bar{h}^{\sigma\tau} - \partial_\alpha\partial_\sigma\bar{h}^\sigma_\beta - \partial_\beta\partial_\sigma\bar{h}^\sigma_\alpha\right) = -2\kappa T_{\alpha\beta} \tag{11.120}$$

These are the working field equations for the linearized theory. We can achieve a further simplification, however, by means of a *gauge transformation*, in a manner which is completely analogous to what we did in electromagnetic theory.

To this end, consider an infinitesimal **gauge transformation** of the form

$$x'^\alpha = x^\alpha + \xi^\alpha \tag{11.121}$$

where the quantities $\left|\xi^\alpha\right| \ll 1$ so they are of the same order of smallness as the $h_{\mu\nu}$. Under this gauge transformation, we find (exercise) that

$$h'_{\alpha\beta} = h_{\alpha\beta} - \partial_\alpha\xi_\beta - \partial_\beta\xi_\alpha \tag{11.122a}$$
$$h' = h - 2\partial_\sigma\xi^\sigma \tag{11.122b}$$
$$\bar{h}'_{\alpha\beta} = \bar{h}_{\alpha\beta} - \partial_\alpha\xi_\beta - \partial_\beta\xi_\alpha + \eta_{\alpha\beta}\partial_\sigma\xi^\sigma \tag{11.122c}$$

In addition, we can show (exercise) that the Riemann tensor and all of its contractions are invariant under the above gauge transformation, and hence, they are *gauge-invariant* quantities. This, in turn, means that the linearized field equations themselves are also invariant under the same transformation.

Upon examining (11.120), we see that the field equations will simplify to *wave equations* if we require that

$$\partial_\sigma\bar{h}^\sigma_\beta = 0 \tag{11.123}$$

which is known as the **Einstein gauge condition**. That is, under the Einstein gauge, it's clear that the quantities in the parentheses on the lefthand side of (11.120) will vanish.

Now, from (11.122c), observe that

$$\partial_\sigma\bar{h}'^\sigma_\beta = \partial_\sigma\left(\bar{h}^\sigma_\beta - \partial^\sigma\xi_\beta - \partial_\beta\xi^\sigma + \eta^\sigma_\beta\partial_\tau\xi^\tau\right)$$
$$\Longrightarrow \partial_\sigma\bar{h}'^\sigma_\beta = \partial_\sigma\bar{h}^\sigma_\beta - \Box^2\xi_\beta \tag{11.124}$$

Thus, if we require that

$$\Box^2\xi_\beta = \partial_\sigma\bar{h}^\sigma_\beta$$

then the Einstein gauge condition (11.123) will be satisfied and we shall have achieved a great simplification. Now, the solution of the last equation is always possible since this is just a wave equation with the source term $\partial_\sigma\bar{h}^\sigma_\beta$.

The above analysis shows that the Einstein gauge condition can always be realized. Now, let's summarize our findings. Upon dropping primes, and raising indices, the linearized field equations will reduce to

$$\Box^2\bar{h}^{\alpha\beta} = -2\kappa T^{\alpha\beta} \tag{11.125}$$

These equations, in turn, are valid whenever the trace reverse satisfies the Einstein gauge condition

$$\partial_\beta \bar{h}^{\alpha\beta} = 0 \tag{11.126}$$

Now, let us assume that the Einstein gauge condition has already been enforced so that (11.126) is satisfied. We still, however, have some remaining gauge freedom. That is, we are still free to perform an *additional* gauge transformation $x'^\alpha = x^\alpha + \xi^\alpha$ as long as we require that

$$\Box^2 \xi^\alpha = 0 \tag{11.127}$$

so that by (11.124), the Einstein gauge condition will continue to be satisfied.

Next, let's see how the solution of the vacuum equation

$$\Box^2 \bar{h}^{\alpha\beta} = 0 \tag{11.128}$$

predicts the existence of gravitational waves. We seek plane wave solutions to this equation of the form

$$\bar{h}^{\alpha\beta} = A^{\alpha\beta} \exp(ik_\sigma x^\sigma) \tag{11.129}$$

where the **amplitude tensor** $\mathbf{A} = (A^{\alpha\beta})$ has constant, but possibly complex components, and it is understood that we must take the *real part* of the righthand side. Here, $(k_\sigma) = (\omega/c, \mathbf{k})$ (where ω is the angular frequency) is the **wave 4-vector;** it points in the direction of wave propagation. Now, (11.128) implies that (k_σ) must be a *null vector*. To see this, observe that

$$\begin{aligned}
\Box^2 \bar{h}^{\alpha\beta} &= 0 \Longrightarrow \eta^{\mu\nu}\partial_\mu\partial_\nu \left[A^{\alpha\beta}\exp(ik_\sigma x^\sigma)\right] = 0 \\
&\Longrightarrow \eta^{\mu\nu}\partial_\mu \left[ik_\sigma \delta^\sigma_\nu A^{\alpha\beta}\exp(ik_\tau x^\tau)\right] = 0 \\
&\Longrightarrow -\eta^{\mu\nu}\left[k_\sigma k_\tau \delta^\sigma_\nu \delta^\tau_\mu A^{\alpha\beta}\exp(ik_\sigma x^\sigma)\right] = 0 \\
&\Longrightarrow -\left(k^\mu k_\mu\right)\bar{h}^{\alpha\beta} = 0
\end{aligned}$$

So, since we are seeking *non-trivial* solutions to (11.128), we must have

$$k^\mu k_\mu = 0 \tag{11.130}$$

This condition, in turn, implies that $\omega^2 = c^2 \|\mathbf{k}\|^2$, so that gravitational plane waves are non-dispersive and must travel at the speed of light.

Similarly, if we substitute (11.129) into the Einstein gauge condition, we find

$$\begin{aligned}
\partial_\beta \bar{h}^{\alpha\beta} &= 0 \Longrightarrow \partial_\beta \left[A^{\alpha\beta}\exp(ik_\sigma x^\sigma)\right] = i\delta^\sigma_\beta k_\sigma A^{\alpha\beta}\exp(ik_\sigma x^\sigma) = 0 \\
&\Longrightarrow ik_\beta A^{\alpha\beta}\exp(ik_\sigma x^\sigma) = 0
\end{aligned}$$

so that

$$A^{\alpha\beta} k_\beta = 0 \tag{11.131}$$

Now, consider a plane wave wave moving in the x^3-direction, so that the wave vector has the form

$$(k^\sigma) = [k, 0, 0, k] \text{ and } (k_\sigma) = [k, 0, 0, -k] \tag{11.132}$$

where $k = \omega/c > 0$. The last two equations imply that

$$A^{\alpha 0} = A^{\alpha 3} \tag{11.133}$$

Since $A^{\alpha\beta} = A^{\beta\alpha}$, the amplitude matrix can have, at most, ten *independent* components; however, the four conditions in (11.133) reduce this number to just six. That is, each of its components may be written in terms of $A^{00}, A^{01}, A^{02}, A^{11}, A^{12}$, and A^{22}. Consequently, we can write

$$\left(A^{\alpha\beta}\right) = \begin{bmatrix} A^{00} & A^{01} & A^{02} & A^{00} \\ A^{01} & A^{11} & A^{12} & A^{01} \\ A^{02} & A^{12} & A^{22} & A^{02} \\ A^{00} & A^{01} & A^{02} & A^{00} \end{bmatrix} \tag{11.134}$$

Now, it turns out that we can exploit our additional gauge freedom, as expressed in (11.127), to reduce the number of independent components to just two. To see this, we consider an *additional* gauge transformation $x'^\alpha = x^\alpha + \xi^\alpha$ such that

$$\xi^\alpha = -i\varepsilon^\alpha \exp(ik_\sigma x^\sigma) \tag{11.135}$$

where the ε^α are constants. It's easy to show that $\square^2 \xi^\alpha = 0$, so that the Einstein gauge condition is not violated.

In terms of this new gauge, we have

$$\bar{h}'^{\alpha\beta} = A'^{\alpha\beta} \exp(ik'_\sigma x'^\sigma)$$

Since, $\exp(ik'_\sigma x'^\sigma) \approx \exp(ik_\sigma x^\sigma)$ to within first order terms, we can rewrite the last equation as

$$\bar{h}'^{\alpha\beta} = A'^{\alpha\beta} \exp(ik_\sigma x^\sigma) \tag{11.136}$$

Substituting (11.135) and (11.136) into (11.122c), gives

$$A'^{\alpha\beta} = A^{\alpha\beta} - \varepsilon^\alpha k^\beta - \varepsilon^\beta k^\alpha + \eta^{\alpha\beta}\left(k_\sigma \xi^\sigma\right) \tag{11.137}$$

Upon recalling (11.132) and (11.134), the last equation implies that

$$\begin{aligned} A'^{00} &= A^{00} - k(\varepsilon^0 + \varepsilon^3) \\ A'^{01} &= A^{01} - k\varepsilon^1 \\ A'^{02} &= A^{02} - k\varepsilon^2 \\ A'^{11} &= A^{11} - k(\varepsilon^0 - \varepsilon^3) \\ A'^{12} &= A^{12} = A'^{21} \\ A'^{22} &= A^{22} - k(\varepsilon^0 - \varepsilon^3) \end{aligned} \tag{11.138}$$

We now attempt to express each of the quantities $A'^{\alpha\beta}$ in terms of A'^{11} and A'^{12}. If we choose,

$$\begin{aligned}
\varepsilon^0 &= \frac{1}{4k}\left(2A^{00}+A^{11}+A^{22}\right)\\
\varepsilon^1 &= \frac{1}{k}A^{01}\\
\varepsilon^2 &= \frac{1}{k}A^{02}\\
\varepsilon^3 &= \frac{1}{4k}\left(2A^{00}-A^{11}-A^{22}\right)
\end{aligned}$$

then (11.138) implies that

$$\begin{aligned}
A'^{00} &= A'^{01}=A'^{02}=0\\
A'^{11} &= -A'^{22}\\
A'^{12} &= A^{21}
\end{aligned} \tag{11.139}$$

Upon dropping primes, we see that the relations (11.139) tell us that only A^{11}, A^{12}, and A^{22} are non-zero, and furthermore, only two quantities, say A^{11} and A^{12} can be chosen independently. Setting $A^{11}=a$ and $A^{12}=b$, we obtain

$$\left(A^{\alpha\beta}_{\mathrm{TT}}\right)=\begin{bmatrix}0&0&0&0\\0&a&b&0\\0&b&-a&0\\0&0&0&0\end{bmatrix} \tag{11.140}$$

in our new gauge, for a wave propagating in the x^3-direction. This new gauge is known as the **transverse-traceless gauge**, or simply, the **TT-gauge**. If we define the **linear polarization tensors** $E^{\alpha\beta}_{(+)}$ and $E^{\alpha\beta}_{(\times)}$ by

$$\mathbf{E}_{(+)}=\begin{bmatrix}0&0&0&0\\0&1&0&0\\0&0&-1&0\\0&0&0&0\end{bmatrix}\quad\text{and }\mathbf{E}_{(\times)}=\begin{bmatrix}0&0&0&0\\0&0&1&0\\0&1&0&0\\0&0&0&0\end{bmatrix} \tag{11.141}$$

then the amplitude tensor may be written as a linear combination of the polarization tensors as

$$A^{\alpha\beta}_{\mathrm{TT}}=aE^{\alpha\beta}_{(+)}+bE^{\alpha\beta}_{(\times)} \tag{11.142}$$

In order to understand the effect of such a wave on matter, we consider two nearby particles, and we employ the *proper coordinate system* of the first particle. According to the equation of geodesic deviation, the connecting vector $\mathbf{V}=(V^{\alpha})$ must satisfy

$$\frac{D^2V^{\alpha}}{D\tau^2}=R^{\alpha}_{\beta\gamma\delta}U^{\beta}U^{\gamma}V^{\delta}$$

where $D/D\tau \equiv U^\alpha \nabla_\alpha$. If we assume small particle velocities, then we have $\mathbf{U} \approx (c, \mathbf{0})$, and $\tau \approx t$. Furthermore, since the connection coefficients vanish in the proper coordinate frame, the last equation reduces to

$$\frac{\partial^2 V^\alpha}{\partial t^2} = c^2 R^\alpha_{00\delta} V^\delta \tag{11.143}$$

Now, in the TT-gauge, it's easy to see that (11.113) reduces to

$$R^\alpha_{00\delta} = \frac{1}{2c^2} \frac{\partial^2}{\partial t^2} h^\alpha_\delta$$

so the geodesic deviation is governed by

$$\frac{\partial^2 V^\alpha}{\partial t^2} = \frac{1}{2} V^\delta \frac{\partial^2}{\partial t^2} h^\alpha_\delta \tag{11.144}$$

In order to solve this equation, we first note that in the TT-gauge, we can rewrite our plane wave solution as

$$\bar{h}^{\alpha\beta} = \bar{h}^{\alpha\beta}_{(+)} + \bar{h}^{\alpha\beta}_{(\times)} \tag{11.145}$$

where

$$\bar{h}^{\alpha\beta}_{(+)} = aE^{\alpha\beta}_{(+)} \exp(ik_\sigma x^\sigma) \;\text{ and }\; \bar{h}^{\alpha\beta}_{(\times)} = bE^{\alpha\beta}_{(\times)} \exp(ik_\sigma x^\sigma) \tag{11.146}$$

First, let's assume that $a = 1$ and $b = 0$ so that (11.144) becomes

$$\frac{\partial^2 V^\alpha}{\partial t^2} = \frac{1}{2} V^\delta \frac{\partial^2}{\partial t^2} \left[E_{(+)}{}^\alpha{}_\delta \exp(ik_\sigma x^\sigma) \right] \tag{11.147}$$

Assuming the two test particles are initially at rest, so that $\dot{V}^\alpha(0) = 0$, we can take $V^\delta \approx V^\delta(0)$ on the righthand side of the last equation, to obtain the approximate solution

$$V^\alpha = V^\alpha(0) \left[1 + \frac{1}{2}\, E_{(+)}{}^\alpha{}_\delta \exp(ik_\sigma x^\sigma) \right] \tag{11.148}$$

In particular, if the two particles are initially at rest on the x-axis, and are separated by a small distance δ, this equation implies that

$$(V^\alpha) = \begin{bmatrix} 0 \\ \delta \\ 0 \\ 0 \end{bmatrix} + \frac{1}{2} \begin{bmatrix} 0 & 0 & 0 & 0 \\ 0 & 1 & 0 & 0 \\ 0 & 0 & -1 & 0 \\ 0 & 0 & 0 & 0 \end{bmatrix} \begin{bmatrix} 0 \\ \delta \\ 0 \\ 0 \end{bmatrix} \exp\left[ik(x^0 - x^3)\right]$$

so that $V^0 = V^2 = V^3 = 0$ and (remembering to take the real part of the solution) we find

$$V^1 = \delta \left[1 + \frac{1}{2} \cos k(x^0 - x^3) \right] \tag{11.149}$$

We conclude that two particles, initially separated in the x-direction, will oscillate in the x-direction as the wave passes by.

On the other hand, if two particles are initially separated by a small distance δ in the y-direction, so that $V^0 = V^1 = V^3 = 0$, then (11.148) gives

$$V^2 = \delta\left[1 - \frac{1}{2}\cos k(x^0 - x^3)\right]$$

So, the particles will oscillate back and forth in the y-direction.

From the above analysis, if we now consider a ring of particles with radius δ lying in the xy-plane, then the entire ring will oscillate in the xy-plane. It will be distorted into an ellipse whose major axis is along the y-direction and then along the x-direction, and back again. This state of affairs is shown in Figure 11.7.

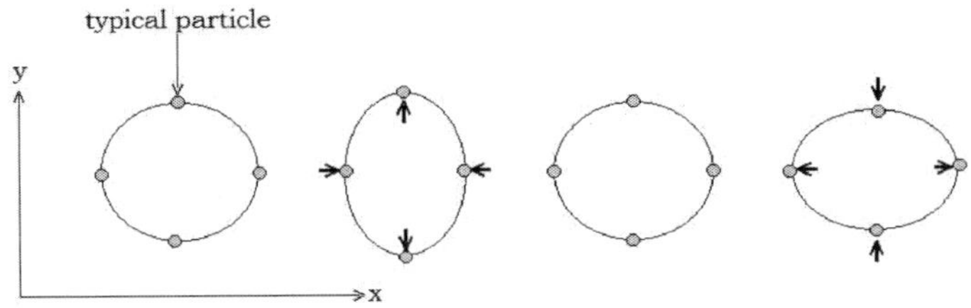

Figure 11.7. The effect of a gravitational wave on a ring of test particles in the xy-plane. Here the wave has the (+) polarization.

In the second case, if we take $a = 0$ and $b = 1$ in (11.142), then it may be shown for the ($\times$)-polarization that

$$V^\alpha = V^\alpha(0)\left[1 + \frac{1}{2}E_{(\times)\delta}^{\ \alpha}\exp(ik_\sigma x^\sigma)\right] \tag{11.150}$$

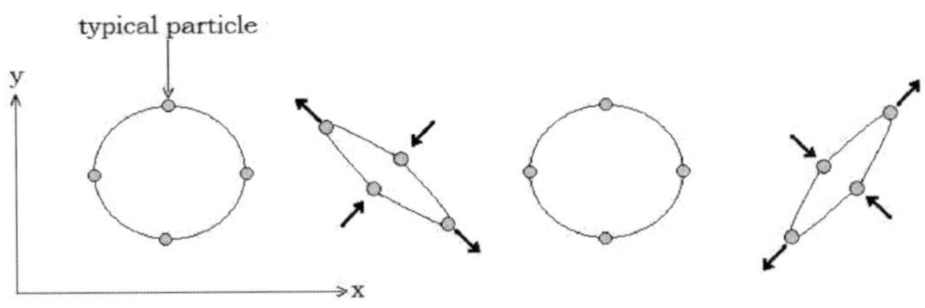

Figure 11.8. The effect of a gravitational wave with ($\times$)-polarization on a ring of test particles in the xy-plane.

For the second case, as depicted in Figure-11.8, if we are given a ring of test particles in the xy-plane, then the ($\times$)-wave will have the same effect as the $(+)$-wave, with the exception that the test particles will now oscillate back and forth on an ellipse whose minor axis is successively inclined at the angles $0, \pi/4, 0, -\pi/4$, and back to 0, as the wave passes by, forming an $\times$-pattern.

Observe that in both cases, whether the $(+)$-polarization or the ($\times$) - polarization is present, the fact that the motion is always confined to the xy-plane leads us to conclude that the waves must be *transverse waves*.

11.8. Exercises

1. Derive the connection coefficients in (11.14).

(Solution)

From (11.13), we have

$$\begin{aligned}\mathbf{g} &= (g_{\mu\nu}) = diag\left[e^{\alpha}, -e^{\beta}, -r^2, -r^2\sin^2\theta\right] \\ \mathbf{g}^{-1} &= (g^{\mu\nu}) = diag\left[e^{-\alpha}, -e^{-\beta}, -\frac{1}{r^2}, -\frac{1}{r^2\sin^2\theta}\right]\end{aligned}$$

so that

$$\begin{aligned}\Gamma^0_{00} &= \frac{1}{2}g^{0\alpha}\left[-g_{00,\alpha}+g_{\alpha 0,0}+g_{0\alpha,0}\right] = \frac{1}{2}g^{00}\left(g_{00,0}\right) = -\frac{1}{2}e^{-\alpha}\left(-\frac{1}{c}e^{\alpha}\dot{\alpha}\right) = \frac{1}{2c}\dot{\alpha} \\ \Gamma^0_{01} &= \frac{1}{2}g^{0\alpha}\left[-g_{01,\alpha}+g_{\alpha 0,1}+g_{1\alpha,0}\right] = \frac{1}{2}g^{00}\left(g_{00,1}\right) = -\frac{1}{2}e^{-\alpha}\left(-e^{\alpha}\alpha'\right) = \frac{1}{2}\alpha' \\ \Gamma^0_{11} &= \frac{1}{2}g^{0\alpha}\left[-g_{11,\alpha}+g_{\alpha 1,1}+g_{1\alpha,1}\right] = \frac{1}{2}g^{00}\left(-g_{11,0}\right) = -\frac{1}{2}e^{-\alpha}\left(-e^{\beta}\dot{\beta}\right) = \frac{1}{2}e^{\beta-\alpha}\dot{\beta}\end{aligned}$$

The calculation of the remaining coefficients is similar.

2. Derive the vacuum field equations (11.16)-(11.20).

(Solution)

Recall that the Ricci tensor is given by

$$R_{\mu\nu} = \frac{\partial}{\partial x^{\nu}}\left(\Gamma^{\sigma}_{\mu\sigma}\right) - \frac{\partial}{\partial x^{\sigma}}\left(\Gamma^{\sigma}_{\mu\nu}\right) - \Gamma^{\sigma}_{\mu\nu}\Gamma^{\tau}_{\sigma\tau} + \Gamma^{\sigma}_{\mu\tau}\Gamma^{\tau}_{\sigma\nu} = 0$$

So, for example,

$$R_{00} = \frac{\partial}{\partial x^{0}}\left(\Gamma^{\sigma}_{0\sigma}\right) - \frac{\partial}{\partial x^{\sigma}}\left(\Gamma^{\sigma}_{00}\right) - \Gamma^{\sigma}_{00}\Gamma^{\tau}_{\sigma\tau} + \Gamma^{\sigma}_{0\tau}\Gamma^{\tau}_{\sigma 0}$$

We now expand each term, *omitting terms that are zero*, to obtain:

$$\frac{\partial}{\partial x^0}\left(\Gamma^{\sigma}_{0\sigma}\right) = \frac{1}{c}\frac{\partial}{\partial t}\left(\Gamma^0_{00}+\Gamma^1_{01}\right) = \frac{1}{c}\frac{\partial}{\partial t}\left(\frac{\dot{\alpha}}{2c}+\frac{\dot{\beta}}{2c}\right) = \frac{\left(\ddot{\alpha}+\ddot{\beta}\right)}{2c^2}$$

$$\begin{aligned}-\frac{\partial}{\partial x^{\sigma}}\left(\Gamma^{\sigma}_{00}\right) &= -\frac{1}{c}\frac{\partial}{\partial t}\Gamma^0_{00} - \frac{\partial}{\partial r}\left(\Gamma^1_{00}\right) \\ &= -\frac{1}{c}\frac{\partial}{\partial t}\left(\frac{\dot{\alpha}}{2c}\right) - \frac{\partial}{\partial r}\left(\frac{\alpha'}{2}e^{\alpha-\beta}\right) \\ &= -\frac{\ddot{\alpha}}{2c^2} - \frac{e^{\alpha-\beta}}{2}\left(\alpha'^2 - \alpha'\beta' + \alpha''\right)\end{aligned}$$

$$\begin{aligned}
-\Gamma^{\sigma}_{00}\Gamma^{\tau}_{\sigma\tau} &= -\Gamma^{0}_{00}\Gamma^{\tau}_{0\tau} - \Gamma^{1}_{00}\Gamma^{\tau}_{1\tau} \\
&= -\frac{\dot{\alpha}}{2c}\left(\Gamma^{0}_{00} + \Gamma^{1}_{01}\right) - \frac{\alpha'}{2}e^{\alpha-\beta}\left(\Gamma^{0}_{10} + \Gamma^{1}_{11} + \Gamma^{2}_{12} + \Gamma^{3}_{13}\right) \\
&= -\frac{\dot{\alpha}}{2c}\left(\frac{\dot{\alpha}}{2c} + \frac{\dot{\beta}}{2c}\right) - \frac{\alpha'}{2}e^{\alpha-\beta}\left(\frac{\alpha'}{2} + \frac{\beta'}{2} + \frac{2}{r}\right) \\
&= -\frac{\dot{\alpha}^2}{4c^2} - \frac{\dot{\alpha}\dot{\beta}}{4c^2} - \frac{e^{\alpha-\beta}}{2}\left(\frac{\alpha'^2}{2} + \frac{\alpha'\beta'}{2} + \frac{2\alpha'}{r}\right)
\end{aligned}$$

$$\begin{aligned}
\Gamma^{\sigma}_{0\tau}\Gamma^{\tau}_{\sigma 0} &= \Gamma^{\sigma}_{00}\Gamma^{0}_{\sigma 0} + \Gamma^{\sigma}_{01}\Gamma^{1}_{\sigma 0} \\
&= \left(\Gamma^{0}_{00}\Gamma^{0}_{00} + \Gamma^{1}_{00}\Gamma^{0}_{10}\right) + \left(\Gamma^{0}_{01}\Gamma^{1}_{00} + \Gamma^{1}_{01}\Gamma^{1}_{10}\right) \\
&= \frac{\dot{\alpha}^2}{4c^2} + \frac{\alpha'^2}{2}e^{\alpha-\beta} + \frac{\dot{\beta}^2}{4c^2}
\end{aligned}$$

Adding the last four results and simplifying, we obtain

$$R_{00} = \frac{1}{c^2}\left(\frac{\ddot{\beta}}{2} - \frac{\dot{\alpha}\dot{\beta}}{4} + \frac{\dot{\beta}^2}{4}\right) - e^{\alpha-\beta}\left(\frac{\alpha''}{2} - \frac{\alpha'\beta'}{4} + \frac{\alpha'}{r} + \frac{\alpha'^2}{4}\right)$$

Consequently, the first vacuum field equation is

$$R_{00} = -e^{\alpha-\beta}\left(\frac{\alpha''}{2} - \frac{\alpha'\beta'}{4} + \frac{\alpha'}{r} + \frac{\alpha'^2}{4}\right) + \frac{1}{c^2}\left(\frac{\ddot{\beta}}{2} - \frac{\dot{\alpha}\dot{\beta}}{4} + \frac{\dot{\beta}^2}{4}\right) = 0$$

The remaining vacuum field equations are found similarly.

3. Derive equations (11.39a) and (11.39b).

(Solution)

Assuming plane polar coordinates (R, θ), it's easy to show that

$$\begin{aligned}
\hat{\mathbf{R}} &= (\cos\theta)\mathbf{i} + (\sin\theta)\mathbf{j} \\
\hat{\theta} &= (-\sin\theta)\mathbf{i} + (\cos\theta)\mathbf{j}
\end{aligned}$$

Differentiating the above expressions, we find

$$\frac{d\hat{\mathbf{R}}}{dt} = \dot{\theta}\hat{\theta} \text{ and } \frac{d\hat{\theta}}{dt} = -\dot{\theta}\hat{\mathbf{R}}$$

Now, the velocity is given by

$$\mathbf{v} = \frac{d\mathbf{R}}{dt} = \frac{d}{dt}(R\hat{\mathbf{R}}) = \frac{dR}{dt}\hat{\mathbf{R}} + R\frac{d\hat{\mathbf{R}}}{dt} = (\dot{R})\,\hat{\mathbf{R}} + (R\dot{\theta})\,\hat{\theta}$$

and the acceleration is

$$\begin{aligned}
\mathbf{a} &= \frac{d\mathbf{v}}{dt} = \frac{d}{dt}\left[(\dot{R})\,\hat{\mathbf{R}} + (R\dot{\theta})\,\hat{\theta}\right] = (\ddot{R})\hat{\mathbf{R}} + (\dot{R})\left(\dot{\theta}\hat{\theta}\right) + (R\ddot{\theta} + \dot{R}\dot{\theta})\hat{\theta} + (R\dot{\theta})\left(-\dot{\theta}\hat{\mathbf{R}}\right) \\
&= \left(\ddot{R} - R\dot{\theta}^2\right)\hat{\mathbf{R}} + \left(R\ddot{\theta} + 2\,\dot{R}\dot{\theta}\right)\hat{\theta}
\end{aligned}$$

4. Derive equation (11.41c).

(Solution)

When written in terms of polar coordinates, the angular momentum is $\mathbf{\Omega} = \mathbf{R} \times m\dot{\mathbf{R}}$, so $\mathbf{h} = \mathbf{R} \times \dot{\mathbf{R}}$, since $\mathbf{h}$ is the angular momentum per unit mass. We can write

$$\begin{aligned} \mathbf{h} &= \mathbf{R} \times \mathbf{v} = \left(R\hat{\mathbf{R}}\right) \times \left[\left(\dot{R}\right)\hat{\mathbf{R}} + \left(R\dot{\theta}\right)\hat{\theta}\right] \\ &= \mathbf{0} + R^2\dot{\theta}\left(\hat{\mathbf{R}} \times \hat{\theta}\right) \\ &= \left(R^2\dot{\theta}\right)\mathbf{k} \end{aligned}$$

Consequently,

$$h = \|\mathbf{h}\| = R^2\dot{\theta}$$

as required.

5. Derive equation (11.43).

(Solution)

Since $R = 1/u$, we have

$$\dot{R} = \frac{dR}{du}\frac{du}{d\theta}\frac{d\theta}{dt} = -\frac{1}{u^2}\dot{\theta}\frac{du}{d\theta} = -(R^2\dot{\theta})\frac{du}{d\theta} = -h\frac{du}{d\theta}$$

Consequently,

$$\begin{aligned} \ddot{R} &= \frac{d}{d\theta}\left(-h\frac{du}{d\theta}\right)\frac{d\theta}{dt} = -h\frac{d^2u}{d\theta^2}\left(\frac{h}{R^2}\right) = -h\frac{d^2u}{d\theta^2}\left(hu^2\right) \\ &= -h^2u^2\frac{d^2u}{d\theta^2} \end{aligned}$$

as required.

6. *(Binet's Equation)* Show that the solution to Binet's equation may be written as

$$u = \frac{GM}{h^2} + C\cos\left(\theta - \theta_0\right)$$

(Solution)

We must solve the inital value problem

$$\begin{aligned} \frac{d^2u}{d\theta^2} + u &= \frac{GM}{h^2} \\ u(0) &= u_0 \\ u'(0) &= u_0' \end{aligned}$$

Now, by inspection, a particular solution is $u_p = GM/h^2$, so the general solution is

$$u = \frac{GM}{h^2} + A\cos\theta + B\sin\theta$$

Using the initial conditions, we find

$$A = u_0 - \frac{GM}{h^2} \quad \text{and} \quad B = u_0'$$

so the solution is

$$u = \frac{GM}{h^2} + \left(u_0 - \frac{GM}{h^2}\right)\cos\theta + u_0' \sin\theta$$

If we define

$$C = \sqrt{\left(u_0 - \frac{GM}{h^2}\right)^2 + u_0'^2}$$

then we can rewrite the solution as

$$u = \frac{GM}{h^2} + C\left[\frac{\left(u_0 - \frac{GM}{h^2}\right)}{C}\cos\theta + \frac{u_0'}{C}\sin\theta\right]$$

Upon putting

$$\cos\theta_0 = \frac{\left(u_0 - \frac{GM}{h^2}\right)}{C} \text{ and } \sin\theta_0 = \frac{u_0'}{C}\text{ , so that } \theta_0 = \tan^{-1}\left[\frac{u_0'}{\left(u_0 - \frac{GM}{h^2}\right)}\right]$$

we find that

$$u = \frac{GM}{h^2} + C\cos(\theta - \theta_0)$$

which is (11.45), as required.

7. *(Gauge Invariance)* Show that the linearized Riemann curvature tensor (11.113) is invariant under the infinitesimal gauge transformation (11.121).

(Solution)

From (11.113), we have

$$R^{\alpha}_{\beta\gamma\delta} = \frac{1}{2}\left(\partial^{\alpha}\partial_{\delta}h_{\beta\gamma} - \partial^{\alpha}\partial_{\gamma}h_{\beta\delta} + \partial_{\beta}\partial_{\gamma}h^{\alpha}_{\delta} - \partial_{\beta}\partial_{\delta}h^{\alpha}_{\gamma}\right)$$

On the other hand, from (11.122a) we can write

$$h_{\alpha\beta} = h'_{\alpha\beta} + \partial_{\alpha}\xi_{\beta} + \partial_{\beta}\xi_{\alpha}$$

Upon substituting the second equation into the first one, we get

$$\begin{aligned} R^{\alpha}_{\beta\gamma\delta} &= \frac{1}{2}\left[\begin{array}{l} \partial^{\alpha}\partial_{\delta}\left(h'_{\beta\gamma} + \partial_{\beta}\xi_{\gamma} + \partial_{\gamma}\xi_{\beta}\right) - \partial^{\alpha}\partial_{\gamma}\left(h'_{\beta\delta} + \partial_{\beta}\xi_{\delta} + \partial_{\delta}\xi_{\beta}\right) \\ + \partial_{\beta}\partial_{\gamma}\left(h'^{\alpha}_{\delta} + \partial^{\alpha}\xi_{\delta} + \partial_{\delta}\xi^{\alpha}\right) - \partial_{\beta}\partial_{\delta}\left(h'^{\alpha}_{\gamma} + \partial^{\alpha}\xi_{\gamma} + \partial_{\gamma}\xi^{\alpha}\right) \end{array}\right] \\ &= \frac{1}{2}\left[\partial^{\alpha}\partial_{\delta}h'_{\beta\gamma} - \partial^{\alpha}\partial_{\gamma}h'_{\beta\delta} + \partial_{\beta}\partial_{\gamma}h'^{\alpha}_{\delta} - \partial_{\beta}\partial_{\delta}h'^{\alpha}_{\gamma}\right] \\ &= R'^{\alpha}_{\beta\gamma\delta} \end{aligned}$$

as required.

Notes

Chapter 1.

1. See Callahan [8], page 87.

Chapter 2.

1. Given two sets X and Y, recall that their Cartesian product is given by

$$X \times Y = \{ (x, y) | x \in X \text{ and } y \in Y \}$$

2. A Mobius strip also provides us with a nice example of a one-sided or non-orientable surface.

3. Recall that a **group** $\langle G, \circ \rangle$ is a non-empty set, together with a binary operation $\circ$, which satisfies the following properties:

a. For any elements $a, b \in G$, the product $a \circ b \in G$.

b. The operation $\circ$ is associative; that is, for any elements $a, b, c \in G$,

$$a \circ (b \circ c) = (a \circ b) \circ c$$

c. There exists an **identity element** $e \in G$ such that for any element $a \in G$,

$$a \circ e = e \circ a = a$$

d. Corresponding to each element $a \in G$, there exists an element a^{-1}, called the **inverse** of a, such that

$$a \circ a^{-1} = a^{-1} \circ a = e$$

If no confusion results, we shall denote the group $\langle G, \circ \rangle$ by G. Furthermore, we say that a group G is **commutative** or **Abelian** if and only if $a \circ b = b \circ a$ for all $a, b \in G$.

4. A set H is a **subgroup** of a group G if and only if $H \subseteq G$, and H is a group under the *same* binary operation as G.

5. Our development of the concept of a continuous group and of a Lie group closely parallels the excellent introduction given in Joshi [27], pp. 110-116.

6. A matrix A is said to be **Hermitean** if $A^\dagger = A$, where $A^\dagger$ denotes the complex conjugate transpose of A, i.e.,

$$A^\dagger = (A^T)^*$$

7. For the purpose of our discussion, we do not include *time translations*

$$t \longrightarrow \bar{t} = t + t_0$$

or *time and space reflections*

$$\begin{aligned} t &\longrightarrow \bar{t} = -t \\ \mathbf{r} &\longrightarrow \bar{\mathbf{r}} = -\mathbf{r} \end{aligned}$$

as elements of the general Galilean group.

Chapter 3.

1. See Bergmann [5], p.p. 173-178.

2. Throughout the text, we use the CGS electrostatic (or Gaussian) system of units.

3. We follow the convention of Ohanian [39], in referring to **B** as the *magnetic field* rather than the *magnetic induction*.

4. Recall that the **divergence theorem of Gauss** states that if R is a region of three-space which is bounded by a *closed* surface ∂R, then

$$\int_R [\nabla \cdot \mathbf{V}(\mathbf{x})]\, d^3\mathbf{x} = \oint_{\partial R} \mathbf{V}(\mathbf{x}) \cdot \mathbf{n}\, dS$$

provided the vector field $\mathbf{V}(\mathbf{x})$ has continuous derivatives, and $\mathbf{n}$ is the outward unit normal to S.

5. Recall that **Stoke's theorem** states that if S is an orientable (two-sided) surface which is bounded by a simple-closed curve C, then

$$\oint_C \mathbf{V}(\mathbf{x}) \cdot d\mathbf{r} = \int_S (\nabla \times \mathbf{V}(\mathbf{x})) \cdot \mathbf{n}\, dS$$

where we assume that the vector field $\mathbf{V}(\mathbf{x})$ possesses continuous derivatives, and $\mathbf{n}$ is the outward unit normal to S.

6. After Griffiths [23], pages 72-73.

7. P.A.M. Dirac was the first to suggest the existence of magnetic monopoles, though thus far, none have been detected.

Chapter 4.

1. The reader interested in a detailed historical account of this time period is referred to Miller [33], pages 18-23.

2. See French [17], *Einstein, A Centenary Volume,* pages 281-293.

3. The question of whether Einstein was fully aware of the result of the Michelson-Morley experiment at that time, and whether it influenced his thinking, has been hotly debated. For a full and fair discussion of this issue, see Pais [42], pp. 127 - 128.

4. See Synge [56], pp. 127-128.

5. Motivated by an example from Aharoni [1].

Chapter 5.

1. We follow an approach which is based upon the development given in Nakahara [36], p.p. 140-142.

2. For $1 \leq k \leq n$, we can define the functions $u^k : R^n \longrightarrow R$ that map each point $q \in R^n$ to its k-th coordinate, i.e.,

$$u^k(q) = u^k(x^1, x^2, ..., x^n) = x^k$$

The functions u^k are called the **natural coordinate functions** on R^n. Thus, the coordinate functions ϕ^k on M are precisely defined by

$$\phi^k(p) = u^k(\phi(p)) \text{ for all } p \in M.$$

3. Alternatively, we can say the a smooth function $f : M \longrightarrow R$ is a **scalar field** on M since f assigns a *unique* scalar (or real number) $f(p)$ to each point $p \in M$. As an example of this idea, one can imagine a material sphere S^2 with $f(p)$ representing the temperature at each point $p \in S^2$.

Chapter 6.

1. Note that if we were to parallel-transport the basis vector $\mathbf{e}_b(\mathbf{x})$ from the point p to the point q, then by symmetry, we would have

$$\hat{\mathbf{e}}_b(\mathbf{x} + \Delta\mathbf{x}) = \mathbf{e}_b(\mathbf{x}) - (\Delta x^a)\,\Gamma^c_{ab}\mathbf{e}_c(\mathbf{x}) \tag{A6.1}$$

The minus sign results from the fact that the roles of the points p and q in Figure 6.1 are now reversed.

2. For relativistic spacetime, we will assume that the connection is torsion-free. Spacetimes with non-zero torsion are addressed in the so called *Einstein-Cartan theories*.

3. Observe that this definition violates the convention for the placement of tensor indices, but this is a standard practice, and it has been done for historical reasons.

4. More generally, we can characterize *any* smooth curve as **spacelike**, **timelike**, or **null** provided that $\mathbf{g}(\mathbf{X},\mathbf{X}) < 0$, $\mathbf{g}(\mathbf{X},\mathbf{X}) > 0$, or $\mathbf{g}(\mathbf{X},\mathbf{X}) = 0$, respectively; *at each point* along the curve.

5. For a *compact* manifold, it may be shown that the solution to (6.202) is both unique and exists for all $t \in R$.

Chapter 10.

1. We follow an approach motivated by Atwater [4], pages 77-79.

2. As pointed out by Hawking and Ellis [24], page 56, there exists an entire *equivalence class* of spacetimes. We can define an equivalence relation on the collection of all spacetimes by agreeing that any two spacetimes $(M,\mathbf{g})$ and $(M',\mathbf{g}')$ are equivalent if and only if they are isometric. That is, there exists a diffeomorphism $\phi : M \longrightarrow M'$ which takes the metric $\mathbf{g}$ into $\mathbf{g}'$ such that g is *form-invariant*. So, when we are working with a particular spacetime, we are really working with a *representative* of such an equivalence class.

Chapter 11.

1. The reader interested in a more detailed history of these developments should consult Pais [42], p.p. 253-256.

References

[1] Aharoni, Joseph, *The Special Theory of Relativity*, Clarendon Press, 1959.

[2] Adler, R., Bazin, M., and Schiffer, M., *Introduction to General Relativity*, McGraw-Hill Book Co., 1975.

[3] Anderson, James, *Principles of Relativity Physics*, Academic Press, 1967.

[4] Atwater, H.A., *Introduction to General Relativity*, Pergamon Press, 1974.

[5] Bergmann, Peter Gabriel, *Basic Theories of Physics*, 1949, Prentice-Hall, Inc.

[6] Bitter, Francis, *Currents, Fields, and Particles*, Technology Press of M.I.T. with John Wiley & Sons, Inc., 1956.

[7] Bose, S.K., *General Relativity*, Halsted Press, 1980.

[8] Callahan, James, J., *The Geometry of Spacetime*, Springer-Verlag New York, Inc., 2000.

[9] Carroll, Sean, *Spacetime and Geometry*, Pearson Education, Inc., (publishing as Addison Wesley), 2004.

[10] Cheng, Ta-Pei, *Relativity, Gravitation, and Cosmology*, Oxford University Press, 2006.

[11] DeFelice, F., and Clarke, C.J.S., *Relativity on Curved Manifolds*, Cambridge University Press, 1990.

[12] Dicke, R.H. in *Gravitation and Relativity*, ed. H.Y. Chiu and W.F. Hoffmann, Benjamin, New York, 1964.

[13] D'Inverno, Ray, *Introducing Einstein's Relativity*, Oxford University Press, 1992.

[14] Dirac, P.A.M., *General Theory of Relativity*, John Wiley & Sons, 1975.

[15] Ferraro, Rafael, *Einstein's Spacetime*, Springer-Verlag, New York, Inc., 2007.

[16] Fock, V., *The Theory of Spacetime and Gravitation*, Pergamon Press, 1959.

[17] Foster, J. and Nightingale, J.D., *A Short Course in General Relativity*, 2nd Edition, Springer-Verlag, New York, Inc., 1995.

[18] French, A.P., (editor) *Einstein, A Centenary Volume*, Harvard University Press, 1979.

[19] Göckeler, M., and Schücker, *Differential Geometry, Gauge Theories, and Gravity*, Cambridge University Press, 1990.

[20] Goldberg, Stanley, *Understanding Relativity*, Birkhäuser Boston, Inc., 1984.

[21] Grøn, Øyvind, *Lecture Notes on the General Theory of Relativity*, Springer Science and Business Media, 2009.

[22] Griffiths, David, *Introduction to Electrodynamics*, Pearson Education, Inc., 1998.

[23] Griffiths, David, *Introduction to Elementary Particles*, John Wiley & Sons, Inc., 1987.

[24] Hartle, James, *Gravity - An Introduction to Einstein's General Relativity*, Addison Wesley, 2003.

[25] Hawking, S.W. and Ellis, G.F.R., *The Large Scale Structure of Spacetime*, Cambridge University Press, 1973.

[26] Hobson, M.P., Efstathiou, G., and Lasenby, A.N., *General Relativity - An Introduction for Physicists*, Cambridge University Press, 2006.

[27] Jackson, John David, *Classical Electrodynamics*, John Wiley & Sons, Inc., 1962.

[28] Joshi, A.W., *Elements of Group Theory for Physicists*, John Wiley & Sons, Inc., 1977.

[29] Kenyon, I.R., *General Relativity*, Oxford University Press, 1990.

[30] Kilmister, C.W., *General Theory of Relativity*, Pergamon Press, 1973.

[31] Kühnel, Wolfgang, *Differential Geometry*, 2nd Edition, American Mathematical Society, 2006.

[32] Landau, L.D., and Lifshitz, E.M., *The Classical Theory of Fields*, Pergamon Press, 1975.

[33] Lovelock, David and Rund, Hanno, *Tensors, Differential Forms, and Variational Principles*, Wiley Interscience, 1975.

[34] Miller, Arthur, I., *Albert Einstein's Special Theory of Relativity*, 1998, Springer-Verlag New York, Inc.

[35] Moller, C., *The Theory of Relativity*, Oxford University Press, 1957.

[36] Mould, Richard, A., *Basic Relativity*, Springer-Verlag, New York, Inc., 1994.

[37] Nakahara, M., *Geometry, Topology, and Physics*, 1996, Institute of Physics Publishing, Bristol and Philadephia.

[38] Narlikar, Jayant, V., *General Relativity and Cosmology*, The Macmillan Press, Ltd., 1979.

[39] Ney, E.P., *Electromagnetism and Relativity*, Harper and Row, 1965.

[40] Ohanian, Hans, C., *Classical Electrodynamics*, Allyn & Bacon, 1988.

[41] Ohanian, Hans, C., *Gravitation and Spacetime*, W.W. Norton & Co., Inc., 1976.

[42] O'Neill, Barrett, *Semi-Riemannian Geometry*, Academic Press, 1983.

[43] Padmanabhan, T., *Gravitation*, Cambridge University Press, 2010.

[44] Pais, Abraham, *'Subtle is the Lord ...'* , 1982, Clarendon Press, Oxford University Press, New York.

[45] Plebanski, Jerzy, and Krasinski, Andrzej, *An Introduction to General Relativity and Cosmology*, Cambridge University Press, 2006.

[46] Pathria, R.K., *The Theory of Relativity*, Dover Publications, Inc., 2003.

[47] Robinson, F.N.H., *An Introduction to Special Relativity*, World Scientific Publishing Co., 1995.

[48] Rindler, Wolfgang, *Essential Relativity*, Van Nostrand Reinhold Co., 1969.

[49] Ryder, Lewis, *Introduction to General Relativity*, Cambridge University Press, 2009.

[50] Schröder, Ulrich, E., *Special Relativity*, World Scientific Publishing Co., 1990.

[51] Sharan, Pankaj, *Spacetime, Geometry, and Gravitation*, Hindustan Book Agency, 2009.

[52] Shutz, Bernard, *A First Course in General Relativity*, 2nd Edition, Cambridge University Press, 2009.

[53] Shutz, Bernard, *Geometrical Methods in Mathematical Physics*, Cambridge University Press, 1990.

[54] Stephani, Hans, *General Relativity*, 2nd Edition, Cambridge University Press, 1990.

[55] Stewart, John, *Advanced General Relativity*, Cambridge University Press, 1990.

[56] Synge, J.L., *Relativity: The Special Theory*, North-Holland Publishing Co., Amsterdam, 1965.

[57] Tourrenc, Phillip, *Relativity and Gravitation*, Cambridge University Press, 1997.

[58] Weyl, Hermann, *Space-Time-Matter*, Dover Publications, Inc., 1950.

[59] Williams, L. Pearce, *Michael Faraday*, Basic Books, 1965.

[60] Willmore, T.J., *Differential Geometry*, Oxford University Press, 1959.

[61] Willmore, T.J., *Riemannian Geometry*, Oxford University Press, 1996.

INDEX

A

B

C

D

E

F

G

N

O

P

Q

R

S

T

U

V